World Free Flight Review

TAFT CONDORS
HOME OF
U.S. FREE FLIGHT
CHAMPIONSHIPS
DON'T USE NEW ROAD - YOU'LL GET STUCK
SAN VALEERS CONTEST
N11127
Challenge

Spirit

World Free Flight Review

Volume 1

1975-1976-1977 Edition

A Pictorial Record of International
Free Flight Aeromodelling

edited by
William R. Hartill

WORLD FREE FLIGHT PRESS

Canoga Park, California
U.S.A.

Library of Congress Card Number 78-64742
ISBN 0-933066-01-5
Address all inquiries to the Publisher.

WORLD FREE FLIGHT PRESS

7513 SAUSALITO AVE., CANOGA PARK, CA. 91307 U.S.A.

Printed in the United States of America by
The Printer, Davis, California

Venture

Dedication

Winged Victory of Samothrace. A statue of Nike the Greek Goddess of Victory believed to be the work of a Rhodian artist, Pythocles, in the early second century B.C., was found on the Greek island of Samothrace in 1863. The work now stands in the Louvre. Winged Victory stood on the prow of a ship, with her beautiful wings extended as if to relinquish the ties with earth and soar aloft. Wind streams her dress against her body, gathering deep shadowy folds, reflecting the purposeful stance. We dedicate this *Free Flight Review* to Winged Victory, a creation of the mythogenetic embodiment of man's aspiration to the freedom of flight.

Discovery

Contents

Color Plates

Acknowledgements

The author gratefully thanks the many people of the free-flight community who have made this *Review* possible. The generous and enthusiastic cooperation in response to requests for drawings, information and help has been most gratifying and appreciated. Approval for use of the drawings has been provided by the original designers and by the various publications that have given them prior presentation. Special thanks go to the following publications.

Aeromodeller	GB
Der Bartabschneider	D
FFN-Free Flight News	GB
Scatter	USA
Model Aviation	USA
MAN-Model Airplane News	USA
MB-Model Builder	USA
Vol Libre	F
Modellflyngnytt	S
Modelflyne Nyt	DK
Alas	RA
Modelar	CSSR
Modellistica	I
AVA	I
Airborne	AUS
East Anglian News	CB
Free Flight Digest	USA
Indoor News & Views	USA

One cannot write of Free Flight without recognizing the basic and pervasive influence of Frank Zaic. No other person has contributed so much to our knowledge, comradeship and spirit of Free Flight. His teachings have sparked the vital perceptions needed to launch the *World Free Flight Review*. May this work help express my gratitude.

Many thanks to Craig Cusick for his very fine detailed drawings, to Elsie Gildersleeve for the meticulous typesetting, and to Doug Galbreath *The Printer* for patiently answering a million questions and doing a great printing job. And most important to this venture; chief typist and source of inspiration and encouragement, my Inza.

—Bill Hartill

Introduction

World Free Flight Review documents the challenge of Free Flight. A compendium that brings a taste of excitement, a view of the craftsmanship and a sense of the beauty of Free Flight Aeromodelling.

Aeromodelling has been with us ever since that time when the sight of a soaring bird first stirred the vital spark of consciousness in prehistoric man to fashion his own free-flight creation from leaves and twigs. This process has been going on ever since, but with precious little documentation along the way to spotlight and report this form of man's dream, this relationship between the psyche and the physical.

We are such stuff/As dreams are made on. (W.S.)

The literature of aeromodelling tends to be characterized by either the boring elementary how-to-do-it approach or by the journals that give us a little of this, a little of that, but not enough for anyone. We Free Flighters need more. Some of us have done something about it. Free Flight newsletters have sprung up in all parts of the world, emphasizing that need. Now we have *World Free Flight Review*, the kind of book we wanted, the kind of book we think you will enjoy. The *Review* reports international activity in the major Free Flight Aeromodelling categories in the areas of design technology, construction techniques, competition skills, organizational direction, history, camaraderie and adventure.

For the uninitiated, a bit of explanation. Free Flight Aeromodelling requires for its psyche, that the model be flown free, with no direct control (radio, control lines, etc.). Upon leaving the hand the model flies free, relying on its design and adjustment to work with the principles and natural laws of physics, mechanics and aerodynamics to soar on the atmospheric ocean of air we live in. The Free Flighter doesn't consider the release of the model to these natural forces a defect or disadvantage, rather he considers this a requisite. In doing so he finds a challenge and beauty of spirit unknown to more mundane sportsmen and hobbyists.

By projecting the true image of Free Flight in the pages of the *Review* we hope to demonstrate and document this challenge.

Publication of the *Review* will be continued at regular intervals to provide a series documentation of World Free Flight activity. Reader comments and suggestions to the editor regarding contribution and selection of future material is respectfully and earnestly solicited.

Roy Collins, Great Britain.

1 Free Flight Review 1975-1976-1977

The View From Downwind

A most outstanding Free Flight World Championship was held in Plovdiv, Bulgaria, August 1975. Although many had expressed some discouragement back in 1973 at the idea of travelling to such a "far away spot", the spirit of competition (and adventure) prevailed and Plovdiv was jammed with competitors, helpers and interested enthusiasts. The Bulgarians put on a very good show and the contestants responded by delivering the largest mass fly-offs yet. In fact the clear conclusion of those present was that indeed fly-offs are the rule, not the exception. The contestants and spectators alike thrilled to the quickening pace, the tenseness of concentration, the *coup de grace* as the winners eked out their close margins, in the closing dusk of the day.

The concept of fly-offs is felt to be in need of retention as an element of the sport. Difficulties in conducting and managing the fly-offs are still apparent and improvements are being proposed. For example, the FAI CIAM Free Flight Subcommittee has agreed to consider procedural changes that should improve the fairness of the fly-off. Another competition "problem" is that of line crossings in F1A. The new hook systems now anchor line to model with little chance of premature release while the circle option has the contestants cavorting all over the field. Line crossings appear to be inevitable under the circumstances. This situation has led some to call for combat rules-no refly-but others point out the inequity of a contestant's model being pulled into the ground by someone else. Putting the burden on the timers to sort our who did what to whom is equally unsatisfactory as the timer is intent on watching for line release, and it is difficult to observe line crossings when the distance of sight may be several hundreds of meters away. The CIAM Subcommittee is struggling with this one too but no clever solution appears at hand. Through it all the towline aficionados will manage to cope, we are sure.

F1C engine runs at Plovdiv were to the old 10-second rule. To those who had been using 7 seconds in preparation for the competitions under the new FAI rules, the flight patterns appeared a bit wasteful. Not all, of course, but a number were stalling and/or hesitant in getting into glide trim following the run. Some of the F1A zoom launches looked far more impressive than some of the F1C power patterns!

Getting to Plovdiv was not without its interesting and challenging moments. Main problem seemed to be the matching up of the realities of the world we live in with the expectations. When these do not match, one must compensate, adjust, or be disappointed. The role of team manager in such circumstances becomes most difficult and, in fact, is a difficult job under the best of circumstances. Selection and training of team managers will be receiving more attention in the future, together with the development of techniques for producing maximum effectiveness from each team member. There is more to it than "making the team!" One idea that has merit is to have a Coach to look after the technical needs and also a Manager that sees to the travel, food and lodging arrangements. Thus, if difficulties arise in regard to travel, these possible unpleasantries are not transferred into resentments between the team members and Coach.

Once having got there. there was little to find fault with at Plovdiv. Although there was considerable question as to why the bar in the hotel always ran out of beer every night at the same time. This is certainly a question that should be taken up immediately at the highest levels of the FAI. In spite of this planning defect, the social activities were warm and enthusiastic and a fitting complement to the World Championships.

The Criterium Pierre Trebod held at Marigny le Grande, France, has achieved unique stature in the Free Flight Aeromodelling World. Each year establishes a new high in unbridled indulgence of international free-flight passions. Yeah man, this is where it's at. Free Flight with a purpose, orgiastic good will!

Coupe d'Hiver International, usually held in France, was switched to England in '75. Held at R.A.F. Halton November 30, the conditions started out with abominable fog and ice. Nevertheless the French once again proved their clear mastery of this event with nonchalant disregard for the British weather, trouncing the field. Who will rise to the French challenge!

In the U.S. Free Flight activity continues strong in a number of classifications. FAI, Indoor, AMA classes, Scale, Vintage. Special interest groups such as NFFS, NIMAS and SAM, spread the word on activities. In the F1A, F1B and F1C classes a major activity is related to U.S. team selection. In addition, a number of large competitions are held around the country. For example, the U.S. Free Flight Championships are held at Taft, California in May of each year. Free Flight events are flown with a contestant entry list of over 300.

In August 1976 the Indoor World Championships were held in the No. 1 airship shed at Cardington, England. A resurgence of interest in Indoor model flying has brought an intensity to this competition that has been unknown in prior years. Bud Romak of the U.S. became the new World Champ with two flights of just under 40 minutes each. This with FAI rules in which the model must weigh one gram. The unlimited indoor world record was smashed in 1976 by Dick Kowalski of the U.S.A. who pushed it up to 50:41. It was not too long ago that 30 minutes was spoken of as a barrier!

Another Free Flight category receiving a build-up in interest and enthusiasm is the Vintage or Old Timer class. The fun of flying the old ships is just as great as it ever was and organized flying sessions and competitions are becoming popular around the world. Here the emphasis is on nostalgia and a look back to the more basic origins of Free Flight. The Custom Cavalier is *still* a beauty!

The year 1977 saw the Free Flight World Championships organization taken over by enthusiastic Danish aeromodellers. They turned out to be excellent hosts for the affair held in Roskilde. And, in fact, their strong driving force, Thomas Koster, emerged as the hero of the occasion, winning F1C in decisive fashion as well as lending a strong hand to the solution of the many organization problems. This was a sweet victory for Thomas in his own country, after coming so close in 1973 when he lost by a few seconds as his model flew behind a tree. And then there was the frustrating experience in 1975, having a clear performance margin with his flapped *Creams*, he suffered a series of crippling crashes that left him with nothing to fly in spite of heroic repairs. Now Thomas is the first competitor to become World Champion in two categories, F1B Wakefield (1965) and F1C Power (1977).

The North Koreans left no doubt in 1977 that their strong showing in 1975 was no accident. Winning the Wakefield Cup in these two years along with team places and high positions in F1A and F1C has made their proficiency quite clear now to everyone. Flying what appear to be conventional models, they now have their competition wondering if the more advanced design concepts have been over-engineered at the expense of some other more basic functions.

Nordic designs are no longer striving for that ultimate 180-second still-air machine. The practicality of such an approach has been definitely discounted. Instead effort is now concentrated on use of the circle-towing hook. Many training flights are needed to gain the easy handling skill needed to snare the lift every time. Zoom launches are in but some work remains to develop an optimum zoom pattern. Auto-surface sophistication is on the increase in F1A to eke out the most climb from the zoom. Climb/glide transition is now a design/trim parameter in towline glider! East Europeans appear to be spending the most time with F1A practice flying and it shows. It pays off in ability to doggedly track the lift, test it, launch in it, and stay in it. Kostadin Abadjiev of Bulgaria, the 1977 F1A World Champion, demonstrated that 'easy' proficiency.

F1C Power models in 1977 are now showing the trends set by such as Urs Schaller and his *Plurabelle* and a number of others. Fully cowled engines, fiberglassed front-ends, glass reinforced tail-booms and wings are becoming the norm. The clean design does help, both in the very rapid climbs and flat glides being demonstrated. Transition patterns in 1977 appeared to be much improved also over 1975. Prop brakes are now universal. A few electronic timers have appeared but they are still under development.

In all a very exciting period in the world of Free Flight. Now if we could only get some good Wakefield rubber. ■

Nordic Glider Development
— by Ivan Horejsi, Plzen, Czechoslovakia

Towline gliders of the A/2 class, nowadays known internationally as F1A, were born back in the 1940's in the north of Europe in Sweden and Finland. That is why they are currently known as Nordic gliders. Originally they were a little bit smaller in size than at present. The total area was in the range 24-26 dm^2, total minimum weight 400 g and the minimum cross section area of the fuselage equaled $L^2/300$ where L was the length of the fuselage.

These Nordic gliders had many good features; inexpensive, easy to build, damage resistant, etc., and were a big improvement over the huge gliders that were the previous standard in those old days. They became very popular with modelers and spread quickly all over the world, replacing those big old birds.

The Federation Aeronautique International (FAI) recognized the new class in 1948 and the new rules were as follows: total area 32-34 dm^2; minimum weight 410 g, minimum cross section $L^2/300$, and towline length 100 m. The year 1955 saw the canceling of the cross-section rule and thus the Nordics received their general characteristics which have not been changed since. It is really an interesting fact that there has been no need to change the model specifications over the many years since 1955. On the other hand, there have been severe changes in other F/F classes — just note the changes in rubber weight in Wakefield. The main reason for this seems to be that we can only attempt to improve the gliding aerodynamics of Nordics while the Wake fliers for example can improve prop efficiency, and find better rubber. Power fliers benefit from more and more powerful engines that are not comparable with the old ones etc.

Though the building rules of the Nordics have not been changed and therefore today's models do not differ too much from the old ones, a top flier of the old days would now probably be an outsider. The reason is that the tactics of flying have been much improved.

During the A/2 development since 1948, there have been a number of fliers identified with bringing the class to today's perfection. Probably the first world-known modeler of this category was **Oskar Czepa** from Austria. He became the World Champion with his model called *Toothpick*. The model, as its namesake, featured a very long skinny fuselage, with the minimum prescribed cross section located back with the tailfeathers.

Next model in interest belonged to the German modeler **Rudi Lindner.** He worked together with his friend **Max Hacklinger** with excellent results. The airfoils and thread turbulators developed by Hacklinger, are still in use today. Nevertheless the better known man is Lindner, as he became twice the World Champion. (1954, 1955).

Lindner's contribution was in the development of tactical flying. He was also probably the first to learn how to profit from wing warps. Lindner's wings were warped in the opposite way to current practice. He washed in the outboard wing panel. A model with that warp flies in broad circles in neutral or down air, while tightening up in lift. This looks very nice. Unfor-

Ivan Horejsi, Czechoslovakia.

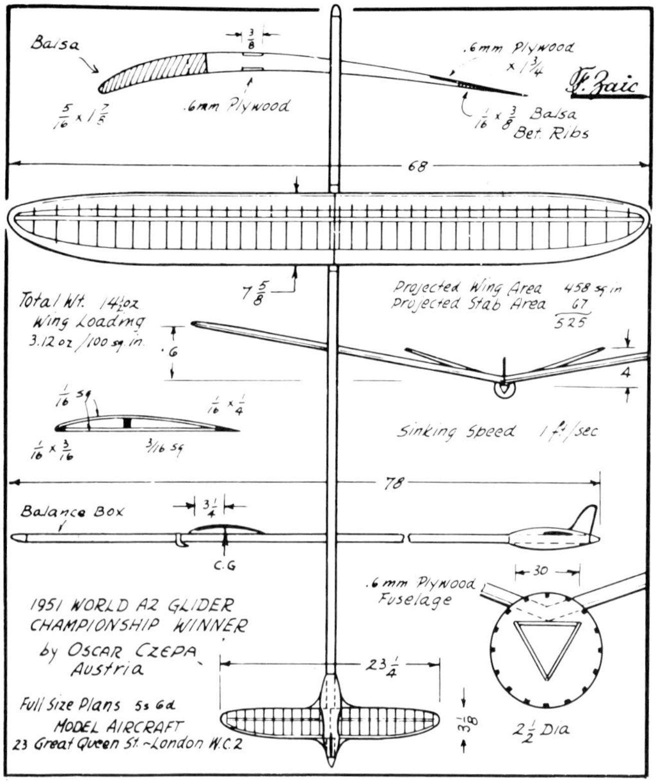

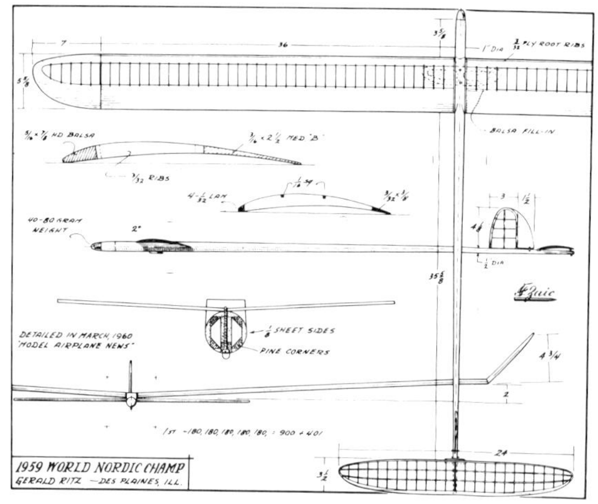

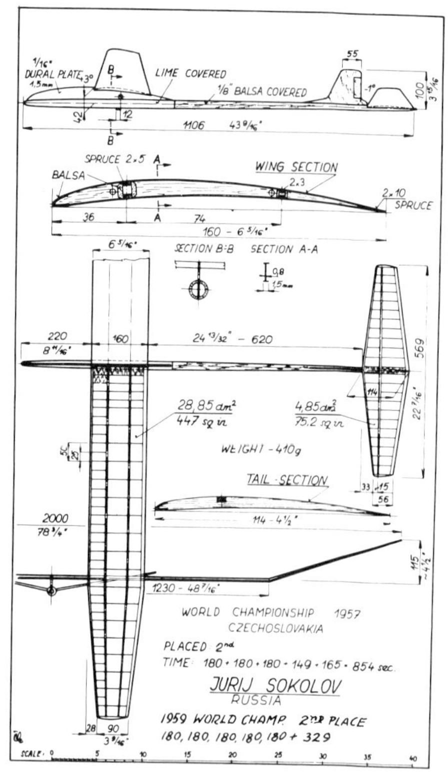

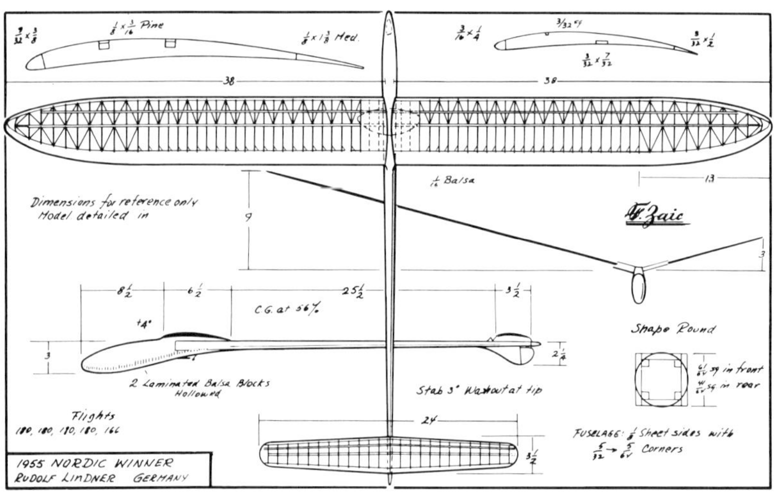

tunately such a model can easily spin out from strong lift. Lindner compensated this tendency with large amounts of dihedral — 13 degrees each panel. This trim method did not retain its popularity but Lindner's work did establish a new performance level. He was also the first to use long tows on line thus feeling for and finding lift. Naturally he was handicapped in those days by the simple towhooks and reliance on fuse DT.

In 1959 **Gerald Ritz** of the U.S.A. won the World Champs with his model *Continental*. This model looked already like a modern Nordic, featuring a high undercambered section, slim fuselage and short nose. Sharply angled-wing tips added a tip-plate effect along with their normal function as dihedral. The fliers of the USSR also did well in that period and the best of them, **J. Sokolovski,** flew a model which became the grandfather of today's Russian models and founded the so-called Russian School. Another Russian flier, **Averianov,** became the World Champ with a similar model two years later in 1971 at Leutkirch, W Germany.

The beginning of the 1960's saw appearance of a great novelty which was of immediate use to the glider fliers. This was the mechanical DT timer. Now it became possible to start the timer just at the moment of release from the towline. Unlimited time of towing now became possible. We did not care any more about burning fuse and did not fear the risk of short fuses. This led to the development of special towhooks. The goal was to make a hook which would allow for unlimited length of towing, even with the line slack.

The first steps toward today's fancy hooks were begun.

There appeared a great variety of "locks" which would keep the line attached even if running through a downer. Very simple and effective devices of this type were developed by 1969 World Champ **Elton Drew** who used a loop of thick monofilament which also creates the pin to start the timer and to deflect the rudder and another device by **Hank Cole (USA)** which was simply a rubber band over the hook.

At the end of the 1960's there was finally a quite new type of towhook which caused a revolution in the tactics of flying. It was discovered that the straight tow is not the only way to tow. It was also possible to make circles on line. The pioneer of this method was **J. M. Berthe** from France. his model *Flamingo* was furnished with an offset hook which allowed circling on line. The line had to be under tension all the time to prevent unwanted release. Berthe flew his model to reach several wins in international contests and his hook was copied by many with some success. The arrangement was often combined with a self-winding rubber motor powered line winder in order to keep the line always under tension. I saw in 1968 such an arrangement used by R. Hofsass of W Germany, known now for his Wakefield successes.

Sometime in that period was born in the USSR a telescopic hook. It did not allow circling on the line but it did introduce the idea of the *zoom* launch instead. Both the systems, the French circling on the line on the one hand and the Russian zoom launch was ripe for combining. The inventor of the combination was **Andres Lepp** from Estonia, USSR. It was not so easy as it now looks at first glance. Among other problems it was necessary to overcome the psychological barrier and to get accustomed with the great line tension necessary to release the model.

The team of the USSR used these hooks for the first time at the 1969 World Champs in Wiener Neustadt and became the team winner even with newcomers on their team. Since that time the new hook idea has spread all over the world like an avalanche. There are very few serious Nordic fliers today who do not use it. The hook enables the flier to reach any part of the flying field with his model on line and doesn't risk going O.O.S. upwind in a straight line which was a frequent problem before. It makes possible unlimited towing time and it is possible to have a rest with model still attached on line reducing fatigue and permitting the flier to think more clearly and remain in command of the situation.

There appeared a new problem connected with the new towing technique; *line tangling.* As more and more fliers used circle tow the situation became worse, but as the rules allowed a new attempt when a line tangle occurred it became possible to take unfair advantage of this rule and make flights until the max is made. It was certainly unfair but that was the rule. The FAI was forced to look for a way out and finally decided to make the flier responsible for his own line tangling. This new rule was used at the 1977 World Champs in Denmark and it proved to be a good rule.

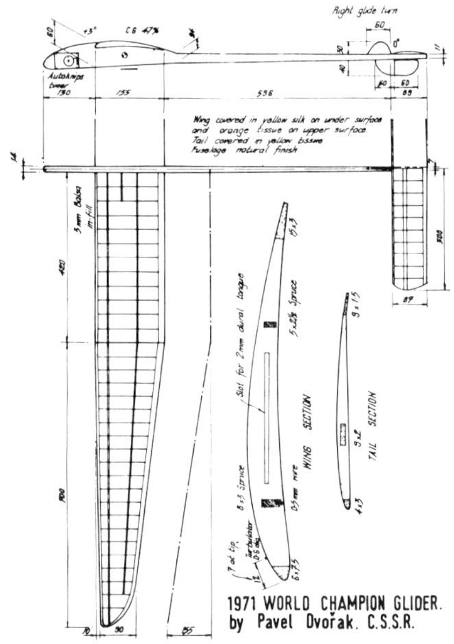

1971 WORLD CHAMPION GLIDER.
by Pavel Dvořak. C.S.S.R.

Another World Champs which is important in the history of Nordics is the one organized by Sweden in 1971. For the first time in history there was an attempt made to fly the World Champs without thermals. As far as the flying was concerned the attempt was unsuccessful. Even worse was the strain on accommodation and sleeping arrangements. We were quartered in barracks and what bothered us most was too short a time to sleep. We flew often till 11 PM and were awakened at 2-3 AM. Try that for a week! It made the event very difficult. The winner was **Pavel Dvorak** of Czechoslovakia. He triumphed over his rival from Finland, Nolo Munnukka, in a fly-off, thanks to a great "zoom". Dvorak flew his *Saper* model with high A/R wing with very long tapered tips. The model made use of hybrid wing warps which were sort of a combination of the conventional and Lindner-type warps. This combination enabled Pavel to zoom his model very high even with a pure conventional hook. The disadvantage of these warps is that it is very tricky to achieve the correct zoom pattern and requires very many trimming flights.

People were told long before the 1971 World Champs that it would be flown in non-thermal conditions and the result was that many developed very high A/R ships which, as was believed, were superior in those conditions. The interest in the big gliders was also due to the report by Peter Allnutt and Kaczanowski which appeared just in time, touting the advantages of high A/R.

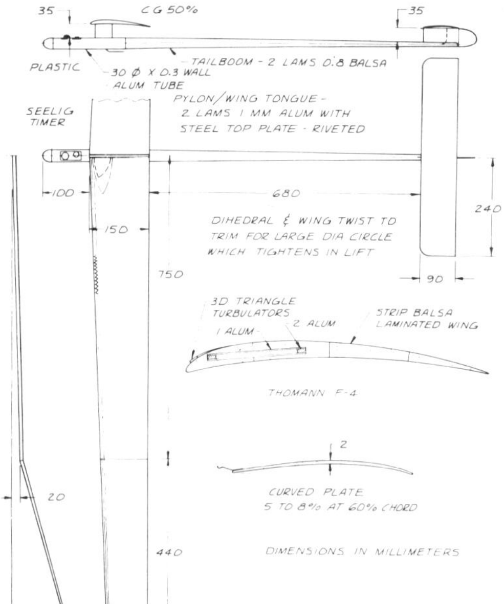

SPL GLIDER
SCHMIDT, PUTTNER & LASCLO
BRYAN SPOONER VERSION

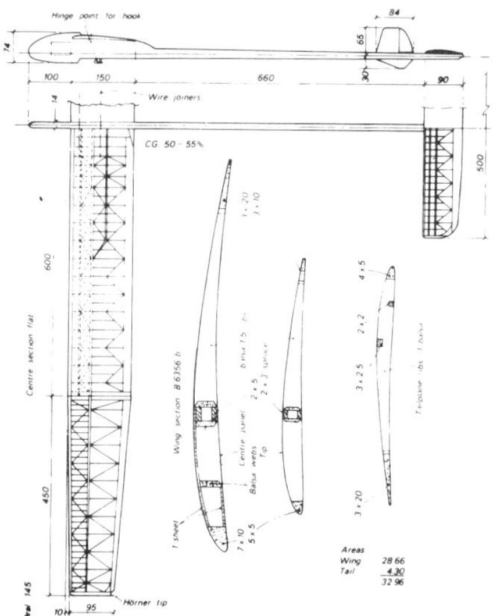

VALERIJ EKHTENKOV's F1A, EVG 1172

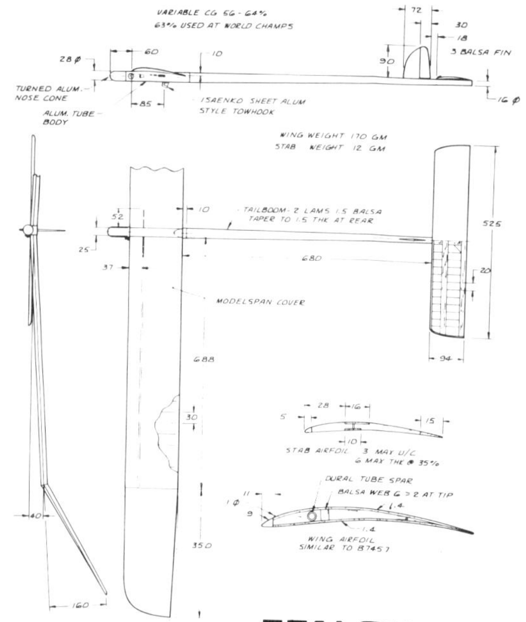

ETALON
1977 WORLD CHAMPION F1A
BY KOSTADIN ABADJIEV, BULGARIA

The best known modeler flying the high A/R models was **Herbert Schmidt** of W Germany. He, with his "SPL" model (A/R = 20), flew great at plenty of contests. The model featured solid balsa wings. Herbert had been flying this model type for some time. Personally, I too had two similar high A/R models, both performed beautifully in the dead calm air and still times were in the three-minute range. Disadvantage was that they were not suitable for the high zoom due to the high-lift, high-drag airfoil employed (Thomann F4). When comparing these models with the conventional ones it was found that they performed equally well since the zoom made up the difference in glide ability of the high A/R models. However, the conventional ships were not so sensitive and could be flown in all conditions. At present it seems the high A/R is a little out of date.

Jean Marie Berthe, France.

Andres Lepp, USSR.

The years 1973 and 1975 saw World Champs which were similar in some respects. The weather for both was so nice that there were 41 in the Wiener Neustadt fly-off and 42 again at Plovdiv. Winner in 1973 was **V. Ekhtenkov** of U.S.S.R. with a superbly designed and built model. Two years later there was the winner **V. Tchop**, again from U.S.S.R. He beat Peter Allnutt in the second fly-off round.

The 1977 World Champs was organized by the Danes and similar to Sweden in 1971, the intention was to fly the event without lift. The organization was simply perfect and the weather for Nordic day seemed good with only slight breeze. But the thermals were there, although weak and hard to detect. Due to these conditions the contest was very interesting from beginning to end. The leading flier was Andres Lepp who was the only one with a full house of maxes after six rounds. Unfortunately for him though, he did not max in the seventh and was beaten by **Kostadin Abadjiev** of Bulgaria.

Abadjiev is not new in the ranks of Nordic enthusiasts and has been flying this class for years. I met him two months before the World Champs at a contest of Eastern Countries. He flew Power there and he told me he got too old to fly the Nordic. He qualified also in Power to fly for his country at the World Champs but because there was no one from Bulgaria to fly Nordics in Denmark, he decided to try once more flying his old class.

The Danish World Champs was probably the best one in many years. All of us have to say thanks to the small group of organizers, led by Thomas Koster. They did a great job.

The winning model by Abadjiev is a modern design. It is worthwhile to note the wing. It is full-sheeted with balsa 1.4 mm thick and is covered with silk. This structural design used commonly on Power models, has now become more popular in Nordic. Along with the zoom launch there has appeared again the problem of torsional rigidity. The period of the zoom, which can be several seconds between release and normal glide speed, is critical for all the flight. It is possible to gain more height which is like gold in the bank. But, on the other hand, it is also easy to lose that height through bad trim or technique. It is necessary to have a model in which the wing warps absolutely do not change from time to time. Also flutter which is often a problem, can really kill off height gain.

There have appeared many wing structures which solve this problem and the full or partial sheeting of wings is one method. Nordic development at present is going to models which are able to withstand the zoom stresses and which are rigid enough not to warp. Here should be mentioned the Russian design of J. Sokolovski which has influenced a number of other designs. His models featured a D-box wing giving good torsional rigidity with low weight. Typical weight is 120-125g. Airfoil is of the B6356b type which has relatively low undercamber and is of low lift, low drag type for better zooms. Wing dihedral is relatively low together with larger fuselage nose side area. The CG is located rearward as far as possible, while retaining stability, to have more speed in the zoom.

At the opposite side, in some sense, are gliders from California designed by Isaacson, Hines and others which are fitted with the CH-407 section. This airfoil has greater glide potential but is more draggy in the zoom. Anyway it is hard to say which is best.

Recently there have been developed devices which improve the zoom transition. The goal is obvious as the transition is too critical and reliability and adjustability are desired. First step was the aileron on one wing. With the model in a hard bank shortly after release, the aileron moves, leveling the model. Unfortunately the aileron is neither easy to build nor operate. Moreover it tries to counteract the unwanted forces rather than remove them. With the model in a hard bank the deflected rudder acts as an elevator. I noticed this phenomena when flying my R/C glider and tried to solve it with the delayed rudder deflection on my Nordic. That was three years ago and it worked. Although I can proudly say that I invented this idea I don't really know if I was the first.

The system works as follows: the rudder is not deflected fully for the normal glide at the moment of release but several seconds later. The dihedral is strong enough to level the model and when all is OK the rudder deflects to its normal flight position. This gadget works well and a number of people use it now. It is convenient as there aren't any extra moving parts.

Another idea tried is the stab incidence change for the transition. I was told that it wasn't good when the decalage decreased during the transition and that better results were obtained when the decalage was increased. Some people also use the stab incidence change to achieve tighter circles on tow.

More fliers now seem to be using the telescoping type of towhook, either made from tube or flat stock. With these, four rudder positions are possible: straight tow, circle on line, zoom position and glide. All positions can be adjusted. The disadvantage of this hook is that if there is a sudden strong increase in tension (gust, strong thermal,etc.) the rudder deflects, the hook unlocks and you must launch. This advantage is avoided by the *Maxaid* hook of Elton Drew which is designed to cope with the gusty English weather, but a good zoom is more difficult to get.

The circling-on-line is a great advantage but it is necessary to learn to use it properly. This takes practice, the more the better. Circling can give some rest to tired legs, it lets you reach any part of the field and finally the piggy-backing is far easier. On the other hand towing all over can be overdone. Circling in the wind is usually useless and dangerous. Circling when the lift is very weak is not good in looking for your own lift as it destroys the continuous feeling in the line and it is hard to detect the small differences in line tension.

What is there to say in conclusion? Some fifteen years ago I was told that the development of the Nordic was finished. I was told that all the gliders were of a similar layout, etc. Some people say it each year since, but when looking back it certainly doesn't seem true. What will the future bring to us? We shall see! ■

Wakefield Techniques
— by Bob White, Monrovia, California

Flying Wakefield successfully can be divided into three parts: the construction and design, the equipment used for detecting the proper air in which to launch the model, and last, the proper equipment for testing and winding the rubber motors. The design of the plane is clearly defined and available to all. After 50 years of evaluation and study, at least the general outlines and wing sections are known and are available to all who would be competitive in the sport.

The motor tube is generally strong enough to stand repeated motor breaks. The tail boom is balsa sheet, either a box or a tapered round tube. It is constructed as light as possible, but with enough strength to stand the rigors of several years of flying.

The front end is solidly built with ballbearings supporting the shaft and thrust loads taken by the front bearing. The nose block is detachable for winding and a separate winding hook or stirrup can be used to transfer the motor from the nose block to the winder. The propellor consists of a hub and detachable blades that can be replaced or adjusted for different pitches or merely checked for tracking and identical pitches. Gauges and fixtures are necessary for this and must be used before any serious flying is done.

When constructing the blades, very hard balsa or hardwood is used since much of the outer portion of the blade is quite thin and would shatter if soft wood was used. One of the two major methods of constructing the blade is to form it over a precise pattern by soaking it in water, then pressing it firmly to the pattern. Three or more laminations are used and epoxy or other very stable bonding agents used to cement the layers together. The other and more traditional method is by carving from a block. The pitch angles are precisely plotted on the sides of the block, and then the blade is carved. The finish work is done with great care. Thickness measurements are done with micrometers or precision calipers. Blades are balanced with a beam, one blade against the other. The finish is usually clear lacquer or epoxy. The use of turbulators on top of the blade has been experimented with and this has been done with some degree of success. True running props with precise pitch and tracking are essential.

As stated before, the body of the plane must be strong and the tail boom light, yet strong enough not to break in normal use. The wing and tail attachment points must be constructed so as to position the wing properly when placed on the body. The tail must also be keyed so that it can't shift or change in any way.

Wing construction must be rigid and resistant to warping. Thin balsa sheet can be used to cover top and bottom of wings. The "D" box method, as used in Nordic A-2, or the Union Jack or geodetic can also be used.

Bob White

Because of the thin wing sections in use today, the wing must resist warps and also resist flutter in the power burst. The most common method of covering, unchanged over the years, is still tissue and nitrate dope, well plasticized.

The use of turbulators is being used more than in the past, especially with the thinner wing sections. One could be located about 5% behind the leading edge and be about .015 of an inch high. The second turbulator should be about 20% to 30% behind the leading edge and about .020 inches high. This has the effect of softening the stall characteristics of the plane and helps in turbulent air conditions.

Wing areas are averaging about 240 sq. inches up to a maximum area of about 250 sq. inches, though this is extreme, and causes problems in turbulent air as the stabilizer is quite small at about 40 sq. in. and this is not enough area for reliable performance. Some experiments have been done on high aspect, small area stabs, which have shown some improvement over lower aspect ratio ones. The modern trend in stabs is to use the same warp resistant structure as the wings, only lighter, but still strong enough to survive. Stab sections are usually low lift and flat-bottomed or slightly undercambered. Areas run from 40 to 55 sq. inches. Tail moments have lengthened and are now usually from 27 inches to 30 inches between the wing and stab.

In most parts of the world the flight pattern of the average Wakefield is usually a right climb and right glide. The notable exception is in the western U. S. and Denmark, where the majority of flyers fly a right-left pattern. The right-right pattern usually requires an auto-rudder on steep climbing planes, and it is thought that with this method, it is easier to put a plane in a thermal with no loss in transition from power to glide. Trim is difficult though, and the launch of the plane is more critical. Also, this flight pattern requires an auto-rudder on faster climbing models. Auto-anything is subject to failure, and, the flyer must use great care in constructing this device. The right-left pattern makes it a little simpler to adjust the plane and the flight pattern is more reliable, due to the action of the rudder and the torque of the propellor. The plane turns to the left and it is thought that it will stay in a thermal better than a right turn.

To get the plane down, a fuse is still in wide use, but is a disadvantage when precise dethermalization is required. The mechanical timer is more precise and has wider use in areas where there are small fields and precision timing is necessary. Mechanical timers are also useful for actuating auto-stab and auto-rudder mechanisms.

Ease of launching the plane is also quite important, as many times it will be impossible to give it a perfect angle when launching. It must be able to cope with a crosswind launch in either direction as well as a too steep or shallow launch. There are many times, in competition, when it is necessary to point the plane in a different direction from the ideal, in order to place it in the proper air.

The power of the Wakefield is always a problem, and the best possible rubber motor is required for the competition flight. Several methods of testing have been developed to cope with the widely varying rubber. To briefly describe the method used — the rubber is stretched about seven times its original length, or until it will not stretch anymore with a moderate force. Then it is wound to its maximum capacity, a torque reading is taken in inch ounces. Then as the motor is unwound, readings are taken, and a graph can be plotted, torque versus winds. The more area there is under the torque-turns curve, the better the rubber. Also the flatter the curve, with the peaks more leveled off, the more useful the motor will be. There are many fine and sophisticated test methods used by different flyers, but they all accomplish the same task, which is: measuring the total energy returned by the rubber motor and the rate at which this energy is returned. As mentioned earlier, it must plot a smooth curve without a sharp bend after the initial power burst.

The test of a particular batch of rubber to take winds is always interesting and exciting. A pre-stretched motor is wound to destruction and the turns, torque, the ambient temperature and proper lubrication noted. If you expect to use the rubber at a temperature of 85°F, you must make the test at that temperature. Temperatures above will lower the ability of the rubber to be wound. Also, at temperatures below 40°F, the rubber will become stiff and fewer turns can be used. The rubber lubricant is very important. Several different lubes are used. Silicon oil, soap and glycerine, and caster oil are the most common. I usually use soap and glycerine, as it is washable and not harmful to the rubber, and its lubricity is high.

The equipment for winding the rubber is quite varied, from an old hand drill and hand-held model, to the specially designed and built, all-ball-bearing, helical geared winders, with digital read-out and turn indicators, and with torque meter attached. Generally, the winder with ballbearing shaft, and a good thrust bearing to take the load of winding, is important as a feel of the rubber can be sensed by the flyer, as he winds the motor. Also, the safety latch is a must to hold the winding loop or stirrup. The winding tube is another safety device to protect the plane when winding the motor. It prevents the occasional breaking motor from damaging the plane and splitting the fuselage, as it absorbs much of the shock.

During the past few years there has been an increasing number of flyers using a holding fixture to retain the plane, when winding. This gives the flyer the advantage of being able to fly without a helper as well as eliminate the "human factors" of the holder. The holding fixture is a rigid stake driven in the ground, and held upright by two strong lines also held by two tent stakes. A torque indicator can also be used in the holding fixture. It gives a continuous reading as the motor is wound and with the continued deterioration in the quality and consistency of rubber, this is quite important.

The power of the plane is now well understood and much of the guesswork is gone. We know how to select good motors of a measured power output and how long the motor can be held while searching and waiting for the "right air" or a bubble of lift to come by. Even though the plane is an efficient design and the motor is properly tested and good quality, it must be put in the proper air conditions. The air must be neutral or rising and not in a downward movement. Indications of this movement can be obtained with an electronic device to measure small changes in temperature. This consists of a thermister on a pole, an amplifier and a meter or a strip chart recorder to obtain a visual read-out. A tall pole 20 to 30 feet high is positioned up wind with a long mylar ribbon to show wind direction and vertical movement of the air. The ribbon and cattails or "fluffies", and soap bubbles, make the air 'visible'.

The system used is to observe the temperature meter and its fluctuations. When the temperature shows a sustained increase and the wind movement decreases,

Bob White's Wakefield stooge has a torque meter built in. Neat idea.

start looking for an indication of rising air, by observing the bubbles or fluffies and also the ribbon. The horizontal air movement should then start to increase and the visual indicators should show a definite rise and possibly a swirl or twisting motion. At this time the wind should increase and the temperature should start to fall. The plane should be launched at this time. It should be noted though, that the exact time of release may vary from day to day and during different times of the year. Also, the horizontal wind velocity is a factor and during higher wind speeds the events occur too rapidly to be observed and the plane must be launched only in the lull of the wind and with a rise in temperature.

The design of the Wakefield, as well as the support equipment may seen to be quite a lot of effort, expense and time to the newcomer but, we must remember that the ultimate goal of this event is world competition with the best flyers from every country. Many of these participants have a lot more at stake than we do and they put all the effort and skill possible into it.

A very happy part of our sport though, is the free exchange of ideas and information. The serious Wakefield flyer can get information from around the globe in the form of newsletters and magazines, where most of the best designs and techniques are described.

In conclusion, the Wakefield has evolved into a refined and sophisticated event in which the newcomer must devote several years of study and building as well as practice at the flying field. Gone are the good old days when the modeler could build a couple of Wakes from a magazine, buy a hank of rubber, and successfully compete in the finals without any serious preparation and training.

Today, it is a long way to successful competition. Not only must you have the finest possible plane, but you must be familiar with how it flies in every situation — light air, strong winds, violent thermals, still early morning conditions, dry, wet, and, above all, the late afternoon fly-off, for this is where the World Champion is made.

There is plenty of tension and stress involved in the tough competition. The flyer must be in good physical condition, and be able to control himself and his emotions during the eight to twelve hours of competition. I know, from personal involvement how exhausting this can be, but as in any sport, winning is a great experience and the privilege of competing well is the best reward one could ask for. ■

Cat tail "fluffies" are floating upward as Bob White launches his Wakefield. Thermal skill is a must. Good ground equipment important too.

Wakefield Perspective
— by Bob Wells, Hornchurch, England

What is recent with respect to Wakefield models development is all relative. Most obvious developments are brought about through the challenge of rules changes and as these have hardly been altered now for over twenty years, model changes have been equally slow. To trace the trends in Wakefield it is therefore necessary to go back to 1954 when for the first time, the amount of rubber that could be used was restricted.

This rule did not cause an overnight change in the models because the amount of rubber allowed was 80 grams and that could be made into motors to fit, with very few exceptions, into existing fuselages. All that was necessary to meet the new rules was to add a couple of ounces of ballast to the old models and as the Championships were then held annually, it was by far the quicker way of getting a model ready. By the present standards, 80 grams of rubber was an enormous amount and even 1947-48 models complete with freewheeling propeller could still be competitive.

However, by 1955, the surplus weight was being built into sheet fuselages, less prone to damage by bursting motors. In 1956 the Russians produced sensationally long tail-moment models and with the dropping of the fuselage cross section rule, circular section fuselages became popular. The folding propeller, already well in favor by 1953 became almost universal with its obvious improvement to glide performance.

It was in 1957 that the old Wakefield finally died with the rubber weight reduced to 50 grams and the World Championships to be held on a bi-annual basis. The Wakefield had become F1B. The first event to the 50 gram rule was held in 1958 and the "new look" models at last came into their own. The British models particularly, with many others, were conservative in design, retaining much the same plan form as before and even built-up tissue covered fuselages in some cases. The winner, Baker, though had a moulded balsa oval cross-section fuselage, longish tail moment and twin fins. Zurad's second place model had extremely long tail moment and Johansson's third place had circular sheet fuselage, long tail moment and sheet-covered wing. The weather was windy so the "freak" models had proved they were not just calm air machines and were here to stay.

Other notable models at this contest were the Hungarian's featuring circular balsa fuselages, very long tail moments and sheet covered wings — they won the team prize. Perhaps though the most adventurous model was the Japanese Nonka's with a 64-inch wing span, geodetic-curved dihedral elliptical-tipped wing and would you believe — an alloy tube fuselage! This machine was flown proxy and did not do very well but was reputed to have a high "still-air" performance.

Bob Wells.

The point about all this though is that any of the shapes and constructional features of 99% of Wakefields built in the 20 years since then can be found in the models mentioned.

The 1959 contest was held in very hot weather which enabled seven competitors to reach the fly-off. This was the biggest fly-off in the event up until then; the very things that all the rule changes were meant to eliminate; just shows what a few thermals will do! The models in the fly-off were a mixture of just about everything that had gone before but of course they all had sheet fuselages and folding propellers. However, a great number of the entries were of the longer, thinner, variety so the previous year's new models had made an impression. Although of modest design, Bob Hatschek's model was notable for its performance and was subsequently copied by the Russian Matveev whose own model was one of the longer ones. A definite trend here started by the introduction of clockwork dethermalizer timers; Joe Bilgri used one, maybe the first.

The contest now fell into its bi-annual pattern with the next event held in 1961. This again was warm and calm but only three made the fly-off. The long tail-moment, slim-fuselage model was now the rule rather than the exception and winner George Reich's model was one of the most "old fashioned" looking, particularly with its open structure fuselage. Comparing it with 1953 models though, it was definitely "stretched". Its other unusual feature, the anhedralled tail plane was not copied.

Clockwork timers had become popular but flatout thermal seeking was not too evident. With most shapes of model now having been tried the efforts towards higher performance concentrated around the propeller and its assembly. A few machined assemblies with large diameter shafts running in journal bearings with adjustable pitch blades were seen in 1961. These were much in evidence in 1963 when Loffler won a four-man fly-off. His model was quite ordinary as was second place Hakansson's but Murani's third place was of the extremely long tail-moment variety and the other model in the fly-off, Wagner's had a sheeted wing so there was no definite trend amongst the top models.

Nineteen sixty-five produced the now usual mixed assortment of designs but most had settled for a moderately proportioned wing and tail on a long moment-arm fuselage. This was the year when thermal seeking really came to the fore with several electronic detecting machines and blowers of soap bubbles. Consequently, twelve made a perfect score and because of the thermal condition prevailing, few further rounds of progressively increasing maxes were necessary to decide the winner. This was Koster flying a straight forward model and second was Matveev flying his derivitive of Hatschek's model of five years earlier.

The two leading models were similar in having tissue covered wings of five-inch chord, 22-inch diameter double blade folding propellers and wing sections that to all intents and purposes only varied in that Koster's had a sharp leading edge entry and turbulator; the winning model also had a longer tail moment.

Rather than alterations to basic model design, this year saw the wide use of precision engineering in the form of fancy propeller assemblies, (Koster's although basically simple, used what is popularly known as the "Montreal Stop"), automatic rudders usually worked off the propeller stop and several variable incidence tails.

It would seem that more emphasis was being placed on getting existing designs to fly better than to try new shapes, sizes, etc. The fact that thermal seeking techniques, used extensively in accommodating weather conditions produced a twelve man fly-off brought the usual panic reaction from those in power who for some reason hate fly-offs and for 1967 rubber weight was further reduced to its present 40 grams. This brought no significant change in the models, the most drastic being a move of motor peg position.

The problems remained essentially the same as those associated with the 50 gram motors but were of course accentuated. The main problem was (and still is) to either extend the motor run by using a relatively thin cross section motor or longer propeller or to try and make more use of the first burst of power associated with a short motor. The initial power run represented a greater proportion of the total motor run. This, of course, was a problem with 50-gram motors but it was possible to extend their run to 45-50 seconds with efficient energy release to climb the model. With the reduced rubber, this sort of run gave very marginal power. Previously, the most successful models had a 35-40 second run but it was difficult to trim the first burst, it was evident however that if the fast initial climb could be trimmed correctly, the resulting over all performance would be better than taking the easy way out and using a long run. With 40 grams, everyone (with a few exceptions) had a first burst whether they liked it or not thus the trimming aids such as automatic rudder and variable incidence tail became widespread.

The 1967 models had all these gadgets amongst them including Koster's which was not fitted with V.I.T. The angle changes were affected either by trigger release from clockwork timers, propeller stops or by torque reaction of the motor. The first was the easiest to engineer and the last subject of most experiment. The winner that year though, used none of these things, just a straight forward model flown every flight in a thermal to top a list of *sixteen* fly-off qualifiers.

Nineteen sixty-nine produced much the same variety of models but conditions were more difficult and only one had a perfect score; again the model had no frills except a simple form of variable pitch propeller. The pattern continued through 1971, '73 and '75 with nothing startlingly new except perhaps the delayed start propeller. Although used by Mike Thomas at the 1973 Champs and used ever since by John O'Donnell to very good effect, it has not become popular which is strange because once mastered, it does give a positive (as opposed to theoretical) gain in altitude. Its one drawback apart from the obvious disaster if the mechanics fail, is that it is not suitable if the propellers fold under the wing as with nose moments becoming shorter, many now do.

Although perhaps not an immediately obvious development this latter feature of short noses has become more apparent in recent years and does seem to give a less critical trim on the first burst of power. It seems to have come about due to a combination of shorter motors, shorter tail moments and the "Montreal stop" making the fold possible.

So what have been the developments over the last few years? It would appear not many that one would notice and it is a fair bet that most of the models at the next World Champs will not be very much different to those of 1958. There has, I suppose been a trend towards the very high aspect ratio wing, sheet covered, but although claims of very high performance have been made, these have not been borne out in practice.

Propellers have rightly come in for a lot of attention with various theories being put forward. Naturally, the more successful fliers are given a more attentive hearing and during the mid sixties it was the Danes, but now it is the turn of the North Koreans. They have recently emerged as top Wakefield fliers and because their models are otherwise quite ordinary, the large diameter hardwood propeller has come in for much attention and it remains to be seen how the inevitable copies perform.

Wing aerofoil sections have been used in such diversity that none can be called the current trend although the Benedek sections have always been popular and there is no sign of this diminishing. Tailplane areas seem to have settled around 25-30% of wing area after a bout of very small units which now seem to have been more of a hazard than an advantage. Most development since the late sixties has been in thermal detecting techniques which is not surprising as even the very best possible model to the current rules must at the very least avoid sinking in the air if the flier is to have any chance of winning.

Still, if we all keep trying perhaps one of us will discover the elusive breakthrough and once again produce that genuine "five-minute" Wakefield. What this discovery will be I cannot imagine; if I did I would be keeping very quiet about it until after the next World Champs — perhaps someone somewhere is doing just that! ■

FAI Power Models Status
— by Bill Hartill, Canoga Park, California

Rules and Specifications. FAI rules and model specifications for all free flight categories have seen little change in the last few years. This has been met with a certain amount of satisfaction by most FAI fliers who prefer as little disruption as possible to their designing and flying procedures. Past rules disruptions, attempting to keep up with the rapidly increasing power of the 2.5 cc engine, were not always greeted with unanimity. Some of these changes were felt to be made without clearly thinking through cause and effect. For example in 1966 nitro-compounds were outlawed in FAI fuel and a standard methanol/castor oil mix was adopted. The reason given for this change was that nitromethane was difficult to obtain (read that *costly*) and so this prevented young modellers with limited resources from being competitive. Instead of achieving the desired result, however, just the opposite happened. Some of the sharper engine builders found that with clever, expensive designing and machining techniques, highpower could be gotten out of this standard fuel. Although now, instead of $10.98 for an OS Max and $2.00 for a can of "hot" fuel, you have to pay around $70 for a Rossi, and $4.00 for a can of FAI fuel. Even with inflation figured in, it's pretty easy to see that this rule change hasn't had the effect that was originally claimed for it. A similar story can be told for tuned pipes. Outlawed in 1970 because of claims that they were beyond the capabilities and availability of beginners, these inexpensive and easily built intriguing aero-thermodynamic devices are denied any consideration.

Engine power has increased markedly in spite of these restrictions but this power has not become "more available" (*read that more inexpensive*). The restrictions are now looked upon as a method of keeping flight performance down to restrict fly-offs and to limit requirements of flying site size. Attempts in earlier years to cope with developing engine power resulted in higher weight loadings, shorter engine runs and shorter maxes. The most recent change, which took place in 1976, was to reduce the engine run from ten to seven seconds. In 1975 fly-off engine runs were reduced from the standard 10 secs to a progression of eight, six and then four seconds. Although the World Championships (Bulgaria) were flown in this manner, most contestants did not favor this approach. Changing timer settings between fly-off rounds was a tricky thing and many also felt that it required different trimming of the transition. Cynics might suggest that what looked like an acceptable transition at 500 feet was, on closer examination at 150 feet, not all that good.

The seven-second engine run now standard for all flights was first approached with apprehension by many competitors. The ability to time the runs accurately and uniformly was doubted. Field tests were conducted by a group in England and reported in the June '75 issue of FFN. Twenty-four timekeepers timed a series of flights. Band width of deviation from the eight second mean varied from - 1.0 sec to + 0.8 sec. Deviations of up to 0.4 sec were numerous.

In the U.S. several gadgets have been proposed that would make a signal on engine stoppage. The most intriguing of these was Dick Lyon's *Flasher.* The flood-off actuator was arranged to close the contacts on a self-energized flash bulb (Magic Cube). Said flash being visible from the ground the timekeeper would, presumably, associate the flash with engine stoppage and punch-off. The weak part of this idea being that the flash and engine stoppage were only indirectly related events that would not be proven linked when the model was 500 feet overhead. Another proposal which has not been tried, is to allow a *total* summation of 49 secs of engine run for the seven rounds. Thus some leeway would be allowed until, of course, the last flight. This method would require the recording of the run time and not just the belief that it was "OK".

As it turned out, the seven-second engine run wasn't too great a problem after all, particularly since the FAI did adopt new fly-off procedures so that only one engine was being checked at a time. This also allowed the same two timekeepers to time all fly-off engine runs. And also the concurrent widespread use of engine brakes reduced timing difficulties since the brake stopped all the burbling and windmilling that complicated the timing.

Periodically a few voices are raised in favor of a lower power *Motorsiegler* class model, something that would move slower and hopefully deemphasize importance of the powerplant. However, no great impetus has grown for development of this class and the current F1C class does have an enthusiastic following.

Design. The very high power available from the modern 2.5 cc glow engine, together with the relatively short engine run (7 secs) has led to a new approach in balancing the ever widening climb and glide flight characteristics. The Zipper showed the way in getting automatic stability by setting turning, rolling and looping tendencies in balanced opposition. This produced the "safe" layout which has served us well. But opposed forces also mean that drag is greater than the potential minimum. Enter at this point, auto surfaces. First used as a trimmer on the conventional pylon layout to reduce looping tendencies, the auto-stab is now being used as a basic design element with maximum advantage being made of its potential. Climb patterns are now straighter and steeper than previously and now require a great deal of attention to the transition. Tom Koster, for example uses a *down* kick on the stab TE to drop the nose from the vertical climb (momentarily) prior to moving to the glide position. Rudder movement also is coordinated to turn the ship into its glide turn. Too soon and a spiral dive will develop, too late and a straight ahead phugoid stall will build up.

The next step beyond auto-surface trim changes is the use of variable wing camber (flaps). Bill Gieskieng of the U.S. and Tom Koster of Denmark have done most of the pioneering on this idea. There is now no question that a performance gain is available with flappers. However a more widespread adoption of this design approach has been inhibited by mechanical difficulties in making the mechanism positive, reliable and goof-proof.

The high power and programmed in-flight trim changes have now led to a real drag clean-up design approach. Exposed flat firewalls are out, cowlings are in. Plug-in two piece wings eliminate draggy gum-band and dowel fixing. Sheeted surfaces are higher aspect ratio than before and sections thinner. Flat-bottomed sections are more popular by far and range from about 9% to 6½% thick. These thin flat sections can produce high L/D but must glide flat and fast. For this reason drag is critical. Verbitsky (USSR) uses a stab section built flat but with high LE entry and reflexed TE to approach a symmetrical section. A number of successful designs use no pylon, with very clean shoulder wing positions no higher than the engine cylinder head. Check the *Design Survey* Book section to see typical area distributions, moments and sections of the top models.

Construction. F1C models have led a revolution in developing new materials and construction techniques. Propellor construction for example has become one of the leading "cottage industries" of the world. It seems that just about everyone has a go at it, and the starting point is the Cox 7x3½. Most still refer to that pattern as the reference with various modifications according to your theories and/or experience. Strength requirements demand that fiber reinforcement be used in molding props. Fiberglass or carbon fiber filaments are laid into the mold around the prop hole (not through the hole, please), and various resin or epoxy mixtures packed into it. The mold is put together, tightened up and baked in an oven. The carbon fiber types come out stiffer and can be made thinner but are more brittle and are prone to damage on landing.

Engine pans cast from aluminum or magnesium are commonly used and are also a "cottage industry" product. The right angle types attached to a vertical firewall are now giving way to the horizontal half-pan type similar to those used in U-control. This approach permits a nice engine cowl to be made integral with the rest of the fuselage. For access to the engine and

fuel system usually only one screw need be undone to detach the pan.

Circular section fuselage tail booms are popular and these are made with either a rolled balsa tube (2 mm thickness) covered with light cloth and resin or are cast up completely with cloth and resin without the balsa. Some have been seen with carbon fiber reinforcement. Forward fuselage sections are now being completely molded in glass cloth and resin by a number of builders. This method turns out smooth, nicely contoured shapes that are not difficult to make once you have a pattern made to your liking. Another approach for the front end is to completely machine it from alloy bits. A very good example of this can be seen on the models of Verbitsky and his Russian teammates.

Wing and stab construction is most commonly done with all sheet covering over internal ribs and spar reinforcement. Light glass cloth and resin outer covering is done also by some to add a lot of strength but care must be taken to keep the weight down. Wing skin is usually 1 mm (1/16) balsa while the stab can be ½ mm (1/32) balsa. Two piece wings are now made to plug into the fuselage (or pylon) in fashion similar to Nordics. A single 3/16 inch dia. music wire joiner can be used with a smaller locating pin at the TE for attachment. A 1 x 10 mm or 1 x 7 mm steel spar joiner can also be used which is fitted into brass box sections fixed to the wing panels. The two panels are hooked together with a tension spring or rubber bands across the fuselage from wing root to wing root. This may not seem too safe in a crash but then nothing is when things go wrong in F1C. Other wing-holding techniques include the use of nylon machine screws run through alloy saddles. These screws can shear on impact.

The Russians showed up at the World Championships in 1977 with aluminum alloy covered wings. A .03 mm heat treated alloy was bonded to the outside of balsa sheeting which was then wrapped around an aluminum tube LE and balsa ribs. A number of modellers in the west are now trying out this interesting idea. These wings come out very stiff with a smooth surface. Attempts have been made to use foam-core wing structure but weight still seems to be a problem as the foam must be sealed with a skin and it must have spar reinforcement at key points. One possibility which has yet to be tried is a .03 mm dural skin over foam.

Engines and Accessories. The Rossi is King and the overwhelming choice. Only serious competitor which has appeared recently is the Cox *Conquest.* The Cox is comparable in power and has been quite successful in the hands of a number of U.S. fliers, but there has not been a big shift away from the Rossi. The stock Rossi *Normale* Free flight engine has gone through few changes, the biggest one being a larger diameter crankshaft. Stock engines all seem to be ready to go with little variation in output or handling characteristics. An ABC version is available; that's an aluminum piston, and chrome plated brass cylinder liner. The idea here is that an aluminum piston is less than half the weight of an iron piston and so absorbs less energy in accelerating and decelerating. An aluminum piston in a steel cylinder liner would have a big problem though in that the coefficient of expansion of aluminum is about twice that of steel. Solution; use a brass liner, whose coefficient of expansion is 18.7 compared to 23.8 for aluminum. By chrome plating the brass you get wear resistance exceeding steel.

Making a Rossi run faster is not an easy task as the basic set-up is so good to begin with. However a careful adjustment of the bore, piston, and bearings fit can be helpful. Head clearance should also be experimented with along with plug type and range for your flying conditions. The use of pressure, a hard tank and flood-off system is universal. Many F1C fliers replace the stock venturi insert with a longer, larger throat version. Variations on the manifolded multiple jet stock system are common and a single orifice seems to work okay. Some of these inserts have been made of Delrin or Nylon to reduce heat transfer to the fuel. Temperature control of the cylinder is also being given consideration with some cowlings offering no entry for cooling air. Others have tried wrapping the cylinder fins with a band of metal or even tape to get the engine operating at its best with less warm-up time.

Stopping the engine is another matter and the most popular engine timer is the multi-function clockwork Seelig. These must be given care and attention as they are a vital part of the flying system. Efforts have been made to develop an electronic multi-function timer as this approach should give exceptional accuracy. Two examples have been produced; one by Thomas Koster (Denmark) and one by Dave Parsons (California). The electronic circuitry is rather straight forward and simple to put together but the mechanical bits to give four functions from the electronic counter requires careful work. Flood-off is accomplished by releasing a squeezed-off line to dump pressurized fuel into the venturi. The engine/prop brake can be timed to go on at the same time or a few tenths of a second later. Brakes were first tried by Eugene Verbitsky (USSR) and his system uses a strong spirally wound claw that is allowed to engage in a slot in the rear of the prop driver. Other types rely on engaging a friction pad or a "strangler" wire gripping the prop driver hub. Bill Gieskieng came up with the simplest and lightest type of all, just a couple of loops of 50-lb. nylon monofilament line pulled tight around the prop driver hub. One of the surprising things about brakes is that they all work so well. Although engines are turning over 25,000 rpm, torque is low and it doesn't take much to stop them. The prop has the greatest rotational momentum of the system and that is what you must stop. A light prop stops quicker. Propellor mounting must be attended to. The Rossi collett mounted prop driver and spinner has a habit of letting the prop loosen up on the shaft. This happens because as the spinner is tightened it tends to distort the back plate so that the outer gripper edges lose good contact with the backface of the prop. Remedy this by countersinking the back of the prop hole slightly and try not to tighten the spinner too much.

Flight Technique and Competition. Trimming the modern F1C model requires a lot of common sense and a certain amount of daring. First of all don't even think about trimming a new ship unless all surfaces are well strapped down to the body. Warps must be removed or dealt with in the trimming session as a permanent design feature of the model. Auto-surface mechanisms must be variable through a range and all adjustment screws must have some locking provision. Hand gliding should be done with both power and glide settings. The power setting will require a pretty hard throw straight out. Do it over soft grass. Take out any left-turn tendency. First powered flight is usually a thriller. Set it up for DT at ½ sec after engine cut. Pray a lot and don't be discouraged. They always fly better after that first crash and repair job. If you are new to the shoulder-wing configuration remember that they tend to hook to the right under power. Wash-in doesn't help much in lifting that right panel and, in fact, seems to cause more drag on that side, making the situation worse. This is where the triple fins can be helpful. The central fin and rudder can be set for a left-turn bias under power to kill some of that right tendency and will be more powerful with the prop blast on it. The outer fins can be neutral or *slight* right. Auto rudder on the central fin then kicks the ship into the right-glide turn. There is no need to play with thrust line changes. Stab saddles must be well anchored and contact surfaces between stab and saddle and auto mechanism must be hardened material such as plywood or aluminum. Movement of auto surfaces (stab and rudder) must be positive and hold their desired positions under high speed flight loads.

What does it take to win in F1C? Common characteristics that are observable in the winner's circle are: reliable airplanes that groove, reliable engines that don't go sour, timers and auto-mechanisms that work right every time (no Mickey Mouse lash-ups), and a good air sense by the flier. A super hot engine is useful but not essential. Exotic construction is also nice but, again, not essential. Your equipment must be ready and dependable. You should have your flight preparation and engine starting routine carefully orchestrated. Anything abnormal then becomes easier to pinpoint and fix. Failure to correct some minor glitch (sick plug, leaky fuel line, loose prop, empty tank, broken brake, auto strings fouled, etc.) cannot be made up by superior performance. Air sense means, pick the right air and right time to fly. F1C models will max off a seven-second run if you don't allow something wrong to happen, and if you don't launch into a downdraft. Thermals are useful but not necessary and besides there is always a downer next to them. These you positively don't need.

F1C is a challenge to the spirit. When you get ready to launch that screaming ship into the air, the feel of the forces you are about to release are very real. The trim had better be right at that moment of truth. ■

Indoor Duration Free Flight
— by Laurie Barr, Berkshire, England

To me the most exhilarating sight in any branch of model flying, is to fly an F1D microfilm model to the limit, in one of the big hangars.

Having crammed every last turn the motor and the model will take, taking off *just* enough turns to drop launch torque to the point where the model will level off and begin the cruise mode, just under the center catwalk (Position "A"!!), count the prop RPM, divide it into the turns, put on the motor, and you know, *provided* it does not hang or collide with the sides of the building, you are about to do your best flight ever. Although most people at indoor meetings seem to be outwardly calm, (the slow walk aids the illusion), I can assure you that most of us are nail biters at heart, if not in practice, during such a flight. Personally I am twitching from the moment of launch, and during a

Laurie Barr, Great Britain.

whole day at say Cardington, even during a fly-for-fun meeting, I will seldom eat anything, but have been known to drink with both hands!

I hope this introduction to what makes modelers become hooked on indoor will arouse your interest to the point of trying this side of our hobby for yourself. Building and flying a microfilm model is not all that hard, and it becomes easier every time you do it. My modelling career has spanned all forms of free flight, radio, and control line, and I have been well rewarded in all aspects, but none more so than indoor.

The current FAI class at 65 cm span and minimum weight of one gram results in a strong model capable of performing very long flights. The structures are simple to make and cover, are amazingly strong, and my indoor models last longer, and survive obstructions much more than any other kind. Most of the building is simplified through the use of jigs, templates, etc. which can be used again and again. Assuming you have all the jigs and your film hanging ready for use, you can build a model in three or four evenings modelling time. Also assuming your design is stable, trimming is simply a matter of raising or lowering the wing posts to change the wing incidence, until you have got it cruising well. Then comes the crucial part of indoor, finding the correct prop and rubber combination for a given flying site on a particular day. If you have hard rubber one would use a large diameter/pitch prop, and on a good day with warm air and not much drift, you would use a positive flare propeller. That is one with more area in front of the spar than behind it, so that the prop goes into "coarse" pitch at the launch, to help "kill" the high torque at the start, and help prevent that hang up. On a bad day, cold and hard to get up through the inversion layers usually present, a negative flare prop will help to pull the model up into the relatively warmer air in the roof.

If you have "soft rubber", you will generate less high torque, but you will get more turns (A La Mode Romak World Champion!) and fly with slightly smaller pitch/diameter props. It is a constant source of debate between microfilmites which is the best route to take. Really good rubber is hard, but not impossible to come by, and in a Wakefield motor, you have enough of the precious stuff to last you for a couple of years at the very least.

Gadgets abound with the indoor flyer. We have: torque meters to wind on, which tell us (for a given site by practice) what launch torque will make you level off at the correct altitude; pitch angle checker for assuring equal pitch angles on both blades, or for "tweaking" more or less pitch by rotating the prop spar. We have tables of data for measuring the cross section of rubber (it varies a good deal) by averaging the length against weight, and also turns per inch your average rubber will take (Bud Romak got within *40 turns* of what his chart and test showed the motor would take!) and most of us have those much prized, and superbly made rubber strippers of the 1974 World

FAI class Indoor Free Flight model. One gram without rubber motor. Endurance capability, forty-plus minutes.

Indoor Champion, Ryzard Czechowski from Poland.

Another facet of the gentle art is steering the model by balloon, filled for safety with non-flam Helium, tied to a 1/32" dia *white* line (so you can see what is happening in the dark roof,) wound onto a quick winding reel, preferably a fishing reel of the multiplier type, with level-wind attachment. In England at least, we are habitual steerers, and will steer just about every flight with confidence at any altitude. Experience has taught us that the best method is to allow the model to fly into the line, between the port wing and the prop. As the model is trying to turn left, with the line resting alongside the motor-stick and up against the leading edge of the wing, whichever direction you walk in, the model will follow. To release (assuming you are not caught up in the prop from behind) (a sharp short tug on the line will usually free it), you just walk faster than the model, on a large diameter left hand path, and the model is free once more and out of danger. Most poor steerers leave it much too late and are not in position soon enough (I constantly monitor the drift of my model when near danger). The way to get plenty of steering practice, is to do what I did, pose as an expert, and do "it" for other people's models! They are usually apprehensive at their own chances with the dreaded balloon, and are easily conned, and if you get them out of trouble without "donging" the model, are usually life-long friends.

The indoor man is a special breed, he will readily impart all his latest technology, help you repair your model, transport it, lend you his best motor, and are usually great corresponders. A lot of underground stuff goes on all over the world by letters, and much discourse takes place about the merits of design and technique.

I hope by now you have been tempted (beware, the affliction is usually permanent!) and I would advise using a known design, but go up approximately 30% on the wood sizes and weights. After two or three models have been built you get progressively lighter structures, and confidence to actually have the nerve to get one of these things down to the exact one-gram lower limit.

Most models are between 1.1 to 1.3 gms. in weight, and flights of 35 to 40 minutes are possible. However it is very desirable to get down to the one-gram weight limit, for the model usually flies on a motor that is 1.3 times the airframe weight. If a model *is* overweight, it follows that its total flying weight with rubber, can be at as much as a 30% weight disadvantage. It flies faster, revs quicker to keep the contraption up, flies well enough, but will never make world-record times.

It was only during my preparation for the 1976 W/C at Cardington, that I built some models right down to weight. One of them during the Saturday practice session, did a flight of 39.39 mins, was "ballooned" six times (it was running out of shed, as the drift down the hangar was strong) and lost 75 feet in height during the process. Having counted the turns left on the motor which would have all been flown off with that missing 75 ft. would have made a 45-minute flight. I am told Kalina, who's rubber it was, that the motor should take more turns still.!

We are lucky to have Cardington so near and available, but it can only be usefully flown in during the summer/fall, as it is far from air tight and it does depend on getting warm calm weather outside, to make long uninterrupted flights inside. (Although Bob Bailey did once make history by flying an EZB model, through a missing roof panel, and was ambitious enough to make the timekeepers rush outside to continue timing, regrettably by this time it had either thermalled away or disintegrated.

As the "bloke" who runs indoor in the U.K for our modelling society (the SMAE), I had the "pleasure" of preparing the 1972 and 1976 W/C at Cardington, and some observations may help others who may stage other major world champ events. First you assure yourself of the hangar. You meet the man at the top in charge, and you meet and become friendly (and get him to see things your way as he will have no other way of knowing what indoor flyers want unless *you* indoctrinate him!) with the man who actually has the task of doing all the audacious things you are about to get him to perform in the *months* before. When meeting with the top man, inform him that you don't suppose he will do all you ask, but for courtesy sake, write him a letter afterwards containing a number check list of ideal things to be done, and also out of "courtesy" send a copy to the site manager. Ring up and preferably visit frequently to gently inquire on the progress — *Are the balloons (full size) being let down, will that or this hole be repaired in time? Are you sure the door closing mechanism (which has malfunctioned all summer long) really shut the door tight? They only weigh 400 tons each! Will they cordon off certain areas to the plan that you just happened to have prepared to make things easy for them?*

Lay on a good meal situation in the hangar and include all whom you are indebted to imbibe with you. Take the hat round for the rigger afterwards, who has risked life and limb to salvage your stupid model. For the banquet visit all the local hotels and get quotes from many. Make it a good meal, impress them with the list of V.I.P's invited, you will get a better deal and more attention. If you use a service station for accommodation and food as we did at R.A.F. Henlow nearby, do similar things, try to involve any modellers on the Base of whatever preferences, to help you get them interested. Above all find the 100 carat characters who are *going* to do wonders for your guests.

Try to think of everything an indoor modeller would want. After all he is coming half-way round the world to fly in *your* shed, so make sure it's like a visit of friends to your home. Delegate little if any planning and control. *Don't* recite your detailed plans to Headquarters. Just assure them all is well, and that way you get no interference. Think it *all* out well in advance, set yourself target dates for parts of the action and *do it*. Live, eat and sleep on all the arrangements — never give up. ■

Coupe d'Hiver Techniques
— by Pierre Chaussebourg, Rouille, France

The first Coupe d'Hiver rules were written in the No. 23 issue of *Le Modele Reduit d'Avion* September 1938 by the editor himself, Maurice Bayet (France). The weight of the model was a minimum of 80g including a maximum of 10g of rubber. The maximum duration of the flight was 180 seconds. In 1959, the rules were modified as follows: no change for the weight but cross section 20 cm^2 and maximum duration 120 seconds.

In 1971, Maurice Bayet agreed to increase the weight to 100g which would give more interest to the formula, avoiding too fragile airplanes. The CIAM adopted Coupe d'Hiver as a provisional series: F1G, with the French specifications. In 1974, an American proposal was adopted in favour of the old rule: 70g minimum without rubber, and 10g maximum for the rubber motor. None of the European Coupe d'Hiver fliers could accept such a change, but this proposal was accepted by the CIAM because of the votes of delegates who don't even know what a Coupe d'Hiver model is . . . ! Since that time, France tries every year to officialize the 100g weight of the F1G series (90g without rubber, plus a maximum of 10g of rubber). It is remarkable that every year, the nations who really fly Coupe d'Hiver in Europe, for a long time, are alone in favor of the heavy Coupe.

Many countries think that Coupe d'Hiver is a rubber model for beginners, but this is wrong. It is not the first step to Wakefield either. A beginner rubber class should be quite different and something like the P30, with a commercial free-wheel plastic propeller, and a flat section airfoil for wing and stab.

Design: There is not a particular airfoil section for Coupe d'Hiver. In fact, the speed is slow during the climb and the weight of the model is low compared to the flying surfaces. The position of the spars may be more important than the airfoil itself. Any thin airfoil, with well-situated turbulators may be a good Coupe airfoil. The aspect ratio does not look to be as important for a Coupe as for a Wakefield or a glider. The area of the wing and stab looks to be more important. With the 80g models, 12 dm^2 was very large. Now, with 100g the flying area is about the same as for a Wakefield and sometimes larger: Dupuis, 17 dm^2; *Trumeaux*, 17 dm^2; *Super Trumal*, 23 dm^2. The

Pierre Chaussebourg writes about Coupe d'Hiver but here he shows how to launch a Wakefield.

moment arm is about 50 cm, the stab area is about 1/5 of the wing area, CG position is around 70%. The 10g of rubber is usually 6 x 1 Pirelli. The motor which was made out of eight strands with the 80g models for a climb duration of about 15-20 seconds, is now made of six strands, for a climb of about 35 to 40 seconds, or four strands for a climb of 115 seconds . . . but there, you need real calm air. The diameter of the propeller is increasing: Louis, 465mm; Georges, 520mm; the pitch is between 465 for Dupuis to 480 for Matherat. The advantage of a long run is to climb as high as with a short run so you get the same altitude, but instead of 15 to 20 seconds you are in the air for a minute. If the model glides for one more minute, it is 80 seconds for the fast model, and max for the second one. Some flyers like to use 4 x 1 or 3 x 1 mm rubber. Twelve strands give about 20/25 seconds, 10 strands give 30 seconds. Eight strands should give about 60 seconds or more.

I don't think the design of the model is so important, except for very cold and calm weather. In winter without wind, it is necessary to climb fast to 60 meters and glide well (Bob White - *Beau Coupe* or Louis Dupuis - *Zebul*) or to climb in 40/50 seconds and also glide well (Bernard Boutiller or George Matherat). With no bad air it makes the two-minute max. In Europe for spring, summer and autumn weather during the day, you must avoid the downdraft to make a max, even if you fly a super design. Or you can easily make a max with a small Coupe which climbs to 20m, if you launch it right in the middle of a thermal.

Construction: The wood must be good; thin and light but strong, 1/32 for the ribs (8/10 mm) is okay, 2 x 2 mm spruce spars for the wing and in the central panels is good too. Diagonal ribs help to avoid warping, especially when the airfoil is thin. The stab and wing must be covered with light Japan tissue with light dope, with plasticiser in it for the last coat.

For the fuselage, there are three ways.

(1) The traditional way: Landeau, *Pamyscaphe* (F.Zaic Year Book 1964).

(2) The *Grenoble* way: Matherat, *Trumeaux* with a thin fibre glass boom. Another version of this, with Jean Wantzenreither, alias *MR 007(!)*, uses a branch of reed instead of the fiberglass!

(3) A moulded tube and a moulded boom made of balsa and tissue (Bob White, Jim Quinn). This can be made out of polyester or epoxy resin with fiberglass cloth.

(4) A moulded fuselage like Louis Dupuis or Jacques Boiziau. Very good and very nice, but difficult to make for a beginner.

The only thing not to forget when you make a fuselage is that the cross section must be at least 20 cm^2. If you make the fuselage out of a tube and use a pylon, this one must be wide enough to satisfy the rules.

Propellers may be laminated and moulded, or carved from a block. Personally, I like this last way better. It is so nice to make blades like that, and it is not difficult. The first one you make will work, and the more you make, the better they are. Nose blocks can be made out of balsa, with a brass tube inside. The other parts can be made out of piano wire. Everything must be simple and work perfectly. Some fliers like to make very sophisticated nose blocks which are very impressive and very difficult to make. They look like Swiss clocks, but they don't work better, particularly as soon as you fly them in the desert dust! It is important to wind without the prop, with a protecting tube inside the fuselage; Bob White's technique is perfect.

Rubber: This is the main problem! Some time ago, Pirelli was making good rubber and everybody was happy. But nowadays, the situation is quite different. It seems that the 3 x 1 mm FAI rubber (Ed Dolby) is performing well and that the light brown Pirelli we have seen at the World Contest in Denmark is going to be a good material too. For Wakefield fliers, it looks possible to make Coupe motors out of old Wakefield motors, but they would be suitable for trimming or practicing flights only. A Coupe needs a good quality rubber to perform well.

Flight technique: As for all rubber models, when everything has been checked okay at the workshop, you'll go to the flying field and will make test glide flights over a very soft grass. You can also do that in a park or on a baseball field. As soon as glide is okay, try the climb. Three ways of pattern are possible:

(1) Right climb — left glide. This is very safe and easy to trim.

(2) Right climb — right glide with auto rudder. It is about the same thing with a little more mechanics, because the auto rudder is operated when the prop stops.

(3) Right climb and right glide with tilted stab like Matherat *Trumeaux*. More difficult to trim but flies well too. Usually when a model is well-built and correctly trimmed you don't have many changes to make; just very small adjustments with different kinds of weather. For instance the CG can be moved forward in case of windy weather.

Performance: When the *French Aeromodelling Federation* (FFAM) decided to increase the weight to 100g, the reason was that there were too many people making the three maxes. So Coupe flyers put 20g of ballast on their models, and it was a disaster. Then some of them tried to make some research in the new formula and they found out some new principles. Most people thought that a small model with a fast climb was able to climb very high, and anyhow, higher than a larger model in the same time. This could look logical. But in fact the new thing was that it is possible to climb as high, but slower, with a longer motor run, and a very much larger flying area, which gives a better glide. So when the prop stops, the model is in the air for almost a minute or even more, and as high as the small model when its propeller stopped, but its gliding time will be longer. So, this new generation of model is in the end, performing better. Of course, there are many more possibilities of research about trimming, airfoils, propellers, surfaces, arrangement of the 10 grams of rubber, etc.

It is not difficult for a beginner to do well with these new airplanes, because the material may be strong enough and the weight will not be heavier than the 90g without rubber. When the CIAM decided to come back to 80g (70g + 10g rubber) the first fliers just threw away their 20g of ballast, but the other flyers, those who took the time to build new airplanes, were also competitive and actually better with larger and heavier models (See 1975 results at Coupe d'Hiver in Halton, Great Britain).

Since that time, with the experience of the larger

models, some very good builders have built the same models at 70 grams which is not a job for a beginner. And these models are fantastic things! The Italian fliers, with their new rubber and their new models have been so good at the last Coupe d'Hiver contest in Torino, that they have decided to increase the weight of their Coupes to 90g (without rubber) as the French fliers have since 1971, whatever the CIAM decision about Coupe d'Hiver weight.

Many European flyers (German, British, Czechs ...) think the same thing, because they have a real and long experience in this series. Something that must be clear for everybody, and repeated if necessary; Coupe d'Hiver is not a series for beginners; it is an expert class which is more difficult and may be more interesting than the Wakefield class. If FAI wants a beginner class, it will be necessary to have a new series, as I wrote before, with flat section, free-wheel plastic prop, and 10g of rubber. This would be cheap and would give the same chance to everyone.

Competition: A good field is important. Windy weather is not so important if there are no hills or trees up wind. Coupe d'Hiver is a small and light airplane which does not like turbulence. It it always very discouraging to see an airplane rolled by the windy turbulences. It is necessary to be patient and wait for good air although, after 10 minutes, the wound motor becomes weak and it is better to unwind. It is absolutely necessary to avoid bad air to start for an official flight. Strong thermals are also dangerous as the models, even dethermalized, often are still climbing and can disappear and be lost. Coupe d'Hiver is now flown in more and more countries. Personally, I think it is easier to begin in this series with the modern 100g design (Georges Matherat - *Trumeaux*, 1977 NFFS best model) than with the old 80g designs. It is interesting to note that the recent best models selected by the NFFS Symposium have been French models. If you want to begin and build a Coupe model, try that French *Trumeaux*. It is a good model, easy to build, easy to trim and easy to fly.

There have been very few international events in Coupe but almost every year since its birth in 1938 there has been *La Coupe d'Hiver du MRA* in Paris and since 1975 the Aeromodeller British Coupe d'Hiver in Halton, near London. For many years also, the Italians have run a Coupe d'Hiver Contest in Torino. Last summer, *Les Journees Internationales du Poitou* in France, had F1G on the same day they flew F1C. Another good thing about Coupe d'Hiver is that it is a small model; wing and fuselage can be made in two parts, packed in a small box and travel alone by airmail! On a Saturday morning of April 1974, I mailed a Coupe model to Bob White; four days later it was in (California) and soon flown at Elsinore and Taft where it has won the United States Free Flight Championships. ■

Free flight World Championships opening ceremony, Plovdiv, Bulgaria, 1975.

2 Competition Report 1975

Amsterdam Cup 1975, by Martin Dilly

The Amsterdam Cup contest is always an interesting one for the British. Competing usually involves an overnight crossing of the North Sea, and properly planned, a day or so in the Netherlands soaking up some of that clean Dutch atmosphere before the flying starts. The 1975 event, held as usual in May, was no exception.

I picked up Mike Warren, who was in the 1974 British A/2 team at the European Championships, and his wife Kate, plus our tents and models, piled into the back of the small Ford van. We drove across north London, luckily just before the evening rush hour traffic started, heading over to meet Ken Faux at Harlow in Essex. Ken is in the British power team for Plovdiv, and its youngest member, at 21. Again we transferred models and equipment to Ken's bigger Ford Cortina, because with petrol at 75 pence — that's $1.75 a gallon — we wanted to keep the costs down. Off again across the Essex countryside — white clapboard houses, thatched roofs, small fields of wheat, still too short to hide a model, and pubs with their signs swinging outside — The Fox and Hounds, The Red Bull, The Ship, The Plough, The Leather Bottle — all turning gold in the evening sun. After an hour's drive we reached the seaport of Harwich (that's Harry-ch in the English pronunciation!) and met up with Jim Baguley; Jim produces those top-notch free-flight kits under the St. Leonards Models label as a spare-time operation.

With the masts and funnels of the ships as a background we again reloaded, this time into Jim's capacious Volvo; soon its vast trunk was full and four Wakefields, two Rossi gas models and ten A/2s somehow were fitted into boxes on the roof rack, together with an automatic bubble machine strapped on top. Up the ramp onto the drive-on ferry for the overnight crossing, where the three wifeless men share a four-berth cabin with an utterly silent, non-modeller fellow traveller during the fairly calm crossing. The North Sea can be really wild and even a stabilized ship like the *Koningin Juliana* will corkscrew its way over for the whole of the ten hours in a storm.

Driving off the ship at 6 a.m. we find The Netherlands covered in coastal mist with a two-hundred-yard visibility. The sun slowly burns through as we drive along E8; if we keep going on this road we will get to the Soviet border, via Berlin and Warsaw. We pass acres of greenhouses, with people already at work in them; their roofs have poly-anhedral. Goats, horses and cows in the misty fields now have a silvery halo of rim lighting from the low sun; as we head eastwards into the dawn the tip of Ken's gas model wing, too big to fit into the box outside, and rubber banded inside the roof of the Volvo instead, acts as an extra sun visor. Silhouetted windmills and clumps of poplars loom out of the mist rising off the irrigation canals. The land here is flat pasture, most of it reclaimed from the sea by the hard-working Dutch; even now there are excavators and heaps of mud as marshy areas are slowly won back.

We cross a ship canal just as a vast barge about the length of a couple of towlines chugs under the bridge; on the deck is lashed the crew's transport, two bicycles and a Datsun car. In North West Europe canals like this are a major part of the freight system.

We reach Arnhem just as the shops are opening and decide to spend the warm spring morning at an open-air folk museum. It is full of old buildings rescued from various parts of the country, all in a vast park with beech woods shading parts of it, spotless and well cared for, not at all phoney or Disneyland-ish. There are windmills, a water-powered, paper-making mill hammering rags into pulp, bakeries, barns full of wooden farm implements, and cottages with interiors like a walk-through van Eyck painting.

In the afternoon we camp at a site in another beech wood, inhabited by woodpeckers, jays and collared doves, sounding off like miniature klaxon horns. A drive up a series of hairpins in what is the hilliest part of the country brings us to the Roosendaalse Heide, an area of heather-covered heathland about two miles square, with a few scattered birches and low shrubs; it is part of the Hoge Veluwe nature reserve with buzzards, goshawks and apparently wildboar resident there.

The wind is blowing about seven knots when we arrive, and a few Frenchmen are getting the feel of the air with A/2s; by the time we have made a few check

Youngest competitor was Julian Masterman, 13, who flew a Jim Baguley-designed Nova, built from a St. Leonard's kit. Father Paul Masterman, co-editor of Free Flight News, holds model.

Van Uden of the Netherlands dropped two flights to place eighth in F1C. He flew an all sheet Rossi model with Hoerner wingtips.

Van Oostrom of the Dutch Hengelo club used a sheet balsa, wire-braced wing and a radio location beacon on his A/2 at the Amsterdam Cup contest. Launcher unknown.

flights on short D/Ts the thought of wading through the twelve-inch-high heather to recover three-minute ones is not too appealing. The stuff drags at the ankles; however the launch area has been mown and at least the early stages of glider towing will not be too arduous. The evening sky is grey and an hour or so before dark it starts to drizzle; we decide to investigate the Indonesian restaurants in Velp, five miles away.

The next morning is still greyish and breezy; as we arrive on the heath the launch poles are being hammered into the sandy ground of the heath, like the setting for a mass execution. Several A/2s are circling, their lines sagging limply beneath them as they drift downwind. Among the British flyers only Brian Baines has a circle tow model he is happy with. He and Joe Barnes have already been here a couple of days, and took second places in A/2 and Wakefield at the Dutch F/F Nationals held here during this week. The rest of us merely hope that the thick heather will discourage the circle towers from waiting downwind of the pack and taking advantage of the models drifting over in lift.

A couple of hundred yards upwind of the launch line the ground drops steeply away in an escarpment about seventy-five feet high. All through the contest this will give problems and fool people into thinking the slope lift is a thermal, only to see their mistake as the 'curl-over" dumps models into the ground after a couple of minutes.

One of the most impressive gliders at Roosendaal is Tom Koster's. Having won the Wakefield Cup at Kauhava at the age of eighteen, and consistently in power fly-offs in the intervening ten years, Tom is now breaking into A/2 flying.

His model shows Soviet influence, with a dural fuselage spine-cum-pylon, long, tapered, wingtip panels and a dihedralled tailplane. The timer and circle towhook are mounted on the dural spine and protected by glass fibre mouldings which form the fuselage back to the flexi-joint. The whole model gives the impression of being well-engineered and Koster has it nicely set-up to circle right overhead on tow; he keeps the excess line in loose bights in one hand and tows without a winch. Once during practice we watch his model sail off with the line still attached, and Koster, who seems to have lost quite a bit of weight

Fourth placing Wakefield flyer Joe Barnes of Liverpool looks for the upwind lift as he waits to stir up the licquorice in his Clio, held by A/2 winner Brian Baines. Model uses v.i.t. triggered by the recoil of the propeller stop engaging in the noseblock.

1975 F1C British team member Ken Faux (left) and West German Gerhard Heidemann of the Celle club, who flew a T-tail, twin pylon model at the Amsterdam Cup.

Piet de Boer holds A/2 by Herman Lammerts of the Dutch Enschede club. Model features all-moving fin, with a wire-braced wing, using 1 mm. balsa skinning over Styrofoam cores, and a 1/64 square bass turbulator 3/16 back from the leading edge.

recently, in pursuit. Maybe A/2 flying could replace jogging and dieting. This model looks a threat.

At noon the first A/2 round starts; at 12:05 the rain starts too. Nobody has flown so the organizers, the free-flight section of the K.N.V.V.L. (the *Royal Netherlands Aero Club*) sensibly decide to call a one-hour hold. Obligingly the rain stops and we head out to our launching poles again. Flyers are divided in groups of four to each launching pole and paid timekeepers are provided, a pair at every pole, equipped with binoculars and neat translucent glassfibre clipboards for the scorecards, with timekeeping instructions laminated into the resin.

A number of typical Dutch gliders are in the first popular patch of lift. They have highish pylon-mounted wings with about 12:1 aspect ratio and elastic outrigger turbulators so the models hum eerily like Aeolian harps, even while being carried. This could probably help to make sure you launch your model at its correct gliding speed, assuming you have a good memory and perfect pitch... Just slow up the tow until the sound is right and cast off—no stall, none of that frustrating 'two speeding' while the model makes its mind up about its gliding speed.

The lift is coming through in large patches, but is not easy to detect; perhaps due to the false indications from the ridge lift, there are a lot of sub-maxima coming in. For the British it is not a good sart to the contest. *Free Flight News* co-editor Paul Masterman has a glider set up to fly tail-high (like a Whitley bomber, for those of us who remember wartime Britain) and he uses a catapult hook system that gives a zoom launch as the hook unlatches with increased line tension. Somehow he has a premature unlatch and the resulting low altitude tight turn puts the model into the lid of an open car trunk. His son Julian, 13, and flying in his first international contest, sheds a wingtip after a tumbling D/T pattern. At the end of the first round 21 out of the 67 flying A/2s have 180 seconds on the scoreboard.

Several of the Dutch and West German models use radio beacons for model location. As far as I can discover they are of rather doubtful legality but with a range of perhaps a couple of hundred yards nobody is likely to object. As the wind swings and models start to D/T into the scattered birches downwind, which all look alike at close range, bleeper owners trot off with their transistor radios to their ears and confident expressions on their faces. The rest of us rely on compasses and frequent checks on back-bearings on the way down the line towards our lost models. Luckily we find there are plenty of sandy tracks across the heather, used by the local riders of mopeds (called *brommas* in Dutch, rather alliteratively), and these cut down the wear on legs.

The various classes are flown in 'laminated' rounds through the weekend, which gives three rounds of each by the time Saturday's flying ends. Only Baines of Great Britain, Weiland and Wilkening of West Germany, Erard of France and Lendering of the Netherlands have treble maxes.

Sunday starts calm and warm, and over half the scores are maxes in the first two glider rounds of the morning. By midday, after five rounds in all, only Brian Baines and Reiner Weiland have five threes. Weiland, however, has lost one model and badly damaged his only spare. Now, inexplicably, he decides to borrow someone else's A/2 and fly that in round 6, only to be spotted by the jury, resulting in the cancellation of his entire score. Baines maxes again and is now 34 seconds ahead of West Germany's Arno Deubel, lying in second place in spite of four sub-max flights. Pete Allnutt, over on a flying visit from Canada, and using a model with circle towhook but no zoom facility, has dropped almost a minute on his second flight. Continuous maxes since then keep him up in third place. Using the Hacklinger airfoil with a 1/16th sheet balsa upper skin and outrigger turbulators, Pete's model has a flexi-jointed fuselage with a dural keel/pylon onto which the swinging towhook is pivotted. He uses a recoil hook on the line to prevent accidental drop-off during circle towing.

The weather for the final round looks murky. Baines must now choose his moment carefully. Too long a wait for good air may bring the rain instead of lift. With the windspeed increasing, he tows and releases the Shoaf-airfoiled model. Maybe the ridge lift fools him, but in 2 minutes 42 seconds the model is down. Even if everyone else maxes now he is unbeatable.

In Wakefield twenty-two flew. After dropping eight seconds on his first flight West Germany's Boeren led from fellow countryman Haacken for the first five rounds. Then he scored 132 and 99 seconds for the last two rounds, while Haacken continued to max after his 156-second first flight to win the contest. The winning Wakefield climbed against left rudder, using extensive right-thrust and down-thrust, and a Montreal prop stop with auto-rudder release; 5° washout on the left wingtip and 2½° on the right one helped to control the motor torque. The propeller blades were turbulated. Haacken's wing had small straight-tapered tips and riblets to maintain the shape of the leading edge portion of the Koster airfoil. Tailplane airfoil was John Gard's.

Second place went to Italy's Gialanella, one of four members from the *Cagliari Club* in Sardinia who were competing at the Coupe d'Amsterdam. They used a thermistor temperature sensor on a tripod a hundred yards upwind from the launch poles, and observed its meter variations through a tripod-mounted telescope. Denmark's Kjell Kongsberg was 12 seconds behind in third place, two seconds ahead of Joe Barnes who watched the bubbles successfully, apart from a first-round score of a minute and a half. Joe, from the *Liverpool* club, flew a Ron Pollard-influenced model with a prop-stop inertia operated auto-rudder similar to that used by U.S. Nats winner Carrol Allen. This relies on the shock of the drop-in pin engaging in the nose plate, which gives a momentary rotary movement to the fuselage, enough to disengage a latch holding the rudder in the climb position.

F1C brought 16 entrants, the majority from the Nord Rhein Westfalen area of West Germany, just across the border, only twenty miles from Arnhem. The power patterns of several of these models appeared rather 'loopy' to British eyes, used to models that turn less than 360° during their climbs. Fully sheeted flying surfaces were almost universal; the *Celle Club* favoured T-tails. As with power flying everywhere the problem was to avoid downdraughts rather than finding lift. Hans Schallenberg was one of four to max out until round six, when he was down for 98 seconds, dropping him to sixth place.

After seven rounds Tom Koster, Josef Brodarac — a Jugoslav now living in Dusseldorf — and Tassilo Schwend had 1260 seonds. All of them maxed off the eight-second run, but in round nine Brodarac was down for 166 seconds, leaving Koster and Schwend to check their timers for the four-second run. Schwend launched first and had a rolly climb which, coupled with poor-feeling air, gave him a time of 1:26. Koster had an almost straight-up climb and a 3.3 second engine run. He used a spring-and-drum propeller braking system that stopped the engine convincingly, but sounded a good deal less destructive than the earlier types used by the Soviet team at Wiener Neustadt. He found much better air and ended his flight twenty feet up a birch tree to max comfortably and win the contest.

Koster had deserted his flappers for this contest, using instead a model with a two-piece shoulder-mounted wing with a flat-bottomed airfoil. The motor — a Rossi, of course, — was enclosed in a close-fitting fibreglass cowl with the top of the head exposed like a C/L speed model and the exhaust leaving the fuselage via a megaphone duct, separate from the engine in order to remain legal. The wing was mounted virtually level with the head of the motor, so the effective pylon height was very low.

Overall the Amsterdam Cup contest was well organized. In 1974 there were problems largely due to the organizers not being specifically free-flight oriented. In 1975 under the efficient control of Jan van Rij of the Dutch free-flight subcommittee the problems were apparently few. One, to which a blind eye was turned, was the almost total absence of evidence of processing of the German models. However, the ever-reasonable Dutch obviously realized that it was hardly the fault of the entrants that they belonged to an inefficient areo club and, as far as is known, no on-the-spot processing was asked for or carried out on the offending models. ■

Young Darko Broderac of West Germany flew this Rossi model with very Koster-like tail group, but retired after four rounds.

ПЛОВДИВ
БЪЛГАРИЯ
'75
ФАИ
FAI

Flying site entrance, Kromovo Airport, Bulgaria

◁ *World Championships Hotel and Headquarters, Plovdiv, Bulgaria.*

World Championships 1975 Plovdiv, Bulgaria

The World Champs exciting preliminaries plus the story of one groups experience in making the trip are described elsewhere in this Volume (Free Flight to Plovdiv on the Orient Express). We've saved the technical details of the competition itself to present here in this report.

The 1975 Free Flight World Championships was organized and conducted by the *Bulgarian Federation of Aircraft Modeling*, Hristo Botev Boul., 48, Sofia. The contest director was Mr. Peter Stankov, Secretary of the Federation. The CIAM Bureau confirmed the jury of:

Mr. Luigi Bovo - Italia
Mr. Yourii Sirotkin - U.S.S.R.
Mr. Petko Petkov - Bulgaria

Pairs of timekeepers were provided in which one was Bulgarian and the other foreign. The Federation did an excellent job in sending out information bulletins to national aeroclubs giving all the details as to schedule, costs, food and housing arrangements etc. A fee of $75 was charged to competitors and managers, while $85 was charged to all others. This paid for six days of food, lodging and transport. Official languages were Russian, English and Bulgarian. Interpreters were provided for each national team and other languages including French, German and Spanish were accomodated. Although the organizors made announcements in the three official languages, often the translations would be tardy and the news would be late, causing some consternation. This is a defect that appears difficult to avoid considering the situation, so Team Managers, be alert (and multilinguistic).

Accommodation for the team members and supporters was provided in a relatively modern 14-story hotel in Plovdiv. Each room was equipped with two beds and bath. There was a small bar on the ground floor and completely inadequate parking facilities outside. Apparently the Bulgarian life style was not geared to cope with the Western commitment and reliance on personal auto transport. In the evening cars and camping trailers were parked everywhere, packed solid curb to curb and on the sidewalks. There was a deluxe fleet of buses at our disposal that ran a shuttle service between the hotel, the dining hall, and the flying field. The dining hall was several km. across town and appeared to be a community center of some sort perhaps in connection with an upper level school.

The flying field was the Kromovo airport located 11 km to the south of Plovdiv. Test flying was not permitted there prior to the official days. An auxiliary site for testing was provided about 40 km away. Use of the auxiliary field was not without its excitement however as some of the American contingent who arrived early found out. Al Vela, Bob White, Bob Piserchio and *retinue* were arrested and detained for a few hours while the suspicious K.A.T. heavies checked their credentials and stories as to what they were doing on that field. They were released without apology, apparently a routine procedure in those parts.

The official buses made the trip with a strident police escort who seemed to take delight in forcing the citizen motorists literally off the road. The locals seemed resigned to such treatment and took our disruption with resignation and good nature. They appeared to be genuinely proud to be our hosts.

The field was quite large, flat and grass-covered. There were no runways and no aircraft except for a few helicopters that were assigned as lost model spotters. The general terrain lay in a wide plain bordering the river Maritza. In the far distance in several directions could be seen mountains rising to over 6,000 ft. Although the field was grass-covered, there were a number of annoying thistles, and outcroppings of pebbles. Although the surface was softer than say Taft, it was harder than what some of the continentals are used to.

Test flying, as well as official flights, were flown from the middle of the field. Available space in all directions was adequate for all but the most flagrant DT failure. This was proved by John Cooper of GB whose straight tow reserve FIA model flew off for 24 minutes and was never found. The weather, after producing some rain the week before, straightened itself out, and gave us sunny skies, and gentle zephyrs of breeze and quite warm days and nights. Really ideal conditions.

Friday afternoon and evening was taken up by model processing at the hotel. It was all well organized and carefully done. The work continued through the night til about 2 AM. All models had been checked (estimated as 750) by Saturday lunch time. Processing included a dimensional check of the models, engraving of identification numbers on engines, checking of rubber motors (which were then placed in the team's sealed container) and a check of team paperwork. Rubber motors were also processed on the field during the contest but that generally took some time and had to be planned ahead so as to not delay the team's flying schedule. Processing at the hotel gave many an opportunity to see some of the models of the competition and to chat with friends. The processing took place on the second floor while the bar was on the first. A continuing flow and replenishment was soon established.

On the Saturday, the last practice day, just about everybody was at the field making those last minute "reassurance" flights. And a few were desperately fighting some trimming problem that had chosen this

U.S.A. Team tries out the field.

moment to surface. A number of wild flights and prangs meant that either things were getting sorted out or new difficulties were only just now becoming evident. In either case new challenges for the keyed-up competitors at the eleventh hour. Included in this group was Thomas Koster suffering a DT failure, Dick Lyons a wing failure from early DT and Bob White a smashed wing and prop through entanglement with a mylar streamer. Roy Collins had a midair break-up of his power mdel when the engine failed to quit as programmed. A number of other flight patterns gave an augury of foreboding to their owners (One might say an augury of augery.).

A number of teams were flying and launching simultaneously to check for a substandard trim. The North Koreans were observed doing these simultaneous launches with quite professional skill. Interestingly they made their final team member selection only after flying in Bulgaria. They seemed to know what they were doing. The Russians could be seen trotting around practicing down wind towing, low snap-releases, high zoom-launches and in general very confident, effortless handling. In fact most F1A fliers were having a great time cavorting about the field, limbering up their muscles.

Sunday, August 17, F1C Power

After a hurried breakfast, and no breakfast for some intent souls, the day dawned gray and misty. The state of the weather to come lay still cloaked in the haze. At the field the organizers were wandering about as if lost, pointing in various directions, trying to decide where to put the line. Contest nerves! Then having settled on a spot, what-ho the wind shifted. After this happened several times it was agreed, what the hell, start from the middle of the field and rotate the line if need be.Breeze was light, only a meter per sec. max. interspersed with calm. With the sun rising, the mist started to burn off creating a few low lying macro-size cumulus only a few meters above us. The official start time, 8:30, came and went without a start, although it was announced that the round would not be extended. The delay was due to the binoculars and stop watches not having arrived. Eventually they did arrive and the green starting flare was fired only 15 minutes late.

Honors for first official flight of the contest went to Aussie wheat farmer Tom Prosser. Team manager Dave Anderson tried to hold him back but that was futile. Tom's model raced into the blue sky chasing the green Very flare climbing adequately but then dropping disappointingly in the first half of the glide, cooling the fervor of the watch and wait crowd. But then Tom's model found good air at lower level and rode it out to a nice max. The flying commenced immediately as the air appeared to be ideal. All up and down the line, spinners were being jammed into whining starter snouts and fliers were agonizing over the screaming engines, tweaking the needle valves, trying for just a bit more, or waving the roaring model around, scaring the timers half to death. Here and there a ship would arc up, the doppler effect making a sudden pitch change in tone as the model accelerated away.

French Team tries out the field, uses rooster for thermal detector.

Siegfried Reda, sheet covered foam construction, stretched out design made fly-off. (BDR).

Dave Rounsaville went the other way, low aspect ratio, had shot-pin prop brake.

F1C winner Lars Olofsson of Sweden shows his winning javelin launch style.

Before the first round got under way several more models had gone berserk. Seigfried Reda had a left pattern and crashed; Frank Schlacta creamed his new model after an auto-stab failure. The greatest drama though was enacted in the Danish camp.

It had been learned that on the way to Plovdiv the Danes and some Americans had stopped off at Weiner Neustadt (Austria) to do some testing on that huge field. Thomas had something go wrong there and so arrived at Plovdiv with little equipment left. During testing at Plovdiv one of his remaining models went off on a very fast climb but with the engine stopping dead at 4 seconds . . . broken conrod! With everything still set for 10 seconds of climb, the unpowered "Zooky Zooky", "Zookied" on in. Thence began the most remarkable field repair in the history of F1C. The entire Danish team appeared to be nothing but elbows and cyanoacrylate, reattaching bits from several of Toms's Flappers that took some imagination to see as part of an airplane. These bits soon formed one new flapper from the wreckage of the two. A hasty test flight caused some minor damage (broken wing tip) but that was soon fixed and Thomas made the first round by 29 seconds. The flight was great with a super straight up climb, a bunt at the top into a flat wide glide circle for a fine max. But then the crazy bird wouldn't come down! The Danes chased under it for 8 maddening minutes before it settled to the Bulgarian turf. It was flown again in the second round. But this time it spun in from the top of the climb for a score of 25 seconds. That was the end of a great effort. There had been every expectation that Thomas would do very well as a number of his practice flights demonstrated an exceptional performance. The flapper concept works but it looks like more effort is needed on the mechanical systems.

Some had a little trouble but maxed anyway. Frank Wolff (US) had trouble starting his engine and launched late in poor air. Engine run was short at 8.2 secs but he maxed with 10 feet to spare. Rounsaville (US) had a 9.2 sec run, good pattern but engine sounded off. Pull-out not too good but got high anyway and maxed. Lars Olofsson had a bad climb pattern but overrun. Second attempt was OK. Verbitsky (USSR) was timed at 9.5 & 10.1 engine run. Team manager protested and protest was upheld.

It was soon apparent that timing engine runs was going to be a problem. With as many as ten models in close proximity it was very difficult to detect engine shut-down. And additionally with only one each of the timekeeper pairs speaking Bulgarian it was often difficult to communicate and hold any kind of meaningful discussion about the fairness of a particular call. Although the Bulgarian timekeepers were well briefed as to the rules, many of them were unfamiliar with the subtleties of timing free flight models. Since Bulgaria is a small country there just weren't enough Free Flight fans available and so RC, UC and other modeling types had been pressed into service. In their zeal to be scrupulously correct a number of overruns were called that no doubt shouldn't have been. For those teams whose managers could speak Bulgarian, Russian, or some close facsimile, mistaken overruns could often be corrected. And for those who could speak the universal language of indignation and disbelief with just the right blend of commiseration with the poor timers plight, by sign language if nothing else, then too, some adjustments in decision could be made. But lo the poor team manager who hadn't learned to cope with this situation! The trick was to browbeat the timers before the time was recorded. Once recorded an official protest would be necessary. And that was when things got hard and inflexible. The smart team would have a manager or assistant manager who could speak the native tongue. All this is not to say that timing was done with deliberate unfairness or prejudice, but to point out the difficulties of using inexperienced timers. Perhaps it would have been better to have F1C flown on the last day?

It looked like this year was the year of the brake. Verbitsky had a brake in 1973 at Wiener Neustadt and now at least half the contestants were trying them out. The most spectacular was that of Rounsaville and Wolff (U.S.) out of Hatschek (the gadget man). This ingenious device used the self energizing strangler spring together with a "Montreal Stop". This infernal device fires a pin into a hole in the spinner a few milli seconds after the strangler spring trys to strangle the prop shaft. The pin stops the spinner like it was impaled with a harpoon. And does it with a bang that can be heard by the timekeeper, startling him into action to punch the stopwatch button. Main disadvantage of this system is that, as can be imagined, the pin suffers unbearably. As a result the Skyscraper design team developed a built-in system so that the tip of the pin can be ground off and advanced to its former position. Grinding is done with a wheel attached to the club starter, all set up with jigs and fixtures. This has to be done after every six penetrations. An additional advantage of this brake is that it stops the prop horizontally.

The British team were pumping away on their bubble machine, a device that must truly be seen to be believed. Considering the great tradition of technology advancement as epitomized by such Englishmen as Stringfellow, Watt, Hiram Maxim and the Lever Brothers, this marvel was to be expected. And with no disrespect intended for our hosts, the British did leave the field a lot cleaner than they found it, and a mite slipperier too.

Well the bubbles said go so Ken Faux did . . . easy max. Roy Collins got ready but discovered a loose timer mount and prudently elected to fall back and repair. Bob Bailey fired up his Night Train derivative and blasted the old but consistent bird into a good max flight. Roy refitted the timer and checked things out on the ground and then made his flight. Climb was fine but the glide was too straight and steep. Help from lift wasn't enough and eventually a spin-in left him with a very disappointing 91 seconds. A post-mortem revealed that the VIT line had somehow got twisted in the repair episode and did not allow full travel.

Plenty of eyes were on the Russians as they prepared to fly. Verbitsky had a new model without flaps, although his reserve models were flapped. The new one, BE34, was a marvel of mechanical ingenuity and workmanship with each piece, gadget and whatever carefully finished, polished, anodized, fitted etc. It seemed there was no limit put on the effort to prepare these models. These were no last minute display models either as it was obvious that everything on

Unidentified Bulgarian flier has big audience.

Roy Collins, GB, gets his ship up straight away but had problems with a loose timer.

Russian fleet awaiting action.

Verbitsky made some trimming flights during the lunch break.

Anxious Cuban team checks an engine run.

them worked well and had been proven in the air. The ships were of conventional layout with moderate aspect ratio, moderate pylon, single aft fin, flat-bottom thin sections, all sheeted and glassed. The port wing had an aileron that worked off the timer but it was used only as a fine-tune trimming device and in fact wasn't needed on this model. The fuselage front end was all machined light alloy. Circular engine pan slipped inside it and was held by the friction of two rubber "O" rings that were squeezed between bulkheads. One axial screw tightened the bulkheads. No metal to metal contact from pan to fuselage. All the gadgets and metal work were turned out with beautiful detail and thought. Propellor was a carbon fibre "special" with wide chord cuffs at the root. Climb was very fast with about 7/8 of a turn in a "safe" spiral pattern in 10 seconds. Prop brake was similar to the one he used at the previous W/C. It consisted of a heavy coil spring wrapped around the engine crankcase. One end securely fastened to an engine mount bolt, the other end with a menacing looking hook that was positioned to engage in a slot in the back of the prop driver. The hook was restrained till the right moment by a wire to the timer. It worked well and once having got the hook shape right and the proper heat treating etc. there has been no maintenance problem. Team mates Sharon and Mozyrsky also had nicely constructed ships similar in some details but otherwise showing somewhat different ideas and original thought.

There were only 10 out of the 73 who flew in the first round who failed to max. Unhappily for the Dutch that included all three of their team, who fell before various misfortunes and bad air. De Luna, Mexico's sole power entry, missed in the second round. He was flying a pair of well built "Night Trains" but the pattern chanced to be a bit off and the air too, to drop him 17 seconds. The two-man Cuban power team perservered with models that looked well flown and of Czech influence but finished in an uninspiring tobacco color. Nevertheless one of them, Valdes, made it through to the flyoff.

It was in the third round that misfortune once again struck the British, this time Bob Bailey. With hawks serenely thermalling at 500 feet over the field, Bob launched with what he thought was a conservative 8 sec run only to have it called at 10.1. Rather than protest Bob chose to take his 2nd attempt, what with the air looking so great. But sometimes these things don't work out. And it didn't. With a deliberately shorter run, Bob put it up but into bad air . . . and there was the chance lost at 133 secs. And with that the gloom got a little thicker in the British camp.

For the fourth round Ken Faux got the nod from team manager Brian Spooner to lead. He did with a 7 sec. run and off on a great max, a morale builder they sorely needed. But then incredulously the two engine timers reported 12 seconds and 20 seconds. With other

Former Champ Franz Bauman (1969) checks the air. Engine angled to side for exhaust to clear pylon. Timer opposite. (BDR).

Frank Wolff checking his ship. Moments later a gust tore it away causing damage.

engines running nearby it was difficult enough, but it was obvious that the inexperienced timekeepers were just unsuitable. Spooner first aborted an attempted international confrontation and then immediately lodged an official protest. It took 40 minutes to petition the FAI jury and receive the decision that the flight would not count and could be reflown. This left Ken with 8 minutes to manage his allowed two attempts. With a short 7 sec run and mediocre air Ken at first appeared to be done in, but defeat slipped from his grasp as the GB stalwarts waved, stomped and cavorted insanely under the model refusing it to land. A max!

The organizors did what they could to improve the timekeeper situation and in fact fired five of them that night and read the riot act (or whatever it is the authorities invoke in the East European countries) to the remainder. At the end of the 4th round 57 competitors had full scores including 11 full teams. Dick Lyons of the U.S. had a close one when his 4th flight was an overrun — and made only 2:10. The second attempt was OK and a max.

At this point a very welcome scheduled two hour lunch break commenced. This period gave many a chance to catch their breath and gather their strength for the following rounds and inevitable flyoffs. Frank Wolff (U.S.) had finished his lunch and was sitting on his model box relaxing a bit when WHAM, a miniature dust devil whipped up behind him and tore his model from his grasp. Pieces flew off down wind with friends grabbing parts. The fuselage broke just ahead of the stab. Quick team work had everything Hot Stuffed back together and Frank added epoxy and glass reinforcement. A test flight was made and it looked OK to Frank. Came the 5th round official however and the Square V rolled out left at low altitude and turned in a disappointing 83 sec. Dave Rounsaville, in unplanned sympathy nearly followed suit in the 5th making it with only one second to spare.

All engines on the field were Rossis except for the MVVS Diesels of the Czechs and an odd Super Tigre here and there on a reserve model. The Czech's diesels were running very well, putting out about 24,000 rpm on the same props that Rossis were turning at 26,000. Good enough to put all the Czech's in the flyoff but not quite the power of the best Rossi examples.

Model design elegance ranged from the very graceful *So Long For Now* by Urs Schaller, to the ubiquitous unassuming model of the Brazilian Eolo Carlini. Eolo's model was his only remaining power model (he was also flying Wakefield) after breaking up his others. The dated design was powered by a Rossi fueled from a huge brass tank; no brake. An inoperative aileron still had the mechanism bits trailing in the breeze. Engine sat on wooden bearers and the timer was home made. The irrepressible Eolo made it into the flyoff and maxed in the 8 sec round. An overrun in round 9 put him out. His efforts and infectious good

Yosef Adlt, Plsen, Czechoslovakia, used MVVS diesels that turned remarkably fast. Team maxed out.

North Korea placed two in fly-off. Conventional Rossi powered models.

nature made him everyone's friend and he was awarded one of three special prizes by the BAMF at the banquet.

Frenchman Denis Fererro used an aluminum strap clamped around the fins of his Rossi. Claims that engine heat reaches and maintains better equilibrium. It did sound very smooth and fast. His model featured a very low pylon and elliptical dihedral. Franz Baumann of W. Germany, and the 1969 World Champion, dropped his chance for a shot at the fly-off when in the fifth round his glow plug blew at 6.2 secs. The short run and bad transition brought the ship to earth in 126 secs. Nielsen of Denmark was also unlucky. In that same round he broke his string of maxes with a 179. N. Koreans were flying what looked like the E. European style model of 15 years or so back. Timers looked home-built, engines were Rossi and props hand-carved wood. Flight pattern was a conventional pylon spiral climb but with rather early use of rudder kick to the glide turn. In the fifth round their Kim Sung Hak garnered their team's only sub-max flight, a disappointing 139 secs. Ken Faux gave his teammates a workout in the the seventh round when he had a DT failure. They chased it off the field and watched it land in a tree. Ron Pollard shinnied up and plucked it from its new-found nest.

The weather stayed good all day and now there were 42 that had maxed out from the 77 entered. Of these, five countries had perfect scores; USSR, Austria, Czechoslovakia, East Germany and Japan. Another ten countries had come very close with only one sub-max flight out of the 21 possible. The organizers were hard-pressed to get the fly-off under way as first there had to be meetings for instructing the team captains and timekeepers, then the starting line was relocated, and then starting positions were assigned. All this took about 45 minutes. Start positions were drawn by lot and the plan was to start at one end and give each man his chance to start his engine and launch. This way it would be easy to check engine runs. The first fly-off round would be with an 8 sec engine run, second 6 sec, third and all additional, 4 sec. Each man was given 2 minutes to start and launch. First off got away at 18:40 in light that was already starting to fade.

Sim Guls ship (PRK) went up in a severe roly poly climb but made his max. The Czech team all maxed in this round. Reda's engine spit and hesitated slightly, giving an anxious moment but it was still a good climb and he maxed. Lustrati had a good glide and possibly good air too but his climb was bothered by a short run and stally transition. He still made it. Rounsaville must have followed through to the left with his throw just enough to screw up the pattern as it flipped out poorly at the top. Engine cut was at 7.5 secs and the flight put him out with 169 secs. Schaller's *Plurabele* looked better in performance compared to two years ago at Wiener Neustadt. This high aspect ratio really clean ship looked great, but in the past such "pretty" airplanes have been difficult to trim. Now it looks like Urs has that detail worked out. Tom Prosser with a more conventional *Night Train* type ship went up with a weak sounding engine. All went well as it floated low for a max. Sharon's ship did an agonizing tail slide in a poor transition but then it went on to max also. Verbitsky's flight looked really great but there must have been something about it he didn't like. He confided that he would change the rudder setting and give it a test flight before the ninth round. By the time the anxious last guy in line got to make his flight (he maxed) it was 19:30. The organizers decided to continue the fly-off on the next evening following F1A. There were now 37 still in as two had gotten overruns and three had flown sub max.

Diesel powered Czech model gets a launch. Gloved hand useful on starting.

Japan had strong effort in F1C. All maxed.

Urs Schaller, of Switzerland, now working in Italy, shows great left-handed form in this launch shot. He designs and builds excellent machines.

With streamers lifting and watches snapping, Frank Schlachta of Canada launches his high aspect ratio pylon ship.

Dick Lyons, USA, launches as Jim Walters waves the signal flag. Teamwork an important element at the World Champs.

Dining Hall at Plovdiv World Champs. Friendly service and good food.

Fleet of deluxe buses were used to transport everyone around at the World Champs, Bulgaria.

The buses were all lined up on the field and they rapidly filled with contestants now that it was known that no final rounds would be flown that night and it was too dark to get in any last minute F1A practice. The bus trip back to Plovdiv and the dining hall was quite an experience. I was driving my rented car in this convoy. Escort was provided by a dozen "KAT" police cars that seemed to relish their duties. Absolutely everything was forced to stop and get out of the way, even the cars in the opposite lanes of the wide highway. They were literally run off the road into fields. A switch locomotive was even stopped from delaying our passage. The Bulgarian drivers all seemed to be highly defensive and wary and gave the KAT's a very wide berth. All this made us feel rather important but uneasy about what the citizens thought of our intrusion into their lives. The ones that we met were very hospitable and friendly.

The buses pulled into the narrow street in front of the dining hall and parked in the rubble. The street was still under construction as was the entrance to the hall. We found a spot for the Zhiguli and picked our way through the boulders, ruts and gravel and on into the large building. There were many tables all arranged restaurant style and each had a flag denoting who was to sit where. The food was brought out by young students, many of whom could speak the languages of the countries assigned to their tables. The food was good, especially if you liked tomatoes. There was a lot to talk about and find out what everybody did that day. Those that hadn't made the fly-off were a bit unhappy but they could also relax and enjoy the occasion. From dinner, our "circus" moved back to the hotel where the trick was to find a place to park.

Tom Prosser of Australia gives us a grin while he waits for Dave and Allan to finish their chow.

Outside the dining hall, the Czech Team seen in the morning mist, load their bus for the trip to the field.

The little drive in front was soon filled with cars parked on the sidewalk and completely filling the drive. Pierre Chaussebourg had brought his travel trailer from France, towed behind his diesel Peugeot. It was parked at the hotel entrance and was used by the French team for storage of their models. Just inside the hotel entrance and to the right was a bar. Needless to say it was filled. The bartender, we learned, had been brought in from some other establishment just for this occasion. Beer was in great demand but we were soon to find that there was to be a perpetual shortage. The government, in its wisdom, had decided how much beer we should have each night. So about 21:30 each night the quota was used up and that was it. This had little apparent effect on the festivities however. A visit through the hotel to various floors found lots of F1A fliers giving their models last minute checks. Most of them couldn't sleep anyway.

British team after a tough practice day. Manager Bryan Spooner, Mike Fantham, John Boone and John Cooper.

The Americans get what-for from timekeeper Paul Masterman. Bob Isaacson, Bob Piserchio, Don Chancey.

Monday, August 18, F1A Nordic

Somehow we survived the night and found ourselves back out at the Krumovo airfield the next morning. Weather was ideal and the early bird fliers declared the lift was there. Ninety-two contestants from thirty-one countries were on the field for the 08:30 start. It was soon obvious that this World Championship would be the one to remember for the wholesale adoption of circle towing. A few countries have been using the technique for a number of years (Russia, France) but it has been slow to catch on. Probably this has been due to the fact that proper use of the circle towhook requires lots of practice and development of technique and strategy. You don't just bolt it on and launch. Additionally the circle towhook increases the chance of error, as gadgets always do, so there has been a reluctance to forsake the simplicity of straight tow. The variety of circle towhooks included all sorts of devices and now the Danish "twanger" impulse release was declared legal and was being used by many.

The reigning F1A World Champion, Valerij Ekhtenkov, had chosen to defend his title independent of the Russian Team. This gave him his own starting pole all by himself and the whole round to decide when to fly. His strategy caught everyone by surprise however as with the firing of the green Very flare, he was off and running. Here was a contestant that had confidence in circle tow and was prepared to tow for the entire hour if necessary. This wan't required this round for in only a few minutes his ship (same as his winner of two years ago) was floating softly in light lift. Everyone else had been putting off as long as possible that nervous moment when they would have to commit to a flight and so this first flight and thermal of the day remained Ekhtenkov's private domain and individual triumph. This flight was all that was needed to break the ice and soon the air was filled with circling models.

This love feast with the atmosphere was only marred, or you might consider it enlivened, by the inevitable congestion of tow lines. With the newly developed tow hook technology and the light conditions it was all too easy to cross lines and get in each others way. Now the engine timing problems of the preceding day were replaced with a new, different situation. The FAI rule book says that the timekeeper must see the line-cross for it to be accepted as valid reason for a reflight. Al least that was the interpretation supplied by the FAI Jury. The long starting line with start positions 7 meters apart made for tight quarters particularly as many chose to do their circling right in front of the line. Add to that the constestants who had their lines stretched out waiting to fly and you had an aerial jungle. Those who broke away and circled downwind (or upwind) had less congestion but also ran the added risk of the timekeepers not being able to see line-crossings if they occured.

World Champ of 1973, Valerij Ekhtenkov of the USSR flew as defending Champion. Dropped one flight.

Paul Masterman, having graduated from *Orient Express* Tour Director, to Official Timekeeper, was able to observe first-hand the intricacies of the line-cross problem. His comments from *Free Flight News* provide an excellent description:

The thermal beaters at work. We even saw one of the timekeepers join the fun.

∞

In the congested flying of an FAI event, with so much *wheeling and dealing* through the use of circle tow systems, line crosses are inevitable. But with 'locked-on' towing systems they are no longer the problem they were. Fliers have the chance to unravel the knitting and go on flying-whilst claiming a line cross. If they fail to max, they have an insurance policy for a second attempt.

A further problem of the line cross rule was soon apparent. In spite of the recommendation that one timekeeper should pay particular attention to line crosses, such seems an impossibility without that timekeeper running along with the tower . . . and heaven forbid, it's like Piccadilly Circus out there already. The tower is often so far away from the pole when he releases the model that binoculars are needed to watch the pennant. Hence, since they are unable to watch the whole length of line (or even *see* it in many cases) timekeepers must rely on competitors, team managers and other interested parties for an opinion. Verification with other timekeepers on other poles proves impossible in the heat of competition — find the competitor, find his pole — all while your next competitor is waiting to get in the air . . . Obviously, the system is open to the worst forms of gamesmanship. Cynics might say that this explains the numbers in the fly-off. ∞

The worst part of the line-cross rule seems to be the unevenness with which it is applied. The flier who is shot-down through no fault of his own certainly feels imposed upon. Even the granting of a second attempt does not restore justice as the conditions are changed and it is later in the round. Why shouldn't the benefit of doubt be extended to the side of the flier? At Plovdiv it seemed to be the opposite and the timekeepers were reluctant to accept claims by witnesses. When pressed by indignant Team Managers they stuck to a strict interpretation of their instructions. Carrying the case to the Jury did little good as it was easier to deny seeing a line-cross than it was to find a witness acceptable to the Jury. In some cases the decision rested merely on the argumentive ability of the Team Manager, which varied considerably, especially with regard to language barriers. Luigi Bovo, a Jury member, suggested that a future rule might consider allowing reflights *only* for the case of a *model* contacting another line or model. This would be easier for the timekeeper to spot.

Once the competitors learned (some the hard way) that line crossings should be rigorously avoided, they spread out more and pressed into service helpers to run with them and help steer them away from trouble. The wind remained very light and the thermals seemed to be very large in diameter. Fliers were now taking more time to pick the place and time to release. Many models were in the air together over a relatively wide area. A popular technique with the circle towers was to disconnect the line from the winch and work with just the line. This gives better feel and quicker reaction in pulling in or letting out line. It also eliminates any danger of losing a flight by throwing the reel.

The Bulgarian bag lunch was very good. Fried chicken, bread, tomatoes, chili pepper, grapes, chocolate and Koka Kola were packed in a plastic sack imprinted with the W/C logo.

Don Chancey, Plano, Texas, U.S.A.

NorthAmerican efforts were spoiled in the third round. Don Chancey of the U.S. circled downwind under four rising models. He was setting it up for the Zoom into the guts of the lift when his line tangled with Andy de Mello's (CDN). The ship veered left and down, unlatched and released prematurely to fly out of the lift and into the down for 92 secs. Andy got 127 secs. Protests were ineffective and in fact the jury decision was not available until two minutes before the end of the round. The Englishman John Boon also had a line-cross problem in an early round and in fact it cut his line. The cut dumped the model at 55 secs as he was approaching a patch of lift, but the timekeepers were adamant that they had not seen the contact. A hasty refly (which maxed) was disallowed by the jury. From thence on the British formed a posse to run with the flier and help him hack his way through, collect witnesses and impress the timekeepers with the visibility of the U.K. towline. The British remained stuck with a 55 sec blemish on an otherwise perfect team score for the day.

Since it was apparent that a big F1A fly-off was in the makings and the F1C business was yet to be taken care of, something had to be done to the schedule. This turned out to be shortening the lunch break from two hours to one. Lunch consisted of a plastic bag filled with goodies passed out at the edge of the field. A large tent was set up with tables and chairs inside. The goodies consisted of a large piece of cold chicken, four slices of bread, three tomatoes, a whole red chile, a big bunch of grapes, a bar of chocolate and a bottle of Koka Kola. The bag was decorated with the imaginative logo of the World Champs. The U.S. Team seemed to be in a crisis session in the lunch tent with a heated recriminatory discussion in progress about team business. This didn't sound too inviting so I went out on the field and enjoyed my lunch watching the Russians test fly. A number of teams had little tent sun shades set up on the field. Extra models and gear was kept underneath while only active models and equipment were lugged from post to post on the flight line.

Lunch break was soon over and the air filled once more with floating, soaring Nordics. The shortened lunch break was a good idea for everyone but Dave Simons of Australia whose model was being proxy flown by Schmitt. A series of timer failures had left him with one airplane which also saw fit to fly off sans DT. The Bulgarian Air Force helicopter went to look for it but didn't trundle it back until the pulled-up 5th round had passed it by.

Hertzberg of Israel showed me his unusual ship. It has a foam core wing sheeted inboard with 1mm balsa and 0.5 mm (sanded to 0.3mm) outboard. Light resin and cloth finish is made by placing a plastic bag over the sandwich of materials and evacuating air from the bag by vacuum pump. This is similar to the technique of Siegfried Reda (W. Germany) in making F1C

Jim Walters, Tacoma, Washington, U.S.A.

Viktor Isaenko, Kharkov, USSR.

Nice model, veteran of '73 World Champs (see stickers) by Cringu Popa, Placed 19th. Wing has ply ribs, spruce box spar and TE.

Martin Dilly, flying proxy for Rosalie Douglass of New Zealand, has made another max by the looks of the fans.

wings. The simple Israel impulse towline release is a 5-inch diameter music wire ring on the line end. When the towline is plucked the ring jumps of the hook.

The sixth round proved to be the nemesis of Valerij Ekhtenkov. Again he had flown early in the round and had released soon in what appeared to be good air. But then after thirty seconds or so it started sinking.Wavingof shirts underneath didn't help and soon it was on the ground at 104 secs, or actually it was caught at knee height by an embarrassed Albrecht Oschatz who thought he was saving a maxing model from damage. That one bad flight was enough to put Ekhtenkov down to a humiliating 56th place.

Jim Walters also ran out of luck in that round. Somehow he got a premature unlatch on his circle hook. He soon released the model rather than risk towing it in to the ground and it found itself in poor air for a 112 sec score. That dumped him to 54th place. U.S. hopes were now on Bob Isaacson. In the seventh round Lepp could be seen circling downwind of Bob. Bob found some good air and released. Lepp then towed back and forth through that air of Bob's, taking its measure, then satisfied as to its quality Lepp zoomed off in the best direction gaining 30 ft. in the first turn without stall. With seven rounds completed there were 42 with all maxes (same number as F1C). This included full teams of Russia and North Korea and many other countries with two each. New Zealand was accounted for with all maxes of Martin Gregory and Martin Dilly who was proxy flying a *Lively Lady* for Mrs. Rosalie Douglas. If only they weren't so short-handed!

Preparations for the fly-off again went rather slow and this time there was a backdrop of ominous weather developing. Thunder could be heard, the sky showed clouds with more form and substance and the breeze started to show more character and uncertainty. The line was moved once again, and some competitors got some quick test flights in. Lepp received a shock on finding two of his models had been crushed under the wheels of a car in all this confusion. The most disappointment and frustration of the day must have belonged to Kimmo Kulmakko of Finland. He missed a max by one sec in the very first round and then maxed all the rest. Denied a spot in the fly-off he was stuck with 43rd place. This remarkable situation brought forth much talk about the effect this would

Ivan Horejsi looking well pleased at this point. Fourth place.

Australian Dave Anderson friendly as ever but with no luck at Plovdiv.

Swedish flier gets his legs shortened a bit by barefoot helper.

Oschatz launching F1A of Lustig, E. Germany; Twelfth place.

Interesting mechanical contrivance of Tammy Thompson, Canada. Offset, swinging circle hook.

have on future rules changes. On the one hand to some it appeared that this would be unacceptable in future Championships; that when you could drop one second out of 1260 and place 43rd, the limits have been set too fine. On the other hand many consider a max a "gate". Either you make it or you don't and sub-max time (under typical thermal conditions) doesn't mean much. Following that reasoning it would seem that fly-offs would be a necessary and vital part of the competitive process.

Finally the green flare gave the go and off they were without as much confusion as had been expected. A number of the strategists immediately towed off into the far corners of the field to be in a safe position for observing the action. Some casualties occured early-on. Martin Gregorie of New Zealand had his rudder go berserk and he spun in for 31 secs. Mordehai Ovrutski of Israel had a mid-air collision, flat-spun-in and attempted to get back in business with his reserve. He didn't make it though as the red flare caught him still in tow. With only four minutes allowed there wasn't much time to muck about. There was good air somewhere as 11 contestants made 180 secs or more but only two had made 240 secs and so requiring a following round to go for 300 secs. The two finalists were Viktor Tchop, U.S.S.R. and Peter Allnutt, Canada.

Now everything was concentrated on the two finalists. The Canadians were in a strategy huddle discussing among other things that there might be some lift right over the assembled crowd of spectators. Mike Thomas was Pete's *runner* who was to come along with a radio and relay observations to Pete from the downwind Canadians. Tchop's *runner* was Eugeny Verbitsky who also had a radio. As it turned out the Russian command post was upwind.

Off went the green Very flare and away went Tchop while Allnutt waited. Tchop ran upwind about 200 yards where he went into a circling holding pattern with Verbitsky at his side. Allnutt had decided not to play that game and opted for the chance that there was lift right here with the crowd. He hooked up the *Thomas Tweaker*. That's a length of monofilament line running from the towline to fit loosely under a rubber band on the right wing panel. At launch it helps nudge the plane into its glide turn. After about two minutes of waiting Allnutt started to run and, with what looked like a somewhat early and low release, snapped it off to the right. Seeing that Allnut had committed himself, Tchop started winding his circle tighter and tighter. Suspense was surging in the crowd and then with a mighty zoom Tchop catapulted his ship into a magnificent 30 foot climb, leveling off in a piece of solid air. There was a gasp and cheer of admiration for this very well done bit of flying. Tchop's model floated on in obvious lift making the 5-minute max with no strain and DT'd at towline height. Allnutt's hoped-for lift wasn't realized and his ship sank for a 131 sec score.

Viktor Tchop readies for the fly-off.

The 1975 Nordic World Champion, Viktor Tchop, a young soldier from Odessa, USSR, is lofted on his teammates shoulders in a victory parade.

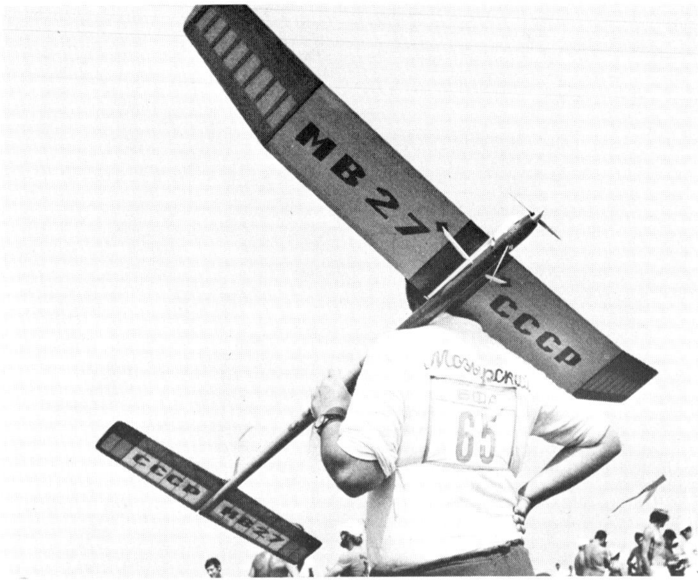

Mozyrsky of the USSR flew this somewhat dated design with strut braced wing. Made fly-off and 23rd place.

That was not the end of the excitement for that day as the stage grips rushed into action shifting scenery and props around for the continuation of the F1C Power fly-off. This the 5 second run fly-off. A thunder cloud had just passed up the valley and then the wind had shifted 180 degrees, threatening to bring the cloud back. The air had that very uncertain feel about it but there was very little breeze. The light was starting to fade and the photographers were trying to guess what film speed would be needed. The fliers had more to worry about.

North Korean power models were simple and unsophisticated but of course carried the essential Rossi, and VIT. Placed two in fly-off and team was 8th.

Procedure was the same as the first power fly-off round. Each flier at a pole on the line (selected at random) was given one minute to start and launch. This was done consecutively starting at one end of the line. Thus a ring of spectators, organizors, KAT Cops, photogrphers and helpers would surround each man as his turn came. Although the flier might not have appreciated all the attention and somewhat crowded conditions, this procedure did give a very exciting stage for the *melodrame vol libre* taking place. As each model was started and launched there were *oohs* and *ahs* with each inflexion to the climb pattern. Of the 37 who were flying only 12 failed to max in this round. This included unfortunate Ladislav Kovacic of Yugoslavia who could not start his engine. Three others overran including the exuberant Eolo Carlini. A special drama was played out by the Russian Sharon. His engine started but ran rough every time he readied for launch. He tried to needle it and get it to run but it just sounded sour. Then the tail popped and he struggled for a moment to reset it, but with frantic signaling from his teammates he tossed the model down and grabbed the reserve. The spinner was tweaked into the starter, it screamed away and with only seconds remaining he launched. At one second the engine stopped dead and the silence was deafening.

By now with darkness settling in, spots of rain and wind and even flashes of lightning appearing, it was obvious that the next 4 sec round could not be started. It was postponed to the following evening. Then with beautiful timing a rain squall hit and there was a scramble for the buses that were standing by to take everybody to the late dinner awaiting us back in town.

We, the working press, hied over to the administration building to attend the press conference, interviewing the F1A winners. The results were still being tabulated but the first three place winners were assembled along with intepreters and officials. The winning models were impounded in an anteroom and were being carefully measured for compliance with FAI specifications. Scores were also being double checked. Finally we assembled in a large room around a rectangular table. Tchop, Allnutt and Ivan Horejsi were placed at the head of the table. At the time it was thought that Ivan was third but it was not until the next day that a mistake in the recording was found and Dietmar Henke of E. Germany was declared third and Ivan Horejsi, (CZ) fourth. Soft drinks were provided everyone and the organizers made a few introductory remarks. Interpreters were available and we were permitted to ask questions of the winners. Viktor Tchop is 21 and a soldier in the Russian Army. He was born in Odessa and began his modelling career in 1969. In 1974 he became the Champion of the Soviet Union. This was his first time to fly at the Plovdiv site and also the first time to fly in a World Championship. His comments included: "It is difficult to become a champion, I will continue to train. I owe my success to my trainers. I don't think my models had any special advantage but I tried hard to fly them right. Our preparation of the team was in three stages of practice and training together. The first in March, then June and finally just before the W/C." Tchop's glider used a thin-walled fiberglas tube tail-boom, 18.5mm diameter at the forward end. Wings had constant chord inner panels and trapezoidal outer panels. Spars were tapered and diagonal bracing more dense inboard. Circle towhook was of the Lepp/Isaenko style.

Pete Allnutt of Canada at second place was both happy and disappointed but with resignation learned from long experience. Pete works as a ticket agent for Air Canada and lives on a farm in Orangeville, Canada and thus has his own flying field. He is also the FAI CIAM Free Flight Committee Chairman. His glider he admits is somewhat dated as he has concentrated on development of reliable equipment, and effective technique. The model has a rolled balsa tail-boom, and constant chord low aspect ratio wing with rounded tips. Short span inner panels are flat. Wing is fully sheeted top and bottom with 1/32 sheet. An elastic turbulator is mounted on ply outriggers forward of the LE. Towhook is a home-grown affair that gives a simple pendulum control to the rudder; release is by spring impulse on the line with no zoom setup. His comments included praise for the Bulgarian organizers and, in response to questioning, thought that the standard of flying is now leveling off and he didn't think a rules change in this class was either imminent or desirable.

Ivan Horejsi was wearing his broad infectious grin and didn't have a whole lot to say, just very happy. He lives in Plzen, Czechoslovakia, that's where Pilsner beer was invented. Ivan is an engineering technician in the heavy industry of the city. He practices flying 3-4 times a week, usually 2-3 hours in the evening when he returns from work, with the logical result that he tows and circles with an ease and self assurance that comes naturally when it counts. His model uses the Czech type circle-hook, a trapezoidal fully tapered wing with tapered spruce spars. Airfoil is a special one that was concocted by combining the B7457d for the top and the *Lindner* for the bottom. A complete construction article for the model is now available in the March '77 MAN.

Before the meeting broke up I was able to ask Tchop one more question. "Did you feel lift while you were circling upwind of Peter Allnutt?" Apparently he understood most of what I was getting at before the translater phrased it in Russian because his face lit up and broke into a grin, *yes*, he said. What a great scintillating feeling that must have been, circling patiently in that thermal waiting for his challenger to punch off in poorer air.

It was dark now and time to get some dinner before it was too late. Martyn Cowley, Ivan Horejsi and I piled into my rented Zhiguli and we drove off onto the now familiar highway into the city. While dodging the pokey trucks and skittering across the worn cobblestones we talked of Nordics, competition and good friends. Outside the dining hall we found a place to park crossways between a rock and a rut and went on inside. Our dinner was waiting and Ivan got a cheer from the late diners. A great day, yes-siree.

Tuesday, August 19, F1B Wakefield

At the dining hall the team helpers were picking up breakfast to go for those Wakefield fliers who were too anxious to stop for breakfast. This, the last flying day, was still putting lots of pressure on the competitors. There were 75 Wakefield fliers from 29 countries. Weather remained excellent. The threatening skies of the evening before had gone away, although they did return later. As is usual on a good weather day, the first round had good solid air with no "tricks" to it. Fifty-four made maxes in that round. One of the exceptions was Jan Zetterdahl. He had some motor breaks and was running out of time. Finally with a wound motor intact and ready to go, there was an announcement in

Typical Plovdiv street scene as Thomas Koster of Denmark hooks tow cable from his disabled German VW to your American editor's rented Russian-built, Italian-designed Fiat (Japanese camera).

Bulgarian, then it was repeated in Russian and then in English; "There is one minute left to the end of the round." Simultaneously with the English translation, the red Very flare was fired up and over a bewildered Jan standing there flat-footed, not able to let go in time!

In Wakefield it doesn't take much of a downdraft to kill a max and there is no running around with it on a string to feel the lift. So you develop your senses, and magic machines to tell you when to launch. First round saw the bubble machines spewing forth frantically, mylar streamers fluttering away, cattail fluffies bounding about and of course thermister antennae were everywhere. There was no chance a thermal could slip through that maze undetected. What you had to do to miss was make a mistake, and if you haven't tried this, believe me, mistakes are easy.

Allan Edwards (AUS) had an elegant looking model with a high-aspect-ratio solid balsa wing, rolled balsa tube fuselage and built-up stab. As a precaution Allan changed his timer before the first round but upon making the flight discovered that the auto-rudder failed to move to the glide position. The resulting straight glide resulted in a 172 sec flight. He found later that a small burr on the trigger wire prevented its release. The Americans all maxed in the first round but all three ships had stally flights which were not expected after much testing prior to the event.

The second round jolted many competitors from the euphoria generated by the seemingly ideal conditions. Mike Woodhouse, keying off of rising clouds of bubbles, climbed to a tremendous height but then glided as if under DT to make only 103 secs. Bob Wells then did the same thing and was rewarded with just 108 secs. Such a phenomenon is common at sites with good weather such as Taft, California. Bubbles can be misleading and there should be other signs to indicate a large enough thermal for a Wakefield. Phasing of the breeze strength and direction must be favorable and thermister readings can help indicate also. For the next half hour no models were seen to thermal and everyone got very cautious. For all of the Swedes and Russians it was too late, none of them maxed. Only the U.S.A., Yugoslavia, and Bulgaria maxed all their men in that round. The tricky air now brought out the 'beaters'. These were hordes of helpers that ran around under descending models waving all sorts of wearing apparel and even bed sheets borrowed from the hotel. The theory was that all this commotion and stirring would break loose a thermal under the model. It does seem to help. Most fliers used winding stooges anchored to the ground with nylon guys and stakes. They had to be pulled up and moved each round. Although inconvenient the non-flinching stooge makes a more reliable anchor than a human sacrifice. Usual procedure is to wind, activate the prop lock pin and then pace around glaring at the opposition and calling to helpers for more bubbles, fluffies (cattail seed pods) and thermister

Tom Prosser is an Australian wheat farmer and a good man to have around when tossing model boxes through train windows. Flew Wakefield and Power.

Allan Edwards also of Australia with his slick looking Wakefield. Noted for his penchant for seeking the aerodynamic truths.

readings. Best air-picker of the lot seemd to be Bob White. His background as an Air Force weather observer in World War II followed by five years with the United States Weather Bureau may have something to do with his skill. He watches bubbles, fluffies, thermister and streamers and of course, looks for other thermaling models.

Yugoslavian Vilim Kmoch was picking some good air also, but what he found in the third round was weird. The model climbed well straight overhead. Then it circled tightly in the glide and drifted to the west. It was coming down. After a minute of that it stopped circling and glided straight off to the east. Then it started circling again but now it appeared to be in strong lift. The thermal had gotten mixed up with a wind shift and the model had first climbed out of it and dropped in the down side outflow. Luckily the circulation fed it back in for another try.

Ron Pollard flies a very nice looking Wakefield. Neat elliptical tips on built-up wing and tail which sit on a tube fuselage; prop hub is faired cleanly with a spinner. Workmanship of these red and black beauties is superb. On his fourth round flight his model collided in the air with another and it fell at 175 secs. On the reflight attempt a motor blew and then there was very little time left. He finally got wound and launched and

Vilim Kmoch of Yugoslavia checks his prop hardware. Placed 11th.

This crowd of Bulgarian young men have gathered about genial Ron Pollard of Great Britain. Ron is showing them all the bits, pieces, whistles and bells. Chap on the left looks a bit dubious!

fortunately all indications showed positive. The ship climbed nicely and off it went in a nice boomer. Later I saw Ron surrounded by about a dozen Bulgarian young boys. He disassembled the model completely to show them piece by piece how it was built. They were probably equally fascinated by Ron's engaging northern English dialect. Mike Thomas also had an interesting dialect that he used while hurling his Wakefield into the air. He charges down the field and with a blood curdling yell lets fly with a tremendous heave. The ship arcs up with its prop blades complacently tucked into the fuselage, restrained by a clockwork timer. At the crucial moment, click, they pop out and start flailing around pulling the model up into its climb pattern. Mike was doing OK, making it into the fly-off and 13th place.

The Swiss models were marvels of intricate mechanisms, variable auto-stab, etc. Something went

Mightier than a speeding locomotive, Canadian Mike Thomas launches his delayed prop release Wakefield. Thirteenth place.

Jon Davis and Groovy Tuna — small stabilizer.

East Germany's Mielitz prepares his Loffler-style Wakefield.

Reiner Hofsass and his Espada are ready for the lift, but friend Sigfried Reda is looking the other way. Beautiful design.

wrong on one of them, however, causing the stab to pivot around to the vertical. With this unorthodox new trim the model layed over on one wing tip and went into spiralling loops. Amazingly it continued to climb in the thermal and maxed. Impresario Pollard had one flight that left no doubt of a max. It DT'd at the usual time but took 24 minutes to come down. Martyn Cowley reported that the happy retrievers running through the vineyards were given huge bunches of grapes by the farmers. Another observation by Cowley was that there was no shortage of lift but some were having trouble climbing to where higher models were marking it. Atlhough the most successful seemed to be the compact low-aspect ratio models of the East Europeans, who had fast punchy climbs that pulled through the lower poor air with no nonsense. The Bulgarians had done very well, maxing through five rounds, showing that they had *this* field figured out. But then in the sixth round two of their men fell by the wayside. Then joining the ranks of the disappointed, Rashkov, Bulgaria, broke his string of maxes in the seventh.

The N. Koreans were doing quite well with two men maxing out. Their models looked conventional and if anything, conservative, with moderate aspect-ratio and fairly compact layout. Some had aluminum motor tubes and some had rolled balsa tubes. Tail-booms were built-up diamond sections with Warren girder truss. All surfaces were covered with dull tissue. Motors were 30 strands of 3mm Pirelli wound to 425 turns. Props had fairly narrow thin blades carved from a material like bass wood. Their flying routine was very effective and efficient. Not much talking, all well rehearsed. If the first man didn't get the signal to launch within 5 minutes from their tactician, he would step down and the timekeepers would be switched to the second man who would start winding. This process was continued until all three had flown. The technique assured climbing with a fresh motor, and in fact the N. Korean's models did climb strongly through the whole run. Only one model, by Kim In Sul, had any auto surfaces and this was a rudder tabworking off movement of the rear peg. The Dane, Peter Rasmussen, is left handed so he thought why not a left hand Wakefield? Sure enough it worked, climbing to the left, probably confusing a number of fliers.

Now the seven rounds were all completed and there were 14 with perfect scores. N.Korea, Canada and Yugoslavia each had two men in the fly-off. Frantiske Rado of Czechoslovakia had the unenviable distinction of being out by only one second. The clouds were now looking a little wet in the distance as preparations were rushed for the fly-off. There was some confusion about getting the correct rubber distributed but with *only*

A fiberglass motor tube and tail boom is used by Ole Torgenson on his Wakefield. Flies from frozen lakes back home in Norway.

fourteen positions to attend to it didn't take long. With the green Very flare in the air, the fliers had four minutes to launch. A blown motor at this stage would best call for use of the ready reserve model. It appeared that those who launched late in the four-minute period missed the better air and could not make the four-minute max. This included the Yugolsavians Kmoch and Pecek and the Canadians Thomas and McGillivray. Also Hugo Benedini, Riener Hofsass and Bert Kroon. In the next round going for a five-minute max all seven fliers made it. Bob White's model appeared to be the highest, having climbed up past the rest at a rapid rate. A max is a max and a five-minute one is a good flight. Now the seven made ready for the six-minute max. This one wasn't going to be too easy. White was down to his last 'good' motor and it was getting tired. Weather was holding with nothing spectacular evident yet. The atmosphere was tingling with suspense and anticipation as the seven wound for this flight. All seven got into the air pretty much together and they seemed to be doing pretty well. White's climb this time didn't look as good as before and now we knew it would be close, and this would be the last round.

Groups of suspenseful spectators gathered near the working timekeepers. Some hurried consultations between the timekeepers and the organizers set off ripples of rumors. Then a burst of excitement from the N. Korean group was quickly accompanied by the word spreading across the field. The N. Korean Paik Chang Sun was first with 303 secs. Bob White was second with 281 secs. Austrian Hans Zachalmel trailed Bob by one second for third place. Following in fourth place was Kim In Sul, N. Korea; fifth place, Albrecht Oschatz, E. Germany; sixth place, Roberto Artioli of Italy and trailing in seventh place but with a very creditable 214 sec flight was Geza Vineze of Romania.

Team victory went to N. Korea followed by Poland second, Canada third and Bulgaria fourth. The N. Koreans were understandably happy. The exuberant team pressed a bouquet of hastily gathered wild flowers into the hands of their Paik Chang Sun, and then tossed him into the air a few times. Bob White congratulated the winner, got his model from the U.S. chase crew and tried to recover a bit from the

Paik Chang Sun of North Korea, winner of the 1975 Wakefield World Championships, Plovdiv, Bulgaria.

Bob White appeared to be outclimbing the competition except for the last fly-off flight. Down to his last good motor, it too got tired and that was that for second place.

shock. He had felt good about his readiness this year and the earlier flights had looked great. Now it was over and time to pack up. On pulling out the fly-off motor it was found to have three broken strands, some indication that it had done its best but was about one flight too old.

The excitement of the Wakefield finale now blended into a quickening pace of preparation for the final Power fly-off. The 21 finalists remaining for the tenth round were hustling their models and gear into position. Team managers and helpers were checking with the timekeepers and organizers to make sure of the procedures to be followed. This would be the 4 sec engine-run round. Light was not too strong but the dusk was not settling too rapidly. Additional maxes at this stage didn't seem likely. There were some moist looking clouds filling in most of the sky at a low altitude. The glow of sunset was filtering through, giving them a slight orange tint. Up into the threatening sky went the green Very flare spitting out a smoky white trail. It's time to fly and get this World Champs over with.

That signal was punctuated by the first man's Rossi leaping to life. A big crowd gathered around but managed to stay clear in a circle of about 30 feet diameter. Everything was concentrated on that screaming engine. That piston is coming to a dead stop about 900 times a second. The fuel-air mixture must be just right to be fired 450 times a second. Will it run clean and fast and stop dead within the four second flight limit? The free flighter knows there is only one way to find out . . . rear back and throw that thing for all you're worth.

As each competitor reckoned with his moment of truth and committed his machine to its own will, the surrounding crowd urged the ship upward with their thoughts, *oohs* and *ahs*. Screaming engines, intense competitors, some wild looking clouds in the offing, wild enthusiasts from all over the world, made a witches cauldron of excitement as each one got ready and fired his ship into the sky. The FAI Jury was used to check all engine runs in turn, starting at one end of the line and working on down to the last man. Surprisingly there were no overruns. In fact most were cutting the runs short at from 3.0 to 3.6 secs, to play it safe. This may well have made the difference in the outcome of the close results. A more precise engine run will surely be one of the items to be closely looked at for next time. Some of the flights didn't seem to pattern correctly causing the transition to be additionally confused and inefficient. This may have been due in part to the crowded conditions and tension upsetting the flyer.

Young Ken Faux of England nervously awaited his turn near the end of the line. He had decided to use his dependable Team Trials model because it had his fastest engine. Changing this engine into his fly-off model didn't seem like too good an idea at this stage because a few test flights would have been needed to check out that combination. However he did fit a special thin blade carbon fiber propellor to this engine. This was his fly-off prop. The whistle blew and an intensely resolute Ken shrieked the Rossi to life. That siren song came in at 29,000 rpm. The prideful U.K. contingent standing by were all smiles. I had been tape recording a number of the launches and I had the machine on for this one too. Playing the tape back later I found that I had captured a very dramatic moment. You can hear Ken's engine start, running clean and very fast. There is a background of voices from the English Team supporters as Ken gives the engine a brief warm-up. Then as he launches, there is an immediate rise in pitch due to the Doppler effect unloading of the prop. And then just about a second later there is an instant change in sound. It is a crazy obscene, cacophony of notes. An incredulous voice (Martin Dilly) is heard saying "O-oh Jesus Christ - Ken's just bust a prop blade." The ship fell from the air with all the grace of a shot duck. Although a shocked Ken Faux had now lost it all in that split second, the English had gained something else. Something that their psyche seems to seek, not so much as an admiration of failure but what J.B. Priestly describes as "an appeal to their imagination, as if they were so many poetic-tragic dramatists looking for a theme, a doomed gallantry, a clear flame of

Ken Faux was looking great in the fly-off until his prop lost a blade one second after launch.

Lars G-Olofsson shows how it's done with his Uncle Remus. Topped the fly-off to become the World Champion.

Eugeny Verbitsky shows off his very technically interesting model. Engine run was a little under. Placed second.

heroism extinguished at last by darkness . . . English history is rich in these apparent failures — appalling sieges, desperate last stands, and the rest — that were welcomed and long remembered by the people as triumphs of the spirit."

And so it was, but now what about the winner? This soon became obvious as we saw this happy Swede being tossed repeatly into the air by a wild group. It was Lars Olofsson. He had flown for 158 secs easing out Verbitsky (USSR) by 10 secs. In third was Mike Burns of Canada. Siegfried Reda of W. Germany and Reinhard Truppe of Austria were tied for fourth place, while Urs Schaller, Switzerland, was sixth.

Olofsson's model, *Uncle Remus*, used a pylon, high-aspect-ratio wing and long tail-moment. All surfaces are sheet covered. Auto-stab and auto-rudder are considered standard these days. Lars used another model, called *Miss Muffet*, which had an undercambered wing section, for rounds 1 thru 5 and 9. *Uncle Remus'* wing section was flat-bottomed. Construction of *Uncle Remus* and *Miss Muffet* was fairly simple compared to Verbitsky's model which had a very complex metal forward-fuselage structure. We had a chance to inspect the disassembled Russian model the next day but more on that in a moment. It appeared that the clean high-aspect-ratio ships had an advan-

Bulgarian organizers held a press conference to kick things off.

tage in the fly-off. Transition efficiency was also very important. Beyond that a few tenths of a second engine run probably made the critical difference for some.

Well, the flying was now concluded and right on cue the rains came. There was a mad dash for the buses and cars as rain drops swirled about, reasserting their claim to the Plovdiv sky. The final press conference was convened this time with the Wakefield and Power winners present. There was some delay in getting this started as a scoring contretemps was set in motion by the organizers. Verbitsky was visibly upset by this misunderstanding and refused to be seated until this was all straightened out. Finally the numbers were checked again and Olofsson confirmed as the winner. This was translated in a number of languages and soon everyone was smiling. Lars Olofsson works as a Fireman. He is a great bear of a guy, very friendly and considerate. His accomplishments include mastery of the English language, and is editor of the Swedish Aeromodelling Magazine, *Modellflygnytt.* During the press conference Lars and Eugeny Verbitsky carried on a lively discussion in English. Verbitsky is a mechanical engineer, lives in the Moscow area and also instructs modelling in the Soviet Aeromodelling Sport Institute. His fame and reputation as a sportsman of great skill and character has spread worldwide over the many years he has been an active and key participant in international modelling. Mike Burns of Ontario, Canada, is 47 years old, married and has two small children. This is his first experience at an international contest. At home he has to drive 80 Km to get to a small test field while a larger 3 min field is 380 Km away.

The Wakefield winner, Paik Chang Sun, appeared to be the older more experienced member of the N. Korean team, and in fact is the Chief Instructor of the North Korean Aeromodelling Institute. He is a teacher in a technical college and has been flying for 15 years. His model had an alloy tube fuselage blended to a long diamond section tail-boom with single fin. It had a prop stop auto-rudder and fuse DT. He wound with prop on with a shield. Motor was 30 strands of 3mm Pirelli. Source of that batch remains a mystery, but it probably was not particularly super. The other PRK Wakefields were similar with minor differences. Teammate Kim Dong Sik placed second in 1973 at Wiener Neustadt, Austria, now was 34th this year. Bob White of the U.S.A. is 52 years old and has been modelling since 1933. He is a supervisor of electrical assembly at Bell & Howell in Pasadena, California. Bob placed fifth in 1973 and third in 1971. An intense competitor, he believes in and practices careful, thorough preparation, taking his modelling quite seriously (any other way seems a waste of time). He recognizes the sport as individualistic in nature and prepares himself accordingly. His models are considered simple (no gadgets) but beautiful in layout; high aspect ratio, elliptical tips, twin fins, high pylon. He builds the best model he knows how, taking no short-cuts. Bob has tried just about every prop shape, aerofoil, rubber and trim set-up there is and has always had the knack to sort out what works best, a difficult trick.

With congratulations all around from those assembled the press conference broke up and we headed for dinner and a night of celebration. At the dining hall the Bulgarian personnel also were feeling the spirit of the occasion and welcomed their new friends that now came streaming back from the final day of flying. National flags at each table identified reserved locations for everyone but of course there was plenty of table hopping. At the hotel the party was soon getting under way. It is doubtful that either Bulgaria or the Free Flight World has before seen such an indulgence in good will and fun. Having everyone together in one hotel made things convenient. However there was no large room for everybody to assemble. Instead the stair landings and corridors on each floor became the arenas for a number of parties. A "togetherness" was managed as a steady flow of people visited the different floors joining in the fun, making it a truly international gathering. Instruction was given in Russian bear hugs, French drinking songs, Cuban dance steps, German polkas, U.S. guitar chords, Yugoslavian liquor processing and Australian farming. This lasted till dawn.

For those who were alive the next day the organizers laid on a bus tour of the outlying Bulgarian tourist attractions. This was spoiled somewhat by the light rain that fell. A few of us stayed behind and gathered in Tom Koster's room where a number of models were dissected and carefully examined. In Thomas' case his model was already dissected under the previous unfortunate circumstances. His ingenious mechanisms and good workmanship could be seen clearly. Verbitsky took a screwdriver to his model and soon had all the bits lying there before us. Prop was a carbon fiber special with wide chord root section notched for the hub. His name was cast into the prop in white letters (class!). Prop brake of two loops of spring wire about 2mm in section was positioned behind the prop driver. One end had an eye in it to fit under an engine mount screw. The other end had a hook ground into it. The hook was pulled away from the prop driver by a wire from the engine timer. On release it went forward and hooked into a slot in the prop driver. Although this sounds like a violent way to stop an engine it does work and there is no breakage problem. Most unusual construction feature was the attachment of the engine pan. This round unit slipped inside an alloy fuselage forward section. It was held in place by two rubber "O" rings that were compressed between these concentric tubes by diaphragms that were tightened by means of a single axial screw. Only thing holding the engine pan in was the friction of those "O" rings. It had a hole through it like a doughnut to make room for the axial tightening screw. Exterior parts were anodized and carried Verbitsky's name and address engraved and filled in with white paint. Timer appeared to be of USSR origin and may be "home-built". It was buried inside with only the scroll protruding. Release arms were hinged from the fuselage alloy tube. Metal work on the arms, hinges and timer start switch was very well done and neat. The engine cowl was retained by a tab locked under one of the cylinder head screws. Two piece wings had a flat-bottom section of a little over 8% thickness. Stab section had an interesting slight reflex at the trailing edge. Stab was hinged to the fuselage by a platform that formed a section of the leading edge. The round tail boom was attached to the forward fuselage by a machined alloy joint pinned together with machine screws. Everything was well thought out and looked very clean with no wear and tear apparent.

The *tourists* came back from their bus ride and everybody got ready for the grand finale that night. It was difficult to recognize many after they got out of their flying clothes and into their banquet outfits. The buses took us first to a large auditorium for the awards ceremony. The stage was set with a backdrop of banners proclaiming the World Championships. Cups and Awards were on display on the right side. Before the awards ceremony got underway we were entertained by folk singers and dancers performing in traditional colorful costumes and to music played on traditional instruments. And then a modern touch was provided by a rock band and some very jivey singers.

The beaming winners were then paraded on and presented with the traditional Cups, Awards and special Medals and Diplomas. From there it was a short walk to the Trimontium Hotel for the banquet. The very elegant Banquet Room featured marble col-

Entertainment at the awards ceremony included traditional folk dancers and this rock combo and singers.

The 1975 Free Flight awards ceremony gets underway.

North Korean table at the Plovdiv banquet shows off the Wakefield Cup.

Dancing in the ballroom of the Trimontium Hotel — "Blue Moon - Stranger on the Shore" — the odyssey comes to a close.

umns, crystal chandeliers and formal decor. Entertainment continued with operatic duets, sword swallowing and dancing. Conversation swung from the recent past results to the prospects of getting to the Pierre Trebod contest in France due to start in three days.

The orchestra swung into a moody "Blue Moon" and then a haunting "Stranger on the Shore". . . We had met a number of strangers on this shore. . . We bid farewell now as friends.

Next day, Thursday August 21 the exodus began with a large number of the participants attempting to get to Marigny, France,for the Classic Pierre Trebod competition scheduled for this Saturday and Sunday. Some were going by train (soon to be labeled the *trans urinal express*) some by airline (the lucky ones) some by private car (driving straight through over a suicide route) and some by hitchhiking. All of these modes resulted in incredible stories that tax belief. Like the one about the train pulling out of Belgrade station with the American Team Manager, looking a bit wretched, standing on the platform with the team's tickets in his pocket.

Martin Dilly demonstrates the preferred method of hot-wiring the Bulgarian rental Zhiguli cars. A procedure needed for the defective switches. The great race to the Pierre Trebod is on!

I drove the rented Zhiguli back to Sofia, accompanied by Edwards and Anderson of Australia. These cars had a nasty habit of jamming their ignition switches. The hot-wiring expert of the Free Flight group turned out to be Martin Dilly. Although nothing in his background would ever lead anyone to suspect. So we motored on having learned Martin's tricks. Then I learned a new one. The gasoline tickets I was dearly holding for authorization of purchase turned out to be the torn half receipts from the previous transaction and were worthless. We refueled but not without learning a few more colloquialisms, both them and us! We got rid of the car in Sofia along with a lot of Levs and rode the free bus out to the airport. Later we found out that the bus wasn't free and neither were we if caught up with. At the Air France counter I was rudely told that my confirmation was *nyet*.. *No one liste etc.* (The phone call made from Plovidv to confirm my reservation must lie somewhere on that great heap of airline hypocrisy.)

We began a great Australian-American-Bulgarian cause celebre, and that seemed to be what they wanted to see, a little spunk. An Air France representative finally gave me a boarding pass after shrugging his shoulders and waving his arms a lot. A Caravelle pulled up to the ramp and we squeezed aboard and got strapped in and firmly immobile. Then came the model boxes! We had with us the entire Cuban team who had brought along a dozen or so pieces of lawn furniture in addition to their model boxes. The boxes got stowed everywhere including under our feet. All this bashing about delayed the take-off a full hour but finally we were off.

It was a good flight, landing enroute at Belgrade, giving a nice view of the city. On the way we discovered Pete Allnutt in the cabin, smirking grandly on hearing my story of my reservations problem. Seems Pete handles that sort of thing for Air Canada. Soon we were circling *La Tour Eiffel* and the great event was in our memories. ■

Fountains Trocadero at la Tour Eiffel.
(Criterium Pierre Trebod 1975)

Pete Buskell of Great Britain fires off into lowering clouds to take seventh place in the sixteen man fly-off.

Criterium Pierre Trebod 1975 Marigny, France

F.A.I.
1975
11eme critérium pierre trébod

The object of the *The Great Race* was to travel some 1200 miles from Plovdiv, Bulgaria (site of the '75 World Champs) to Marigny, France in just two days. That many made it in time to attend this great Free Flight tradition is a tribute more to the gracious French hosts than to the unfortunate schedule. And of course a fine tribute to the stamina and mobility of the modern free flight enthusiast. Of course many P/T participants this year did not go to Bulgaria and they were already at Marigny awaiting news of the World Champs. Your editor and Australians Dave Anderson and Allan Edwards arrived from Bulgaria via Air France Caravelle — neat! We rented a Renault 16 at Orly and drove off to Sezanne/Marigny/Pleurs.

We didn't arrive until about 11PM and unfortunately didn't find our hotel open. We did find a number of English enthusiasts milling about in the streets of Sezanne. They pumped us of all the information they could and then we continued our search through the byways of France searching for lodging. This proved to be futile and we decided to sleep in the car. A friendly spot under a tree was picked and we piled up the model boxes outside. Stretched out inside as best we could and with a night-cap of Air France Brandy we spent our first night, *En France*. Later we learned that not a few others spent that same night so encamped. The next day we found our hotel, had a great English breakfast, and checked in at the 11th Criterium Pierre Trebod at the Marigny le Grande military airfield (It is unused with only a few military caretakers and guard dogs in residence.)

Genial Pierre Trebod Contest Director, Yves Olard.

Rounsaville, Collins, Bailey, and Faux quiz Buskell about something. Paul Bond lost interest.

The friendly efficient P/T director, Yves Olard, checked us in and took care of changing my registration. Having lost my "Tuggy Buzzard" F1A at Bassingbourn in England two weeks previously I now signed up to proxy fly Australian Dave Simons' F1A (Dave's model was on its way home after being proxy flown in the World Champs). This gave the Australians a full team. I became the team manager also so that if anything went wrong the "foreigner" would get the blame. So in a relaxed jovial mood we checked out the field and the arriving competitors.

Make no mistake about it, this is a *major* free flight happening and as some of us claim, the most enjoyable international free flight competition of the year. This year, which is fairly typical, saw 148 in F1A, 78 in F1B and 54 in F1C. That's a total of 280 compared to 244 for the World Champs. *Eighteen* countries were represented. It is difficult to describe the friendly atmosphere of this competition without getting carried away. It is like a rival World Championship in which *anyone* can enter and the organizers only know three things; free flight, competition and friendship. They really put it all together. Our advice is, if you wiped out at your local team selection lottery (there are more of us who did wipe out than didn't!) plan on getting to the Pierre Trebod next year. It's not the World Champs, but in some ways it's more enjoyable. Air Charter flights, LAX to Paris and return cost about $430.

I digress. Back to the field. Finally after talking with many old and new friends we decided we ought to do some test flying. We got all the stuff out and I was just about to try a flight with my new charge when the field was suddenly hit with a downpour of rain. We scurried about shoving stuff back into the Renault, and sat there steaming up the windows, not able to drive off as the rain would come in the half-opened rear hatch if we turned around. We finally struggled around pulling stuff and boxes in, battened down and made own way back to the hotel.

Swedish camp.

Dave Hipperson rummages in the weeds for parts of his Wakefield. Tenth place.

Ernst Neumann, Munich, W. Germany, edits the Bartabschneider newsletter.

Pecchioli of Italy had turbulated all sheet wing, alloy tube.

Allan Jack of Great Britain flew a Pollard type Wakefield.

Keith Proctor, G.B., dressed in his "Welleys" ready for a jaunt through the weeds.

H. Braud of France. Timer operated A/R, finger release for prop.

Robert Allais of France dropped a bit in the last round to place sixth.

Jon Davis, USA, lofts the Groovy Tuna skywards.

Hugo Benedini from Argentina tied with Allais for sixth.

L. Manche of Holland, elegant machine.

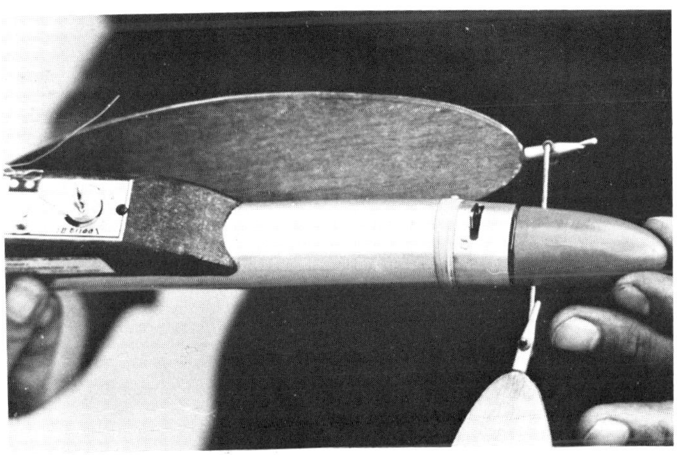

Seelig timer on cabin forward of wing for DT, nicely finished prop hub.

Ian Kaynes of Free Flight News fame.

Pierre Chaussebourg almost ready for takeoff.

The next day the bad rain was gone, clearing skies and little breeze greeted us. F1A was flown on Saturday with F1B & F1C flown together on Sunday. Fliers were grouped in a 'clutch' of up to seven and this clutch shared a team of two timers for the round. It was up to the competitors to decide flying order. A certain amount of reason and cooperation between contestants prevailed and this system worked better than it sounds. The good weather helped and there wasn't much waiting. Seven of the 148 managed seven maxes each.

Tammy Thompson of Canada chose to circle down wind for long periods. He is quite good circling on his own away from the crowd. The seventh round saw him drop two seconds in air that was good but not good enough. Pete Allnutt found his lift further upwind, making the fly-off. Only two, Peter Seelig (Hans Seelig's son) and W. Kraus of Austria made the four minute flight. Kraus had a low-aspect-ratio sheeted wing model, short coupled. He had a good zoom launch technique. Peter Seelig was flying a higher aspect-ratio design with light built-up structure wings, fiberglass tail boom and molded fiberglass fuselage front end. He used a resonant spring release mechanism on the line. In their flyoff Kraus gained a bit on a zoom launch. Seelig launched flatter but eventually seemed to be in slightly better air, nosing out Kraus by 13 seconds. Allnutt placed third. The fresh air and exercise seemed to agree with your editor as the Simons/Hartill effort managed 15th place. Now if only we had a bit more practice!

The weather the next day was right on. More of the journey-weary stragglers from Plovdiv showed up and wandered around like Zombies. The Great British

Greaves launches on his winning flight.

team had returned by Masterman Express to Oostende, Belgium where a number had parked their cars and then drove to Marigny. Never give up! Wakefield saw some excellent flying but with only two in the fly-off. Bob White missed it by two seconds in the first round for third place. Greaves of Great Britain and Lepage of France flew the 8th round. It was near the end of the day with the air rather still. Greaves flying a long box fuselage ship had an excellent glide making 203 seconds. Lepage's climb pattern looked a bit off trim and he managed only 156 seconds. The English went ape. The Argentinians picked up 4th, 6th and 37th, the Cubans were 9th, 11th and 12th with Americans Piserchio 14th, Smitz 20th and Davis 33rd.

Alain Landeau and "Boom-Boom" placed fourth.

The happy victor.

Frank Wolff made the fly-off but had an overrun.

Bob Bailey looks a bit mean here, must be after dropping out in the first fly-off round.

Buskell showed the young kids a few things, Seventh place.

Ray Monks, a favorite with the P/T gang.

Very nicely done streamliner, designer unknown.

Barbaro of France had flaps on his ship with short unflapped tips. Seemed to work well for ninth place.

Paul bond launches away nicely.

Dick Johnson, G.B., didn't have any luck at all.

Reda launches his high aspect ratio ship into fifth place. Surfaces balsa skin over foam.

Richard Schley, Germany, had fun with his Danish looking model except one bad flight.

Hans Seelig gives Dave Rounsaville's prop a smash.

Dave Pimm of Great Britain missed fly-off by 5 secs. Low aspect ratio multi-spar braced with diagonal glass fibers.

Sauer of Germany gets some muscle into the launch. clean ship but no luck this time.

M. Jean's rather ordinary but potent model and extraordinary assistant. Overran in fly-off.

Graux of Belgium trumpeting up a thermal.

F1C gave us a better look at some new airplanes that we weren't able to examine closely at Plovdiv. Also there were some new ones that hadn't been at Plovdiv. Hans Seelig, and a number of his German clubmates, had similar shoulder wing unflapped models with high aspect ratio foam-core wings. The forward fuselage section was molded from fiberglass and formed a single structural element including the engine cowl, wing root fillets, engine exhaust duct, timer mount and main body. Tail boom ws molded balsa and glass; single aft fin, auto surfaces. Climb was straight out at about a 60 degree climb angle. Easy transition to a flat glide. Seelig had some kind of trouble in the 4th round. I believe something happened to his engine. No fewer than 17 fliers maxed out, to produce a very exciting fly-off. As usual the weeding out of these contenders was taken care of by the inevitable mistakes. Embarrassing but that's the game.

Ray Monks had a VIT line break on launch and the resulting power-trim glide was a mite fast. Bob Bailey, with everything looking good, D/T'd early. Frank Wolff had an overrun on the 5 sec round, as did Michel Jean. Jack Allen had his engine gobble itself on the way up. In the 4 sec engine run round there were seven left. Buskell and Collins both went out with short runs while Reda and Landeau were in poor air. Three maxed. In the final 4 sec round Austrian Werner Kraus made up for his second place in glider to win. Seydel and Truppe came in second and third respectively. The air had gotten worse in the gathering gloom and that was the end of the maxes and the great Pierre Trebod '75.

That evening a banquet was held in the little restaurant in Gaye, nearby the airfield. It was jampacked. After a round of food and several rounds of wine we adjourned to the outdoors awards presentation. Short on ceremony but long on sincerity the affair was soon consummated and there was Rounsaville and his guitar again. A very nice time was had by all! ■

Sigurd Sydel of Germany placed second. Immaculate model.

Austrian Werner Kraus having lost Nordic to Peter Seelig the day before, made up for it by winning F1C.

Fully streamlined model by Sauer, Germany, with sheeted plug-in full taper wings, built-up stab, similar to Seelig design.

Unidentified power flier has his engine run ground checked by an accommodating timekeeper. Starting points were rotated and identified by lettered cards on back of the chair.

Landeau of France and "Boom-Boom", fourth place.

Le Gayon got a bit crowded as several hundred party-mood fliers descended upon it.

Jan Zetterdahl surveys the wreckage inside Le Gayon.

Allan Jack checks the action.

Aeromodeller Coupe d'Hiver trophy.

Ian Dowsett (UK) designed and crafted the Aeromodeller trophy.

COUPE D'HIVER INTERNATIONAL 75

COUPE d'HIVER INTERNATIONAL 75
R.A.F. HALTON GREAT BRITAIN
NOVEMBER 30, 1975
By Ron Moulton

THE GHOSTLY COUPE, due to a number of difficulties was not staged in France in 1975. It looked like the XXXIst running of this great French affair would be aborted. But then the great mass of world interest came to bear and the English proposed to Maurice Bayet, creator of the class and publisher of *Le Modele Reduit D'Avion* that the event could be held in England. "Tres bien" was his immediate response, and so off to organize.

The Royal Air Force proved a friend indeed and offered the facilities of R.A.F. Halton. The Coupe d'Hiver enthusiasts throughout the world were alerted and began their preparations for appearing in person

Ron Moulton, Aeromodeller director.

French Coupe d'Hiver group visiting the Royal Air Force Museum at Hendon.

or in shipping their prideful efforts to anxious proxy cohorts. The weather turned quite miserable and now they'll never believe us in Europe that London fog is a thing of the past!

We tried our best to convince the French equipe that the grey mantle of cold dampness that smothered the U.K. through the last day of November was exceptional but the message couldn't get through. They'll say we were trying to excuse a tradition. On the European Continent, everyone expects to find fog in England at this time — and we weren't letting them down! Actually, it rained so much on November 28th that the whole country swelled up like an over-leavened loaf of bread and hit cloudbase. Many were the tales of ice-faired aerials and wing mirrors on cars that loomed out of the clag at Halton's airfield gate, from parts as far away as Yorks, Lancs, Somerset & Kent and it says much for the rose-tinted optimism of aeromodellers that many were actually on the apron waiting for the start at 09.00 hours.

Visibility at that time was 50 meters. Exploratory checks with chuck gliders showed that the air was surprisingly bouyant. When a couple of R/C electrics were put together, hopes were raised for discovery of the vertical visibility and sure enough the *Kloudrider* could be seen for all of a hundred feet upwards. Yuck! After cavorting around, the whirring soarer ditched itself into the frosty grass — frozen out of the sky, with great globs of ice all over its wings and the elevator gap filled solid. Brhh!

Meanwhile our French friends and just a few British diehards were showing a mixture of impatience and blind hope by putting a few Coupes together. As precious daylight hours slipped by, and the field became alive with scattered bodies chasing strange silent shapes above, the visibility improved to all of a hundred meters by 11.30.

From nowhere but the model box and small knapsacks, the French extracted tents, Welly boots (Napoleons?), 20 ft. thermistor masts, floating drift streamers and all the paraphernalia of professionals. They'd seen the new trophies. Maurice Bayet's Coupe de Paris for 100 gr. and the Aeromodeller Trophy commissioned from Ian Dowsett for 80 gr. models. It's such a beautiful piece of art that it possibly accounted for all the over-wound broken motors of the day. Ian made a scaled down Coupe in silver-plated metal, designed a mould, produced a title plaque and had the job set in clear acrylic. The result is something to treasure — superb honor for the winner to keep for the year. And quite a few eyes glowed at the thought of taking it home!

By 11.45 the barbecue had been set up, so that at least there was to be some action, if only to raise a man-made thermal. General bustle gave an impression that the contest might be viable, so control was established for a start — 3 hours late, and about 300 meters out into the eerie grey of a seemingly endless void. To cope with the conditions, timekeepers were allowed to follow under the model. Drift was at most a mere 3 mph, and tests had shown that 90 second flights were easily followed from launch to touchdown. So at noon, the first official flight was made by Peter Cameron and when the timer came back with the card it was a 'max' — two minutes upwards until DT'd. Second flight by Len Ranson went the same way; but the DT failed — and within the first quarter hour, we had seven maxes on the score cards!

So it went on, with just over three hours of the weirdest, yet most pleasant contest flying in ages. Out of 266 flights recorded, almost a third were maxes and despite the visual limitations, only three models remained lost at the end. One of these was Claudine Dupuis', the only lady competitor, whose husband dominated both classes. One might say the British were 'duped' by this dedicated pair from Chatelleroux. Louis (like all others from France) was deliberately flying at over 100 grammes weight in the 80 gr. class to ridicule the claim that the 20 grammes

Louis Dupuis and Bernard Boutillier of France, winners of the British Coupe d'Hiver International '75.

makes any difference. In fact, he won the 80 gr. class with a 107 gr. model, came second in the flyoff and fourth in the 100 gr. class to total up 10 maxes in 11 flights. He was beaten only by French C.d.H. Champ Bernard Boutillier with a better flyoff decider and even then with his own motor which he sportingly loaned.

Ice on propeller blades, frozen fuze, soggy tissue and masses of rubber breaks (one entrant broke 14 motors) some spectacular climbing episodes, gallons of coffee and pounds of sausages and beefburgers consumed, will be remembered as part of a true Winter Cup story. By 1600 hours it was dark and the entourage moved off to the Sgts. Mess annex where warmth and prizes awaited the first 28 in 80 gr. and all 26 who flew in 100 gr.

Results had been posted as soon as the last flights were made, so that as each prize winner was called he was given his flight record and allowed to take his choice from a ladened table. Thanks to the generosity of MAP, Ripmax, Solarbo, Humbrol, KielKraft, Henry Nicholls and Ron Irvine, the awards were very well received as a practical souvenir of a fine experience. The contest proved many things to myself as organizer, not the least of which is that there are still many kind souls around who are prepared to give their time for others to have their fun, and to them we all owe thanks:

R.A.F. Halton:	Brian Tew, Stewart Funnel
Timers:	G. Lynn, Aylesbury & St. Albans Clubs. Some competitors.
Chief Cook:	Evelyn Barrett and of course the STATION COMMANDANT, No. 1 School of Technical Training, Royal Air Force — plus a great bunch of Coupe fliers.

Thanks to all of you. See you next year. ■

Ian Dowsett preparing his Coupe model. Linstrum's Coupe "The Spirit" waits its turn.

Spotlessly clean engine area on West German Sigurd Seydel's Rossi model with immaculate tissue decoration.

3 Competition Report 1976

Holland International 1976, by Martin Dilly

The 1976 Dutch Spring International — now called the *Holland International*, rather than the Coupe d'Amsterdam, and well-run by the *Arnhemse Luchtvaart Club* — had the biggest A/2 entry yet — 87, making it 13% up on 1975. Eight nations had flyers on the heath at Roozendaal (Austria, Belgium, Denmark, France, Great Britain, The Netherlands, Sweden and West Germany), and they faced the worst weather conditions for many years, with rain and stiff winds, as well as the more usual large patches of gentle lift and warm sun.

With the *Dutch Free Flight National Spring Meeting* a couple of days before the weekend contest, a couple of dozen British, French and West German flyers decided to take in both contests. Britain's Steve Marriott from the Biggles Free Flight Team, and Alan Jack from Tynemouth in Northern England, won Wakefield and A/2, while F1C went to West Germany's Borczewski.

In Britain the Biggles group's rapid rise to free flight success has all happened within two years. In 1973 several clubs around the Northampton area were becoming increasingly R/C-minded, with the resulting conflicts of interest. Trevor Payne, Martyn Cowley, Andrew Crisp and Steve Marriott formed the nucleus of the new club and concentrated on attracting younger flyers. The name 'Biggles' comes from the hero of a series of between-the-wars flying books by W. E. Johns, much read by British schoolboys before television set in, all very stiff-upper-lipped and handlebar moustached. Unusual for Britain, where parkas and rubber boots are usually *de rigeur*, the Biggles F/F Team has tee-shirts — yellow ones with a Charles Schultz Snoopy figure sitting on his doghouse roof, saying "Rats, Woodstock's maxed out! . . .". At the *Holland International* the tee-shirts spent most of the contest three layers down and soggy wet.

A 1 p.m. Saturday starting time left long enough time for morning trimming, and there was already quite a gathering on the heath on Friday evening. The light drift was varying through about 120°, making glider towing tricky and leading to frequent line crosses as circle towing aircraft drifted in different directions, trailing lines round each other as their flyers were slowed by the heather and the occasional fox-holes and weapon pits, probably dating from the Allied parachute landings in 1944. Steve Marriott, who had finally removed the Wakefield gold medal from his neck after his earlier win, was doing some trimming with an A/2 and had a D/T failure. After a

Martyn Cowley of the Biggles F/F Team reels out the line as Page Ayers holds his well-used A/2; model has rod blank fuselage all the way to the nose and turned brass nose weight. Page is an American studying at Oxford.

thirty-minute flight into the high overcast, Steve gave up watching through binoculars and started trimming out another glider. At dusk, as he packed the box to head off to the Indonesian restaurants of Arnhem, one of the park rangers stopped nearby to check on what a British car was doing out on the heath so late and there, in the back of his Land Rover, was Steve's A/2, picked up four miles away by a strolling couple and handed in to the rangers. In 1975 I had rather similar luck myself after a D/T failure. When I arrived home in Britain after that year's *Coupe d'Amsterdam* there had been an international phone call from the Dutchman who had found the A/2 in the woods and called my home on the other side of the North Sea the same evening. Many of the people in the Netherlands speak better English than a lot of the British do and, being a lazy nation linguistically, this does little to improve our Dutch.

Thomas Koster was as usual, doing a lot of precontest trimming, flying his unflapped *Cream* with a fibreglass helmet cowl, just exposing the top of the Rossi's cylinder head. Ken Faux, who was on the British F1C team at Plovdiv, forgot to trip the timer as he launched, the resulting 30 seconds of climb putting the all-sheet model high enough for everyone on the heath to watch its attempt to glide with its stabilizer on the climb setting. It took less time coming down than going up, whistling impressively as it went into the heath vertically. Pulled out of the peaty, sandy soil, the only damage was a broken prop, but the large divot of Dutch countryside that encased the Rossi meant a complete strip for the engine.

I had travelled over to the contest with Jim Baguley and his family and now we Volvo-ed back to the

Thomas Koster won F1C at the Holland International with his unflapped Cream.

campsite, where we were comfortably installed in a vast, multi-coloured tent in a beechwood, like a tribe of forest-bound Bedouin. Like the British, the Dutch are an ex-colonial nation, and they have the benefit of a variety of restaurants specializing in Indonesian food — spicy, rather like a cross between Chinese and Indian — and we spent a happy evening in nearby Velp, eating our way through the menu like a swarm of locusts.

With the late arrivals registering through Saturday morning, the first A/2 round was 70 minutes long, presumably to allow the stragglers time to settle down. The line of launch poles gave about 150 yards upwind before trees and bushes made towing all but impossible. An early wind swing took maxing models towards another wooded area which looked denser than it was, as the birches and scrub pines opened out as you approached them. While this helped to reduce the number of treed models it still made it very hard to follow an accurate line during recovery. Radio beacons helped here, and are becoming quite common on the Continent.

Six flyers were allocated to each pole, with contestants from the same country being grouped together in most cases. A 'five minutes at the pole' rule existed, but was rarely invoked, so reasonable cooperation was needed between people on the same pole so each had a fair share of the round in which to fly. Most people flew alphabetically or let those who had maxed choose when to fly. People who have to make a second attempt inevitably cause problems with this system, and on some poles there were several. On one of the 'British' poles a second civil war was narrowly avoided after Brian Baines waited for well over a quarter of an hour before moving to the pole to start his official five-minute period. This left the remaining four people about six minutes each, including their flight times.

With light drift and large patches of gentle, and thus hard-to-spot lift, 61 of the 87 flyers maxed in the first round, including Baines, who was hoping to repeat his 1975 *Coupe d'Amsterdam* win. His A/2 was double covered in light orange Modelspan, had twin timers coupled so that if one failed the other could still release the D/T line, and a Baines-designed circle hook exposed through an access cut out in the fuselage side.

One of those who failed to max on the first round was Klaus Salzer from Austria, who finally took second place. He made 162 seconds on both the first two rounds, flying a model with a flexi-jointed fuselage of balsa and a four-piece wing. Airfoil was Isaakson 73508, using a D-box leading edge. But the interesting thing was the jointing method. Klaus had two wire

Interesting West German A/2 used MVA 123 airfoil and "buzzard" wingtips, made from thin plasticard, with thin steel reinforcements strips to set the twist of the "feathers".

Austrian A/2 flyer Klaus Salzer used Isakson 73508 airfoil on a four-piece wing, held together with two wire joiners, one steel and one brass.

West Germany's Heino Fleischacker uses a moulded fiberglass nose on this A/2, and an Italian-made boom, tapering from 2 cm to 1 cm at the tail.

"I fly Condor", says Carsten Hohls' tee-shirt, but he actually flew this neat A/2 with Benedek 6536 airfoil, dihedralled stabilizer and fibreglass nose, housing circle towhook, timer and radio location beacon. Note beacon aerial in front of fin. Hohls is from the West German Niedersachsen group.

Henri Errard of France launches A/2 for Dutch flyer Gerard Brinks at the Holland International. Model has dural pylon, D-box leading edge to wing and outrigger elastic turbulators.

joiners attaching each tip panel to the center ones; the front one was steel and the rear one of brass wire, about 1/16" diameter. This rear joiner was flattened slightly at the outer end which increased its diameter locally, allowing it to 'pop' into a suitable diameter hole drilled into a piece of nylon sheet let into the facing tip rib. The brass joiners could be bent to vary the washout on the tip panels; the point being taped to maintain the correct setting. Stabilizer was Mylar-covered and Klaus used a self-spooling winch for towing.

While some of the German A/2's used large diameter rolled balsa fuselages or detachable push-on noses of light alloy tubing to reduce drag and keep the dust out of the timeer and hook, the Dutch, who usually fly over heathland rather than dirt, had circle towhooks fully exposed and hung beneath their glider fuselages. A trend with several of the German A/2's was the use of moulded fibreglass front ends. Hanno Fleischacker and Carsten Hohls of Niedersachsen both used these, Carsten's containing a radio beacon as well as timer and circle hook. The moulding gives enough strength to tie the wings and tailboom together, keeps the dust out of the works and still has a low enough wall thickness to keep the frontal area to a minimum. Overall result is neat, but unpainted versions give the impression of one of those translucent tropical fish, swimming around with its systems visible.

Several of the Dutch, French and German flyers used 'towing sticks' a couple of feet long, with the winch at one end and the line running through a loop at the other; this acts as a longer arm and allows a quicker response to gust disturbances of the model. About the only kit A/2's were *Sapers,* — Pavel Dvorak's taper-wing model that won the World Champs in 1971, now kitted by the Czech Igra company. Very popular, and cheap in the Netherlands, their wings, using a composite spruce and music wire mainspar, were deflecting alarming but safely in the gusts on tow.

West Germany's Beate Kappe was the only woman flying in the Holland International; her A/2 used a high plywood pylon/pod and a rod-blank fuselage. She towed with a rubber-driven winch.

John Williams from the British Stanstead club flew this Lively Lady at the Holland International.

Jim Baguley of the Norwich club flew a Bob White Wakefield with a Laurie Burrows glassfibre fuselage, Wife Beryl holds while daughter Sue looks on.

John Bailey of Bristol & West club flew Wakefield; low pylon model is held by Korsgard of Denmark, Fuse D/T.

Louis Manche of the Netherlands was a strong contender in Wakefield, flying this Hoerner-tipped model with alloy motor tube, auto-rudder and wing featuring long, tapered tip panels.

Henk van Hoorn was a member of the Dutch team at Plovdiv; his Wakefield uses an alloy motor tube, Hoerner tips and an alloy noseblock and hub with a neat wire loop serving both as drive pin and hold pin for the Montreal stop.

Jack North was thermal spotting at the Holland International, rather than flying, but then he has flown for Britain in all three F/F classes at World Championships. Here he adjusts the thermistor shroud on his chart recorder lift spotter, while Wakefield winner Dave Greaves from Birmingham wonders how the twisted liquorice is getting on in the wooden fuselage.

Toyn Proctor holds for father Keith, of the York club in Britain. Prop blades flipped past the correct position, turning inside out on the power burst, spoiling his chances in round one.

Thomas Heidemann of the West German Celle club launches all-sheet Rossi model.

Ian Kaynes of Croydon club, took time out from co-editing Free Flight News to take third in Wakefield. Rolled balsa fuselage, two-piece wing. That heather drags at the feet.

After the long first round, the remaining rounds were 50 minutes for A/2 and 40 for the combined Power and Wake rounds. Twenty minutes into Round 2 the drizzle started and the C.G. shifts as stabilizers picked up the raindrops produced a crop of stalls and only 36 maxes for the gliders. Meanwhile the 22 Wakefield and 18 Power flyers had made their first-round flights. Several of 1975's high placers were missing from Roozendaal this year, among them Gialanella from Italy who took second in Wake, as well as Denmark's Kjell Kongsberg and Britain's Joe Barnes, who were third and fourth. A newcomer to the Heath was Dave Greaves from Birmingham, who had the Pierre Trebod Wakefield Cup on his mantle shelf from 1975. Flying a triple-finned model with a square sheet balsa fuselage, Dave is a monumental mason by profession, and builds immaculate models, stylish and 'sensible-looking' without looking too frilly to be reliable.

For Dave and the rest of the Wake contestants the third round — the only one on Day 1 — was the big let-down. Nobody maxed, and overnight only 34 seconds separated the top eight places with Dutchman Louis Manche on top and Hans Haacken, who won for West Germany in 1975, in sixth place. Overnight there was a 180° wind swing which involved the hard-working organizers in an uprooting of launch poles and a new line to set up; unfortunately the new line that gave the maximum area of heath over which to fly, also had scattered birch trees a few feet from the launch poles and a shallow slope facing into wind — now about 20 knots — two or three hundred yards upwind of the poles. At 8 a.m. we started again, with the turbulence from the trees combining with the curl-over from the slope to set the lift-spotters and the glider towers quite a difficult task.

The rain produced trim changes for Greaves, still flying A/2 as well as Wakefield. All through Day 2 he was adding modelling clay to the nose of the Wake as the stabilizer got slowly heavier, and dropped over a minute on his fourth-round flight. In this first Sunday round, local man ter Velde was the only one to max in Wake, although Ian Kaynes missed by only two seconds. After seeing the distances that aircraft were flying on Sunday morning, the max was reduced to two minutes for Rounds 5, 6 and 7, and this, interestingly, made no difference to the number of glider maxes, 25 being scored in the three minute round and 25 two-minute ones in the subsequent round. As the gusts gusted and the ground turbulence boiled over the heath, the impression, commonly held by the British, that Continental flyers were only at home in the calm was shattered. Steve Marriott ended up the top British A/2 placer at tenth.

Steadily maxing through both days was Ton van Dijk, towing with a stick extension and flying a model that looked well-used and had a flexi-jointed balsa fuselage, with a sheet covered wing with outrigger turbulators. Ton finished over a minute ahead of Austria's Klaus Salzer for a convincing win in poor conditions. Although several of us thought that the two-minute max would lead to many people holding

Mike Warren, who writes some of the Free flight columns in Aeromodeller, dopes patches on Dave Greaves' wingtip, after the winning Wakefield shed a tip on its sixth flight. cyano-acrylate made it possible. . .

onto their overnight positions, it was only Dave Greaves and Holland's Henk van Hoorn (who was on their 1975 team at Plovdiv) who managed three twos in Wakefield on the Sunday. Dave used 'feel' for thermal spotting, rather than the thermistor and chart recorder that Croydon's Ian Kaynes favored to place third. Certainly on the second, more turbulent day, when the wind was gusting around 15-20 knots, Ian had a couple of very disappointing flights.

After the sixth round Dave's model came back with a wing tip off. The rough air discouraged any check flights after repairs, so Mike Warren and I, who had decided that our efforts did not warrant further A/2 flying that day, persuaded Dave that this was no occasion for the five-minute epoxy and set to with the cyano-acrylate. With a 50-minute glider round to work in the three of us made a good job of the repairs and another convert to instant skin-bonding was born. On the last flight Dave's climb was fast and a trifle loopy, but high enough at the end to max safely and maintain his 36-second lead over Henk, who also maxed into second place.

British entrant Fred Chilton of the Crookham club was out of luck with this tissue winged model, using glass strands for wing warp resistance, and Laurie Burrows glass fibre fuselage.

This A/2 sized flying wing by Bos of the Haarlem club won its class at the Dutch Spring F/F Nationals.

Alan Jack (left) won A/2 at the Dutch Spring meeting; here he and fellow Tynemouth member Ron Pollard, with Vitar, eye the lift at the Holland International. Lattice tower in background is for firewatchers.

The rather small entry in F1C (18 flew, but only 11 survived till the second day) were led by Dutchman Bert Huijben and Denmark's Thomas Koster. Josef Broderac — a Jugoslav now living in West Germany — dropped 43 seconds on the first round but held onto his third place from then on. For Thomas the hairy round was the sixth; launching fairly late in the round, the Rossi over-ran, with the Danish recovery team in pursuit. Unfortunately the spare model was in the car,too. After a hectic chase in another car along the rutted paths of the heath Tom was back with a model seven minutes after he launched and with two minutes to spare in the round. Working with a certain urgency, he assembled the wings at a run, fuelled, set the systems, offered the spinner up to the starter and fired up the tightly-cowled Rossi. Halfway through the seven-second power run, the signal pistol marked the end of the sixth round.

Koster having maxed again on his sixth flight and again on the seventh, meant that both he and Huijben were to fly-off. The Dutch model was unusual, with elliptical dihedral on both wing and stabilizer, which

Ken Faux of the British St. Albans club surveys his peat encrusted Rossi after his model tried to glide on the power settings from a 30 second overrun. Only damage to the all-sheet model was a broken prop.

Sigurd Seydel from West Germany and copiously tissue-decorated Rossi model.

must have given Bert a few headaches while applying the compound-curved upper sheeting in the jig. In the eighth round with the five-second run, Koster did a thorough overhaul of his aircraft's systems, v.i.t., auto-rudder, D/T, brake and engine run. Apparently worried that the Rossi was not holding its needle setting, he tightened the clamp nut before flying the unflapped Cream. As it happened, a couple of seconds out on the climb the revs sagged sharply, and the model was at barely towline height when the engine cut at 4.9. Here was where the endless trimming paid off; the transition was perfect and saved Thomas. With an A/2-like glide to help he was lucky to hit some weak ridge lift at about 50 feet after a minute and a quarter, and this was enough for the two minutes needed. Huijben, flying second, was less lucky. Although he had no engine run problems, the model managed to miss the lift off the downwind ridge, finally gliding behind it for 1:55.

The Holland International was not a weekend of many major technical innovations, but one that we did spot was a two-function electronic timer, built by a German whose name we missed. Weighing 80 grammes with batteries, it used a neat switching block, about an inch by one-quarter inch square, containing 10 miniature rocker switches, used to alter the delays. Actual output was in the form of two triggers on relay armatures, projecting from the timer faceplate and retracting flush, to release the operating lines hooked to them.

The contest advance information told us that ". . . a pleasant evening has been arranged . . ." for the Saturday night. This certainly turned out to be the case, and we moved en masse into the bar of a local hockey club to carouse and watch a couple of films by Ewout Reuss, one shot at the 1974 European Championships and the other from Plovdiv. Definitely better than most amateur movies, Ewout had some great shots. One in particular sticking in the memory is that of a gas model climbing against the sun into a pearly overcast with the camera rotating as it held the model. To compensate us for the poor weather, Jan van Rij told us that the organizers were providing "Drinks on the house" as well. Another fine Dutch treat. ■

Author/Photographer/Free-Flight raconteur, Martin Dilly

Reception - Sangria

Criterium Pierre Trebod 1976 Marigny, France

The 12th annual running of the popular *Criterium Pierre Trebod* took place on August 21 (F1A) and August 22 (F1B & F1C) just a week before the Indoor World Champs in England. The concrete of the huge runway stretches on for almost 10,000 feet on this ex-NATO airfield located about 70 miles east of Paris near Sezanne. Unfortunately the runway direction and prevailing wind direction do not coincide. Consequently most of the flying and chasing has to be done by slogging through high weeds. The field is nonoperational and has only a small complement of military caretakers. Apparently there is nothing in the French Military Budget to spend on mowing machines for unused airfields. These weeds also have a number of sharp thistles which are far more painful than they are visible. In spite of these few discomforts the P/T is very popular. This year there were 136 in F1A, 64 in F1B and 40 in F1C. Countries represented included Argentina, Austria, Belgium, Canada, Denmark, France, Italy, Holland, Spain, Switzerland, United States, West Germany, Great Britain and Portugal.

A large number of the competitors camp out at this meet. This year the designated campground was sited in a freshly plowed field. Rather than put up with the dust, the tents were pitched along the road and in odd spots in the woods. A few camped in the commercial campground at Sezanne which boasts a swimming pool. Those of us without tents stayed at the hotels in Sezanne and Pleurs. Our favorite is the Cheval Gris. Mme. and M. Laureaut are gracious hosts and put on a fine dinner table, drawing a big crowd each night from the tent city.

Roy Collins, Great Britain

J. Godinko, France

Switzerland's Fritz Gaensli did well except for the first round, twenty-third place.

Fritz Gaensli's stabilizer was sheet covered and super thin. Ribs are plainly visible.

Christoph Chaussebourg holds for Dad's F1A. Offset hook gives tight circles in tow.

Brian Baines of Great Britain with typical British model. Uses two DT timers in parallel. Placed second.

Mlle. Challine launches for Remy Lepage, France.

Nordic of Fritz Wilkening, Germany

Sheeted wing Nordic, H. Prison, Germany.

John Guffens of Belgium with his long tipped Nordic. Used a simple straight towhook.

The shirt of J. Abrew, Portugal, says it all.

John Cooper, Great Britain

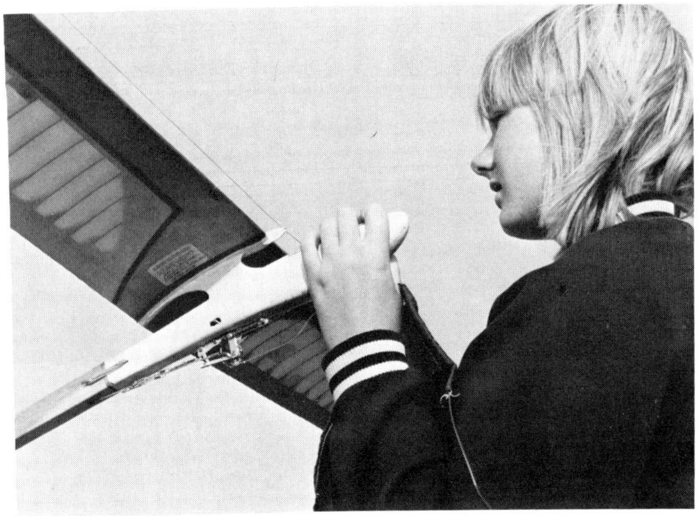

F1A by G. Brinks of Holland. Had shock mounted tail boom, pylon, outrigger turbulators and full-house hook.

Tom Koster contemplates offset twang hook gadgetry of fellow Dane, Dave Rasmussen's Nordic.

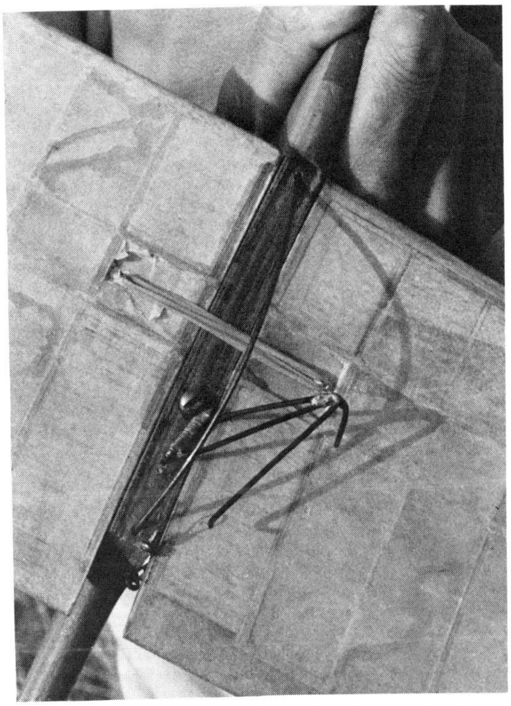

Boutillier's low-wing Nordic with offset towhook.

Vaclav Horcika from Austria, flew F1A. Tows without reel. Winds line on stick. Was F1C World Champ in 1973.

A. Hacken, Netherlands.

Flying commenced at the 07:30 start time with the usual P/T timekeeping system. Contestants were placed and assigned in 19 groups. These groups with a maximum of eight (some had fewer) had to share the two timekeepers at the assigned starting position. It was completely up to the contestants how the flying order would be selected and how much time each could spend flying within the round. This system simulates somewhat the World Champs procedure but the added numbers sharing the round makes it all very tight. The self-imposed selection of flying order is considered by some to be unfair and arbitrary, particularly since one must depend on the voluntary cooperation of the eight. On the other hand other solutions to the timekeeping assignments are not without problems too. With tactical flying becoming so important, choosing the right moment to fly becomes crucial.

Four maxes only were recorded in the first round, in which the conditions were calm. As the air warmed up light lift became more noticeable near the end of the round. From the second round on the breeze freshened a bit and carried straight across the runway. Thermal activity was good as was the strength of the accompanying downdrafts. Tony Cordes (GB) for example was stuck with a particularly bad round time slot ending up with a crushing 61 secs. On the other hand if you hit lift you were in for a good chase. I enjoyed one such peregrination off through the weeds and thorns chasing my *Tuggy Buzzard* in the sixth round. The breeze carried it faster than I could keep after it, stumbling through the brambles and daisies. Besides I had to stop and sample the luscious blackberries. It DT'd OK but far enough beyond to have me worried. After reaching a field of low-lying crops I glimpsed a wing tip. Coming closer I saw a Scat Bird decal on the wing. Hooray! Then finally arriving, discovered it was not *Tuggy Buzzard* but Pierre Chaussebourg's F1A! By the time I picked it up and got out of that field I saw Pierre further downwind with my ship.

The French fliers favored towing with offset hooks and fast loopy type tows. An unfortunate few were seen to lose control on the downwind swoops. Observing line tangles didn't seem to be as big a problem as at the World Champs; perhaps the timekeepers were more experienced. The 1973 F1C World Champion, Vaclav Horcicka, from Austria, was there flying a 15-year-old airplane (just one year younger than my *Tuggy Buzzard*). He had been relatively inactive recently, having been severely ill with heart trouble. We compared our first round scores; both were 145 secs.

Eventual winner was Per Grunnet of Denmark. His *Jessi* had maxed out. This ship sports the current Danish look, tapered round tube fuselage with alloy spine-cum-pylon. Wing has flat center-section with longer outer panels tapered to rectangular tips. Root chord is the ever popular 150 mm. Wing section is a design by Arne Hansen. Stab section is that used by Hacklinger and Lindner. (HA13 and Kria). Per has experimented with a 6% thick flat-bottomed stab but found the HA13 to be superior. Tail boom is flex-jointed to the forward fuselage and it weighs 25 gr. Wing spars include two full depth webbed "I" beams. The forward one closes off a boxed-in leading edge. Model is very similar to the Koster model detailed in FFN 76/03.

Per Grunnet of Denmark, 1976 Pierre Trebod Nordic Champion.

Nordic front end by winner Per Grunnet. Note springmounted tail boom, swinging modified Czech hook and dural spine.

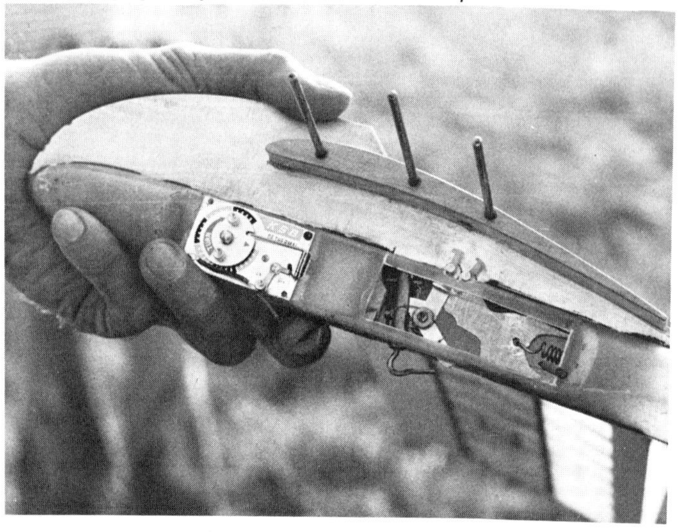

Karl Sauer's very clean new German style power model. Glassed front end.

Front end, Sauer F1C

Didn't know Martyn Cowley of Great Britain could move that fast, but look at that action.

Tom Koster and Square Cream

Tom Koster's dad enjoying the fun.

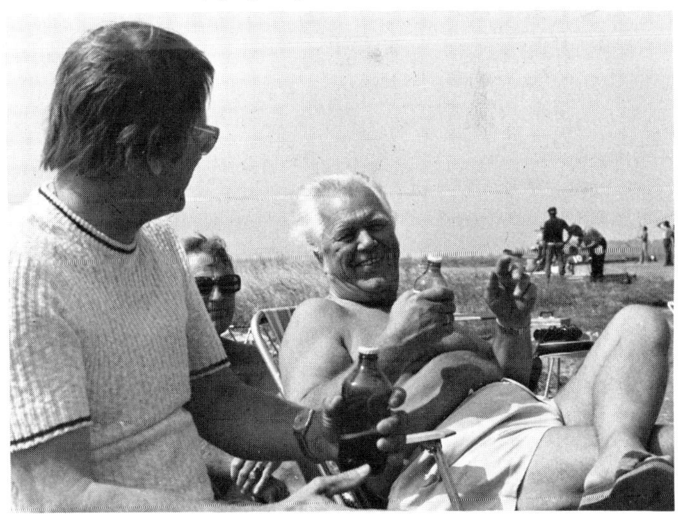

Martyn Cowley's F1C front end

F1C model by Ken Faux, Great Britain.

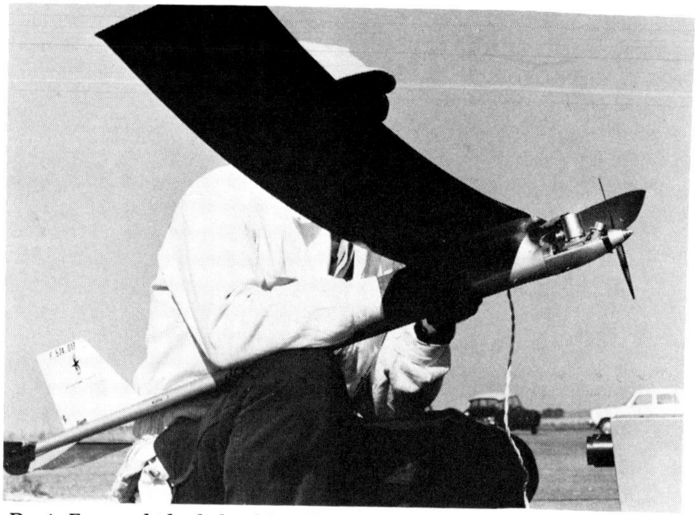

Denis Ferrero hides behind his elliptical dihedral F1C

Paillu (France) had been matching Grunnet max for max until the last round when he collected a disastrous 134 secs. Brian Baines (England) and R. Spann (Austria) followed up with maxes also but it was their first round scores that settled the issue. With no fly-off there was time for some trimming flights of Wakefield and Power models. This included Ken Faux of England flying barefooted.

The dinner and socializing at the Cheval Gris that night was, as always, a high point of the occasion. Our beaming hosts welcomed us and soon the delicious food and wine was flowing to our table. Dinner and wine came to $2.75 while the room for two was $5.70 per night. The next morning Power and Wake got underway with the Power fliers electing to start right off while the Wakefield people kept checking for lift. Miraculously the wind shifted in the A.M. to blow straight down the runway helping the chase regimen considerably.

All eyes were on Thomas Koster to see what he would come up with after all his troubles in the World Champs the previous years. His new ship, *Zooki Zooki* has a nicely cowled front end and shoulder wings without flaps. Usual circular tail boom and triple fins are nicely proportioned. Flight pattern was very impressive, going up almost vertically then going over on its back and rolling out downwind at engine cut. Thomas also told us of plans to host the 1977 World Champs in Denmark, a good piece of news. Koster's flying was very impressive but in the sixth round old devil downdraft got to him for 100 secs, placing him ninth.

Meanwhile Ken Faux, Ray Monks and Hans Stetz proceeded to garner maxes, that is until the last round. Stetz and Faux flew into some bad air to drop while Monks ignominiously DT'd early from a good altitude. This turned things all around as Ferrero (France) and Hubler (W. Germany) made their maxes thus placing them first and second by virtue of their second round scores of 169 and 168 secs respectively. Denis Ferrero's model had a low aspect ratio wing with elliptical dihedral. The single piece wing sat on a low pylon. Fuselage was short, stab rather large and Monokote covered. Rossi was fitted with a metal heat retaining muff. No landing skid was used (fairly common in Europe). Engine ran very fast and very smooth.

Ferrero shows his winning form-very rapid.

Time out for a little kissy-kissy by Roy and Jill Collins, Great Britain.

Haven't figured out how Hans Stetz, Germany, keeps his model so clean and unblemished.

Illan Espinosa, Spain.

Trevor Grey, Great Britain.

P. Rasmussen, Denmark.

Louis Dupuis has the turns cranked in, now for the thermal. He found all but one, for second place.

J. Delcroix, France, Jedelsky wing.

C. Costa, Portugal

C. Bastos, Portugal

H. Helmbrecht, Germany

Pierre Chaussebourg and son Christoph proxy flew Bob White's model to thirteenth place.

Steve Marriott, quick, look behind.

The cautious Wakefield fliers managed to find 17 maxes in the first round including Hugo Benedini, that great Pierre Trebod fan from Argentina. Pierre Chaussebourg was flying Bob White's Wakefield by proxy. And so Klaus Wetterburg of Denmark proxy flew Pierre's Wakefield. I helped out a little and since all three of us were wearing Scat shirts it looked like a real team effort. Klaus kept us all in stitches and I am sure we had more fun than anybody else there. Klaus speaks English with an Indian accent having been tutored in Africa. A very resourceful guy, he came back to the Cheval Gris after the Banquet and discovered he had left his key locked inside the room. We heard a lot of crashing and banging accompanied by a vociferous running description, in Danish, of his predicament (we heard the noise and didn't know what the hell it was). The next morning we discovered he had found a ladder somewhere which was propped against the building allowing him to climb up to his second story window. At least he said it was his room. Considering that some didn't make it back from the Banquet until the next day, Klaus did quite well.

Some of the French Wakefields were unusual by U.S. standards, but they were certainly effective as they took the top three spots. The winning Wakefield by Emile Gouverne of France, *Mini Ostrogoth* features a high aspect ratio solid balsa wing. Section is his own with a single surface thread turbulator. Wing weight is 75 g, root chord 112 mm, span 1540 mm. Propellor is 560 x 750 mm, driven by 14 strands. The prop was held by a finger-operated bell crank so that it did not start to rotate until the instant of release. Motor tube is balsa with spiral wound epoxy glass fibre reinforcement. Tailplane uses a thin flat-bottomed section with five small spars. Area is 3 dm^2 and weight is 5 g, tissue covered. A modified Graupner timer operates an all-moving fin forward of the stab. The stab sits up on 2½-inch high wire stilts.

Emile maxed out in the seven rounds easing by Louis Dupuis who dropped 15 seconds in the fourth round. J.Delcroix with a Jedelsky wing yet, was third, having dropped 33 seconds in the first round. Dupuis' model had a molded balsa fuselage, plug-in shoulder wings, single underslung fin and small stab. Prop run was only 15 secs.

With all flying completed by 5:15 p.m. it was truly miraculous and a tribute to the organizers that promptly at 7:30 p.m. that same night printed results were handed out to everybody in the courtyard of the restaurant in Gaye. Another P/T tradition is the inexhaustible supply of Sangria served in the courtyard. About a half hour of Sangria and socializing prepared us for the awards presentation which took place on the outdoor stage. Colorful lights and banners were strung about to accent the occasion, as the happy winners collected their trophies, kisses and applauses. Then a mad dash for the Banquet inside the restaurant Le Gayon. The fitting of such a large group into the small restaurant was not without its feeling of togetherness and intimacy. The scene, as the benches were filled with the happy crowd, resembled Brugel's *Wedding Feast*. Another great Pierre Trebod was soon ensconced in our memories. ■

ROYAL AIRCRAFT ESTABLISHMENT
CARDINGTON

Inside Cardington Air Shed No. 1.

Indoor World Championships 1976 Cardington, England

The Ultralight free flight model.

F.A.I. '76 CARDINGTON
INDOOR MODEL WORLD CHAMPIONSHIPS

The Ultralight Indoor Model is a unique device that operates in an aerodynamic regime generally ignored both by nature and modern man's technology. And because of this it has produced an irresistible challenge to a number of aeromodellers throughout the world. The Federation Aeronautique Internationale, FAI, has established a set of simplified rules to use in international competition. These specify that the model must weigh at least one gram (a mere 35 thousanths of an ounce) without the rubber motor and have a wing span no greater than 25.61 inches (65 cm).

The World Championship competition for these models was held August 27 through August 30, 1976 at Cardington, England. It was organized by the Royal Aero Club of the United Kingdom, and Society of Model Aeronautical Engineers (SMAE). A beautifully prepared Souvenir Programme was made available to the participants and we reproduce here, with thanks to the SMAE, the introductory essay, that describes the art and sport of indoor free flight.

Souvenir Programme

When you see Indoor models fly, it is as if you are in a world of slow motion. Everyone moves slowly—those on the sidelines lest they create turbulence equal to a tornado for these feathery aircraft, and especially the competitor as he carries his model to the chosen spot for launching lest he crumple a wing by creating an air speed greater than that for which it is stressed. And once the aeroplane is launched that same slow motion action is, if anything, emphasized. Ever so slowly does the aeroplane move forward, barely more than just one mile per hour, and the propellor does its part in adding to the illusion—ticking over at about one revolution per second. Nowhere else in the world of aviation is there a sight like this to be seen.

The airship sheds at R.A.E. Cardington provide an ideal setting for a World Championships for Indoor models. While any reasonably air-tight building may be used to fly these models, the tremendous height of the airship hangars allows the competitor to exploit the full potential of his craft which, in turn, taxes the flyer's know-how to the utmost.

Indoor models for World Championships are judged for duration—how long they stay in the air from the time of launch until the model returns to the floor (or until the flight is 'permanently' stopped from hitting an obstruction). When you consider that the models fly by pre-adjusted control surfaces—set for circling flight—and many flights will be in the 30 to 40 minute range, it is obvious that any appreciable amount of air drift would push the model into the side of the building in only a few minutes, aborting the flight. A certain amount of steering is permitted (by a long pole or by a string attached to a helium-filled balloon) but this usually is done only as a last resort since the possibility of damaging the model is very great.

The models are constructed from the best grades of balsa wood, the traditional model building material. So critical is the choice of the wood, cut into the thinnest of strips, frequently less than 1/64", that probably less than one-tenth of one percent of all the balsa wood available has sufficiant straight grain and strength-to-weight ratio to be considered for championship-calibre models. Even with the best of wood, the structures are usually braced, à la bridge construction, with wires thinner than a spider's web and which are all but invisible.

The flying surfaces of these models (wing, stabiliser and vertical fin) are usually covered with a substance called "microfilm"—and so is the propeller. This transparent skin is made by spreading a special liquid on water to form a sheet of surface film which, when dry, is lifted from the water and placed on the structure. Microfilm can be made in various thicknesses and strengths, which can be judged by the colours of the spectrum created from refraction of light rays. That which does not readily refract and appears clear is the strongest and heaviest, followed by (in order of increasing weakness and lightness) mottled green/red, violet, gold and silver. Typically the microfilm covering adds little to the craft's structural integrity. Most often it is applied loosely to minimize the chances of its warping the delicate structure.

Power for these Indoor models comes from a loop of rubber about one and a half foot long and about as thin as that found inside a golf ball. It's the twisting of this rubber motor which produces the power, and for all-out competition flights the motor typically may be wound for 2,200 turns. Virtually all Indoor flyers use geared winders so that the motor may be wound in just a few minutes. Coming increasingly into general practice is the use of a torque meter as a winding aid; in conjunction with previous tests, the torque meter gives a strong indication of when the rubber motor will have the proper twisting force to perform as per expectations.

The rules for the competition require that each model, without the rubber motor, weigh at least one gramme—a mere 35 thousandths of an ounce! The only other model restriction is that the wingspan must not exceed 25.61 inches; most models are built very close to this limit in order to maximize lifting surface area.

WHAT TO WATCH FOR. Typically the better performing models will reach the apex of their climb in about eight minutes at a point very near the ceiling rafters and then begin a very slow descent. To reach the ceiling too soon runs the strong risk of the model being irretrievably caught in the building structure, and in any case expending the rubber motor's power too soon. To not reach the ceiling does not allow sufficient height for a long "cruise", that portion of the flight when the power fall-off of the rubber motor is very slow and during which there is a large percentage of the total flight time.

Competitors are allowed to change models, propellers and rubber motors—in whole or in part—as they wish. This gives them maximum flexibility for recording their best performances. Since the rules allow six flights, the longest two to count in the official scoring, it is typical for the earlier flights to be "safe," and the later flights to be all-out efforts—producing greater risk for the aeroplane to be caught in the building's framework, but also greater possibility for a long flight if no bad luck is encountered.

Each competitor is entitled to a maximum of six official flights, the better two of which are totaled to determine the World Champion and other places. Onlookers are cautioned to stay on the sidelines and to not distract competitors or timers either when a model is being prepared for or is in flight. Do not touch the models under any circumstances!

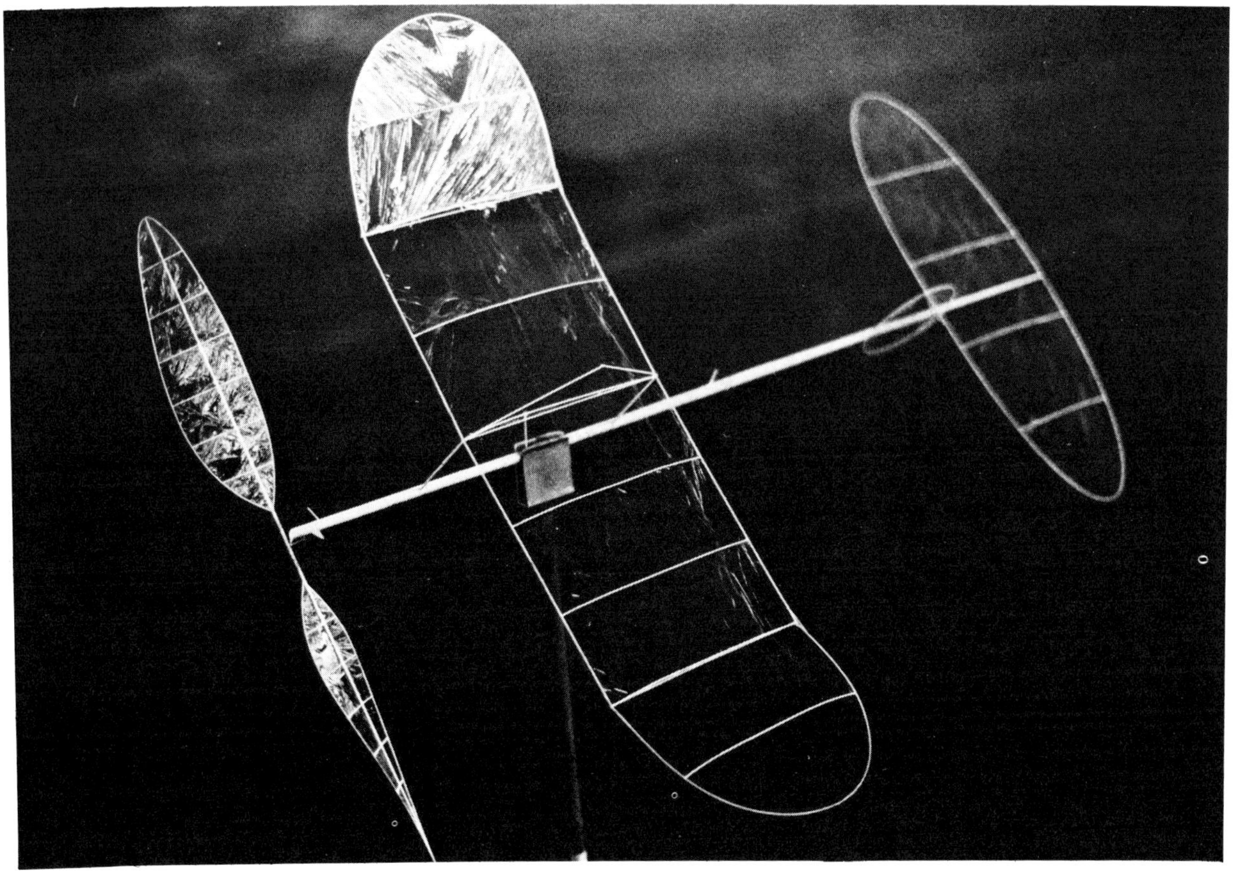

"Rafter Banger", by Laurie Barr.

John Blount, Great Britain.

KEEP CLEA
KEEP

The site at Cardington was the Airshed No. 1 at the Royal Aircraft Establishment. There are two identical sheds side by side. No. 1 was built in 1918 and No. 2 was built in 1930. They were designed to house the British rigid airships R 100 and R 101. With the unfortunate crash of the R 101 in France in 1930 the sheds have found little use in their intended role. What is more fortunate is that the SMAE has been able to obtain their use, on a frequent basis, for Indoor model flying. Shed No. 1 has become one of the most popular and useful such sites in the world.

Cardington is situated near the pleasant town of Bedford, about 60 miles north of London in scenic Bedfordshire countryside. Competitors were housed in barracks at RAF Henlow a few miles away. An enthusiastic crew headed by Laurie Barr and George Lynn of the SMAE did a tremendous job of organization and seeing to all the problems and details of this affair. This included the registration of 41 contestants from 15 countries. Rescuing lost teams and helping cart the large awkward model boxes about was only a small bit of their activities.

Some early test flying got underway on Friday (official rounds to be flown Sunday and Monday). Conditions inside the shed weren't too good as there was a pretty stiff wind blowing and it was hot. This caused turbulence and drift in the shed that the veterans didn't bother trying to cope with. But since there were a number of contestants who had never before flown in such a large site, there was plenty of activity. Official practice got underway Saturday and then there was no holding back.

The conditions didn't get a whole lot better but they did get different. After a crippling drought had been plaguing Great Britain all summer it had now turned cold and it was raining. And with the few holes here and there, it was raining inside too. Some hasty movement of models and boxes was seen to hurriedly to prevent their destruction. The practice flying started with a certain amount of caution and the use of half motors (the other half replaced by a stick). Some basic adjustments were still being made by those with new models and new experiences to learn. Dieter Siebenmann and his Swiss team were trying to get their new ultra-long models sorted out. Apparently with some success as he did a 37.53.

Poland's Ciapala did 38 minutes as did Ron Green, the "beginner" on the GB team. Many flights were ballooned down before they could demonstrate their full capability — and penchant for sticking to the rafters. It was enough to check the climb. Near the end of the day conditions seemed to have improved and it appeared to be time for some strategic one-upmanship. Impressario Laurie Barr did just that with a 39.39 which turned out to be the longest flight of the entire meeting, but of course unofficial. This flight was ballooned once, forcing it to dive and lose 50 ft. before continuing on.

Heise and Siebenmann (Switzerland) strip some rubber for another flight.

Bucky Servaites (USA) repairs a prop.

Polish team members Eduard Ciapala (1), team manager (c) and defending World Champion, Ryszard Czechowski (r).

Ron Green (GB) selects a propellor.

Andy DeMello (Canada) has a brew at the Bonanza Club.

Jim Richmond, U.S., stuck one in the catwalk at 5 mins . The indispensible Rigger of the RAE dislodged it only to have it lodge further down the shed where Jim finally rescued it. Romak had a "lost" model returned by Czechowski who had been up wandering the catwalks himself. Finally the light grew dim and at 7:00 p.m. or so we made our way back to Henlow and the "Lower Ranks Mess" for the official sustenance. The food was good and the companionship of that fine group made it doubly so.

Sven-Olav Linden and wife; Sweden. A 34.06 flight.

It was a short walk over to the Bonanza Bar and recreation hall where this reporter was taught a number of new pub games. There was also a showing of two films: *Wings and Things* and *180 is Max.* A number of those assembled had to grimace their way through seeing themselves portrayed as movie stars.

The next morning our English breakfast seemed to set things straight and it was back to the shed to continue our worries about photo technique in that unusual place. There was an overall darkness throughout the shed that you sort of got used to but then there were windows part way up the walk that gave hot spots at right about the angle you were shooting. There was nothing to bounce flash off and you didn't dare move fast to get in position. My solution was to try everything. Tri-X at ASA 400, 1200. Flash, no flash, tripod, hand held, Kodachrome, Ektachrome, the whole bag. Nothing worked, or should I say everything worked about the same. Trickiest part was shooting from the top catwalks while trying to keep my balance in the pigeon dung without dropping the Nikon through somebody's wing. With the kind Rigger's permission I had the dubious pleasure of traversing the entire catwalk system up and over and down the length.

The conditions on this, the first contest day, were not exceptional as there was still turbulence at lower level. It was cold and gray outside. There appeared to be some confusion that the contest would not be flown in rounds but that three flights could be made each day but no more than one team's member at a time could be in the air. Official period was to be from 10:00 a.m. to 6:00 p.m., but this was delayed 30 minutes. The next day the 30-minute delay was not used, to the consternation of some who did not get the full import of the schedule decisions as they were made. This due to some laxness in making non-English announcements. This seems to be a natural hazard of *all* international meetings.

Sven Pontan of Sweden started the show off with an underpowered flight of 17.07. Ron Green went up and had to balloon and drop it a few feet away from the girders to gain a nice 33.10. The outgoing Canadians then put on quite a show with their Mike Thomas. His model climbed fast and then proceeded to investigate the truss-work in the dark area in the corner of the roof. It slithered, walked, backed up, turned around, rolled over, and sometimes actually flew, all the time raising great clouds of dusty pigeon dung which was dislodged to rain down on the crew below. The model eventually puttered out into the center again to fly for 32.19.

Bud Romak's first flight set things on edge a bit when it was found that the two timekeepers' watches differed by two minutes. A careful check showed the watches to be operating properly. What to do? The rule book was consulted and the CD declared the average of the two should stand (34.54). There was some reason to suspect that the lower time was right and the U.S.A. Team Manager Pete Andrews volunteered to accept the lower time but this was not allowed.

All eyes were on Jim Richmond as his ship went up. The distinctively shaped wing and stab tips made it easy to spot. It climbed too fast though and promptly hung up at 9.30. Servaite's model also climbed to the top, bumped several times and tail slid 15 feet. From there it cruised the length of the shed and landed at 34.57. Czechowski, the reigning World Champion, got off to the side and hung up early. Ciapala got off to a very good start recording a 34.54 and then a very good 36.18. The Poles were not going to be left behind!

Siebenmann launches his long tailed model.

One of the surprises of the meet was the strong showing of the Canadians. After cleaning off the pigeon dung from his wings Thomas put up a neat 35.32 with De Mello close along with a 33.22 and a 34.33. McGillivray then polished things off with a 29.50 and super 36.06. It looked like they were getting the range.

Laurie Barr made his first 'insurance' flight a respectable 34.30. Using the same motor but adding a bit of wing incidence and more turns he went again for the best flight of the day, 36.54. The Americans weren't having much luck. Romak tried again with the same model using .057 x 18.5 inches of Pirelli, 2100 turns on a 18.5 x 35 inch prop. The extra torque took the model straight for the side, some fast action by Bud with a balloon saved it and got it in the clear but all that dragging around killed the climb and it settled in at 29.23. Servaites tried with a few less turns and yet again clobbered the roof. Rigger to the rescue. A cautious Richmond then played things conservatively and made a 31.32. The Swiss, with their interesting long models weren't doing too well after their promising display in test. More practice and test, more airship sheds! The Czechs were right up there with a strong team showing, with times well into the mid-thirties range.

Model of Guy Cognet, France.

Servaite's model.

Czechowski climbing on low torque.

Mike Thomas did well with this model.

Brothers Ferdinando and Pierluigi Migani both bettered 30 minutes. Team placed seventh.

Bucky Servaite's model.

Small vertical fin on Servaites model.

Model of Poland's Ciapalo.

Another view of Serviates model.

Czechowski climbing to the rafters.

Around 5 PM the drab misty conditions outside added to the gloomy darkness inside and visibility started to become a problem. A model was lost by the timekeepers OOS. Some of the teams realized they were running out of time and had to hustle. Permission was given in some cases to have more than one team member in the air at once, a compromise of the rules that was resorted to as the lesser of two opposing unsatisfactory situations.

Romak tried again for his third flight with a different model. A fair climb but turbulent air resulted in *only* 32.48. Bud felt somewhat frustrated but also seemed confident that he now had it all figured out in his mind what rubber, prop, model, turns, etc. would be best. Only it would remain to be seen tomorrow if this feeling was justified. While Bud was in the air Servaites processed his model and launched as soon as Romak had landed. Andrews was doing a great job of managing his team and, although there were many frustrations,Pete kept a steady hand. Servaities' model proceeded to bedevil its owner by heading for the side. A good steer got it right only for it to wander off again. An opportunity to steer again with the balloon did not present itself and the model hung up at 10.15. Although unfortunate for Bucky, this incident gave time for Richmond to get into the air for his third flight. Jim used a Pirelli strip .067 x 19 inches on a 21 x 36 inch prop at 2150 turns. Everything went well to about 110 feet and then some wild air currents were hit. There was a stall and flopping about and a loss of altitude to about 60 feet. it didn't look good, but the lower level of air must have been OK, for the ship cruised for about 30 minutes at 60 feet, landing at 36.29, the best time for that period.

Australian Boyd Felstead had been dreaming for years about flying in Cardington *sans-proxy* and here he was doing very well with two flights over 31 minutes. Although he was the sole Australian entry he did have a first-class Team Manager, Reg Parham, helping out. Old friend Manny Raddoff was also there to cheer him on. The Jugoslav team of Kmoch, Strazberger and Gabrijel was also doing well with a number of flights over 29 minutes. The relatively inexperienced Japanese team was doing a good job considering the brief time they had to let their models out in that big shed. Yasutoshi Banba made a flight of 30.14, the first time ever for Japan over 30 minutes.

At the end of the day GB was in the lead with Canada second, Czechoslovakia third and U.S.A. fourth. A somewhat surprising outcome. But then the expected 40-minute flights had not yet materialized and there was another day's flying to do. The big guns would be brought out then so we thought. For now the English and Canadians were in joyful shock and heading for the nearest pub. The rest were reluctant to give up and stayed in the shed glueing things back together, stripping rubber, and even doing a little flight testing.

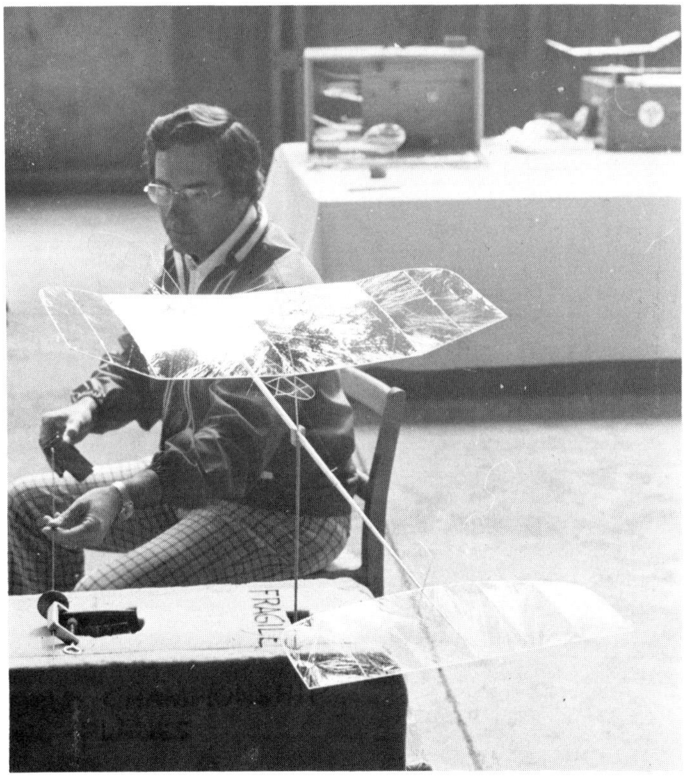

Jim Richmond, from North Carolina, was the 1968 World Champion. Several potentially good flights spoiled by bad air.

Australian Boyd Felstead had a great time, and best flight of 35.45.

Cardington Air Sheds.

We drove back through the damp green countryside and small villages away from those giant gray Mastodons looming out of the swirling mists. Those strange structures left over from a different era, an almost forgotten episode in aerodynamic technology. Perhaps their real purpose in being was only now to be realized. Who should be the judge? Should we ask the Canadians whooping it up there in the *Dog and Duck* (or was it the *Wild Goose*?) What do you suppose the Czechs, Finns, or Italians were thinking? Or how about the Japanese, Americans, Germans and Swedes? And there too, the Dutch, Jugoslavs, Australians, French and Swiss?

Yes, definitely all to a good purpose we mused. *Hail to the RAE, hail to the Rigger!*

We tooled along, finding our way back to Henlow and the Lower Ranks Mess where we soon were joined by happy confederates.

The *bon cuisine* at the Lower Ranks Mess was once again consumed with gusto and not without some wonderment as to what it was. But that is all part of the mystique and camaraderie of a geographically diverse group. Adjournment to the Bonanza Club once again brought us into a comfortable sociable gathering. The cinema attraction turned out to be *Those Magnificent Men in Their Flying Machines*. Jiri Kalina, having seen it 12 times already, was excused to shoot some billiards with Bob Bailey. The film's spoofing of some of our nationalistic traits was taken good naturedly by the assembled representatives of those being spoofed and there were no ugly incidents.

As you may recall the film had an American western hero. Was this an omen of what was to come the next day? Sure glad it wasn't a King Kong movie.

Convenient dining facility at RAF Henlow.

The "roast joints" refers to dinner, not the smirking trio of Laurie Barr, Butch Hadland (U.K. Team Manager) and Bud Romak, caught waiting here in the chow line.

Monday dawned (somewhere) as we were drawn back through the mists to that bulking edifice that was referred to as the world's largest tin hut. Nobody seemed anxious to start till the obvious press of time raised the adrenalin level to a critical point. That was about 11:00 a.m..Ron Green, who had shown such promise, ran amuck in the air as his model stalled badly over the body heat from a large group (that should have dispersed). It tail slid and Ron caught it only to have the motor stick collapse taking most of the model with it. It was his No. 1. A number of others had trouble with collapsing motor sticks (tubes actually). This was laid to the damp air which was thought to soften the wood. That plus the need for more power to overcome the added moisture weight was sufficient to cause trouble.

Ciapala stalled out and bumped things askew. Surprisingly he chose to nudge everything back in place and launch again. Something wasn't right unfortunately and he was down again in a mere 46 secs.

Romak blew a motor while hooking up, causing that model to look a bit ravished. Well he didn't really want to fly that model anyway. Richmond was pressed into the breach and made a disappointing flight of 20.53. What was going wrong? Romak was all set again and started up with a dodgey climb, stalling slightly for two minutes but eventually climbing to near the top catwalk. Sometime later Bud was observed still staring up at the roof. Yes it was still that same flight and now he had a tense audience. The model had that smooth slow look about it as the cruise, with obvious reluctance, wore off. With much applause Bud landed at 39.22. It was time for that sheepish grin.

In this period Werner Wetzel of West Germany came through with a nice 32.17, his third flight here over 30 minutes. The Czech Karol Rybecky put one up for 34.08 right in line with his previous efforts. The Poles only seemed to be gathering even worse luck.

Laurie Barr cranked away on his motor which refused to cooperate by promptly exploding. The released Pirelli bomb wreaked vengeance by ricocheting through Laurie's model which was parked a "safe" distance away. A few vital pieces were disintegrated so Laurie had to pull out another model. He put that one up safely, if not spectacularly, for 32.05.

Lunch was conveniently served at the hanger.

Starting up to the cat-walks.

Ron Green and model.

Mike Thomas and Mrs. Andrews relax at the Air Shed bar.

All this called for some fall back and regroup maneuvering. As it was now lunch time we fell back to the magnificent buffet served in the shed by the Henlow people. Other much appreciated services available were the bar and tea house set up in the shed. How civilized to buy a round of grog as your microfilm flimsy is wafted about there right before your peers, jauntily carrying you to victory!

Laurie's model made 36.54, third longest flight of the meet.

Servaites up again.

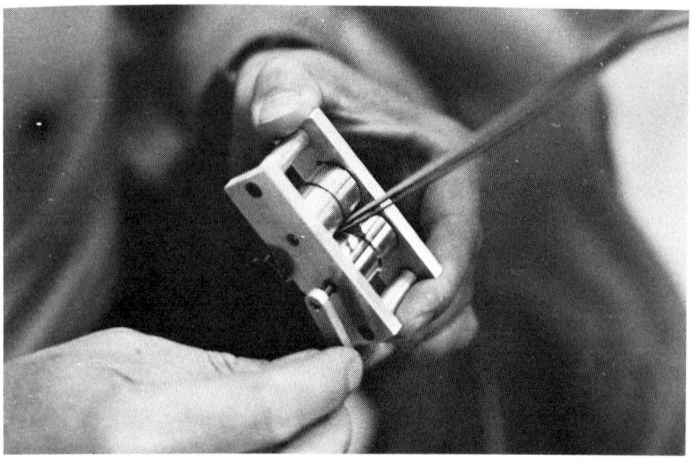

Popular rubber stripper manufactured in Poland.

Romak bites his lip as his ship goes for the title.

Heise, Butti and Siebenmann of Switzerland.

Peter Allnutt guards Mike Thomas' model. Canadian team was third.

Jim Richmond's model. Very nice work.

Japanese transparent-side model box.

Everyone's favorite, Laurie Barr.

Ferdinando Migani, Italy.

Sylvester Kujawa, Poland.

Mike Thomas, another view.

Bucky Servaites was next to go for the Americans and he put up a useful 32.34 bolstering the team's position. The Dutch had not been having much luck at all although it was clear they had plenty of enthusiasm and smarts. Much admired was their covered wagon which was used to transport the ready model over to the processing room and back. The closed box kept off damaging air currents as they walked it along. Their Edmund Liem put up a very nice 34.12, no doubt giving something for the locals back home to shoot at. The Swedes had struggled mightily and had gotten little in return. But then their national record stood at only 11 minutes. Mike Thomas gave Sven Linden one of his motors and, *Voila*, he did 34.06. That should shake up things back in Stockholm. And then Teddy Strazberger of Yugoslavia, upped his sights from struggling for 30 minutes all weekend and got a walloping 36.51.

All eyes were now on Romak as he made ready for his fifth flight. He used the same number of turns as before and held the model back for about 90 secs. The torque indicated by the prop flare looked okay to him so off it went. The climb wasn't quite so ragged this time, although still right to the edge. A little bit of nervous steering was necessary at one point and even some strategic movement of people provided a fine tuning of the tin hut meteorology. It was another good one, 39.36. There was little doubt now that we had witnessed the proving of a new champ, and the spontaneous congratulations from everyone were gratefully received by a real happy guy.

In the meantime Barr was up in an all-out effort. A nice climb took him to the center girder which proved sturdier than the model's fin which fluttered down to the deck. The model continued on in somewhat foolish fashion but showing no sign of wanting to come down. He had to balloon it down so he could get on with the next attempt. Now with time getting short those hoping to catch Romak had to do something spectacular. These attempts loosed a fusillade of models against the roof with the hope that nothing would stick. Between *that* lost hope and the rapidly deteriorating air quality, no threat to Romak's standing could be mustered.

McGillivray in his last attempt smashed into the roof. The wreckage produced a kamikazi attack on Siebenmann's model which was speared as it climbed up through the 50-foot level. Oh dear! Boyd Felstead was babbling over with sweet satisfaction having managed the high time of the last and final period: 35.45. His model would have done better if it had not descended on some piled-up obstructions at one end of the tin hut. Romak did not bother with a sixth flight as the placings were now set and he was feeling no pain. So ended a great World Championship.

Bud Romak became the new World Champion, and individual high-time winner. Team honors also went to the U.S., a clean sweep. Eduard Ciapala of Poland was second, Laurie Barr third, John Blount fourth and Mike Thomas fifth. Teamwise Great Britain was second, Canada third, Czechoslovakia fourth and Yugoslavia fifth.

Bud Romak with winning models and smile to match. FAI weight 1 gram without rubber.

Romak winding.

Distorted prop and stab show torque effects on Romak's winning flight.

A happy Bud Romak with the Ernest Kopecky Trophy-longest flight. Also received C.S. Rushbrooke Trophy as winning individual.

The Banquet and presentation of awards took place at the *Angels Reply* in the small town of Hitchin a few miles from Henlow. There was a friendly reception where we had a chance to talk with some of the organizers who had made all this possible. One of those was F/Lt. Richard Bould, the R.A.F. Liaison Officer. He saw to our housing and meals and coordination with the Royal Aircraft Establishment at Cardington and Royal Air Force, Henlow. The chief Timekeeper, the guy who kept score, was David Stapleton. Technical Director, Peter Freebrey, checked all the models and Contest Director was Mr. George Lynn. This was a very capable cast. The behind-the-scenes orchestration and strong driving force was led by Laurie Barr who also found time to place third in the competition, a whirlwind of activity. He had a few good jokes to tell too!

After a very good dinner, toasts and speeches by the Directors and Patrons of the World Championships, the awards were presented. The C.S. Rushbrooke Trophy was awarded to Bud Romak of the U.S. This trophy is a large baselighted rectangle of lucite in which a large dragonfly replica is embedded. It is awarded to the winning individual. Romak also won the Ernest Kopecky Trophy for longest flight in the Championships. This trophy features a replica of an indoor model embedded in a block of plastic. The Langley Cup awarded to the highest scoring team, went to the U.S. fliers, Romak, Servaites and Richmond led by Team Manager Pete Andrews. Federation Aeronautique Internationale, Gold, Silver, Bronze Plaquettes and Diplomas were awarded to winners and runners-up. And that was the conclusion of this very enjoyable meeting of *Those Magnificent Men and Their Flying Machines - On Gossamer Wings.* ■

WORLD CHAMPIONSHIP
F1D 1976

Lady Harland presents certificate to Laurie Barr at the concluding Banquet.

Lee Hines shows off his Nordic winner at the 1976 Western Free Flight Association International Challenge.

WFFA International Challenge 1976 Taft, California, by Jim Wilson

The 1976 WFFA was held at Taft on November 20 and 21. *SCAT* managed the contest again and the usual three events, F1A, F1B and F1C, were flown. The *Western Free Flight Association* has been more active in other years but is now reduced to this one contest. Perhaps now would be a good time to re-energize WFFA as a way to give more clout to FAI modelers' voices. Being that the Team Selection program fell on its ear this cycle, someone should redirect it and that someone might as well be a group of active modelers instead of hangar colonels.

Coming over the Grapevine on Saturday morning we drove into some delicious fog that filled the entire San Joaquin Valley and set the weather for the weekend. When the fog burned away it was replaced by low clouds and cool 60-degree still air. The 2-5 MPH ground winds were seldom higher at towline height so the Nordic fliers got their miles in this weekend. Bill Bogart was Contest Director this year

Jim Wilson

Bill Gibbons, Denver, Colorado

Gary Baughman, Colorado Springs, Colorado

Hector Diez, Simi, California

Juan Livotto and "Creeper" design. Has auto-wing-wash control from timer.

Creeper mechanism externals.

Creeper mechanism internals.

Ed Wogulis and wife Marty ready Nordic.

Don Zink uses "twanger" hook.

Not WFFA but Bob White and Pierre at the Taft "Pierre's Classic."

and he did an excellent job, unruffled as usual. A different timekeeping wrinkle was tried with the contestants rather than the organizers providing timers and it worked well. Many fellows teamed up in "an I'll time you if you'll time me" deal. I heard no protest over this scheme and since it's simpler it will probably stick around. Total entry was down this year from last but for the first time in a long time there were more power fliers than Wakefield or Glider. Models were mostly like last year's although several sported detail changes. Les Dewitt's Wakefield is highly developed. Reid Simpson uses a moulded fiberglass front end on his power ships. Gary Baughman of Denver had a very responsive glider; *"Prowler MK2"* with small 62 sq. in. stab, steep tip dihedral on short tips, and wrinkled tissue covering.

Finally — the flying. In general flying was improved this year. Partially due to the weather but also due to improved skills particularly in Glider and Power. No doubt that the 20 FAI contests we've had in driving range of Southern California this year encourage the contest work needed to master the craft.

Power: what a contest! Very high level of flying as the scores show. Hardy Broderson also commented that the level of flying was very good compared to other sections of the country. And Hardy allows himself the extravagance of traveling around and contesting in all areas. So, good show! Most rounds the air was helpful or neutral, say above 300 ft. so with a good run and no mistakes on launch a max could be had. Usually, the power fliers would wait for a glider or rubber ship to show that the air was at least neutral then a gaggle of power ships would go off together. Cox engines placed very well and Ron Young only broke his string of maxes with a left launch on the tenth round. Winner Ed Carroll was the only flier in the total entry to max out and this must make Ed feel good after permanently losing his best ship to the Minnesota corn at the team finals. That loss also broke a string of maxes for Ed and took him out of contention.

Dick Gildersleeve with "French Connection."

Wakefield, another tough contest. Several big names including two-thirds of this year's team wound up in the middle of the pack. But, other big names anti-gravitated to the top. Winner Les Dewitt has the best Wake in the area now and the credit must go to Les as an innovator and modifier because there is no way anyone could start with a blank sheet of paper and come up with Les' ship on the first try. Mostly, the flying difficulty was caused by lack of signs of lift at ground level. None of the usual indicators consistently worked. Many times a good launch and climb would result in two minutes. Just no help. There were a few out-of-trim flights so it wasn't lack of preparation that hurt people. There did seem to be an advantage to a high climb even if the glide was sacrificed some in order to go get up where the air was more favorable. The last flight Saturday included the now traditional timed-to-ground champagne round. Les Dewitt was all smiles with 267 seconds until Bob White landed at 271. The four-minute weeny lives! As an aside on that fifth round, I counted 20 ships of all types launched at about the same time early in the round. The sight was beautiful . . . who says free flight is a dead dinosaur?

Jim Quinn, Escondido, California, put a Bob White wing on his Wakefield, placed third.

Les DeWitt won Wakefield with his out-of-the rut design. Vee dihedral, swept tips, own-design airfoil is flat bottomed with LE & TE Cranked down.

Glider: again, tough competition. Most rounds there was little indication of lift at line height. Calm winds allowed towing in any direction and a lot of flights were tactical even if you had to tow downwind to get to the marker model. On my own log I show eight of the 10 flights piggybacking someone and five of those eight having to tow downwind to do it. This manueverability is the real advantage of circle towing.

Lift, when present, was weak at low altitude and very gentle at all times. I only saw one monster thermal on the two days. That was one that took Bill Hartill's power ship to great height in the eighth round. Unfortunately for Bill, on the next flight his auto didn't or it decided to be an awful-rudder or an otter-rudder or something and allowed him to retire and drink beer.

The right ship for this weather would have been a good zoomer to get up high, along with being very responsive to light lift. Winner Lee Hines was the only repeat winner at the contest. Saturday night we had a great finger food and light booze party catered by Inza with Bill Hartill's slides of *Indoor W/C*, *Pierre Trebod*, and touristing in England. The meeting was a pretty good rap session as many people have ideas for improving the FAI program. Unfortunately, neither our own VP or FAI committee member showed up, but Hardy Broderson, District 7 VP did. That way we were able to learn some about the give-and-go at the Executive Council level. Altogether a super weekend and definite must on next year's calendar! ■

Bob Piserchio of San Diego, California flies a Bob White design. U.S. Team member 1977.

Dave Bogdani, Northridge, California, builds a very neat all-sheet F1C model.

Bogdani's front-end details. Wing joiner is 3/16 in. dia. Brake added since photo.

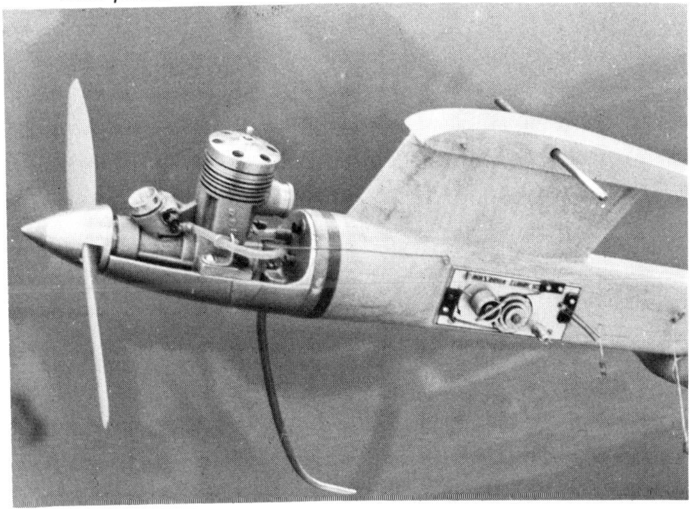

Hardy Broderson all the way from Michigan. Loves Taft but ran out of luck in fly-off. Believes in long-is-beautiful.

Rear-end details. Parson type auto-stab mechanism used.

Nicely made glass moulded front end of Al Vela's power ship. He does beautiful work.

Ronny Young's Cox Conquest set up in a Vela pan with Galbreath brake. A good runner but Rossi still king.

Bob White gives V for victory salute as his Wakefield departs into the sunset.

Vela and two different styles of his F1C front-ends.

Doug Galbreath taking time out from his book printing chores to show he still is a threat in F1C.

Beautifully trimmed ship by Bruce Hannah.

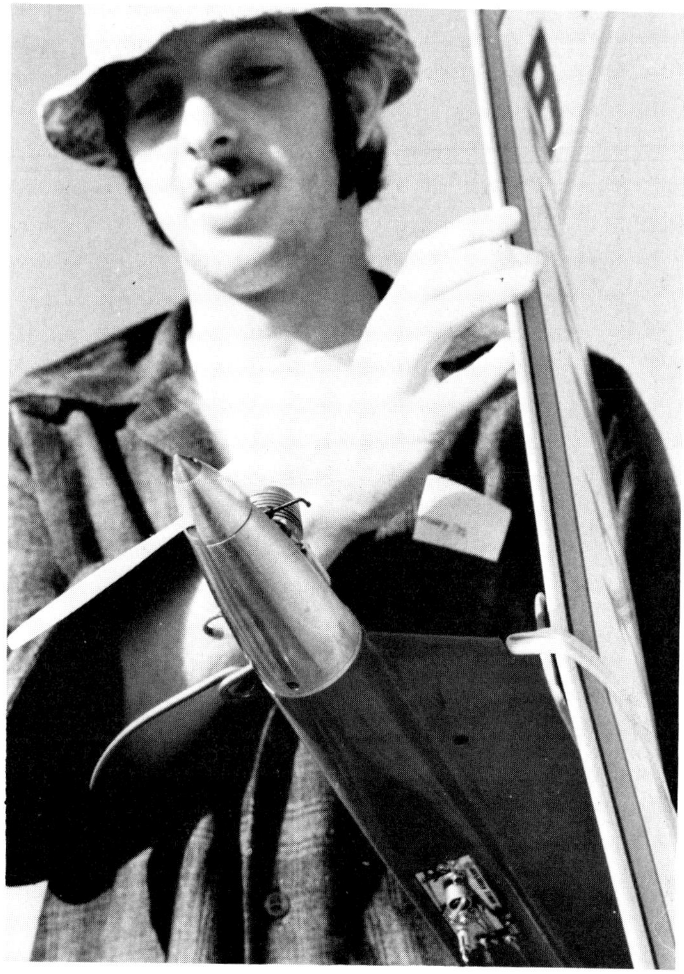

Neat Rossi installation on its side using Oliver pan by Bruce Hannah.

Ken Oliver has stuck with built-up structures even with his high aspect ratio design. Flies well but Ken still working on improvements.

No brake here but added since. Pressure line goes in through engine mount lug. (Bruce Hannah.)

The three WFFA perpetual trophies together with the engraved glassware ready for presentation. Awards designed and built by the SCAT Club.

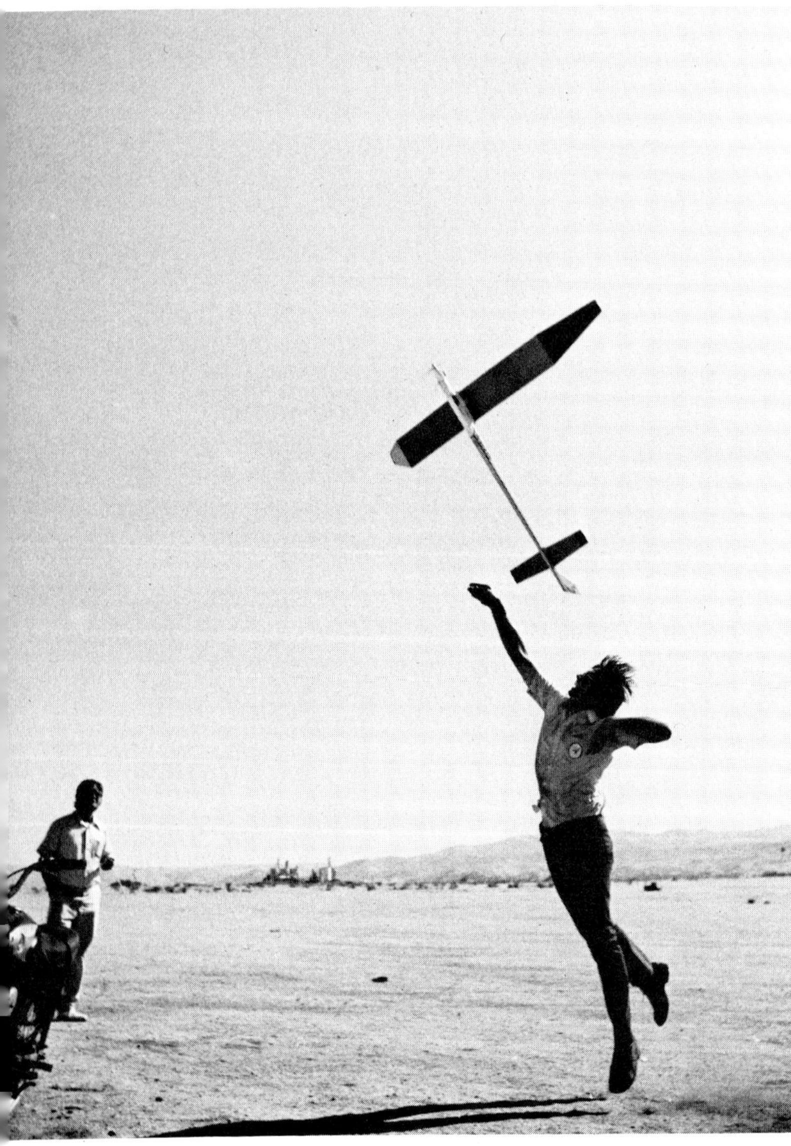

F1C winner Ed Carroll of Riverside, California lets go with a mighty heave.

Ed Carroll needles the Rossi on his "Moon Raker". All sheet design has been doing well.

Jim Wilson and Sam get their ration of competition Sangria from Contest Director and accredited Biggles Representative, Bill Hartill.

Egyptian Team, interest picking up.

North Korean Team.

British Team outfitted with smart uniforms provided by GOLA.

West Germany.

Team Members, Soviet Union.

Kiwis Ron and Diane Magill, Martin Dilly (proxy), and Martin Gregorie.

Serious business here, food line stalled at the Skalstrup Air Base

Happy time at the bar.

Pretty announcers kept the information flowing in several languages.

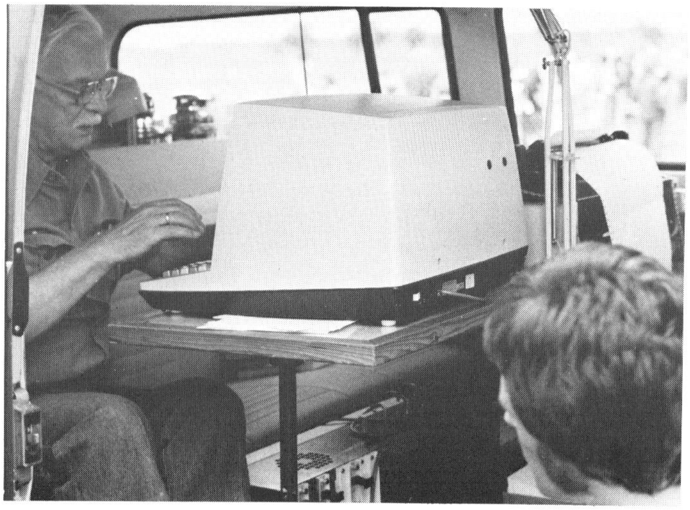

Wang computor and a line printer were used on the field to keep results current.

Meal arrangements were necessarily unusual because of the schedule. A box breakfast (yogurt, boiled egg, tomatoes, open faced sandwiches) was passed out at the school at 2:00 AM. After flying closed down at 9:00 AM everyone adjourned to the cafeteria of the Skalstrup Air Base for a hot lunch. This was located adjacent to the Roskilde Airport. The serving bogged down with very long slow lines. Evening meal was another box lunch (more open faced sandwiches) passed out at the school around 15:30. Those Skalstrup boiled potatoes and Viby Masonite sandwiches will be remembered and will go down in the record books with Finnish cucumbers, Austrian hard-boiled rolls and Bulgarian tomatoes. A great tradition!

The opening ceremony held at the airport at 15:00 on the day before the start of competition was brief and to the point, with color supplied by the many flags of the competing nations and a Danish boy's military marching band. From there we went right to the last evening's practice session. These practice sessions at World Champs are always very pleasant affairs, with lots of old friends to meet and new models to see and this session was idyllic. Weather great; field of neatly clipped green grass crossed with long concrete runways and the red sun taking its sweet time lingering on the horizon. The strong East European Teams and N. Koreans were going through methodical testing of all their models. Their flying routine seemed to be as well organized in testing as later to be shown on the competition days.

The next day, Friday July 8, saw 93 Wakefield contestants from 31 countries scurrying around a wet uninviting field at 04:00 wondering what happened to all the good weather. A 15-mph wind beat a steady light drizzle of rain into us and there seemed to be a general reluctance to get the models out and get on with it. After all, who would want to deliberately fly in such stuff? The flying line was soon assuming the normal look of hopelss confusion with thermal streamers, thermisters, bubble machines, odd plastic rain shields, signal flags, radio antennae, winding stooges and grim contestants scattered about.

Sandy Pimenoff of Finland, FAI-CIAM Chief, takes his ease.

That first round saw only 12 maxes, a real surprise as most felt that the weather was disappointing but hardly disastrous. Flight patterns varied from a wild spin to the ground for Siebenmann (36 secs), through a lot of weak mushy climbs for the majority, to a fairly fast stable climb for the N. Koreans. They had two of the 12 maxes and a 171 by 1975 World Champion Baik Chang Son. It was clear then that they were going to be tough to beat. Piserchio of the U.S. snapped a wing on launch. The reflight with a spare netted him only 110 secs. Smitz did little better with 116 and Ghio, complaining about the damp affecting his trim, managed 167. The British, accustomed to this sort of weather, weren't doing any better.

Round two was pretty much a repeat except that the rain stopped and the breeze picked up. The N. Koreans dropped slightly and now Japan had two men with two maxes each. Italy got all maxes in this round, starting to show their strength. At the end of the third round there were only three with all maxes; Samokish, USSR; Nimptsch, D; and Armesto, RA. Most of the contestants looked like they were in a state of shock as their models failed to perform as expected. Mike Thomas summed it up succinctly, saying: *If we knew what we were doing wrong we would have stopped doing it!*

In round four only Samokish stayed on with all maxes. British efforts were not hitting the mark with all sorts of indecision as to what to believe; thermister, lulls, streamers, bubbles, intuition. Even with everything reading *right* the air had the last word and it was *wrong*. On top of that, Peter Williams' ship hooked and dragged a mylar streamer. This improved its visibility remarkably but did precious little for its performance.

The N. Koreans were using their very effective wind-and-wait system taught them by the Russians. At the first sign of the green flare their first man winds and holds. If the coach doesn't signal him to launch in the first five minutes, the next man winds and the timekeepers are switched to the second man. This continues all through the round until all have flown. Flights are made on freshly wound motors, at least for the first two fliers. Surprisingly, very few other teams used this system. But then team-flying precision varied greatly among the 31 countries represented. N. Korean models looked similar to the straight forward designs they had two years ago, and in some cases were the same models. They used aluminum motor tube, lightweight tail booms (some were built up diamond sections). Nothing extreme in length or aspect ratio. Recent Italian rubber of 3mm made up into 28 strands took 360 turns. Props had relatively slim blades carved from a material similar to bass wood. The story was that the N. Korean Trade Mission was asked to purchase a batch of rubber for the team. All that was available at the time was the narrow 3mm strip that everybody else was avoiding. This stuff looked good in the climb although a few samples examined by Western experts did not appear to be exceptional.

Round four also saw the first maxes for the U.S. as Smitz and Piserchio finally climbed into some better air. Round five still brought no maxes for the 1973 World Champion, Joachim Loffler. His teammate, Albert Oschatz the 1976 Champion was wiped out in

Peter Williams with the help of Bryan Spooner look for some lift.

Lapinski of Poland in the wet.

Second place Wakefield winner Sergey Samokish of USSR

Something's gone askew there.

Tim ten Veld, Netherlands.

Delayed prop release launch.

Argentine Team readies for a flight.

Willard Smitz, USA, with his Gard type model

Baik Chang Sun of N. Korea, defending World Champion, looks a little cold here.

Cassi of Italy, placed seventh.

Mid-wing Wakefield by Andre Graux, Belgium.

Ron Pollard, Great Britain.

this round, dropping 71 secs after doing well earlier. Leader Samokish dropped 15 secs but almost didn't get that as there was a mix-up in flight cards. A high level investigation resolved the situation in his favor but now he was tied for second with N. Korean B. C. Son. The Japanese thrust faltered in this round as Masabumi dropped eight secs but did not lose fourth position, while Isav lost 36 secs to drop down six places.

It was now 09:30 and time for lunch, as we all packed off for the cafeteria on the other side of the airport. The warm food was welcome after trying to pick air in that wet cold sink hole. Adjournment to the school and dormitory gave an opportunity for sleep for those who could. In no time at all though it was 16:00, time to get the supper-in-a-box and get back to the field for the last two rounds.

This proved to be anticlimactic as a rain squall had passed through clearing the air and apparently taking away all the miserable previous conditions. The leaders retained their positions while the second place tie was resolved by Biak Chang Son, N. Korea, the 1975 World Champion when he dropped five secs putting him third. Many managed two maxes in this late afternoon period including Neglais, Rado, Cassi, Mauri, Liwenborg, Zachalmel , Pollard, Ghio, Loffler, Boutillier and others. Twenty contestants didn't get a single max the whole day

Team honors obviously went to N. Korea, running away from the field. Down 226 secs in second place was Italy followed closely by Japan and USSR. An interview of the winning N. Koreans revealed that Kim Dong Sik has flown four times at World Championships and made second place in 1973 at Weiner Neustadt. Their entire team is chosen from only 15 candidates from several districts in N. Korea although

Edy Mauri, Italy, jiggles in the nose plug. Tenth place plus second place team.

it was stated that the sport of model flying is very popular there. They were not really used to flying in the kind of weather we had this day, although they do a tremendous amount of practice flying. The organizors had obtained the use of computing and printing equipment to use right on the field to track the scores. A VW bus held a Wang 2200 minicomputer, CRT input/output, floppy disk data-storage system and a line printer. A 3M Dry Photo Copier prepared copies for distribution after each round. This permitted a quick ceremony right on the field for presenting the winners. Great idea!

Kim Dong Sik, Wakefield winner, was seven secs short of all maxes. N. Korea Team experienced and capable.

F1A glider was flown the next day with 96 contestants from 30 countries. The drive out to the airport took us over picturesque country lanes past neat farms and little villages that weren't awake yet. There was still plenty of mist but it wasn't raining. At the airport the breeze was up and there was a solid overcast. The green flare lit up the scene and with that a number of the hot-shots were off and running to take up circling positions downwind. The ground was smoother there on the runway, taxiway and the edges. The N. Koreans chose to do it differently though. They circled upwind. That had the advantage of giving more room but there was more wet vegetation to stumble over and the position didn't facilitate getting under a thermaling competitor.

The teams of USSR and Israel maxed out in the first round and it appeared that figuring out this day's air was going to be easier than it was the day before, as there were 24 maxes and it was still only 04:50. Second round saw the Russians continue their string while the Israelis fell by the wayside. The U.S. got its first max (Sifleet) and a close one by Markos of 178. Walters was showing great form with good launches but still hadn't got the right air.

It appeared that the great majority of fliers were circling. Most used 50-lb. test monofilament line with no reel. A loop in the end was tied to the wrist and it was played up and down hand over hand. Best technique was to keep the model high, making flat turns, without a lot of line tension. The idea that a circle-tower doesn't need stamina and agility is only true to the extent that he doesn't hope to be competitive. The adroitness and confidence of the top fliers was demonstrated by the Russian team in the first round. Isaenko's towline had become tangled and knotted. As Viktor continued to circle-tow with what was left of the untangled part, Lepp and Tchop ran alongside and tended to the knitting. This was done with unruffled deliberateness without any outward heroics. Line crossings occurred inevitably but there seemed to be much less difficulty here than in Plovdiv two years ago. The fliers seemed more agile and the timekeepers more observant. Zoom launches are in and the best set-up is to use a timer to hold almost straight rudder for two or three seconds during the speed of the zoom.

Andy DeMello of Canada doesn't look too pleased at this point.

Partial view of the Nordic line. Telephoto lens makes trees look close.

Good looking Nordic by Gino Stranieri, Italy.

Viktor Isaenko, USSR, makes a change in tow adjustment.

Then a kick into the glide at the top sets you up there 20 to 30 feet higher. Aileron control was found to be unnecessary and caused too much spiralling. Since lift was available but quite weak, a good zoom made a lot of difference in the outcome.

The Russian sweep faltered in the third round when Isaenko released under a thermaling model only to find himself down at 103 seconds. So much for sure things. The Czechs Horejsi and Dvorak both broke their string of maxes in the fourth round. Horejsi had a line-cross that went unnoticed by the timekeepers. The premature release cost him 15 secs. Dvorak, the 1971 World Champion, dropped nine secs. At the end of the fourth round it looked like Tchop and Lepp could now divide this all up between them. Then in the fifth round Tchop slipped up and lost a crucial 10 secs. Lepp made his max and relatively unnoticed outside the spotlighted "stars" were Thormann of E. Germany and Abadjiev of Bulgaria. They also had five maxes. Kostadin Abadjiev had come to Denmark to fly F1C Power. Someone had persuaded him to bring along a Nordic glider at the last minute, as there was no Bulgarian glider team selected. A fateful decision as it turned out.

The midday break came along to let everyone give in to total exhaustion. While we rested the weather started to turn warmer, dry and sunny with only a moderate breeze. The rest was only a tease for it was time to go back at it again. The weather was a tease too, the cloud cover came back.

The sixth round got underway with Lepp chasing the green flare. Perhaps it was creating a thermal? Fifteen minutes of towing were enough and he catapulted beautifully gaining 30 feet and away to max. Abadjiev made a very good flight but it was three seconds short

E. German towhook by Dietmar Henke.

Nordic has circle towhook and shock mounted tail boom.

Jim Walter's model.

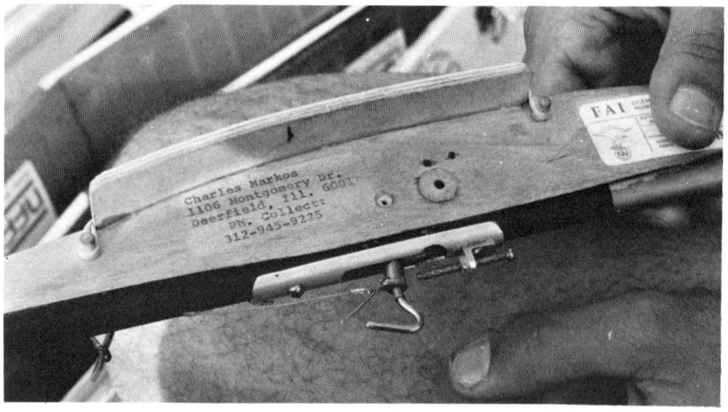

Russian style towhook by Chuck Markos, USA.

French style offset towhook.

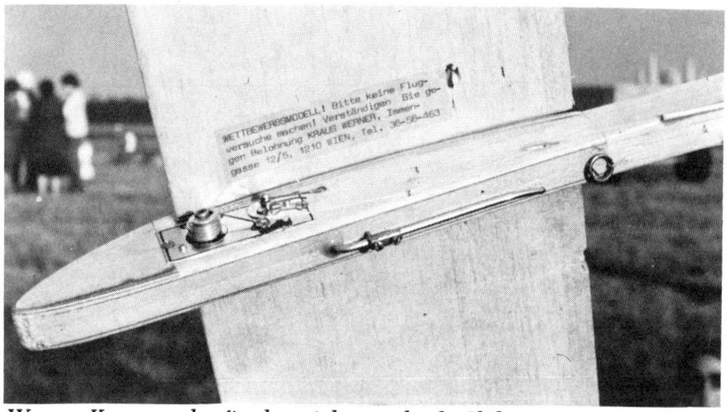

Werner Kraus used a fixed straight-tow hook. Sliding stop with set-screws changed ring position. Third place.

Modified Maxaid unit.

Swinging impulse type hook permits circling, by Finn Bjere,

Large loopy wire on towline gives impulse release, no circle tow.

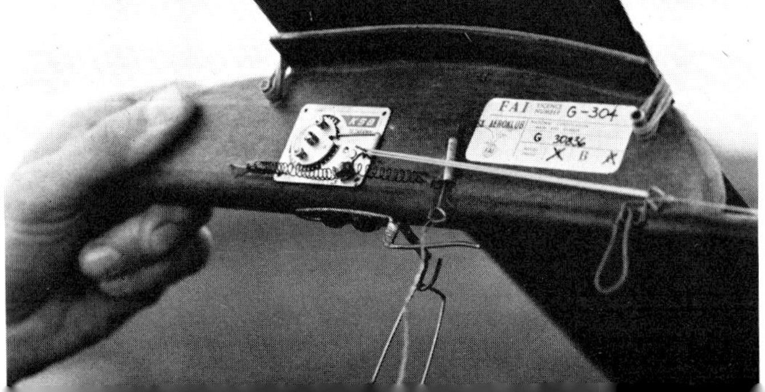

Andres Lepp's hook and timer details. Has rudder delay by timer

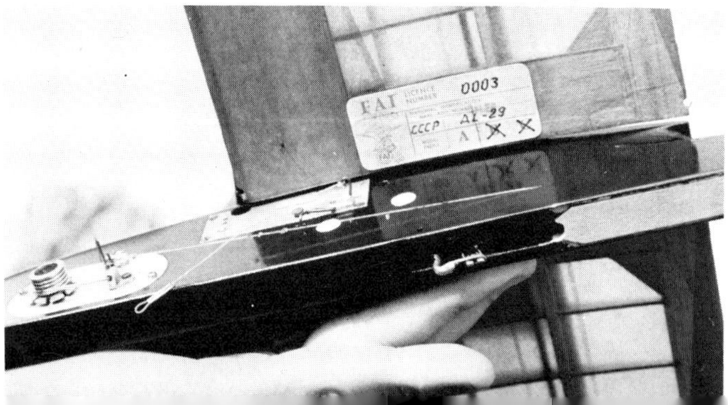

Master of circle tow, Ivan Horejsi, calmly looks away from the glider.

Isaenko looks one way, helper Lepp looks the other and model is yet in another direction in full control.

Viktor Tchop's wing structure and repairs visible.

Danish model of Peter Otte.

Team of Mexico, strength developing fast.

Per Grunnet makes ready while Koster lectures.

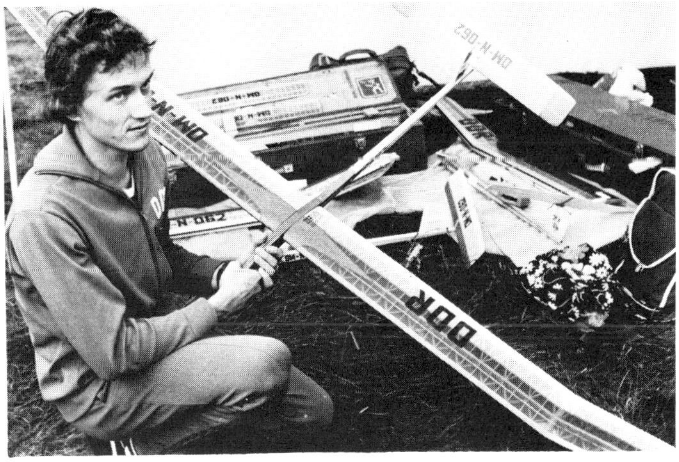

East Germany's Dietmar Henke was third in '75, dropped to 30th.

Lepp's model held by Team Manager while Tchop psyches himself for his turn.

of a max. F1A fliers know how hard it is to get that kind of time. Either it's a max in lift or it's 160 secs at best and a lot less at worse. Now it was up to Klaus Thormann. He had started towing at the first of the round and towed for 45 minutes, his team giving him every possible opportunity. He had no luck finding air, and eventually scored only 100 secs.

Last round started at 18:05 with plenty of light but not much lift. Only 10 contestants would max. Tchop looked miserable having come so close to making it a repeat win. There was still a chance but someone else would have to make a mistake. As it turned out the fickle thermal gods deserted the Russians. Lepp, possibly unsettled by a discussion with his timekeepers about the functioning of his towline release system, failed to find lift and scored 164 secs. Sympathy of the crowd had been with him all the way for his persistent and skillful pursuit of that elusive crown. He could be seen wandering around alone downwind, overcome by the shock of disappointment.

Now it was up to Abadjiev. For almost a half hour he worked his model over the area and then released. It was a good one and that was the gold medal. Andres Lepp was the first to congratulate the ecstatic Bulgarian. Tchop proceeded to add to his own misery by bombing out with a 137. Final results showed Lepp to be second while Werner Kraus of Austria hung in after making 154 secs in the third round, to come in third place. N. Koreans Ri Sang Chan and Kang Jong Sik came in fourth and seventh respectively, while Braud of France was fifth and Horejsi, CZ, sixth. Interestingly, Kraus was one of the few who straight-towed. Team spot went to N. Korea, of course!

The Bulgarian winner had wings of moderate aspect ratio and blunt tips with sheet construction. The sheet was covered with silk and it was turbulated. Stab was built-up and tail boom was a rolled balsa tube. Forward fuselage was a round aluminum tube. Wing sat on top of this tube. Towhook was of the Isaenko type built up from sheet metal bits. Seelig-type timer was partially enclosed in the fuselage. Lepp flew his AL 29 which is a few years old and showed a number of neat repairs under the tissue. He uses the older style Russian hook but modified with timer-controlled zoom rudder function. The Russians also had interesting looking fly-off models which were not called into service. They were stretched out ships with enclosed timers using a radially spiral scroll instead of the familiar Seelig axial scroll.

The brief ceremony recognizing the winners was hardly over when the sound of screaming Rossis announced that the last-minute check flights for tomorrow were already underway. It was another idyllic evening of flying. The weather had now settled down to reasonable, almost calm conditions. One could wander around from one group to another on the short grass between the runways and see one model after another from all the corners of the world. Their

Dave Simons, Australia.

Finn Bjere shows off his elegant fully tapered wing.

Abadjiev shows off the winning Nordic.

Andres Lepp.

Lepp, Abadjiev and Kraus, second, first and third.

Werner Kraus of Vienna, Austria.

Victory acclamation, Lepp, Abadjiev and Kraus.

owners were checking them over, wiping them down, setting timers and auto-functions, starting engines and firing them into the spectacular sunset-fringed sky. The typical FAI Power flier goes through his flight readiness and launch sequence automatically. Any hesitation, confusion or forgetfulness is automatically paid back very quickly with a Rossi tree.

This evening the Russians were going through their entire fleet, checking everything several times. Verbitsky's models were being proxy flown by Igor Ziljberg, a Wakefield flier. He admitted to being a bit apprehensive but soldiered on nevertheless. Eugeny Verbitsky, the very popular and well known-mainstay of the Russian team, had been bitten by a poisonous snake during a contest in *Yerevan* (in Armenia - USSR) near the biblical mountain *Ararat.* The story told was that he had been very ill and was in a hospital for some time. Although on the road to recovery now, his convalescence did not permit him to attend this year in Denmark. A number of the fliers and fans gave cards and souvenirs for the Russian team to carry back to Eugeny.

The Verbitsky models from two years ago were there along with some new ones sporting even cleaner lines and nicer cowlings. A modern touch was the use of 0.03mm dural skin over the 1.5mm balsa sheet wing covering. This was on the inboard panels. The dural was a hardened grade and came from a helicopter blade application. SergeySharon had somewhat similar models, which many thought were the nicest looking ones on the field. Not only was the workmanship and engineering clearly outstanding, the finish and trim showed remarkable artistic sophistication, with use of airbrushing. Mozyrsky, the third team member, also had nicely done similar models. There seems to be a trend of teams flying similar designs.Part of this could be the use of common molds to turn out a number of pieces of complex parts. In any event the Russian models showed an apparent strong team effort in their design and manufacture.

The Danes had an admiring audience also. Koster, Agner and Christensen had similar models of the Danish school (but no flaps this time). They were putting on a very impressive show. Engine note was very painful on the ground and on launch picked up even faster. Koster was turning in the vicinity of 29,000 rpm. Part of the change in engine note on launch was due to prop flexing on the ground. As soon as he launched the blades smoothed out. This was with the plain glass and epoxy props. Another prop he used in the contest used Verbitsky carbon fiber, was cooked in Tom's oven, trimmed by Dave Rounsaville and balanced by Bill Gieskieng. There seemed to be some uncertainty which prop got the ship higher. Pattern was just about straight-up. Transition was helped by a quick shot of *down* elevator and right rudder at the crucial moment (the Koster Bunt). The Square Cream plan doesn't show this device. The flights that evening showed that Koster had the performance potential edge. But could he hold things together? The Russians,

Ziljberg proxied for hospitalized Verbitsky. New models have dural sheeted wings.

Sharon waits with assistants. Trim and lettering air-brushed.

Mirko Bjelajac of Yugoslavia with newstyle streamliner.

Rounsaville tending Koster's Speed Cream.

Koster pushes away.

Howard Costelow, Australia.

Steen Agner, Denmark.

Barabella, Italy.

Walter Nutini, Brazil.

Niels Christenson, Denmark.

Sergey Sharon.

Jose DeLuna, Mexico.

Steen Agner makes ready.

Abadjiev flew F1C also.

Mike Burns, Canada.

Martyn Cowley, G.B.

Urs Schaller talks it over with Ken Faux.

Michel Iribane, France, high aspect ratio triple fins.

McLaughlin, fast climb.

Result of attacking a beer truck; Reda went on to place sixth.

Ziljberg and Mozyrsky.

Lots of detailed engineering on the USSR fleet.

Bruno Fiegl placed tenth for Italy.

Martin Dilly attacks the box breakfast.

For some, the pace was just too much.

Thomas Koster.

Beauty (Sharon) . . .

. . . and the beast (McLaughlan).

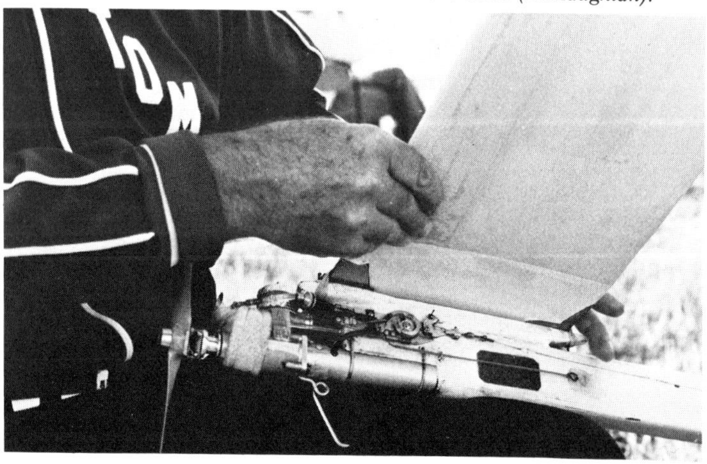

Urs Schaller, Italians and Meczner among others were looking very promising also.

The next day was upon us awfully early. The spirit was willing but the bodies were straggling several paces behind. This was Sunday, July 10 and there were 83 contestants from 25 countries ready to fly. Weather had improved considerably and there was a real sun. Wind was still 10-15 mph but with the warm sun it seemed much more civilized. Past Champion Lars-G Olofsson chose to make the first flight. Several things seemed to be wrong with its trim but he made the max and the next four for the A.M. contest. Martyn Cowley (GB) was another that had some bothersome-looking climbs and stally glide but still maxed out. Martyn had not been able to keep his building program up to speed due to a savage attrition/survival rate and was now still in the trimming stages of his *Hot Rats* and *Hot Ritz* designs out of *Biggles England*. And now on his first flight it was lost in the wheat. There had been strenuous warnings from the organizers not to tramp down the wheat and most attempted to comply with restraint. But as reported by Free Flight News; *Some people, it was alleged, were driving bikes (and in the case of the N. Koreans, the embassy's black Mercedes limousine!) in some very weird countryside locations.* The walkie talkie airwaves were saturated with all sorts of languages and static, making their use somewhat frustrating and uncertain. Signal flags proved to be better for signalling a launch.

The morning's flying went well and performance of the majority was impressive. At the end of five rounds there were three countries maxed out: Italy (always a top team), Canada, and Great Britain. The U.S. had dropped only one flight. Bissonette had the misfortune to have his stab hold down bands weaken and cause a stally glide for 139 secs. The routine of the midday break (if you can call 08:10 to 17:05 midday) was upon us again but now for the last time. Could we survive? The weather stayed fine and now gave a wind shift to blow lightly down the length of the other runway.

The eighth round started to undo a number of hopes as Charlie Martin (U.S.) went down along with Truppe (A), Lutini (BR), Johnson (GB), Engelhardt (DDR) and the defending champion Lars-G Olofsson. Lars had an overrun and then on the refly the VIT line broke smashing the model in 5 secs. Al Vela (Mexico) had some beautiful ships nicely cowled in the latest fashion, but also had some back luck. One flight went straight into the runway from a great height, engine turning fantastic revs. A perfect impression of the engine and wing leading edge tip to tip was made in the pavement. Canadians lost a man too when Frank Schlacta dropped in the seventh round with 148 secs. He joined Roux of France and Krycer of Czechoslovakia in bombing out after a string of six maxes. The Russian chase crew got an unexpected view of the *Northern Lights* when Mozyrsky's model drifted over and DT'd down onto high-tension wires. The aluminum covered wings lit up with several thousand volts. Holes were burned through and for all we knew, half of Copenhagen was now without lights.

It was now time to check on what the fly-off would produce. There were 22 to fly, with only one full team, Italy. Six countries had two men in. They were Denmark, Russia, Canada, Great Britain, W. Germany and would you believe N. Korea? It now appeared that some characterizations could be made of the surviving model designs. Most were high aspect ratio and had full or partial engine cowlings. Most pylons were either very low or non-existent. Most had two-piece sheeted wings. All had VIT, auto-rudder and engine brake. Tom McLaughlan of the U.S. had ships like his *Swinger* of 1973 but now the wing didn't swing. The folding prop shaft and open air detailing of the plumbing and structure could only be described as extraordinary. N. Korean models looked dated but still were well flown.

The Danish organizers had prevailed upon the CIAM to permit the use of the newly proposed but not quite official, fly-off rules. This is the increasing by one-minute-at-a-time the fly-off max with engine run held constant at seven secs. Also used, was a 15-minute launch period and a second attempt permitted as in preceding rounds. The first fly-off round knocked out nine including Sharon, (close with 237 but no cigar) McLaughlin (short run) and Cowley (the stall caught up with him).

Koster's good model was still in the trees after the four-minute flight so he got out the reserve to make a quick check flight. It went very quick up over and down splat. Back to the woods, boys! Soon they came racing forward with *Speed Cream* all in typical Koster tradition. Siegfried Reda also saw fit to make a check flight of his high aspect-ratio foam-core wing ship. It went up in a huge high-velocity arc and came on down seeking out a VW beer van which it impaled. The wreckage bits were brought back in a beer case. These disasters seemed to be hardly noticed as the fliers turned quickly to their reserve models.

The tenth round going for six minutes looked like that would settle it. There were now eight competitors left. The top N. Korean had missed it by five secs. Schaller was one of the first ones up and he got very high. But then a flop over at the top brought him down a little. Koster's flight looked comparable to Schaller's but Thomas had a nice transition. The sky was getting a bit gray and timers had to strain a little to make sure. Reda was down in 253 behind trees. Sugden looked good as did Meczner and Verbitsky (proxied by Ziljberg). In fact, it was a very thrilling finish. You could see the tension building around the Danish camp. Then there it was. Thomas had done it! *First time ever* for a World Champion in two classes.

There was universal admiration for this tremendous achievement. Not only had Thomas Koster developed a technically superior machine, and managed the temperamental beast to a hard-fought win, he had pushed this great meeting into being and marshalled the efforts of his countrymen in its support.

No, the N. Koreans didn't take the team win. They fell back to third with Italy first and Canada second. The parties that night fully measured up to the FAI standards and the earlier fatigue was soon forgotten. The total involvement of Koster in the affairs around us came to light when it was pointed out that he had been helping arrange parking lots earlier that day. And also in earlier years Thomas had been a student at the Peder Syv school where we were now happily toasting his health. The next day, for those who could interrupt the partying, there was a bus tour of Roskilde laid on. Highlight of this was inspection of a Viking ship now restored after being discovered in the vicinity. A number of others took this occasion to drive in to Copenhagen where the main attractions were *Tivoli Gardens*, the harbor and *Little Mermaid*, and the most fun-loving city in Europe.

Banquet and awards ceremony took place at the casino of the Ringsted horse race track. The long and narrow seating arrangement made it difficult for everyone to see the proceedings, but otherwise all had a great time. The banquet service was first class. Menu included shrimp cocktail, cucumber salad, grape salad, roast beef, boiled potatoes, peas and carrots, sweet white potatoes, red wine, white wine, soft ice cream, fruit, coffee and Port wine. A bar was available to keep the glasses filled. The unfortunate N. Koreans had no place on their table to place their glasses, it was so covered over with trophies and awards. As proof that nobody was really tired, the dance floor was filled with gyrating free flighters dancing to the live band whose name translated was "When the Farmers are Making Love".

Somewhere around 2:00 A.M or so the buses carted us back to Viby, where the bar there was reopened and festivities continued. The organizers had provided some 6,000 plastic cups for use in the bar. These had the contest logo imprinted on them and they were gratis with the drink. Needless to say they are now all gone, scattered to the four corners of the world. But we'll wager they're being put to good use now, toasting the memory of this great world meeting.

Next on the grand tour, the *Pierre Trebod* just four days hence and 733 miles to the south, en France. On our way we met half of the happy W/C group waiting in line at the Puttgarden car ferry for the crossing to Germany. Martin Gregorie was there with a full crew of hitchhikers loaded into his Land Rover. He was up from New Zealand and now was planning a drive to India to visit his sister and have a little adventure. See you downwind at Marigny! ■

Andres Meczner, Hungary, has been fighting for that top spot for many years, this time second place.

World Champion Thomas Koster casts a long shadow. Superb performance.

Ziljberg (for Verbitsky), Koster and Meczner receive their recognition, a few minutes after the final fly-off, Roskilde Lufthaven, Denmark.

Koster family perched on VW to view the ceremony.

Thomas Koster, Wakefield World Champion 1965, F1C Power World Champion, 1977.

Igor Ziljberg, a Wakefield flier, proxied Eugeny Verbitsky's complex model to third place, well done!

Canadian group prepares to launch Peter Allnutt's model.

Criterium Pierre Trebod 1977 Marigny, France

THE WORLD CONTEST
MARIGNY 77
13ème CRITERIUM PIERRE TREBOD

The great exodus from the World Championships at Roskilde, Denmark began on 12 July with many of us heading for the 13th annual, fabulous, *Pierre Trebod*, just 1200 km to the southwest. Our first stop was at a pleasant little campground in a rural area outside Bremen. This also marked our first go at trying to erect our *Malaga* sports twosome rent-a-tent that we "hired-out" in London. A few anxious minutes were spent before we got the upper hand and escaped the clutches of the aluminum octopus framework. The additional gear that came with this outfit made for a comfortable camp and we were soon enjoying our *gourmet diner un terrain*, or rather, *festmahl auf dem landweg*.

Not wishing to overexert our matronly VW, we toodled along and stopped the second night at a Rhine River campground near Remagen. We enjoyed our drinks there, with the river splashing on the rocks at our feet and the big diesel barges going *ka-whump, ka-whump* by our camp in a never-ending parade. Next day, after inadvertently investigating an *eingang verboten*, as is often my wont, we made our way upriver to Koblenz then headed southwest on the *autobahn* arriving at Trier, the oldest city in Germany. Trier is situated at the confluence of the Saar, Ruwer and Mosel rivers, having been established some time ago as the northern capitol of the Roman Empire. A few km from Trier we came to the frontier border crossing to Luxembourg. After picking our way through dozens of huge trucks that were stopped for customs check we were waved on by a smiling *inspector*. It's a tiny country but crammed full of interesting things. Too bad we had so little time. The ancient *citadel* with its 53 forts

linked by 16 miles of tunnels and casements hewn into solid rock gives Luxembourg City the appellation, *Gibralter of the North.* The local food combines German heartiness with French finesse and the *Luxembourgeois* are proud of the old saying: "One Luxemburger, a rose garden; two Luxemburgers, a *kaffeeklatsch;* three Luxemburgers, a band."

Just outside the city is a large, immaculately groomed, American cemetery. Here, lying with his fallen comrades, is General Patton. And just down the road a piece is a German military cemetery. All is peaceful here now with only the long rows of crosses to bring back the memory of these men caught up in that grinding struggle.

Driving west we crossed over into France as the road twisted and climbed through the mining region around Longwy, then as we headed south the country flattened out into pleasant farmland. Here and there could be seen remnants of the old *Maginot Line* now used in growing *champignons.* Our journey brought us to Verdun for lunch and then we realized that all the local festivities so apparent were due to this being July 14, Bastille Day.

To the south is the village of Bar-le-Duc which is surely remembered by the *Flying Aces* Phineas Pinkham fans who enjoyed the uproarius exploits of that fictional WWI character. Our meandering took us on and eventually that evening we arrived at *l'Hotel Cheval Gris* in Pleurs, our home at the Pierre Trebod. It was good to be with our friendly hosts and flying companions once again as we settled in and enjoyed another great dinner in the *country Francais* tradition.

Next day out at the field we noticed that the on-site campground had been located on the field next to the back gate making access to the site much easier, a great improvement. Flying surface was the same as before however, long scratchy weeds. Too bad this stuff couldn't be mowed.

The afternoon before the F1A Glider event I was having so much fun I just had to make one more flight with my old *Tuggy Buzzard.* Off it went in a nice thermal and what-ho ! It failed to DT. A thorough search downwind turned up nothing except a great increase in my intimate knowledge of what half of France looks like. Oh well, old TB had a good look at England two years before, now she was off to see *La Belle France.*

The competition was again up to high standards, the weather stayed very cooperative and everybody had a great time. Maxing Nordic in the first round was a little difficult as the thermals were a little slow in making their appearance. Nevertheless 10 managed to reach the fly-off and Herman Motsch of West Germany (Saar Club) emerged the winner with a convincing 199 sec flight made in fading conditions. Motsch claimed this ship was capable of 170 secs with a "soft" launch and 190 secs with catapult which is the method he chose for the final flight. A Lepp type circle hook is used. Brian Baines of GB and Leloux of France had tied for second place at 143 secs and so had their own personal fly-off. Leloux eased out Baines by four secs.

That night we enjoyed the usual revelry at the Cheval Gris — the place was bursting at the seams.

Power and Wakefield fliers were out in force the next morning and now the breeze was taking the models off in the short direction toward Marigny. Martyn Cowley of GB and *Biggles* fame shepherded his *Hot (Ritz, Rats, Lips)* F1C model to a win in a four-man fly-off, in spite of a rather ragged launch. The Snoopy Special has a superb glide. Relaxed World Power Champion Tom Koster was content with sixth place and pleased to see his good friend Dave Rounsaville come in fifth.

A few Wakefields were lost in the wheat off the field including those of Chaussebourg and Sebenmann. Winner was Jean-Claude Neglais of France topping Keith Proctor of GB in a fly-off that saw Neglais get a nice four-minute max while Proctor was down in poor air at 159 secs.

The British were feeling quite rambunctious having done very well. They were once again the winners of the *Jean Magniette Challenge Trophy* for the best total of the top placers from each country. That night they received their rewards and recognition along with the other winners while the rest of us joined in the fun, helped along with the generous supply of Sangria at the *le Gayon* cafe. The dinner was much improved from the year before and it was a truly marvelous gathering. Eating, drinking, dancing, story telling — What more could a World Free Flighter ever desire? ■

Yves Olard mans the starting gun.

Start-line scene, Nordic day.

Yves Olard and Robert Guilloteau, Contest Director and Manager.

Lunch time and the story telling continues.

A cockpit for the timer, by Rotteleur, France.

J. Godinho, France, dropped some time in first round.

Concrete aprons are useful but high weeds make towing and chasing difficult . . .

What's this, Ron Pollard with a Nordic? A Rony-Tube no doubt.

Cubans did not fly at Roskilde but did at the Pierre Trebod and had a good time.

Lebee, France, sorts out the twanger hook bits for Boutillier's model.

Martin Gregorie, N.Z., tried a very short tail moment. Flew OK but no luck.

Debra Collins, G.B., and Peter Seelig, Germany.

Danes check the score board.

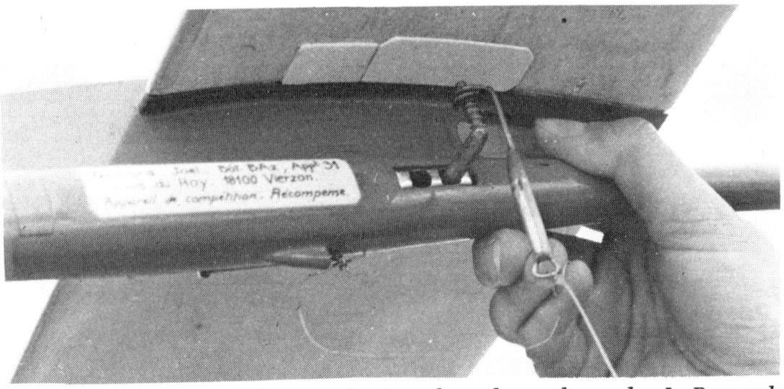

Interesting variation on the impulse release theme by J. Besnard, France.

P. DeBoer, Netherlands, circle tow with cable braced wings.

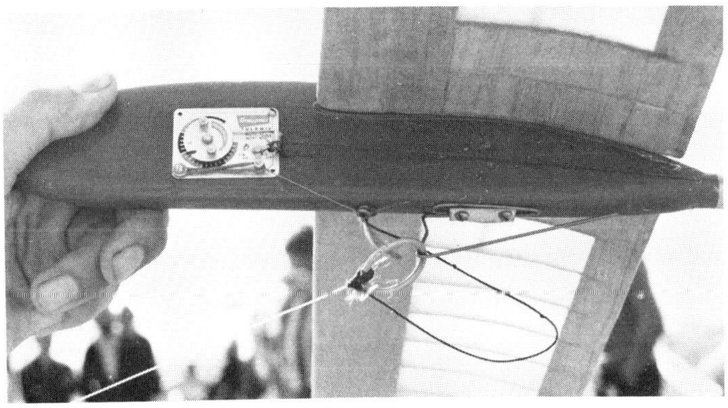

Simple hook system of P. Lafeuille, France.

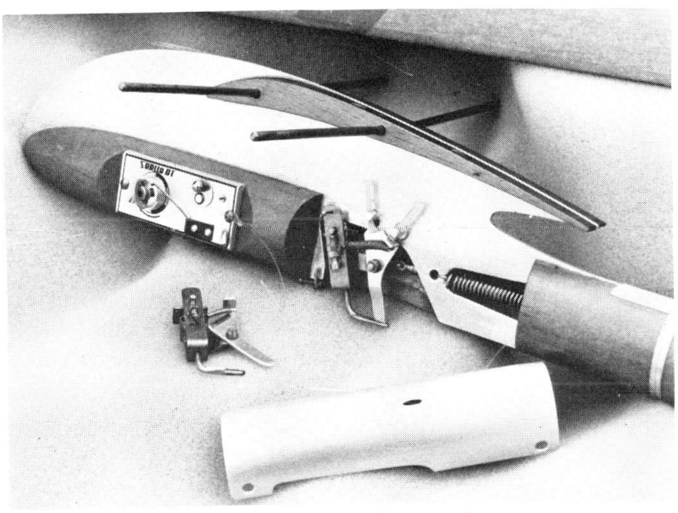

Rasmussen's Danish style hook and front end arrangement.

L. Reynders, Belgium.

M. Bazillion, France, circles with offset hook.

Van Dijk, Netherlands, with impulse hook, braced wings and recovery beeper.

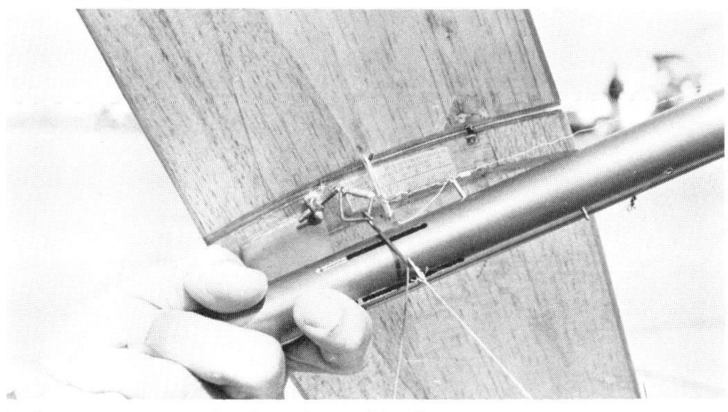

Siebenmann, Switzerland, with novel hook system.

P. Laneurie, France, hinged wings tied with rubber.

The new look in zoom launches.

Allais (F), repairs damage after hitting car. Had one bad flight. ◁

Bob Hatschek patches tissue. ▷

Godinho, France.

Herman Motsch, Germany, first place. ▽

Maiworm holds Motsch's winning Nordic.

Robert Champion ready to launch.

Ian Kaynes, G.B.

Now that's better, Pollard with Wakefield.

Jon Davis, USA.

Dieter Siebenmann lost this model.

Handy bar manned by French Air Force.

All the conviences of home.

Free Flight tent city was on the flying field this year.

P. Lepage gets some thermal advice from the local meterologist.

Lunch was just a few steps to the campground.

Pim Ruyter, Netherlands.

Pim Ruyter details. Prop-stop autorudder, autostab from timer, two string turbulators.

Keith Proctor, G.B., readies for the fly-off.

Proctor sets nose plug.

Proctor launches away, placed second.

Winning Wakefield of J.C. Neglais, France. Four minutes in fly-off.

Jan Zetterdahl shows Ken Faux which way is up.

Koppitz holds Neglais Wake, 'Soft Machine'.

Dick Johnson, G.B., uses a circlip for prop brake. Plenty of hose there.

Mauricio Zito, Argentina.

Frank Schlachta, Canada, has been trying higher aspect ratio.

While Alain Landeau does well with low aspect ratio.

Ken Faux's new model.

Neglais gets tossed into the lap of the thermal god.

Martyn Cowley did no wrong to max all the way with "Hot Lips"
Biggles takes the Pierre Trebod!

Participants begin to gather in the Le Gayon courtyard for the trophy presentation. A wonderful Country French Extravangaza.

Dinner at Le Cheval Gris.

English fliers 'apres diner'.

Wakefield Challenge, CPT.

Power Challenge, CPT.

Reception at Le Gayon.

Pollard peeks at the camera.

Dave Simons checks the wine.

Nordic Challenge, CPT.

Jean Magniette Challenge Cup.

Old friends at Le Gayon again.

Linda Aker and the U.S. Free Flight Champs Grand Sweepstakes Award, the Chuck Broadhurst Memorial Trophy.

Billy Hunter readies for a go at the Taft thermals with his big Satellite.

US Free Flight Championships 1977 Taft, California

The USFFC is the biggest free flight contest held anywhere in the world. How does one write about it? With great difficulty! It must be experienced personally. All I can do here is offer a few statistics, anecdotes and some images of this great orgy.

The USFFC began in 1971 on the May 30 Memorial Day weekend and has played that date every year since. It is organized by the FFMAASC group and the *Northern California Free Flight Council*. FFMAASC is the *Free Flight Model Airplane Association of Southern California*, formed by all the free flight clubs in the area (about eight). It was found some time ago that the only way all the competition in Southern California could be fitted into the calendar, without too much overlap and confusion was to work together as an association and plan the entire year's calendar at one shot. That idea worked. Then the opportunity came in 1971 to organize a West Coast Nationals. That didn't quite make the grade, conflicting with other people's vision of what and where a *Nationals* should be. Undaunted, the FFMAASC went ahead and organized one big F/F shebang.

The 1977 meeting saw 32 F/F event categories with registration of 305 contestants and 756 entries, with the following division:

Nordic A-2	23	Coupe	26
Nordic A-1 (Sr,Op)	25	P-30	19
Nordic A-1 (Jr)	6	Rocket	10
Outdoor HLG (Sr,Op)	38	Payload	11
Outdoor HLG (Jr)	13	½A (Sr,O)	70
Wakefield	20	½A Jr.	11
Un. Rubber (Sr,O)	21	A Gas (Sr,O)	71
Un. Rubber (Jr)	8	A Gas (Jr)	13
O.T. Rubber	24	B Gas	39
E.Z.B.	9	Indoor HLG (Sr,O)	21
C Gas	40	Indoor HLG (Jr)	6
D Gas	19	Scale (Gas, Rubber)	20
FAI Power	19	A/B/O.T. Pylon	19
Night Flying	22	A/B/O.T. Cabin	12
30 sec. Antique	13	Peanut Scale	23
.020 Replica	28	Sweeps/Team	26

Indoor events are held under 20-ft. ceilings in two gymnasiums of the Taft Junior College in the middle of Taft only about five miles from the outdoor site. Top indoor hand-launched glider time is around 63 secs with about 10 mins for EZB. Outdoor events are held on the Taft Condors field east of town. It is a flat spot in the desert with a large cleared hard-packed area surrounded by miles of nothing but ideal chase area. There are low-lying gently sloped mountain ridges to the north, west and south that are well beyond range of anything but a 40-minute flight. These ridges serve to shield the site from wind although beyond the ridges to the east there isn't any wind problem anyway. The one best feature of the site has to be the great weather. A little bit of rain occurs occasionally in late winter/early spring but outside of that *it never rains*. And since the wind doesn't blow either, one just doesn't worry about planning a flying weekend at Taft. You just go and do it. The mid-summer months are very hot with temperatures as high as 125 degrees F during the day. It cools right down without the sun and is quite comfortable in the evening and morning. Humidity is extremely low, being around 5% in the summer. Not many contests are scheduled in July and August because of the heat. Taft regulars use motorbikes for model chasing. The Honda 90 trail bike is the most popular as it is light enough to horse around and handle with one hand while holding the model with the other. It is also big enough to ride two and power through some of the softer outcroppings. A large number of fliers camp overnight right on the flying site so as not to miss any flying time. Besides, the clear desert air gives a great view of the stars , and a hearty cook-out over glowing embers adds a fine flourish to the spirit of the whole thing.

These ideal conditions have led to another unique Southern California free flight institution: night flying. Flown with the larger class AMA gas ships, five minute maxes are not difficult and night thermals are not an unusual occurrence. Small lights attached to the models keep them in sight and timing and retrieval are not nearly as difficult as would be imagined.

Probably a first for any free flight competition was the wedding that took place on the field during the '77 Champs. Terry Welch and Duane Ripley of Bakersfield walked down a red carpet, beneath Starduster and Shocer wings, to exchange vows. The groom prefaced his honeymoon by flying C gas.

Old-Timer (Vintage) events are popular at the Champs, and Megow Commanders, Zippers and Buccaneers can be seen buzzing around with that apprehensively uneven staccato peculiar to those long stroke ignition engines. And they have honest-to-gosh balloon-tired wheels! These guys seem to be having more fun than the all-out competition fanatics, if that's possible.

The US Free Flight Champs draws fliers from all over the US, Canada and Mexico. Hopefully more will make it from Europe, South America and the Pacific in future gatherings at this great site. ■

Al Bissonnette and Bill Chennault get a Big Mama Pearl ready.

Walt Ghio's Wakefield is surprisingly similar to the N. Korean layout.

Dick Myers and Lucky Lindy with ST65.

Al Hotard.

Indoor Champ Bud Romak with T-tail Wakefield.

A scaled-down Playboy maybe?

Sharp B job, Bill Burke, Dallas, Texa.

Wayne Drake, the flying Dentist, with all sheet Wakefield. Issacson looks incredulous.

From Mexico, Jose Luis Ramirez and A-1.

Al Hotard jokes with Bob Isaacson.

Lee Hines looks pleased.

While Jon Davis checks thermal with tongue.

Roger Gregory sticks his Wakefield in a hot thermal.

Kathy Grissom, Dallas, Texas, and homemade 6-inch reflector telescope at Taft.

Ed Wogulis ready to launch.

A first for Free Flight — Terry Welch and Duane Ripley of Bakersfield were married here at Taft during the USFFC competition.

That's Eva Allnutt, Peter Allnutt's mother, there on the left in attendance.

Bob White and Unlimited.

Bill Bogart, all knotted up.

Peter Allnutt, Canada.

Mike Taibi (rt.) cuts and sells some fine balsa.

Clean ship — unidentified.

Isadore Suarez, Bakersfield, Texan 600 with ST23.

Jose Luis de Luna of Mexico.

Floyd Aker & Tartar 610.

Free Flight strategy session.

Bob White cools off.

Jim Parker's model box.

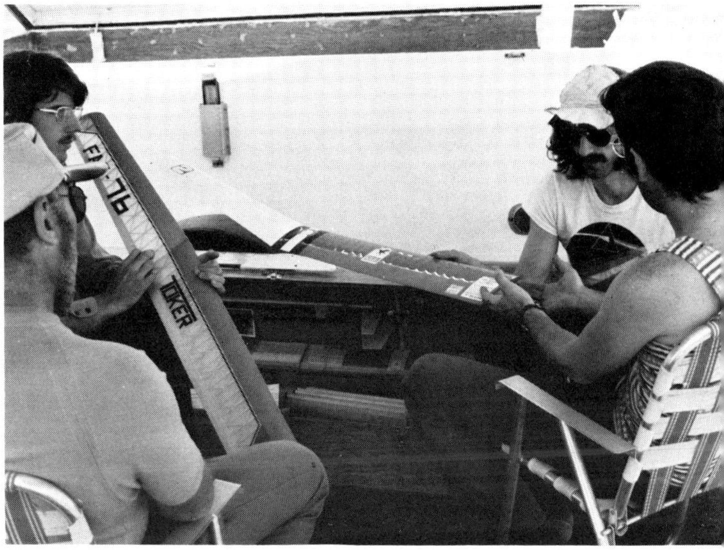

Let's see what you've got there, Jim.

Irv Aker works at it.

And so does Bob Piserchio.

Night flying at Taft is a surrealistic image.

Parker's towhook filed from alloy bits.

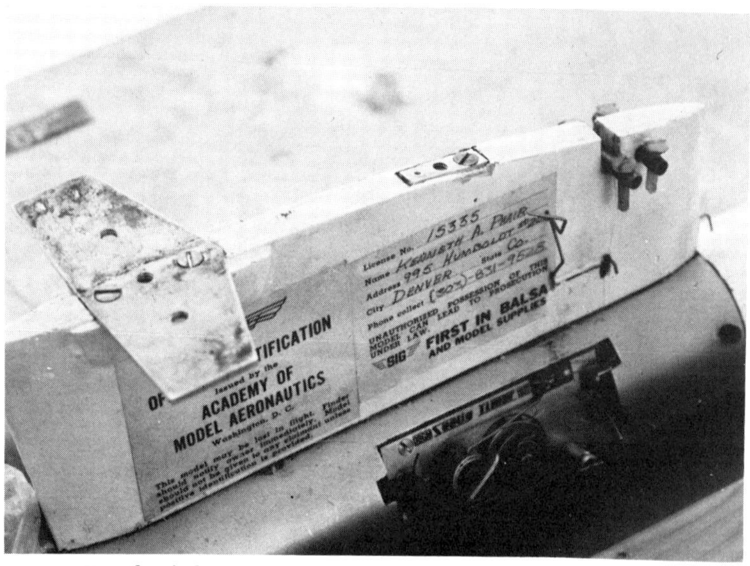

Ken Phair's flapper wing mount. Nylon screws hold wing.

Randy Secor.

Wakefield fans from Mexico.

The early rising moon looks in on the Taft night fliers.

George Batiuk Jr., Denver.

Gieskieng heaves away with flaps up.

Mel Schmidt fuels Shocer.

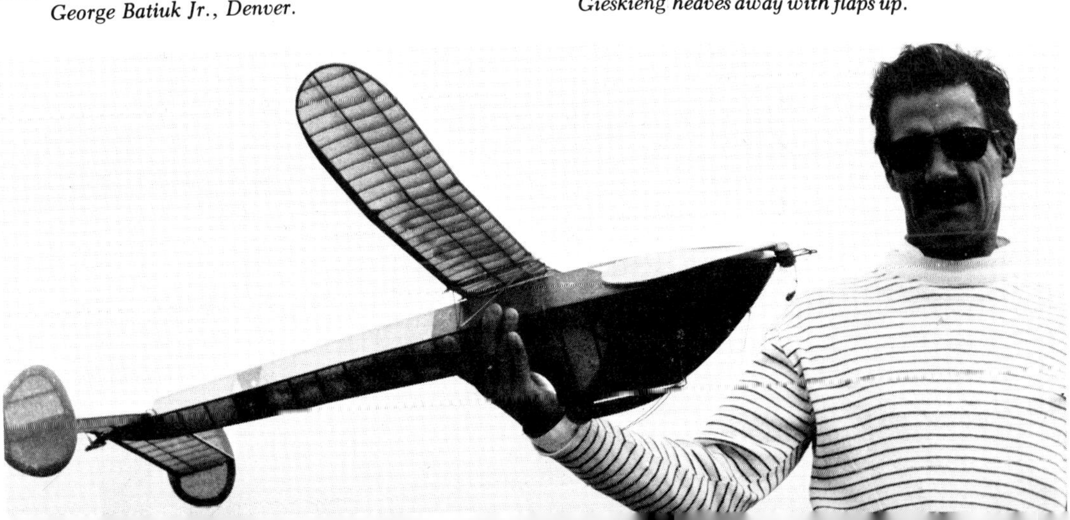

Vince Costanzo and his very neat "California Champ."

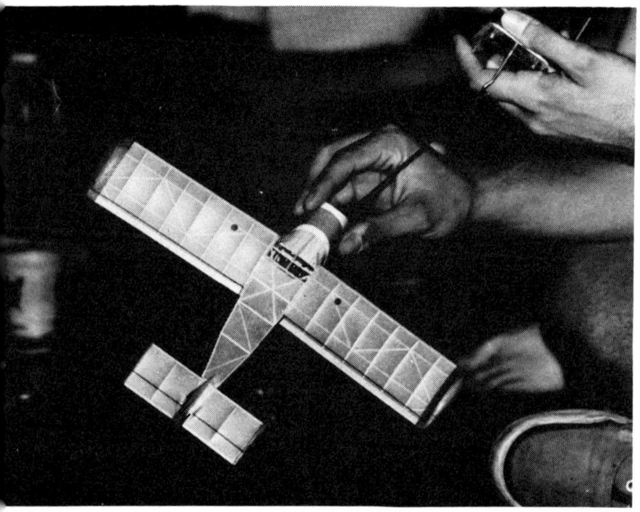

Indoor rubber scale scene at the USFFC, Taft.

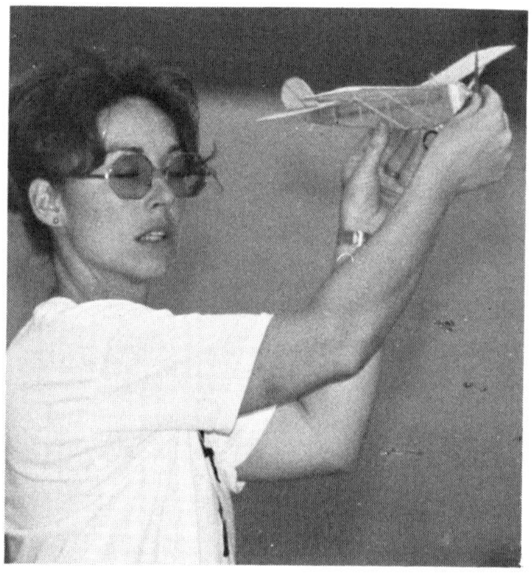

On the Taft gym floor to try out some hand-launch gliders.

Nan Taibi shows us Sal's Forestor 99 powered "Power House."

The young Hunter lad is oblivious of admiring glance by the pretty girl in background. Toö busy sorting out that Satellite.

Z. MACH PHOTOS

F1A fly-off group pictured at the 1975 Cup of Bohemia.

F1C winners at the 1975 Cup of Bohemia; J. Kaiser, C. Patek, K. Houcek.

F1A winners at the 1975 Cup of Bohemia; J. Pokorny, I. Horejsi, L. Vaculik.

F1B winners at the 1975 Cup of Bohemia; J. Nemec, H. Pernica, J. Krajc

Cup of Bohemia 1977 Wakefield winners, J. Loffler, 5th; J. Klima, 2nd; P. Chaussebourg, 1st; E. Mielitz, 3rd; J. Libra, 4th.

Cup of Bohemia 1977, Czechoslovakia, by Pierre Chaussebourg

This FAI contest is organized every two years, same years as World Champs. The competition was scheduled on August 26, 27 and 28, 1977. But, in fact the first day is reserved for model processing, practicing and opening ceremony. The third day is for departure of the contestants. So all three series are flown on the same day, Saturday the 27th. Accommodations and meals are provided in a school in Tabor, two miles from the flying field. One-hundred forty-eight competitors were registered, from CSSR, DDR, Poland, Bulgaria, Austria, France, Sweden and West Germany; most of the competitors being from Eastern countries. Seventy-four flew F1A, 32 F1B and 29 F1C. I was the only one to fly two series, Nordic and Wakefield. It did not cause too many problems, as the weather was beautiful and the flying site great. Also, the contest organization was perfect. There was allotted one hour and 15 minutes for each round and only five or six competitors in each group with two timekeepers equipped with binoculars. Also there were five minutes between each round to move (three poles), first round beginning at 7:00, last round finishing at 17:00.

Weather conditions being exceptionally good many maxes were scored. Nevertheless, it was necessary to be very careful about thermals; sometimes the models were going very high in the first minute, and you could expect a good max, then it was coming down like DT'd! Joachim Loeffler missed the third flight and scored 142, also Ivan Horejsi scored only 151 in the fourth round. Everything was going pretty well for the little western team, particularly for Bo Jansson (Swedish, 22nd in Roskilde) flying F1A, also Hans Lindholm, flying F1C. Personally, I was the first one to score seven maxes in the middle round as I had four maxes in Wakefield and three in F1A. It was going like that up to the end of the fifth round. There were 31 with a full score in F1A, six in Wakefield and seven in Power.

In the sixth round, I flew Wakefield first, and maxed. I had a long chase on this flight, and a lot of holes in the wing as the model DT'd in the straw. I preferred to concentrate on the last round in Wakefield, and patched the holes, put in a new motor, and then flew Nordic at the end of the round. I was flying the same model I placed 13th at the Pierre Trebod this year;

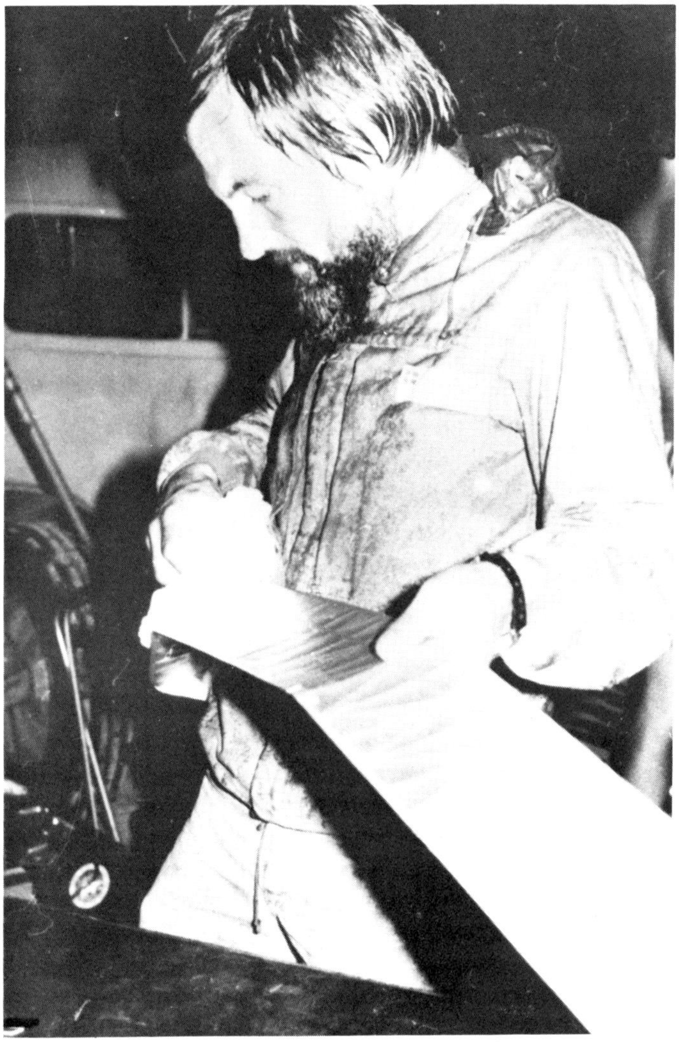

Pierre Chaussebourg waiting in the hanger for the rain to stop, after the first Wakefield fly-off (1977 Cup of Bohemia).

long tips and Paul Crowley tow hook. The air was cold and I didn't find anything and released just before the end, scoring only 126 secs. I understood the weather was changing, a storm was coming and I was very careful about picking good air in the next round. There were clouds, no more wind, only cold air. In this last round only 13 succeeded in making the seven maxes in A2, four in Wakefield, and three in Power. I had some problems to make this seventh flight. The Wakefield was flying well and was going to max, when at 15 seconds from the end of the three minutes, it was smashed to the ground by a glider tow line. The prop hub was bent, one fin broken, and many holes in the wing and stab. I had only 25 minutes to repair, remove the motor and wind, before the end of the contest. Well, I decided to repair; my wife Josette helped me, holding the stab and the fin when the epoxy was hardening, and I could make some tests to check the prop tracking and the prop pitch. Everything looked OK, so I wound and was ready to fly about eight minutes before the end.

The air was cold and Hans Lindholm, standing beside me confirmed my feeling. I waited for a while. My helper, Josette was already gone for retrieving, she didn't want to bother me at this very exciting moment . . . Well it seemed to me that the air was getting a little bit less cold, and the mylar ribbons were moving quite OK. I thought it was now or never! I launched, good launch, good climb, but no lift . . . The model was going down very slowly, and the time was going . . . but downwind, there was a very big tree, and the model was going that way, just to disappear behind that tree, at 2 mins. 58 secs . . . reappearing at 3 mins. 4 secs, and landing at about 3 mins. 30! I had even the time to fly Nordic. I zoomed it so strongly that I bent the wires and it was almost spiral diving to make 56 sec!

Pierre sets the D-T timer for 5 mins. in preparation for the fly-off with Joseph Klima (1977 Cup Bohemia).

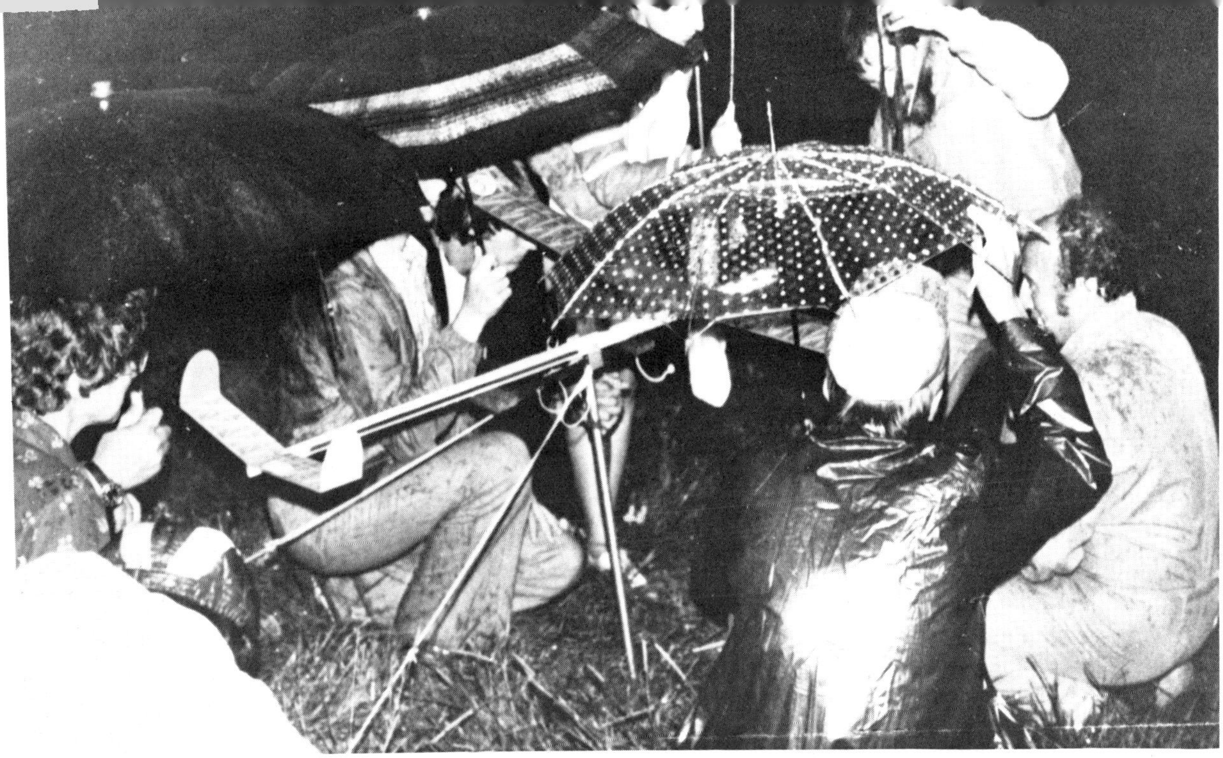

A tense moment as Pierre changes a broken motor. Prop in the mouth is good as, (a) it is safe from damage and, (b) it is impossible to talk.

The three fly-offs were flown together (F1A-F1B-F1C) and I had no time to see anything but my model. The storm was coming pretty fast, with big flashes and thunder. Klima and I flew the four minutes, so did Hans Lindholm in Power. Dvorac and Korczak (Poland) were in a second fly-off for 2nd place, while 1st place was won by Preuss, DDR; Bo Hanson took a very good 4th place. The 3rd Swedish on the field, Bjorn Soderstron, was unhappy for 2nd and 3rd rounds and came in 19th in F1B.

It was raining, and the organizers decided to wait for better conditions for the second fly-off. We stayed in a hangar for about half an hour. Then it looked better and we went out. The jury gave us a start and the rain started again! Klima was the first one to go, as I had to remove the motor; one strand had broken on the very first turns. My model climbed well and began to glide and was just staying in the air for 207 seconds. I was running when Ivan Horejsi who was chasing for me told me, *you are the winner*, I could not believe it!

That night, we had a banquet and a prize-giving with gold, silver and bronze medals, prizes and diplomas for the five top people. The prizes were superb, crystal from Bohemia, glasses, etc. Also, traditional band, waltz and polka . . . and very good beer! Plzen beer!

It was a very good moment when someone came to the French-Swedish-Czech table, holding Hans Lindholm's power model! He had lost it when he flew four minutes in the fly-off and had not found it. A small group of Czech modellers, ignoring the rain and the storm, finally found it: how about that! Hans and I had to sign many programs, and posters of the contest. A good moment also, when we popped the bottle of Champagne from France!

A great thank you to the organizer Vladimir Kubes and his team; also many thanks to Jiri Kalina, the President of the Jury, and the timekeepers. It was a great contest in a beautiful country.

I should like to mention the sportsmanship of the Czechoslovak modellers when they came around me with umbrellas, protecting my plane from the rain. When I was winding, not one drop of rain touched my rubber. Also, when I was coming back, after the last fly-off, with Ivan and a young boy who had been the first one on my plane, somebody came from the starting line, telling me I had won, and also to help me with an umbrella. This is something I'll never forget.

Josette and Pierre receive the congratulations of Joseph Klima after winning the Wakefield fly-off, 1977 Cup of Bohemia.

92 London / Amsterdam – Cologne – Belgrade – Athens / Istanbul

↑ 101	A	B	C	D	E	F	G	H	J
	R	R	R	R	R	R	R	R	R
London Vict. . . d							11 00	11 00	14 45
Oostende . . . . . . d	. . .	. . .	. . .	8 51		. . .	a17 20	18 51	21 03
Bruxelles Midi . d	6 42	. . .	. . .	10 04		. . .	18 42	20 04	22 19
Köln . . . . a	9 10			12 49			a22 05	23 01	1 35
London Liv.Str.d								10 20	
Hoek van Holl. . d	. . .	. . .	. . .	. . .		. . .		19 31	
Rotterdam . . . . . d	. . .	. . .	. . .	. . .		. . .	3)17 40	19 57	
Köln . . . . a							20 42	23 05	
Amsterdam . . . d				9 17			17 50	Austria-Express	
Den Haag C.S. d	. . .	. . .	. . .	9 03		. . .	17 34		
Utrecht. . . . . d	. . .	. . .	. . .	9 50		. . .	18 29		
Köln . . . . a				12 35			20 42		
Köln . . . . d	9 18			13 07			a22 10	23 29	1 45
Stuttgart . . . . . . d	14 21		17 21	17 46	18 26	19 37		4 20	6 19
München . . . . a	16 57		19 59	20 30	21 25	22 16	6 09	7 30	9 02
München . . . . d	17 18	17 30	20 15	20 50	21 38	22 51	6 21	7 56	9 22
Salzburg . . . . a	18 58	19 11	21 56	22 37	23 22	0 34	8 17	9 39	11 12
Salzburg . . . . d	19 20	19 30	22 24	23 05	23 50	1 01	8 42	9 55	11 40
Villach . . . . d	1)22 20	1)22 57	1)1 37	1)2 25	1)3 06	4 35	12 20	1)12 58	15 30
Jesenice . . . . . . a	23 06	23 45	2 33	3 14	4 02	5 29	13 17	13 47	16 17
Ljubljana . . . . . a	0 29	1 14	4 00	4 52	5 31	6 58	14 46	15 15	17 46
Ljubljana . . . . d		Orient / Tauern			7 00	8 09	15 18		
Opatija . . . . . . a					9 22	10 04	17 18		
Rijeka . . . . a					9 33	10 15	a17 28		
Ljubljana . . . . . d	0 41	1 24	4 20	5 08	5 58	7 53	Rijeka-Express	15 25	18 15
Zagreb . . . . a	2 51	3 32	6 34	7 25	8 10	10 12		17 36	20 27
Zagreb . . . . . . d			7 05					Akropolis	21 05
Split. . . . . a			15 10						6 50
Zagreb . . . . . . d						11 05			Tauern-Express
Ploče . . . . a						23 23			
Zagreb . . . . d	3 11	3 48	Dalmacija-Express	8 05	9 00	Mostar-Express		17 56	
Beograd . . . . a	8 32	8 40		12 58	13 55		. . .	22 46	
Beograd . . . . d	Jugoslavia-Express	9 25		14 00	Istanbul-Express			23 16	
Niš . . . . a		12 55		17 36			. . .	3 01	
Gevgelija HEC a		20 00		0 20				10 10	
Gevgelija HEOr d		21 40		2 00				11 50	
Thessaloniki . . . a		23 50		4 40			. . .	13 42	
Athènes . . . . . . a		9 40		15 10				22 05	
Beograd . . . . . . d		9 37		Hellas-Express	14 55				
Niš . . . . a	. . .	13 12	. . .		18 35				
Dimitrovgrad a		15 25			20 48				
HEC d	. . .	16 00	. . .		21 15				
Sofia . . . HEOr a		18 23			23 33				
Sofia . . . HEOr d	. . .	18 50	. . .		23 59				
Svilengrad . . . . . a		23 44			4 26				
Istanbul . . . . . . a		8 25			13 37				

Column B: Athènes-Exp. (Beograd–Athènes); Marmara-Express (Beograd–Istanbul).

TEE

Trans-Europ-Express

Two doubled-up steam high-wheelers, the working remainder of the grand and mysterious Orient Express, roar through the outskirts of Plovdiv.

5 Free Flight Adventure

Free Flight to Plovdiv on the Orient Express

The gold lettering on the side of the royal blue *Wagon-Lits* coach spelled out, *Compagnie Internationale des Wagon-Lits et des Grande Express Europeens.* Struggling with the model box and three pieces of luggage I made it along the platform of the *Gare de l'Est a Paris.* This must be it, I thought, as I headed for the coach entrance steps. Yes, there it was on the destination board. *Simplon Orient Express - Trieste, Ljubljana, Belgrade, Sofia, Istanbul.* A high-pitched scream from the locomotive up forward announced I had better get with it, so up the steps I scrambled. The loco's light flashed on, reflecting down the double ribbon of steel pointing eastward and at exactly 7 P.M. we smoothly accelerated out of the station and off into the night, aboard the fabulous, mystical, *Orient Express.*

That's the way it was going to be, I mused, reading over the latest letter from Paul Masterman. The '75 Free Flight World Champs were scheduled for Plovdiv, Bulgaria — Where the hell is that? And the English were organizing a group travel plan from London to Plovdiv. A number of memorandums and surveys had been circulated by Paul to determine what travel plan could be agreed upon. A chartered bus was vetoed with visions of sitting on some lonely mountain pass with the bus's rear end strewn about. Air fare looked a bit steep and so that was ruled out. That left the train. But then what a train! *The Orient Express.*

I started making my plans which included the trip to Plovdiv, the French *Pierre Trebod* Contest, and a tour of England. In no time it was July 28 and I was at Los Angeles International Airport boarding an Air France 747, TGC charter. Take off was at 11 P.M., and at first the cabin seemed crowded and seating close. But after a few drinks and dinner it seemed quite comfortable.

Melon au Porto
Brochette de Boeuf Bordelaise
Riz Pilaf Salade DuBarry
Fromage Patisserie
Wine, Champagne and Cocktails
were complimentary.

A French film was shown, a comedy. *The Mad Adventures of the Bouncing Betty,* le petit Baigneur. Before we knew it we were landing at Montreal to

AIR FRANCE

Fountains Trocadero from La Tour Eiffel.

City of Light, par La Siene.

refuel. (For the Wakefield fans a "Montreal Stop".) Back soon into the morning air we were on our way to Paris. Breakfast, lunch, more refreshments, stereo music and some sleep kept us pretty well-occupied, staving off cabin fever till the bitter end.

At 7:50 P.M. we landed at the vast, very modern Charles De Gaulle Airport, and I stuffed the model box, camera bag and two satchels and myself into a bus aimed toward Paris. After transferring to a taxi, I found a small hotel in the Madeleine district near Place de la Concorde and crashed, nursing my jet-lag until the next afternoon. I spent a few days in Paris photographing the delightful scenes and visiting such places as the Louvre, Musee de L'Air and French AMA. But that's all another story. My journey then took me by train to Calais thence by ferry across the Channel to Dover and train to London. I spent another five days knocking around England in a rented Mini 1000.

On July 12 the plan was to rendezvous at Victoria Station for the big trip. I drove up and found the English group waiting outside along with a huge pile of model boxes and luggage. I added my stuff to the pile and raced away to turn the car in. Train time, 2:30 P.M., was rapidly approaching so after struggling through the red tape of the car turn-in, I jumped in a cab and beat it back to Victoria Station.

Outside, no English Team Members, no luggage. Where were they? How about that Pub over there? After two, or was it three, pints of ale I felt a bit more relaxed. From the Pub terrace, looking across the busy and anachronistic but essential edifice I could see the sleek cars of *The Night Ferry* at track 2. It's a train, not a boat, and a British/French institution.

Here it was almost 10 P.M. so I finished my ale and made my way on out to step on board what is considered a most civil way to journey to the Continent. I swung aboard, found my compartment and noticed I could make the trip in my very own bed. I wasn't ready for that, however, so I stepped into the corridor to be greeted by the English Conductor; "Good evening, Sir". As the train pulled out and rolled through the humpty-dumpty London surroundings I made my way to the dining car. "Bonsoir monsieur" says the French waiter and it is dinner time through the quiet soft green Kent countryside. At Dover the cars are shunted onto a ferry where they are chained and anchored down for the sea journey to Dunkerque. At the Dunkerque dock the cars are rolled onto the main-line again, recoupled and the train continues off to Paris to connect with the *Orient Express*. There are no stops, no people getting on or off, no customs delays, just calm, sensible and considerate transportation. I finished the dinner in unhurried leisure and sipped my pint of ale. Now, where were those English Team Members? What could have happened to them?

I paid for my ale, tossed off my daydreaming about *The Night Ferry*, and looked up and down the Victoria Station platforms. What's that mob over there; yes, that's our group. I recognized Australian Tom Prosser, he's easy to spot, and there's Martin Dilly, Brian Spooner, Paul Masterman and all the gang -some 24 in all. The boxes and luggage were stacked on some carts that had been commandeered and we were in the forefront of a flying wedge ready to assault the barrier and get on the train. All that was lacking was any certainty as to which was the correct train.

This was to be the first test of our group leader Paul Masterman and, as it turned out, we got everything and everyone on board. Not without some consternation from the conductor and baggage chiefs, but that was soon to become a standard pattern. This was also our first try at stashing the model boxes where they wouldn't be stumbled over. We soon found that Bob Bailey's F1C box was going to be a problem. It wouldn't go through the window, around corners or fit in a coachette. I think it ended up in the baggage car, with Bob visiting it periodically to see to its safety.

The train rattled its way along and in two hours we pulled into the Dover Port Station. This was our first go at unloading all the gear and it didn't go without a bit of confusion and crashing about. We were not permitted to carry the model boxes on board the ferry. They were loaded onto wooden pallets, nets were thrown around them and they were hoisted aboard like so much simple cargo. We stood about trying to exercise some control over the luggage loading but that was rather futile as the stevedores seemed a bit single-minded about doing it their way. The ship was the *Princess Paola* of Belgian registry and we were headed for Oostende, Belgium. Paul checked us all aboard and up the gang plank. He had all the tickets so we didn't want to lose him.

Heavy vibration of the deck signalled that we were under way and we eased out through the opening in the breakwater. Now you could see the white cliffs stretching away to the eastern flank of Kent. In this region are located the *Cinque Ports;* Sandwich, Dover, Hythe, Romney and Hastings. These five ports grouped together for mutual protection in the period preceding the Norman Conquest. The fishing, trade and smuggling carried on from these ports became the nucleus of a Navy which eventually came under the control of the British Crown. At Dover a great castle, built on the bluffs by William the Conqueror, overlooks the harbor. We could see it easily as we sailed out into the Channel. The earliest part of the structure dates from the Roman period. Of interest to free flighters is that Louis Blériot in 1909 made the first aerial crossing of the Channel, flying from Calais, landing in the meadow behind Dover Castle. Bleriot held the very first license issued by the Federation Aeronautique Internationale.

This late afternoon the Channel was calm and most of the group found the top deck invigorating with the fresh breeze and salty tang in the air. Seagulls soared along taking advantage of the deflected slipstream around the ship to fly with a minimum of effort. The camera was soon in action with my 70-210 zoom lens trying to fill the frame with these, genus *Laurus*, artists of free flight. Certainly a beautiful sight, showing off their mastery of flight, about which we really know so little. They looked so proud. The activities of the day and the fresh nautical atmosphere stroked our appetites and before long most of us were in the dining room on the restaurant deck. The food was first class and although we were unaware at this time, this was to be our last such fine meal on the journey.

Nearing the French coast the course took us past Dunkerque. The Gothic church tower of St. Eloi, 295 ft. tall and built in 1560, was easily visible as was the lighthouse at the breakwater and canal entrance. The dramatic rescue of the British Armies from the Dunkerque beaches in May 1940 is well known. Remnants of the German fortification from WWII and WWI are visible in the dunes. The sun was a red ball

Entering the harbor at Oostende, Belgium.

Oostende quay.

City spires.

Free Flight companion.

in the sea back in the direction of England as we approached the harbor entrance to Oostende. The Captain pirouetted the ungainly mass of *Princess Paola* through a 180 and proceeded to back into port through a break in the *Digue de Mer*, the sea wall that stretches out for 9 miles to protect the coast of West Flanders. We were soon trooping down the gangplank looking to the retrieval of the luggage and model boxes.

The train platforms were only a few short steps away from the dock and Masterman appeared to have no difficulty finding the ***Tauern Express*** a major element of TEE, the *Trans-Europ-Express* system which has emerged as the successor to the *Compagnie Internationale des Wagon-Lits et des Grande Express Europeens.* We clambered aboard shoving the model boxes wherever they would fit. There was some question as to what cars we should be in, but that would be sorted our later. First rule is to get on board. At 21:03 the show was on the road humming and clicking off into the eastern darkness. A number of the group preferred standing in the corridors of the coachette cars, hanging on the open window sills. The closeness of the world flashing by, the cooling breeze, the noise and motion gave an immediacy to this adventure. Who could sleep?

First stop was Bruxelles which only took a few minutes. Rolling through the outskirts we saw a large circus/carnival set up in a square. Lots of lights, music and laughter from the crowd even at that late hour. Further into the countryside we were treated to pungent stretches that made it clear that the local *ferme cochon* were quite prosperous. The border to Germany was crossed at Aachen and the train followed the Rhine through Koln, Bonn, Koblenz and Mainz. The conductor came through the cars selling beer which was quite acceptable. Eventually most of the group decided to try out the coachettes in the sleeper version. They make up into six bunk-type beds and the conductor supplies a sheet, pillow and blanket. We weren't crowded so the model boxes went on the unused bunks.

I woke up around 07:30 to find we were about halfway between Stuttgart and Munich. At 09:02 we arrived at Munich and were backed into the large *Bahnhof.* I jumped off and got a large sandwich from a pushcart vendor and while eating this breakfast managed a few photos of the station. In thirteen minutes we were off again toward Salzburg. The scenery at first was misty and wet looking but this soon changed to craggy mountains, alpine meadows and sunshine. At Salzburg we learned we were to change trains, and the schedule only allowed a few minutes. Model boxes flew out windows, carts were commandeered and then we were told we had to go through customs check. Also the porters wanted six Austrian shillings for each bag and box. We finally settled for a group rate and Masterman talked the

Martyn Cowley surveys luggage carts.

Out of the night, Orient Express departs Oostende.

Change cars once again.

Gregorie and Bailey run with the beer at Munich.

Munich Bahnhof.

Bavarian countryside.

customs people into letting us go right on board without check. We scrambled aboard without a moment to spare. Needless to say we were in the wrong cars so had to move forward a number of cars banging the model boxes through the heavy sliding doors and across swaying-wiggily plates between the cars.

We found compartments to our liking, settled down to catch our breath, congratulating ourselves on passing that hurdle. The conductor came through collecting tickets. We found that the solution to this dilemma was to point to each other and say *sportzen-groupen*, and then point to the next car and say, *groupenfuhrer*. A cup of coffee on this train cost 2.10 Marks. There weren't many takers. Hunger was getting to us and it didn't look like a dining car would be added on. At Bischofshofen we stopped for 10 minutes. An old lady with a pushcart of food, flowers and drink, hawked her wares on the platform. I tried to make her understand that I wanted a boiled egg which occupied a prominent spot in the cart. It wasn't till I gave an imitation of a chicken that she caught on and we completed our transaction through the window as the train rolled out.

My lunch was made up from some emergency rations I had brought from England. A can of pate, bottle of Schweppes Bitter Lemon, English biscuits, a banana and the boiled egg. Delicious. We passed through some very pretty country with the train clinging to one side of a mountain pass. The view across the valley was magnificent. Now we were smack in the middle of the Tauern mountain range in the Austrian Alps for which this train got its name. Gross Glockner, at 12,457 feet is Austria's highest peak and was visible from our rolling club house. Suddenly the lights go out and the confined rattle and roar gives everyone a start in the second or so it takes to realize we are in a tunnel.

Jill and Debra Collins catch the refreshing breeze.

The beer is running low!

Faces in the night.

Bob Bailey guards lunch with his British umbrella.

It was quite long, as we lurched through several thousand feet of tunneled rock, bursting out again in the bright sunshine on the southern side of this great Alpine barrier.

Spittal-Millet to Villach it was all downhill and the brakes got a real workout. On arriving at Villach we sat in the station for some time. The wheels were red hot, smoking, snapping and crackling from all the braking they had gone through. We watched a maintenance crew walk along the train with long handled hammers. They would tap each wheel in turn. The wheels gave off a bell-like ring which seemed reassuring, but then we heard one wheel go *thunk*. The men kept moving along though with no concern for that one. I guess the wheel would have to break and fall at their feet before any attention was needed.

With the wheels and brakes cooled off we rolled out onto the mainline to cross the Austrian/Yugoslav border a few miles south. The right of way was bordered with patches of wild strawberries alternating with delicate mountain flowers. Our direction took us along the edge of a long green valley feeding the Sava River. Stops were made at Jesenice and Lesce Bled. That name sounded familiar. Yes, in 1951 the Swedish Cup Competition (F1A) was held in Lesce Bled. Oscar Czepa of Austria won with his unorthodox but cleverly designed *Toothpick*. The fuselage was 78 inches long, 10 inches more than the wingspan. Fuselage cross-sectional area, as required in those days, was taken care of with a pod at the tail end. The next year Bora Gunic went to the '52 competition in Graz, Austria and brought the Swedish cup back to Yugoslavia. The resulting competition in '53 resulted in a win for Hans Hansen of Denmark, who maxed out. That was back in the days when towlines were 100 meters long, 300 secs was max and three rounds were it.

The Lesce Bled flying site is the sport aviation center located near picturesque Lake Bled. Mountains surround the site on three sides, medieval castles and churches are perched about. Some of the non-flying activities included riding in the horse-drawn *droshkys*, sailing in the lake, sampling *sljivovica*, the national institution. In '53 the English Team had traveled to Lesce Bled by the very train we were now on. They had some bad luck, the model boxes disappeared and never made it to the Championships. Hopefully the phantom of the baggage car would not strike again.

Another thirty miles down the mainline we pulled into Ljubljana, the Capitol of Slovenia, one of the six Republics making up Yugoslavia. This region is the birthplace of Frank Zaic, the grand old man of free flight. The Zaics emigrated to the U.S.A. when Frank was a boy. But apparently the grandeur of these meadows, mountains and sky had some subliminal effect on Frank, sharpening his questioning mind and creating an awareness of the challenge of the atmosphere.

The stop at Ljubljana gave us a chance to stretch our legs a bit on the platform. With all the changes taking place in the rail system, equipment and management there had arisen a certain amount of confusion as to which is the real *Orient Express*. Some of the sign boards on our train said, *Direct Orient-Paris-Lausanne-Milano-Venezia-Beograd-Sofia-Istanbul*, others said, *Tauern Orient Express-Munchen-Ljubljana-Beograd-Athenes*. Also the printed time table lists the *Marmara Express*, *Istanbul Express*, *Direct Orient*, *Simplon Express* and *Tauern-Orient*. Very confusing. *Simplon* refers to the *Simplon Pass* and tunnels in Switzerland that were the key to opening a direct line, Paris to Milan. This route is the one more generally associated with the *Orient Express* as glamourized in literature and the cinema.

Origin of this unique institution is found with the Belgian, Georges Nagelmackers, who in 1876 formed the *Compagnie Internationale des Wagon-Lits*. He was able to put together a remarkable coalition of politicians and railroad men. He managed to transcend nationalistic rivalries, jealousies and unmatching track gauges to eventually link all of Europe with silver bands of steel *giving flight to great chariots thrust forward upon the vapours of boiling cauldrons*. The inaugural run, Paris/Constantinople, began October 4, 1883, at the Gare de L'Est. The panting 2-4-0 locomotive trembled to be away on this great adventure; clouds of steam wafted over the two sleeping cars, dining cars and two baggage cars. Everything gleamed in the royal blue enamel, polished teak, brass and steel. Nothing was spared in seeing to the comfort of the passengers as so much hinged on the success of this journey. There needn't have been any fear of that though as success and acceptance came in abundance in the following decades.

The inaugural run bypassed the Alps by swinging north through Germany, Austria, Hungary, Romania and Bulgaria. Disruptions and prejudices stemming from World War I and the plans put forth by the Allied Governments, post-Versailles, put pressure on opening a new more direct route across northern Italy. This plan became reality when German and Swiss engineers blasted a 12⅓ mile tunnel through the *Simplon Pass* near Brig, Switzerland. The*Simplon*

Gross Glockner.

Tunnel was opened for traffic in 1905. At about the same time work was started on the huge railway station in Milan which boasts a marble and granite main concourse 700 feet long. With this new alignment, travel time was cut from 96½ hours to 56 hours, Paris/Istanbul, by way of Milan, Venice, Trieste, Zagreb, Belgrade and Sofia.

Our grand tour organizer, Paul Masterman, had determined that the *Simplon Orient Express* would have been a bit dicey for us as it meant transferring from Paris (Nord) to Paris (Est) railway stations. That would have created problems in getting the model boxes carted across Paris and we might have lost valuable team members to the lures of Montmarte and Pigalle.

Arriving at Zagreb at 20:27 we got the word that there would be a 1½-hour wait for our train connection. Through some sort of railway higgery-jiggery we would be integrated into the *Simplon Orient Express* eastward bound. With great military precision we off-loaded the model boxes and luggage through windows and doors. Soon we had this humongous pile of bizarre looking stuff taking up most of the Zagreb *stanicu ulaz*. I devised the brilliant plan to split our forces in two groups. The first would stand by and guard our precious baggage while the second would have dinner in the station restaurant. There was no question that we were famished so this brief interlude in Zagreb looked like a lifesaver.

Soon a clutch of our British friends emerged from the restaurant with a horror story. The menu was in some strange foreign language and completely unintelligible. Their courage failed them and they were afraid to take a chance. We had been drooling, thinking of our chance at eating (anything), so took their place in the restaurant. We tossed aside the menu, found our waiter and discovered he spoke excellent English with superb modulation and inflection. Needless to say we had a quite agreeable repast of veal, chips, salad and fine Jugoslavian white wine. We handsomely rewarded this fine servant and rejoined our timid comrades on guard duty. We ex-colonialists took the opportunity to point out to our friends the disadvantage of a social system that had not prepared them for coping with starvation in the midst of plenty. We topped off the round of jest by discovering a beer kiosk on the platform. A triumph of American ingenuity, improvization and know how!

By this time with much whistling, bumping and grinding of cars in the dimly-lit train yards, our train was ready. The boxes and baggage were carted aboard and we found that this was a *Deluxe Wagon Lits (Schlaufwagen)* with very nice sleeping compartments, not the inelegant coachettes we had ridden in so far. Loading was completed when there then ensued a spirited discussion between Brian Spooner (British Team Manager) and railway officials. Seems that the model boxes violated the sanctity of this fine train in some manner and this would not be permitted. They must be put on some other local train with the peasants, goats and chickens. Apparently these officials enjoyed yelling and shouting about all this and,

Stop at Villach, Yugoslavia, to cool the brakes.

Exciting panorama.

of course, we were equally determined that the boxes stay with us. After fifteen minutes of intense argument we were forced to compromise. After all, they had guns, and they wouldn't let the train start.

Brian Spooner and Aussie Tom Prosser volunteered to go with the model boxes on the other train. They ended up sleeping on top of the boxes, off into the night to some questionable destination. We climbed back into our deluxe compartments and barricaded ourselves in, suspecting that this really wasn't our train and we would be thrown off in some Serbian backwater. The compartments had telephones, wash sink, individual reading lights, crank-up window, coat hangers, soap, towels, three bunks per room, etc. There was little opportunity to enjoy this comfort for we awoke from a sound sleep around 6 A.M. and saw we were coming into the Belgrade station.

That was the end of the fancy train and we had to move everything off and on to the platform while Paul Masterman disappeared somewhere to deal once again with the authorities. Eventually we were reunited with Spooner, Prosser and the model boxes and were on our way once more, now nearing Bulgaria. A few hours out of Belgrade, Customs men poking about in the luggage discovered a cache of bundled up team uniforms in the British kit. Spooner was accused of importing clothing into Bulgaria without a license. Brian's explanation that these were the uniforms of the nine-man sports team representing Great Britain at the Free Flight World Champs was accepted by the Customs Agents. Although to the rest of us it seemed a bit implausible.

Change again at Sofia, a weary group tries to recover.

At Dimitrovgrad on the Bulgarian frontier, we stopped and were shocked to see model boxes being *thrown* out of the baggage car on to the platform. Some of our group raced forward to stop this madness and a big argument developed. One of the baggage handlers jumped on to the model boxes after throwing them to the platform. This had an immediate reaction from those on board and box owners were jumping from the train and running to the scene. The officials had to show who was in charge of this situation, however, and they herded everybody back on board and locked the doors.

Just why these boxes should be put on another train was beyond comprehension. Apparently some government bureau dictated that luggage of a certain size rode on certain trains, period. The lackeys charged with carrying out these rules were not allowed to exercise any judgment or common sense as far as we could tell. The impasse was broken finally when we learned that with payment of some fees the boxes could go along with us. Miraculous! With boxes back on board, fees paid and passports checked we trundled on down the line and across the border. It was marked by a double steel fence stretching off to the horizon. Either side of the fence, the land was cleared flat to make a no-mans-land. The train ran parallel to the main highway here and at the border crossing check point you could see a long line of cars and trucks waiting to be admitted to Yugoslavia. We learned later that we were following the major overland trade route between industrialized Europe and the Middle East.

Yugoslavia is now planning to build the key link of a four - lane major superhighway linking the highways of Italy and Austria with Greece, Turkey and certain Middle East countries. This is long overdue as the two-lane road built shortly after World War II is generally regarded as the most hazardous and poorly maintained road on the continent. Those that drove to Plovdiv confirmed that opinion with wild tales of their experiences.

In a few hours we arrived in Sofia, the capitol of Bulgaria. We had some time to spend so Ron Pollard (the British Wakefield flier) and I wandered around looking for some breakfast. We found a sidewalk cafe a few blocks from the station which had a good view of a wide street and square. We started with some coffee. It turned out to be Turkish, heavily sweetened and served in tiny cups with a mineral water chaser. It tasted like valve grinding compound. That was followed with a plate of sausage and pork bits, bread, and two big cold chiles. This was all washed down with a half litre of cold beer.

The streets were busy with a lot of battered heavy trucks bouncing along belching black smoke. The pedestrian traffic was heavy with the going-to-work crowd nimbly making their way across the cobblestone streets through breaks in the traffic. Many were stopping at kiosks to have a *Banitsa*, a very sweet and sticky pastry. This close look at our host country had to be cut short as it was time to get back to the train and be off to Plovdiv.

Loaded again into our fine reserved coachettes, we were soon rattling along through the Bulgarian farmland. There were many small plots of corn, cabbage and other vegetables, all rather flat and green country but with some mountains guarding the horizons. Our route took us through the valley of the *River Maritza*. To the north was another parallel valley feeding the *River Strjama*. That region is known world-wide as the Valley of Roses. This sun-flooded valley nestled between the gentle slopes of the Balkan Range and the Sredna Gora Mountains is literally covered with roses that were first brought here centuries ago from Persia. They are now grown here as a commercial crop. The roses are used in a variety of products including perfumes, soaps, and wines. This is a major tourist attraction in Bulgaria and there are many interesting folk programs and pageants that celebrate the traditions and customs of the people.

Here and there we spotted fields that would make good free-flight flying sites. The weather seemed to be holding and everyone was getting anxious to get there and tend to the flying. All of Western Europe had been sweltering for months in a heat wave. Crops were withering and everyone, except us, had been anxiously hoping for some rain. The pessimists, of course, felt that the deluge was simply awaiting the start of the competition.

This train looked more like a Local than an Express as a number of stops were made at unlikely looking spots. For a while we had a band of Gypsies riding on the platforms between cars. They were sitting sprawled all over the place and were travelling *sans-ticket*. They disappeared somewhere along the way having decided apparently not to field their team and compete with us at the World Champs. A pity, as the party afterwards would have been a real deuzy.

Ron Pollard has a Balkan breakfast.

Way station outside Sofia hosted our train for thirty minutes in wait for priority traffic.

Sofia, Bulgaria, train terminal.

As we anxiously looked at the sign boards of each station, a question arose. How will we know when we get to Plovdiv? All the signs are in the Cyrillic alphabet and we haven't seen what it looks like. Some of our more intellectual types tried to come up with the answer by attempting to re-invent the Cyrillic alphabet from scratch just by starting from basic reasoning and logic. The heroic attempt failed and the resulting word structure had to be suppressed for reasons of obscenity. No doubt our wordsmiths had been overly influenced by some of the interesting examples of literary profundity scrawled along the right of way for our attention and reading pleasure.

Well we did finally arrive and we did recognize our destination simply because Australian Dave Anderson happened to be standing on the platform. Of course it could have been a terrible mistake as this might have been Adelaide, but we took a chance and got off. We were 55½ hours of solid travelling out of Victoria Station across all of Europe on the *Orient Express*. Now it was time for a party.

The Bulgarian organizers had a deluxe bus, interpreter and pretty young hostess to greet us. In fact they met all the trains. We were carted off into the night through the shadowy streets of Plovdiv and arrived at a twin-tower, fourteen-story hotel blazing with lights. It was set by itself in a little park and was reserved exclusively for the World Championships. At reception we were assigned rooms which slept two and were equipped with toilet and full bath. My preoccupation at this point was with the hotel bar, which was filled with many old friends. The party was enjoyable and I don't believe it ever ended for the duration of the competition.

The next morning we found that a fleet of a dozen deluxe buses was at our disposal and their first task was to take us to breakfast. Our communal meals were taken in the dining room of a college in Plovdiv about two miles from the hotel. It was a new, modern building but the street outside and landscaping had not been completed. Inside, tables were set with national flags identifying our places. Young girls and boys, volunteers from the schools, brought us our food quickly. Their appearance and spirit was fresh and genuine, providing a most hospitable atmosphere. Table assignment of the young students was based on the particular languages that they were studying so no table had a language barrier.

After breakfast I went to the Trimontium Hotel in the center of the city and got my Zhiguli rental car, a Russian-built Fiat. A large number of these cars had been brought from somewhere to satisfy the Westerners' obsession with private transportation. Now I didn't have to depend on the buses to get around. Next stop was the flying site. With some Australian hitch hikers we set out following the map. First surprise was the sight of so many flags, banners and posters proclaiming the Free Flight World Championships. In the city they were hung from every lamp post. Along the highway to the field, they were never out of sight. The field was Krumovo airport, 11 km to the south of the city. It was a large grassy area of several square miles. There were no paved runways and no full-size aircraft around, except for a few helicopters. The grass hid a few thorny thistles here and there. This site had been used a number of times for the East European Championships but this was the first time western fliers had been here.

Some of the earlier arriving Americans got themselves in trouble when they tried to locate the alternate practice field. They were arrested and questioned by the police for over an hour. On release there were no apologies or explanations given by the authorities. The incident was considered *routine*. The only difficulty I had was on the dirt road entrance to the Krumovo field. I stopped to take a picture of the area and a gnarled old man of about 80 years or so pedalled up on a rusty old bicycle, making threatening motions and sounds. He had a rusty rifle slung over his shoulder also, the barrel of which looked about eight feet long. I wanted to get a picture of him too, but he seemed to think I should be intimidated, so I obliged.

After the morning practice session I drove back into the city to attend a press conference. This was put on by the contest officials, city officials and Bulgarian press officials, and consisted of a welcoming and background briefing. From what we heard it was apparent that this World Championship was being promoted and supported by the highest levels of Government. They saw in this meeting an opportunity to promote a spirit of international cooperation in the framework of a competitive sport. Considering all of the other more depressing ways that international cooperation, or rather the lack of it, tends to be drawn, this approach by Bulgaria was greeted by us with much enthusiasm. Of course, it was recognized that by being such a generous and friendly host, little Bulgaria would benefit from the respect and admiration shown by many of us from other countries.

After this meeting I had a chance to walk around the city and see the sights. Plovdiv, the second largest city in Bulgaria has a population of around 700,000. The oldest portion of the city, founded several hundred years before the birth of Christ, is perched on the cliffs overlooking the Maritza River. A citadel was built here by Philip of Macedonia (Alexander the Great's father) to guard his Thracian gold mines. He called it Philipopulis. For 23 centuries this city had been a commercial half-way meeting place on the main caravan trail between Europe and the Turkish land bridge to Asia and Africa.

This area had been subject to a number of incursions of contentious tribes and nations over the centuries. The most recent example is that of 1944 when the Soviet Union declared war on Bulgaria. A Communist coalition gained control of the govern-

View of Plovdiv, Bulgaria, from the World Champs hotel.

Director of the Championships, genial Peter Stankov, Secretary of the Bulgarian Federation of Aircraft Modeling,

Deluxe bus service for participants.

ment as Soviet forces were marching into the country without meeting resistance. Since that time the country has been ruled by the Bulgarian Communist Party. Bulgaria is considered to have the closest ties and friendship with the Soviet Union of any nation.

Although the economic conditions and standard of living of the country have been rapidly rising, these levels are visibly lower than the Western industrialized nations. Principal economy of Bulgaria is agriculture which accounts for 25% of industrial output. The country is not densely populated and has the highest ratio of arable land to population in Eastern Europe. Plovdiv itself is an important trade and agricultural center. It is the site of a major international trade fair in which East and West trade is given a two-way exposition annually in September. The Sofia University has one of the richest paleontological collections in Europe. One of the most interesting exhibits is a Deinoterium which was a kind of 'harmless' but giant cousin to the elephant. It was found near Plovdiv, long since expired of course, and it is the only completely collected specimen in the world. The city of Plovdiv has been called Eurmolpia, Phillippopolis and Trimontium in ancient times. Hence the name for the premier hotel in Plovdiv, the Trimontium. The stores had little on display compared to Western standards. Goods in plentiful supply included tobacco, embroidery, woodcarving, pottery, furs and, of course, attar of roses.

While walking across an open plaza, a military helicopter flew over and dropped pamphlets to shower down and swirl around everywhere. It was an announcement of the *Free Flight World Championships* and invitation to the opening ceremony to be held the next evening (Saturday, August 16) in the Plovdiv Sports Stadium. After another evening of partying and meeting free flighters from around the world and then another day (the last one) of practicing and checking of models we were ready for the opening ceremony.

After an early dinner everyone got dressed up in their team uniforms and boarded the buses again for the short trip to the stadium. The convoy was shepherded through town by a number of police cars making generous use of their lights and sirens. The pedestrian citizenry stood in awe at this procession, giving us all quite a thrill to be a part of it. Soon the buses made their way to a little park at the rear of the

USA Team ready for opening ceremonies.

Pretty Bulgarian hostess.

British Team.

French Team enter Plovdiv Stadium.

All teams lined up in the Sports Stadium to reaffirm the Sportsman Oath.

sports stadium where we unloaded and formed up into national groups. Each team was given an escort of a bevy of cute young things in mini-skirts who also were to carry a placard stating the team's nationality. There was a lot of photo snapping and good natured banter going on as we waited for the signal to enter the arena. We were hoping this wasn't to be a re-enactment of the Christians and lions bit.

The reason for the wait was that a "warm-up" show was about to be put on for the 40,000 or so spectators that were packed into the stands. It started with three combat models flown from the center of the arena. They went after each other's streamers with a vengeance and the lively flying had the crowd on edge. This was followed in rapid succession by three control-line stunters and then a model rocket demonstration. With precision timing the show phased into full scale sport aviation as a pair of high performance sailplanes appeared in the sky above on tow. Then a pair of Zlin Acrobats, Czech-built aerobatic sport planes, performed in formation directly over the stadium. Following closely in the program a Russian Antonov An-2P cabin biplane flew over, disgorging a dozen parachutists. They trailed colored smoke flares on the way down and all landed in the center arena. A number of the jumpers were women, noticeable by their shouts to the crowd as they descended to burst spot landing balloons in the center.

At that climactic moment the band struck up the Bulgarian equivalent of the *Entry of the Gladiators*, the entry gate was thrown open and the procession of teams marched onto and once around the inner periphery of the stadium. The dimming twilight was bolstered by banks of high-powered floodlights mounted high above the stands. They caused a strange surrealistic magic lantern effect as the shadows and light rays ricocheted across the green grass. The scene, as the phalanges of national groups formed up in the center of the stadium, produced an extravaganza, *phantasmogoria*.

After a few mercifully short speeches by Bulgarian dignitaries hundreds of young boys and girls ran forward and gave carnations to everyone standing in review in the arena. Selected team representatives then stepped forward and read the Federation Aeronautique International *Oath of Contestants*. The lights then switched off and the competitors were whisked away to the stadium exit under cover of a great fireworks display touched off with split-second timing. The effect of all this on the competitors can only be described as, *electrifying*. The universal thought was, how can more people be brought together and share such a feeling?

The skyrockets faded away into the blackness of the night sky to be replaced by the stars. My musing brought to fore the image once again of the *Orient Express* driving through dark nights and strange lands, bringing together East and West as we rode that magic chariot. ■

World Free Flight Championships, Plovdiv Bulgaria, 1975.

The way it was (and still is), a proud, four-square, flying machine.

6 Vintage Free Flight

Old Times - Good Times, by Ken Sykora

Fredrick Lewis Allen, or some equally profound social historian, could likely explain the conditions which brought about the hobby within a hobby which in the last 18 years has become known as Old Timer (OT) modeling. Perhaps it was the social and technical climate in the 1960s. The long war in Asia, political assasinations, civil demonstrations, the individual's struggle against computerized stupidity, and other such factors constantly reminded us of the turmoil and change in our daily lives. Such conditions have been known to drive people to seek comforting distraction in recreating nostalgic elements of younger and (seemingly) more happy times. Such profound analysis we will leave to the sociologists. However, there was a factor which we should record that non-modeler researchers might fail to comprehend.

In the 00's there was still a large body of active modelers who had spent their adolescent years in modeling during the 1930's and early 40's; which was the so-called, "golden age" of free flight development. By the mid-60's these modelers were in varying levels of middle age. They had developed careers in business, and some were even beginning to plan early retirements. Their children were more or less grown, and many other factors of their busy earlier lives were either accomplished or in-hand. As a result, they began thinking in terms of regular building sessions, and once more attending contests.

The Retreads Look: In surveying the contemporary model scene however, these "retreads" found things much changed from their youthful memories. The AMA national organization had enjoyed tremendous growth ... in both numbers and remoteness. Its emphasis was on radio control activities, with less emphasis on building and more emphasis on spectator-oriented flying. This was where the buck was to be found and "big" business (by model airplane standards) was calling the tune. What really hurt was that the modeler's organization had taken on all the warm, personal communication of an IBM or a GMC. Greater and greater numbers of faceless marchers... all in lock step.

Digging deeper into the purely free flight activities around the country, the middle-aged modeler looked for current alternatives. He examined the international FAI events; and found them about as casual as America's Cup racing or Formula auto competition ... and requiring roughly the same amount of time, money and travel! He also surveyed the contemporary FF scene and found it less regimented than FAI, but still involving a race of constant technical and mechanical refinements.

Two Real Reasons: There was also another factor inherent in both of these approaches which did not jell with the budding old timer prospect. That was the deadening factor of monotony in design planform and general concepts in modern models. Wild, unusual and highly individual approaches, all directed toward a common goal of stable performance and good duration were part of the Golden Age of modeling development. With the definition of sound design patterns, these variations largely died out in post-war modeling and as the future OT modeler viewed the accepted modern concepts he realized that a common monograph planform could be laid over any contemporary design (FAI or AMA) with little deviation required . . . and from a little distance, there was no difference at all!

This factor of expressing individual design tastes, along with the more leisurely, low pressure approach to organized competition, may be the key reasons behind the unexpected interest in OT modeling, here and abroad. It is certainly clear that a variety of OT designs are selected for building and competing each year. And, among the more serious OT'ers, the search goes on for the unusual and for long forgotten subject . . . with which to surprise, entertain (and perhaps defeat?) fellow flyers.

In retrospect, it is difficult to back track all the individuals and events that made OT modeling bloom in the 1960's as a popular "unofficial" national movement. However, we can identify a major factor which unified nationwide pockets of interest and activity. This was the formation of a national organization for OT'ers based on a statement of objectives, a common definition for OT models and a single set of rules for OT competition. The story of how this organization came about also involves many individuals and events, but we can isolate some blocks of people/actions which were key factors in getting the job done.

The Warbucks Factor: Running like a rampant thread throughout the fabric of Old Timer development is a package of boundless energy named John Pond. His smooth dome, shockwave personality and earth-moving organizational capabilities have earned him the affectionate nickname, "Daddy Warbucks."

During the first trickles of the movement, John got into a running discussion with a fellow Northern California modeler, Bob Bowen, on the performance merits of pre-WWII models. Bowen contended that many of these ships could perform as well as contemporary designs, and that it would be fun to have regular contests for them as a distinct class.

These conversations aroused the interest of other flyers in the area and as a result the Stockton GMC held its first Old Timer Annual in 1961. Around this time other clubs throughout the country were also producing OT advocates, and some competitions with widely varied regulations, but there was a unique outgrowth of the Northern California activities and it was to become one of the true foundations of the movement.

The $1.25 Dream: John Pond, in his zeal to get the California OT contests off to a good start, volunteered to draw full-size plans, on order, of the kit and magazine published pre-war models. The tab was $1.25 for a blueline print of the original vellum, and from the casual beginning the demand for these plans mushroomed. More and more modelers seemed to want to re-build that first "real" flyer from their youth, or to finally get to that dream ship that the rush to adult responsibilities had not left time for.

Today this demand for old plans continues to expand, with no end in sight. Even the most obscure designs continue to be unearthed in this country and literally from the four corners of the globe. John's old drawing board has grown into the Old Time Plans Service with a four volume catalog (Gas, Rubber and two on Flying Scale) now in its ninth edition update.

The Collection Agency: With the beginning of scheduled club competitions and the available working drawings from Pond, the OT'ers had two-thirds of their requirements. Their remaining need was old ignition engines; and that presented a problem. Even high production rate pre- and post-war makes, such as the Ohlsson, had long been off the public market by the beginning of the 1960s.

Some builders solved this dilemma by using modern glo engines. However, there were a couple of things wrong with this that kept the problem from going away. First, some of the older designs weren't happy with the increased power of the modern engines; and second, many OTers were of the purist silk and since they were resurrecting old designs, nothing less than three-and-one fuel and an internal ignition system would do. Some of these flyers had hoarded their old engines, but most had not.

The solution to this requirement began in an unusual way because it involved a group of largely non-flyers and even many non-modelers, in the active sense. Roughly in parallel with the OT flying activities in the early 60's, a modeler named Joe Wagner decided the rich history of model *engines* was also an important aspect of old time modeling, and should be preserved. Toward this end he literally single-handedly began to publish a newsletter called, *The Engine Collectors' Journal.* Joe struggled along with his dream for many months. He completed a massive alphabetical listing of common and rare engines, and published much of the old engine histories and technical data gleaned from his personal files.

Joe's efforts gathered a small but strong following of enthusiasts who shared his fascination with the old engines. But, like so many labors of love, one-man-band efforts, in the end Joe had to give up his publishing operations. However, the seed had been sown, and others came along to continue the work Joe had begun. In August 1963, Tim Dannels of Denver began publishing the *Journal* again, and from that point on this branch of the hobby began to gather momentum, Regular collector "swap meets" began to spring up, and an ever increasing wealth of history and how-to information began to fill the pages of the *Journal*.

For the average OT flyer/competitor, this engine collector activity had two distinct benefits. First, the intense activity of locating old engines by the collectors brought to light many non-rare and otherwise unattractive "chipped" engines. These soon found their way into the hands of those seeking "runners" to power their growing fleets of OT models. Second, the collector's newsletter would provide a nationally distributed grass roots publication which could document and promote the widely scattered and largely independent building and flying activities of OT enthusiasts across the country. This convenient circumstance was to set the stage for the final chapter in formalizing the OT movement.

Let's Play It Again, Sam: In October, 1963, John Pond and another OT modeler, Lee Freeman, were sharing a ride to the Stockton GMA's 3rd Annual OT contest. During the trip, Freeman suggested that what the OT movement needed at that point was a national organization; some focal point for administration and rules development that would unify the current widely varied activities. The benefits to the individual modeler would be a source of information exchange with like-minded hobbyists, and a common platform for those interested in OT competition on both local and national scales. And, of course, uniform contest rules would be the salvation of the poor local contest director!

John agreed wholeheartedly with Freeman's ideas, but when Lee further suggested that John was just the man to organize such a plan, Pond begged off. At the time John was heavily involved in AMA politics as well as several local model organizations, and just could not find the time for another undertaking. Freeman let the matter drop, and decided it was closed. Like so many others, however, Lee had underestimated the Warbucks Factor!

In the Spring of '64 Freeman was somewhat taken aback to read in Dick Black's VTO column in *Model Airplane News* that he (Freeman) was forming an Old Timer League and interested modelers should get in touch with him for details! (One guess who'd leaked this tidbit to Black!) But, being a congenial fellow, and seriously concerned with helping the OT movement along, Freeman accepted the task. The result was a staggering pile of incoming letters and outgoing replies over the next three years, as OT'ers all over the country began to participate in hammering out their own organization.

"Colonel" Ken Sykora checks for spark on his Super Cyclone powered "Rambler," a Gil Sherman design, 1939 Flying Aces.

By the winter of '64, Tim Dannels had made a going thing out of the *Engine Collectors' Journal*, and in the Jan/Feb issue John Pond presented the first of his series of 3-views and descriptions of famous OT designs.

The Mar/Apr '64 issue also started his famous "Gas 'N Oil" column which he would produce for several years. This feature of the Journal presented many photos of OT'ers and their ships, news items, results of contests, gossip, and technical tips and data.

Dannels further extended his courtesy to the active OT'ers in the Sept. issue of '64 by providing Lee Freeman with space for a column to promote, and provide an opinions soundingboard for, his continuing efforts to establish an OT League which by now had acquired the name, *Antique Model Association*. (In a classic demonstration of their adroit group handling of a crisis, the founding members of the movement, some eight months later, discovered that their initials might well be confused with another group of old fuds, namely; the American Medical Association. In response, Freeman asked for alternate name suggestions. Bob Stallick of Albany, Oregon, provided the eventual winner: **Society of Antique Modelers**. . . and "SAM" was born.)

Founding Fathers; In his first column, Freeman proposed an idea that was to finally let the penny drop on moving forward with formal national organization. He proposed the formation of Chapters, made up of five or more OT'ers, from a specific club or area. These chapters would establish basic regional units to form the national organization, and provide the voting system (majority vote of each Chapter's members represented one national vote) to elect national SAM officers, decide national rules and determine other Society business matters.

By the fall of '65 the chapter idea had brought into the fold over 100 registered SAM members, representing a group of 15 Chapters scattered across the country. These first groups, and the sparkplug individuals who worked to organize them, represent the true "founding fathers" nucleus of SAM:

1. Denver, Col. — Tim Dannels
2. Erie, Pa. — Vic Didelot
3. Canoga Park, Ca. — Art Swift
4. Albany, Ore. — Bob Stallick
5. Overland Pk, Kan. — R. Schroeder
6. Fostoria, Ohio — John Peck
7. San Francisco, Ca. — John Pond
8. Vienna, Va. — M. Schindler
9. Boulder, Col. — Charles Wendt, Jr.
10. Tiffin, Ohio — Jim Deats
11. Litchfield, O. — Ed Snider
12. San Diego, Ca. — Larry Boyer
13. Garden Grove, Ca. — Bud McNorgan
14. Pontiac, Mich. — Bill Shirley
15. Albuquerque,NM — R. Schellenbaum

At Last. . . Rules! Acting as the coordinating point for these groups, Freeman was able to assemble a multiple choice set of basic OT rules and published them, for the Chapters to vote on, in the Jan/Feb 1966 issue of the *Journal*. When the next issue of the *Journal* came out that Spring, Lee published the results; and after three years of hard work by many, many people, SAM had its first nationally recognized rules!

These rules consisted of three basic areas:

I. **Definition Of An Old Timer:** An "OT" was defined as a model which was designed, kitted, or had plans published prior to December 31, 1942. An "Antique" had the same qualifications, with the exception of a cut-off date of December 31, 1938.

II. **Modifications:** This section stated, generally, that it was not permitted to play around with the basic structure, airfoil shape, or to make radical changes in thrustline vertical position. Minor beefing up of weak areas was permitted, and modification to allow DT installation was OK.

III. **Power & Classes:** Both ignition and glow engines were allowed, with an engine run handicap for the glows. (Currently this rule has been modified to allow only pre-1950 ignition and diesel engines and glow engines which have been converted to ignition operation.) The class sizes were divided thus: A-.000/.200, B-.201/.300, and C-.301/1.00.

Officers Call: In the Fall of '66 Freeman had to give up his SAM coordinator job due to the press of his own consulting business, and the task was taken over by another Los Angeles area modeler smitten with the OT bug, J. G. "Bud" McNorgan.

McNorgan, like Pond, was an enthusiastic, tireless organizer (as well as a craftsman builder and fierce competitor) and he played a major role in guiding the final basic organization operations for SAM. Building on the accomplishments of Freeman and the founding father Chapters, Bud instigated the concept of issuing serial numbered membership cards to each member. The purpose was to refine the existing voting method so that the Chapter majority no longer provided a single vote, but rather, the simple majority of all members would be individually tallied. This proved to be a popular idea, and it was soon adopted. Bud also laid the ground work for the election of SAM's first national and regional officers, and the addition of some detail rules (engine runs, weights, flight limits, etc.) to round out the basic three-part regulations adopted the previous year.

In the summer of 1970 the 4th Annual SAM Nationals was held in Taft, California. (The growing national character of the membership led to rotation of the sites, as was done with the AMA Nats.) At the SAM members meeting held during this meet the chairman, John Pond, presented a slate of nominated officer candidates, for voting on by the entire membership, by mail. Some rules modifications/additions were also placed on this ballot, as well as a "Preamble" introduction to the rules to document the aims, goals and spirit of the Society.

The results of this national vote installed the recommended rules expansions, and elected the first SAM officers. They were: President, Bill Ladner; Secretary-Treasurer, Woody Bartelt; and Vice Presidents for Eastern, Central and Western Regions, respectively, were: Gene Lapansie, Tim Dannels and Gene Wallock.

In May 1971, the first official SAM newsletter was produced by Woody Bartelt, and the circle was finally complete. SAM had regulations, officers and a voice, that would be heard. Recognition by the AMA, and inclusion of OT rules in the AMA Rule Book, would follow. The rest is a history of growth in both interest and activities. Local club and regional contests thrive throughout the country, and in 1978 the 12th Annual OT Nationals will be held. Membership is currently around 1700, representing all fifty states and many foreign countries.

Cliff Silva just keeps turning them out. Here's an oldie, the Cleveland Cloudster.

The turning point in Free Flight Gas, Carl Goldberg's Comet Zipper.

As with any national organization, ideas and opinions from regions with diversified local interests continually flow through the SAM organization. Some have been quite "unique", and have quickly burned out. Others have triggered tempers and resulted in hot debates. And some, which met with generally favorable response, have been incorporated.

One of the major new directions in the Society has been the incorporation of a radio control branch of SAM activity. Meeting the needs of those with a lack of flying sites (and/or too decrepit to chase?) this activity has a large following, and a special set of rules which were developed for RC classes and competition.

McNorgan's re-establishment of the Texaco Trophy fuel allotment, unlimited flight even on the West Coast has helped speed the popularity of the "Antique" class of pre-1939 designs. These large "pioneer" type models now compete in two categories: a ¼-ounce of fuel per pound of model weight event (Timer follows the flight), and a 30-second engine run, 5-minute maximum flight event.

Another "special" event, mutually developed by the SCIF and SCAMPS OT clubs in Los Angeles, is the "020 Replica". This class is for scaled down (or sometimes scaled up!) copies of legitimate OT designs. The Cox .020 glo engine is used, and flight rules are: 20-second engine run, ROG launch, and a 3-minute flight limit. This class was intended originally as a simple "fun" event, requiring minimum building time and suitable for small field operations. However, its appeal has been universal and its popularity has promoted a couple of manufacturers to produce lines of kits for the class. It is frequently a hotly contested event, and is included at the SAM Nats and many local meets across the country.

"Commercial Rubber" is a unique event also developed by the prolific promoters in the SCAMPS club. It too was intended as a small field event utilizing the old "$.25-type" rubber models which were so prolific in both kits and model magazine plans during the pre-war years. The simple rules specify a maximum 36-inch wingspan, fixed landing gear, and a free-wheeling prop shown on the plan. This class is frequently included in West Coast meets and has found favor in other areas as an ideal short-field event . . . although some of the models are rather potent performers!

Other special events are also flown. Among these OT Scale (for both gas and rubber models), Precision (based on the pre-war contests promoted in England by Colonel Bowden), and OT towline glider. These categories are sometimes included, for variety, in the larger contests. They draw small, but enthusiastic, groups of flyers.

The original concepts and organizational aspects of the OT movement have weathered the test of time pretty much intact. Various influences have assaulted its structure and officials with varied opinions, both popular and obscure, and frequently highly vocal! Yet the OT movement, as a major "special event" in the highly diversified field of American Free Flight modeling continued to flourish, with no greater discord than any other AMA category.

In seeking the answer to this reasonable stability in a national organization of some 1700 members, perhaps the clue is back at the beginning. As a true independent, grass roots modeling movement, the Old Timer phenomenon was the result of much voluntary hard work by many *active* modelers. They planned well and laid a good foundation, with much arbitration, but always with specific, defined goals.

If this assumption is correct, the simplest answer could be in the Preamble on the first page of the SAM Rule Book. It was written in 1970 by SCIF clubman Bruce Chandler, and adopted by the membership along with the first officers. It goes like this:

> **"PREAMBLE:** The competition flying of free flight model aircraft of vintage design is intended to be casual, enjoyable and interesting for both competitor and spectator alike. It is neither desired to advance the state-of-the-art of aeromodeling, per se, other than to increase participation in the sport generally, nor to reprove again that which is already recorded in aeromodeling history books. The intent of these rules is to categorize the basic types of vintage models and establish an equitable and simple framework of regulations for competition purposes. Therefore, model designs that revolutionized free flight competition and necessitated the formation of two basic classifications, 'ANTIQUE' and 'OLD TIMER', are expected to compete in the Old Timer Events." ■

Two great Sal Taibi designs, the Power House (above) and the Brooklyn Dodger (below).

Rudi Lindner shows his World Championship Nordic Glider.

Free Flight 1955 — The World Championships

The '55 Free Flight World Champs was the first such competition in which the popular F1A, F1B and F1C FAI categories were combined in one grand super championships. Logistical problems of such a large gathering were handled by combining the resources of two countries, the U.S.A. and Germany. Joint responsibility for the organization was shared by the A.M.A. and the Deutscher Aeroclub. The U.S. Air Force, Europe, provided facilities at its headquarters in Weisbaden to house, feed and transport some 350 contestants, officials and supporters. A large contingent of AMA officials and members were brought over with the U.S. team on NAA arranged air transport to help out with organizational duties.

My personal thoughts of this meet, the first of the great Free Flight Extravaganzas, were sharpened by my having won a spot on the power team that year. In hard-fought trials I had nosed out some very good fliers in the U.S. eastern zone, so I was hoping I could carry off my responsibilities with some success.

There were only a few weeks separating team selection and leaving for Germany and I spent that time practice flying at two New Jersey fields, Solberg and Hadley. The *Thunderbirds* club gave me a lot of help and support and soon I was packed and on a DC-6 heading out across the Atlantic. There was a refueling stop at the Azores and we got a short tour of that fascinating spot. Landing at Frankfurt, the USAFE had buses waiting which took us on to Weisbaden, about 10 miles away. We soon met many other contestants who were arriving in a steady stream. A delay in setting up meal service at the base caused everyone to turn to the PX for sustenance and a meeting place. The PX exchange medium was military script and since I had the good luck to get some on the way over I had the pleasure of treating a number of newly met European friends.

The next day we got right to some test flying in a field just outside the Weisbaden base. We carried our model boxes out there as it wasn't far. The field was covered with dried wheat stubble that was a few inches high, causing some discouragement to a few who weren't anxious to do a lot of tissue patching. I got one of my *Spoofems* out and tried a flight. It looked good and soared away in a nice thermal. The DT fuse had been lit, but it didn't DT, as I chased after it. Eventually after about a two-mile hike I found it sitting casually there in a potato field. It was then that the uniqueness of this situation (for me) became apparent. Here I was on the other side of the Atlantic Ocean

USAF PHOTO

Gunner Kalen, Sweden, has his Nordic checked by meet official Pete Sotich of the USA. Model modern looking except for long nose.

USAF PHOTO

U.S.A. Team Members; standing, Harry Gould (P), Jean Schaap (W), Manny Andrade (W), Herb Kothe (W & N). Bob Champine (W), Hank Cole (N), kneeling, Bill Hartill (P), Otis Goss (P), Joe Harris (N), Jerry Kolb (N), Ernie Shailor (P).

Unusual Nordic of Ossi Czepa. Austrians great experimenters.

Harry Gould and Hank Cole. Long noses were common to keep weight down.

Czechoslovakian Team.

thousands of miles from home, deep in the heart of a foreign land where I couldn't speak the language, chasing a model airplane that almost got lost first time out of the box and here it was, safe again. A strange feeling but nevertheless an exhilarating thrill, and the World Champs hadn't even started yet.

The first contest day, Saturday, was for F1A Nordic (generally called A2 then). The buses took us 12 miles from Weisbaden, through Mainz across the Rhine to Finthen, location of the contest flying site. It was an all-grass sod airfield situated in the French-controlled district of Allied Powers administrative jurisdiction. Weather was overcast with an occasional light drizzle at first but this cleared up later. Circle towhooks hadn't been thought of yet and tactical flying was still in its infancy. Most A2 fliers concentrated on eking out the best still air performance they could. Stabs were big and heavy by today's standards and many models were overweight. Long noses were common as the CG could be located properly with minimum effect on increasing the total weight. There were a few exceptions to this by a few forward-thinking A2 fliers. Gilroy of GB had an extremely short-nosed ship with which he placed second. Herb Kothe of the U.S. also had a short nose and small stab on his ship. He placed seventh. Flying had started at 8 AM with five rounds of two hours each scheduled. This gave one-half hour to fly for each of the four-man team members. That first round proved to be the nemesis of a number of contestants as the scores show. Improving conditions later allowed more maxes.

At the end of the fourth round there were six who were in contention for the top spot although Rudi Lindner of Germany was the only one with four maxes. Thomann (Switzerland), with a ship that was quite advanced for the time, dropped mightily in the last round (130). Hagel (Sweden) almost maxed but was three secs short. (Hagel later became F1C World Champ in 1971). Giusti (Italy) and Gilroy both maxed out putting pressure on Lindner who would have to get 161 secs to ease ahead of Gilroy. Rudi had held back in that last round and now there was only a half hour left. His German teammates proceeded to put up a number of "pilot" models to test the air. Although there was controversy over this move there were not protests made as it was apparently not denied by the rules. With the light now starting to fail, Rudi launched his last flight into a smooth flat glide. The "pilot" models had not helped any as there didn't seem to be any lift or turbulence. The model circled gracefully off to the edge of the field where the grass sloped down a little and it finally touched down at 166 secs, edging out Gilroy from the top spot by six seconds.

Earlier that day I witnessed one of Lindner's flights that certainly made an impression. Under tow he trotted along at just enough speed to keep the model in the air. He had no circle-tow capability, of course, but he did weave a bit across the field. Then, obviously feeling lift he gently cast off with no stall, reeled up the line and trotted under the circling model. The drift was very light and at three minutes he was still on the field. Then it DT'd and without any strain he caught the model. He pulled the wings off their tongues, tucked them under his arm and stepped over a few feet to the edge of the field where the organizers had set up an outdoor beer garden. Sitting down at a flower decorated table he ordered and finished off a stein of Lager and a sausage sandwich, then headed on back up the field. Cool!

Our bus ride back to Weisbaden wasn't too joyful at first considering that the U.S. hadn't gotten much glory, but on the other hand we did have fun because a few of the Czech team members had hitched a ride with us. We stopped at a beer garden near the Mainz *Bahnhoff* to relax a bit and commiserate with our new friends.

That night I didn't get much sleep thinking about the power event the next day. I kept dreaming that I heard someone trying to start an engine. On finally waking fully I discovered it was someone snoring in the next bunk.

The weather that morning looked a little gray but it was dry and there wasn't much drift. Engine timing seemed to be a problem and a number got zeros for double overruns. Also there were a few that had trouble with takeoffs even though huge plywood platforms were provided that were elevated above the grass. The eventual winner, Mike Gaster of England, was flying with VTO, sparking a lot of admiration for the beautiful climb pattern his Oliver Tiger powered *Gastove* was getting. My own efforts with *Spoofem* were uneven in the result. Given a good engine run close to 15 secs and barring any downdrafts, I could max. But my Green Head Torp 15 wasn't all that steady and those with the new Oliver Tiger diesels were outpulling the Torp anyway. *Spoofem* had the MVA342 wing section and aspect ratio of 12.9. Weight was 19 oz. (yes, my son, it could *glide*).

The fifth round saw three drop after having maxed the preceding four rounds. They were Buskell (GB), Vidossich (I) and Hajek (CZ). Buskell's drop was due to a short eight-second engine run, just not enough to get the altitude. However there were three others who had maxed all the way. The concept of fly-offs was still very new in those days and it was considered, by the organizers at least, as a troublesome thing only necessary to see which of the "joint" winners would get to hold the cup for a year. As such, fly-off time was unlimited thus requiring only one extra round.

The three subject winners were Mike Gaster (GB), Fransisco Stajcer (RA) and Bryant Jones (CD). They set about getting ready to fly-off in the gathering dusk. It was 6:45 P.M. when the models were launched all within 15 seconds of each other. Jones had the misfortune of an over-run, a great disappointment. Stajcer's

Andrade gets his Wakefield off the board, gear retracting.

Wakefield scoreboard. Four-man teams in 1955.

Gilroy, second place, G.B.; Lindner first place, Germany; Rolf Hagel, third place, Sweden.

USAF PHOTO

United States Wakefield Team; Schaap, Kothe, Andrade, Champine, (Manager, Cole). Variety of designs, Team placed third.

Shailor anchors as Schaap cranks in the turns.

USAF PHOTO

USAF PHOTO

Bryan Jones of Canada, placed third, used Oliver Tiger, had overrun in fly-off.

USAF PHOTO

Herb Kothe's Nordic up and away, placed seventh.

ship glided in at 175 secs while Gaster's was still at a good height, now the obvious winner. *Gastove* subsequently landed at 313 secs, a great flight.

That night I learned that I was going to be the Wakefield proxy flier for John Upton of New Zealand. Well talk about short notice! In looking in the New Zealand model box I discovered there was everything I needed for one Wakefield — except rubber. Along about 8 P.M. the English *equippe* stopped by and dropped a skein of fresh Pirelli on my bunk. That was to be the New Zealand motors. Such is the way World Champs are won I mused as I spent half the night making up and breaking in motors. The dormitory doorknob made a convenient hook for anchoring the rubber as I wound, but things did get a bit hectic as several water fights with fire extinguishers erupted around me between some of the national groups. All in good fun but there wasn't much sleep to be gotten.

The weather the next day was similar to the previous day, a light overcast sky with little drift. With the help of Joe Harris of the U.S. Nordic team I made some quick trim flights on the Upton Wakefield and it seemed to be in reasonable trim. Biggest problem with this model turned out to be the loose prop blade retainers. These consisted simply of little plastic "O" rings that were slipped over the 1/16-inch wire blade hangers. They kept slipping off and the blades would just fall off when the model landed. Launching was tricky, trying to hold the blades on while maintaining contact with the retractable take-off peg (ROGing a hot Wakefield from a flimsy "representation" of a landing gear is a thrill that our modern-day Wakefield enthusiasts will never know). The ship was capable of maxing as it proved in the first round, but I couldn't keep up that pace. One following flight might have maxed had the ship not landed on the roof of a hangar. Joe and I had the additional help of a young German boy who attached himself to us as our "official" chaser. He took his position quite seriously and really scrambled after our New Zealand entry. After the meet Joe gave the boy his Nordics.

The U.S. team started off in fine shape with all maxes up through three rounds. Then in the fourth, Champine dropped one second, Kothe 22 secs and Andrade 32 secs. Schaap maxed but dropped 34 secs in the fifth round. A disappointment after such a good start but the U.S. did retain third place in team score.

Seven competitors had maxed out, for those days an unusual situation. They were, in the order of their eventual fly-off scores, Samann (D), Hakanssen (S), Scardicchio (I), Altmann (D) Fresl (Yu), Fea (I), Muzny (CZ). The air was getting dark as the fly-off began and there was some concern about being able to see the models to the ground. Samann was in the air for 315 secs, a very good flight.

A banquet and prize-giving ceremony was held that evening in the airmen's mess at Weisbaden Airbase. The facilities were taxed to the limit by the unexpectedly large assembly but somehow things were managed and the American-grilled T-bone steak dinner went well with the local German wines. This meeting established a new popular format for Free Flight World Championships that has served as a fine example for many great World Champs in the intervening years.

The Wakefields in 1955 showed a leaner more compact look. The cross section requirement had been dropped and rubber was limited to 80g (twice the current rubber weight rule). The rubber limitation

Editor Bill Hartill was busy at the 1955 World Champs flying his "Spoofem" F1C ship and also proxy flying this Wakefield of John Upton, New Zealand.

eliminated the very long fuselages and the geared motors. Many were still using built-up truss-work fuselages but a few sheet-box fuselages were seen, as for example the winner Samann. Winding tubes were not being used but then motors were being wound to a greater margin of safety than is the case now. Samann probably pushed that margin a little closer than most since his fuselage did look pretty tough. Samann also used the Benedek 8356b wing section, leading a new trend away from the thicker foils of the earlier traditions (RAF32, NACA 6409, etc.).

Rudi Lindner's win in A2 with a ship very similar to his 1954 World Champion winner (it used the same wings) set the Nordic glider world on its ear and made a great impact on glider design for a number of years. The "Lindner" wing section, a thin well undercambered section, became a big favorite. The lines and clean details were very graceful and inspired many a similar configuration. The construction too, showed the way with a new level of precision that admittedly took more work but gave much more satisfaction in pride of craftsmanship. In light of today's experience we know that the steep vee dihedral tended to make banked turns tighten up too quickly. The vee dihedral stab looked nice but it was very sensitive to keying. A slight misalignment and you got a lot of turn. The narrow-boxed tail boom with pine or spruce longerons amounted to structural and aerodynamic overkill as it was much heavier than the current standards. Unfortunately Lindner didn't have a fiberglass boom to work with. In spite of these "complaints" made with hindsight, it was quite a machine.

The Lindner glider made such an impression that a very important lesson was missed there on that Finthen airfield. That was the concept of tactical flying. Lindner did an excellent job of reading the air and observing progress of other models near him. It helped. But then the next year (1956) in Florence, Italy, a highly antiquated model cleverly handled by Marcel Brems, humbled the field of contestants who seemed to be caught off base, thinking too much of "performance" and not enough of the man/model system.

Mike Gaster's power winner was also a very neat and well constructed model. The wing aspect ratio looks awfully low and the stab looks awfully big by today's standards but these features were effective (in the days before VIT) in keeping the climb pattern very steep with little spiralling. This ship got very good height, with that Oliver swinging a nine-inch diameter by five-inch pitch prop.

Summing up, it was a great year and a great World Champs, and it appeared to this observer that a new world awareness had now developed. The top models showed a new maturity and attention to detail, the participants found a comradeship that stretched around the world. Free Flight had found itself and it *mattered.* ■

USAF PHOTO

United States Power Team; Hartill, (Manager Champine), Shailor, Goss and Gould. More variety but all with Green Head Torps. Team fifth place.

Top Wakefield Team from Sweden. Anders Hakanson, Malte Blomquist and Ragnar Ahman.

USAF PHOTO

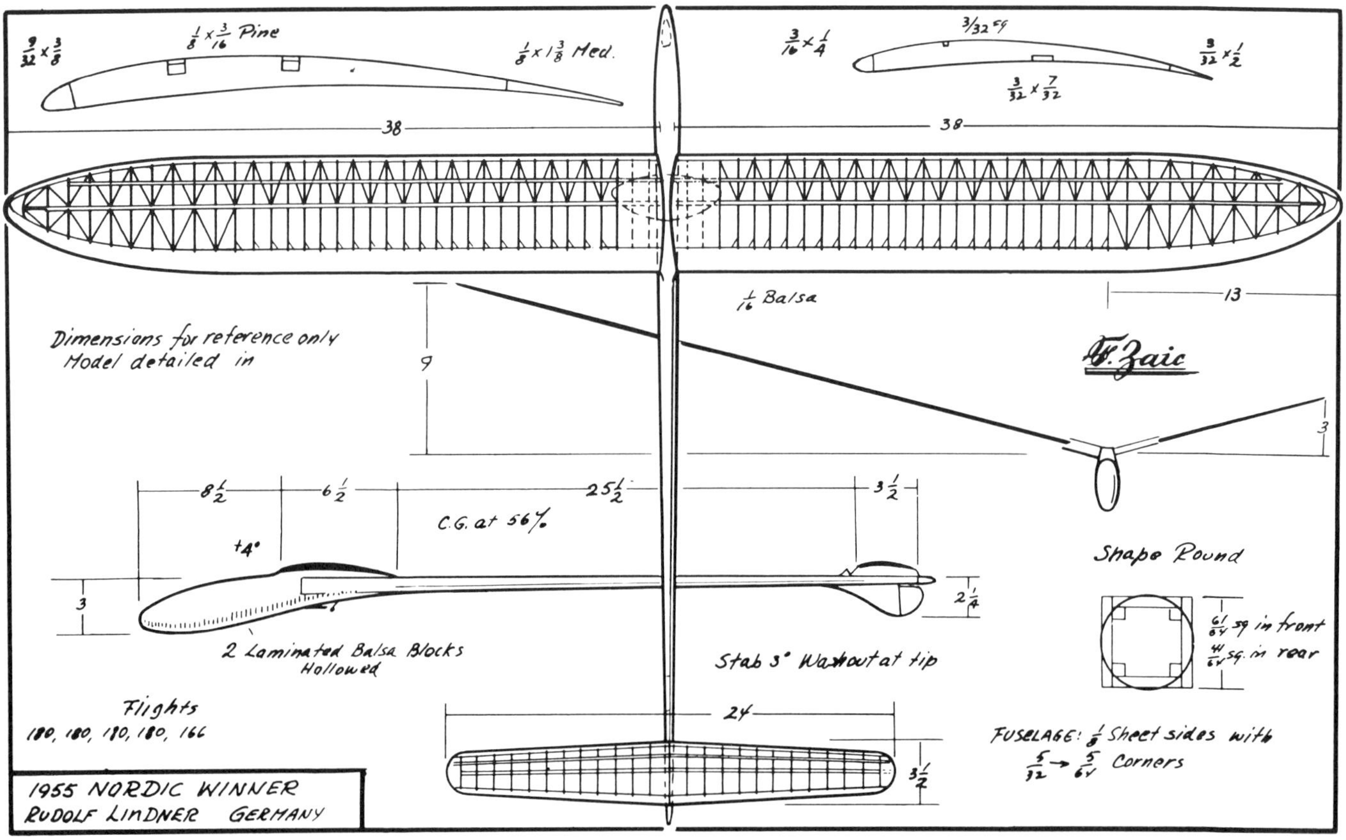

INDIVIDUAL RESULTS FOR THE SWEDISH CUP

			1	2	3	4	5	Total
1.	R. Lindner	... Germany ...	... 180	180	180	180	166	886
2.	R. Gilroy ...	... Gt. Britain	... 160	180	180	180	180	880
3.	R. Hagel ...	... Sweden ...	... 176	180	180	164	177	877
4.	E. Giusti ...	... Italy ...	... 156	180	180	180	180	876
5.	J. C. D. Esvelt	... Netherlands	... 163	180	137	180	180	840
6.	H. W. Thomann	... Switzerland	... 166	180	180	180	130	836
7.	H. Kòthe	... United States	... 143	180	145	180	180	828
8.	V. Horyna ...	... Czechoslovakia	... 180	180	133	180	152	825
9.	H. Hansen	... Denmark ...	... 180	180	158	106	180	804
9.	M. Vilchair	... France ...	... 118	180	180	180	146	804
11.	H. Ege ...	... Switzerland	... 174	116	180	144	180	794
12.	C. Varetto	... Italy ...	... 130	180	180	180	114	784
13.	C. Goetz ...	... France ...	... 135	100	180	180	180	775
13.	B. McElwain (P) Kurth	... New Zealand	... 104	180	131	180	180	775
15.	G. Overlaet	... Belgium ...	... 147	180	180	180	85	772
16.	L. Murtagh	... Ireland ...	... 138	180	180	93	180	771
17.	A. Cavlevski	... Yugoslavia	... 171	165	128	125	180	769
18.	L. Gustafsson	... Sweden ...	... 141	180	180	90	168	759
19.	J. O'Donnell	... Gt. Britain	... 96	180	180	180	114	750
20.	L. Feron ...	... Belgium ...	... 150	139	180	180	97	746
21.	D. Mackenzie	... Canada ...	... 130	99	180	156	180	745
22.	V. Spulák ...	... Czechoslovakia	... 166	95	180	116	178	735
23	Eduardo Vich	... Argentina ...	... 110	128	180	147	167	732
24.	F. Sussdorf	... Saar ...	... 157	180	159	127	106	729
25.	S. Pedersen	... Denmark ...	... 128	180	139	101	178	726
25.	L. Olsson ...	... Sweden ...	... 114	180	72	180	180	726
27.	R. Berthe ...	... France ...	... 108	144	180	180	110	722
28.	P. Petrovski	... Yugoslavia	... 180	124	125	108	180	717
29.	C. Boscarol	... Italy ...	... 140	146	158	178	122	716
30.	J. Lock ...	... France ...	... 126	180	90	172	143	711
30.	A. C. LeBreton (P) Mussig	... New Zealand	... 165	180	180	80	106	711
32.	J. Harápát ...	... Czechoslovakia	... 143	180	180	87	114	704
33.	W. Etherington	... Canada ...	... 180	151	72	180	115	698
34.	B. Jones ...	... Canada ...	... 105	180	101	180	129	695
34.	H. Rau ...	... Saar ...	... 133	180	180	127	75	695
36.	P. Nironi	... Italy ...	... 180	84	180	70	176	690
37.	J. Fraquelli ...	... Argentina ...	... 132	147	180	153	75	687
38.	M. Newnham (P) Barth...	... Australia ...	... 110	109	99	180	180	678
39.	M. Vuletic ...	... Yugoslavia	... 150	180	78	100	167	675
40.	R. Knoll ...	... Saar ...	... 92	180	128	180	83	663
41.	W. Woerle	... Germany ...	... 121	180	69	145	143	658
42.	P. Smith ...	... Ireland ...	... 87	180	112	71	180	630
43.	H. G. Wachter	... Germany ...	... 116	180	83	96	152	627
44.	F. Menc ...	... Czechoslovakia	... 125	103	180	109	108	625
45.	H. Nielsen	... Denmark ...	... 80	180	178	98	88	624
46.	C. de Coslo	... Mexico ...	... 124	180	179	86	54	623

			1	2	3	4	5	Total
47.	R. Melzer ...	... Germany ...	... 113	95	180	47	180	615
48.	C. Aubertin	... Monaco ...	... 104	176	76	74	180	610
48.	A. Klaver ...	... Netherlands	... 139	172	82	180	37	610
50.	M. Zito ...	... Argentina ...	... 94	87	62	180	180	603
51.	J. Turk ...	... Austria ...	... 98	180	30	180	96	584
52.	O. Czepa	... Austria ...	... 161	161	135	52	67	576
53.	J. Kolb ...	... United States	... 95	128	180	56	116	575
54.	H. Glavitsch	... Austria ...	... 131	123	176	20	118	568
55.	H. Cole ...	... United States	... 121	175	113	93	63	565
56.	H. Sayar ...	... Argentina ...	... 107	104	180	106	64	561
56.	H. Schnabel	... Switzerland	... 138	180	40	82	121	561
58.	R. B. Lester	... Canada ...	... 106	180	83	120	71	560
59.	D. Yeabsley	... Gt. Britain	... 134	101	72	180	67	554
60.	H. Weintraut	... Saar ...	... 102	173	66	101	47	549
61.	B. Hansen	... Denmark ...	... 132	117	180	53	64	546
61.	G. Lefever	... Gt. Britain	... 90	167	137	55	97	546
61.	F. Bachli ...	... Switzerland	... 137	180	96	86	47	546
64.	J. Maes ...	... Belgium ...	... 178	180	75	29	83	545
65.	G. Kalen ...	... Sweden ...	... 115	117	133	75	77	516
66.	M. Walsh ...	... Ireland ...	... 131	86	166	82	44	509
67.	J. Wastl ...	... Austria ...	... 116	67	45	156	100	484
68.	R. Molinari	... Monaco ...	... 47	70	180	90	86	473
69.	A. King ...	... Australia ...	... 108	114	62	180	—	464
70.	L. R. G. Ackroyd (P) Denzin	... New Zealand	... 119	119	56	68	96	458
71.	A. A Teunissen	... Netherlands	... 114	86	91	64	89	444
72.	R. Aubertin	... Monaco ...	... 90	133	84	81	49	437
73.	J. Harris ...	... United States	... 122	90	77	77	56	422
73.	L. Pinter ...	... Yugoslavia	... 63	76	99	61	123	422
75.	V. Adamski	... Belgium ...	... 103	83	117	83	35	421
76.	J. Thompson	... Ireland ...	... 87	114	92	54	53	400
77.	H. B. M. Luykx	... Netherlands	... 126	35	60	58	63	342
78.	P. Carter (P) Boy ...	... New Zealand	... 120	45	27	—	—	192
79.	W. Malcolm	... Australia ...	... —	180	—	—	—	180

(P) Indicates proxy flown.

TEAM AGGREGATES FOR THE BELGIUM DAUMERIE TEAM TROPHY

1.	Italy ...	... 2376	8.	Yugoslavia	... 2161	15.	New Zealand	1944
2.	Sweden	... 2362	9.	Denmark	... 2154	16.	Ireland	... 1910
3.	France ...	... 2301	10.	Canada	... 2138	17.	Netherlands	1894
4.	Czechoslovakia	2264	11.	Saar ...	... 2087	18.	Austria	... 1728
5.	Switzerland	... 2191	12.	Belgium	... 2063	19.	Monaco	... 1520
6.	Gt. Britain	... 2184	13.	Argentina	... 2022	20.	Australia	... 1322
7.	Germany	... 2171	14.	United States	1968	21.	Mexico	... 623

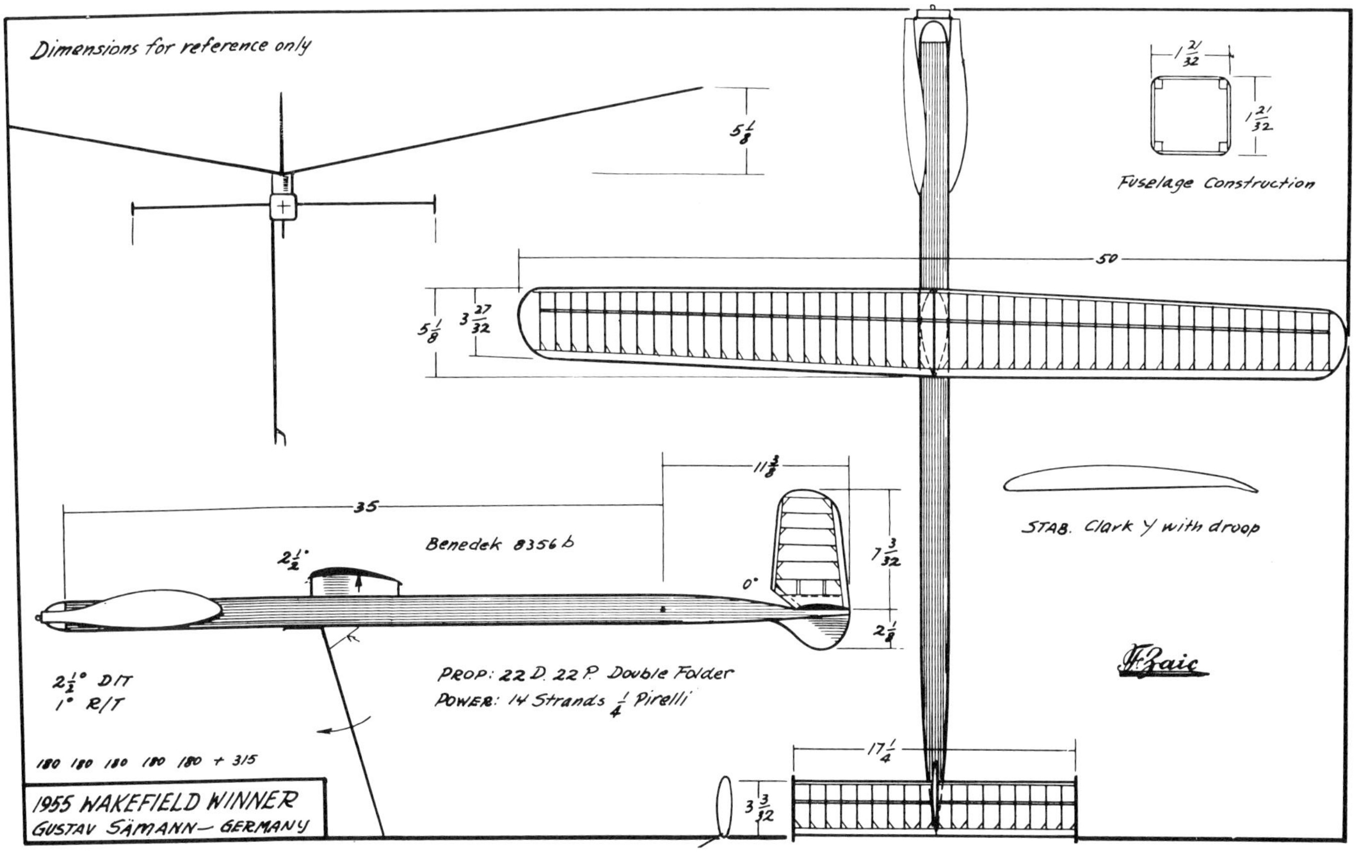

INDIVIDUAL RESULTS FOR THE WAKEFIELD CUP

	Name	Country	1	2	3	4	5	Total	Fly-off
1.	G. Samann	Germany	180	180	180	180	180	900	315
2.	A. I. Hakanssen	Sweden	180	180	180	180	180	900	289
3.	C. Scardicchio	Italy	180	180	180	180	180	900	286
4.	J. Altmann	Germany	180	180	180	180	180	900	284
5.	E. Fresl	Yugoslavia	180	180	180	180	180	900	270
6.	Guido Fea	Italy	180	180	180	180	180	900	213
7.	L. Muzny	Czechoslovakia	180	180	180	180	180	900	169
8.	M. U. Blomquist	Sweden	180	180	180	172	180	892	
9.	K. E. Widell	Denmark	180	180	180	180	170	890	
9.	R. G. Ahman	Sweden	180	170	180	180	180	890	
11.	F. Holland	Gt. Britain	180	180	180	180	160	880	
11.	R. A. Champine	United States	180	180	180	179	161	880	
13.	H. H. Kothe	United States	180	180	180	158	180	878	
14.	F. Mursep	Argentina	164	180	180	173	180	877	
15.	H. O'Donnell	Gt. Britain	180	180	156	180	180	876	
16.	E. Balasse	Belgium	180	180	180	149	180	869	
16.	C. R. de Vries	Netherlands	180	180	180	159	170	869	
18.	M. D. Andrade	United States	180	180	180	148	180	868	
19.	G. Maibaum	Germany	180	180	147	180	180	867	
20.	G. J. Schaap	United States	180	180	180	180	146	866	
21.	A. S. P. Balogh van Galantha	Netherlands	180	143	180	180	180	863	
21.	O. de Bare	Belgium	180	180	180	143	180	863	
23.	V. Kmoh	Yugoslavia	141	180	180	180	180	861	
24.	R. Cizek	Czechoslovakia	178	180	132	180	180	850	
25.	H. J. v.d. Geer	Netherlands	148	180	180	177	158	843	
25.	H. Toersen	Netherlands	180	125	180	179	179	843	
27.	E. Knudsen	Denmark	180	174	136	168	180	838	
28.	R. K. E. Johansson	Sweden	180	180	117	180	180	837	
29.	D. Prandini	Italy	180	180	180	114	180	834	
29.	J. O'Donnell	Gt. Britain	180	180	114	180	180	834	
31.	D. R. Mackenzie	Canada	180	158	133	180	180	831	
32.	R. McGlashan (P) Etherington	Canada	175	172	180	172	180	829	
33.	M. Bodmer	Switzerland	162	180	160	160	165	827	
34.	E. Gerlaud	France	105	180	180	180	180	825	
35.	I. Pietralunga	Italy	180	115	180	148	180	803	
35.	R. Miyahara (P) Schulz	Japan	156	164	173	180	180	803	
37.	P. W. Read	Gt. Britain	138	165	136	180	180	800	
38.	L. F. Murtagh	Ireland	125	180	180	180	129	794	
39.	B. R. S. Baker	Australia	180	154	114	180	165	793	
40.	L. Nesic	Yugoslavia	103	180	157	167	180	787	
41.	E. Nienstedt	Denmark	180	144	180	98	180	782	
42.	N. Corwell	Ireland	108	180	180	180	133	781	
43.	M. Chevrlot	France	180	180	180	123	113	776	
44.	R. E. Conzalez	Argentina	180	173	121	144	153	771	
45.	J. C. Parnisari	Argentina	180	151	125	180	127	763	
46.	J. Hemola	Czechoslovakia	149	180	70	180	180	759	
47.	J. Morisset	France	180	172	145	134	127	758	
48.	O. Hyttrek	Germany	173	156	110	122	176	737	
49.	A. Leong (P) Hertsch	New Zealand	118	180	131	115	180	724	
50.	J. Upton (P) Hartrill	New Zealand	180	151	70	156	163	720	
51.	Y. Mikami (P) Tonn	Japan	180	145	180	104	110	719	
52.	C. Goetz	France	78	180	122	180	155	715	
53.	V. H. Ure (P) Strattner	Canada	180	113	127	180	114	714	
54.	N. W. Sorensen	Denmark	155	131	180	104	143	713	
55.	J. A. Djorde	Yugoslavia	171	144	180	114	89	698	
56.	M. Kimura (P) Sabel	Japan	110	180	137	140	108	675	
57.	G. Lippens	Belgium	109	180	121	125	121	656	
58.	Z. Mach	Czechoslovakia	147	180	—	155	172	654	
59.	R. E. Bird (P) Ehmann	Australia	170	180	48	60	180	638	
60.	V. J. Rizzi	Argentina	27	109	106	180	180	602	
61.	K. Miyoshi (P) Bathel	Japan	180	98	103	48	158	587	
62.	L. J. Walter (P) Naegele	Canada	125	47	180	128	105	505	
63.	A. D. King	Australia	180	151	170	69	—	576	
64.	L. R. G. Ackroyd (P) Heidmuller	New Zealand	—	180	114	180	—	471	
65.	A. Gordon	Ireland	—	103	99	64	145	411	
66.	B. R. McElwain (P) Ruedle	New Zealand	—	133	113	89	—	337	
67.	D. R. Boughton (P) Lichte	Australia	—	—	—	180	137	317	
68.	H. Houtrelle	Belgium	92	—	64	68	91	315	
69.	J. D. Thompson	Ireland	—	38	104	47	57	246	
70.	R. M. Aubertin	Monaco	—	10	—	—	—	10	

(P) Indicates proxy flown.

TEAM AGGREGATES FOR F.M.A. CUP

	Team	Total		Team	Total		Team	Total
1.	Sweden	2682	7.	Yugoslavia	2548	13.	France	2359
2.	Germany	2667	8.	Denmark	2510	14.	Japan	2197
3.	United States	2646	9.	Czechoslovakia	2509	15.	Australia	2007
4.	Italy	2634	10.	Argentina	2411	16.	Ireland	1986
5.	Great Britain	2590	11.	Belgium	2388	17.	New Zealand	1918
6.	Netherlands	2575	12.	Canada	2374	18.	Switzerland	827

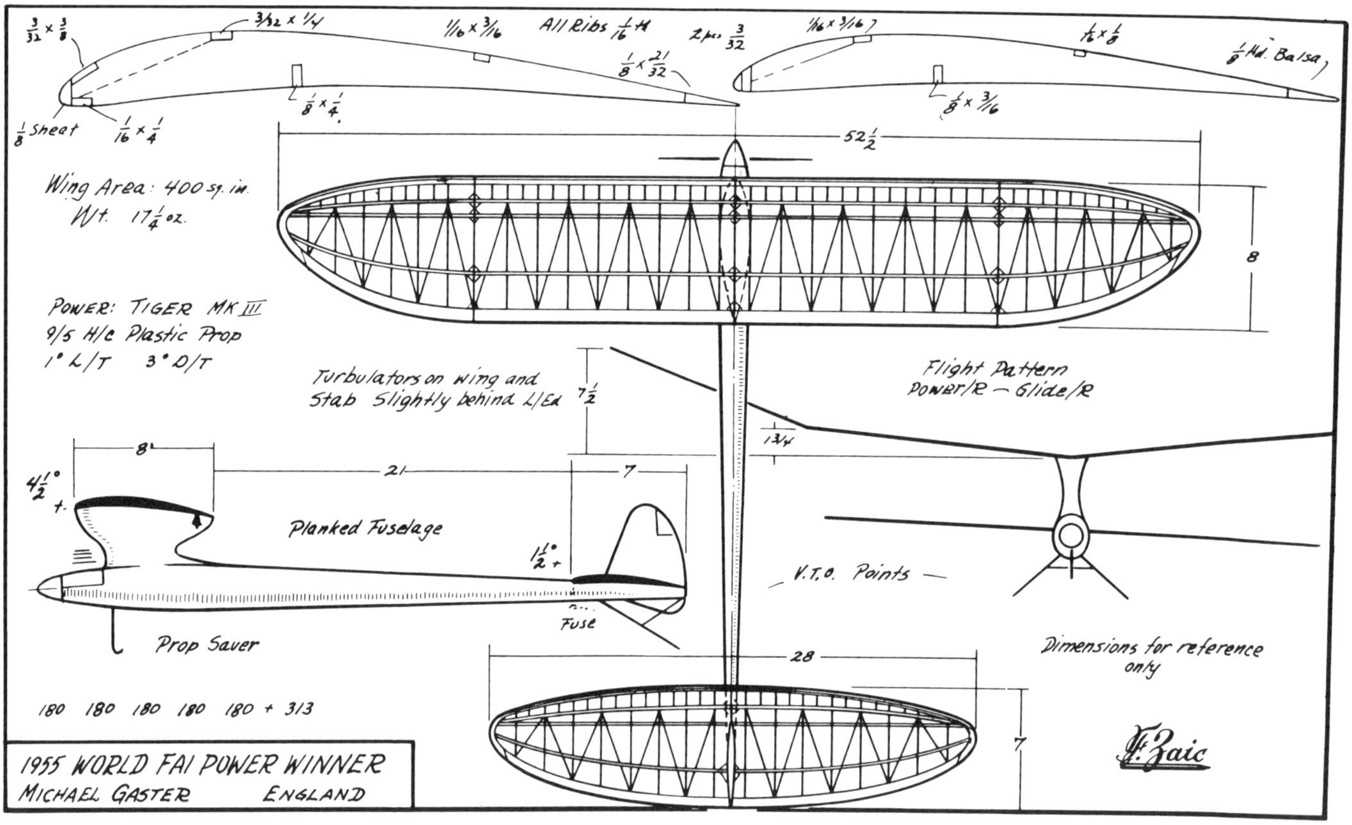

INDIVIDUAL RESULTS FOR THE F.N.A.F.O.M. CUP

			1	2	3	4	5	Total	Fly-off
1.	M. Gaster	... Gt. Britain ...	180	180	180	180	180	900	313
2.	F. Stajcer	... Argentina ...	180	180	180	180	180	900	175
3.	B. Jones	... Canada	180	180	180	180	180	900	000
4.	V. Hájek	... Czechoslovakia	180	180	180	180	166	886	
4.	L. Mangino	... Mexico	166	180	180	180	180	886	
6.	P. Buskell	... Gt. Britain ...	180	180	180	180	151	871	
7.	G. Vidossich	... Italy	180	180	180	180	150	870	
8.	M. Rudolph	... Germany ...	179	180	166	180	164	869	
9.	P. Goss ...	... United States ...	180	180	148	180	178	866	
10.	L. F. L. M. Bausch	Netherlands ...	160	180	180	180	127	827	
10.	Antonio Podda	Italy	170	142	180	180	155	827	
12.	J. Partinen	... Finland	132	180	158	180	167	817	
13.	H. Gould	... United States ...	180	180	142	180	130	812	
14.	R. Bacchi	... Italy	180	180	180	174	87	801	
14.	B. Gunic	... Yugoslavia ...	180	180	81	180	180	801	
16.	J. Parrott	... Gt. Britain ...	180	180	102	180	143	785	
17.	J. Heidemann	... Germany ...	120	180	173	180	176	779	
18.	G. Hormann	... Austria	180	169	133	180	102	764	
19.	O. Lucas	... Argentina ...	162	180	60	180	180	762	
19.	J. Thompson	... Ireland ...	150	127	125	180	180	762	
21.	M. Zito	... Argentina ...	180	155	111	180	134	760	
22.	S. Davila	... Mexico	180	125	129	166	157	757	
23.	F. Aiken	... Ireland ...	180	154	180	165	74	753	
23.	E. Johansen	... Denmark ...	157	132	104	180	180	753	
25.	G. Rupp	... Germany ...	169	108	109	180	180	746	
26.	E. Fresl ...	... Yugoslavia ...	180	130	95	151	180	736	
27.	H. Fries ...	... Sweden ...	144	180	160	180	69	733	
28.	L. Nesic	... Yugoslavia ...	147	76	147	180	180	730	
29.	G. Lippens	... Belgium ...	136	152	180	180	81	729	
30.	P. Schmitter	... Switzerland ...	154	180	123	180	91	728	
31.	W. Hartill	... United States ...	135	100	180	180	113	708	
32.	B. Baker	... Australia ...	180	153	101	114	158	706	
33.	H. Buhr ...	... Switzerland ...	118	90	180	180	137	705	
34.	A. Mussell	... Gt. Britain ...	180	156	—	180	180	696	
35.	R. Cerny	... Czechoslovakia	159	99	143	110	180	691	
36.	J. McMillan (P) Benkert	... Canada	79	180	136	153	139	687	
37.	A. Lundin	... Sweden ...	102	132	180	180	90	684	
38.	T. Morelli	... Ireland	115	137	180	142	109	683	
39.	H. Entzeroth	... Switzerland ...	148	119	180	103	104	654	
40.	W. Etherington	Canada	173	49	143	156	131	652	
41.	R. Schenker	... Switzerland ...	137	133	67	117	174	628	
42.	J. S'Jongers	... Belgium ...	172	180	103	71	95	621	
43.	G. Guidici	... France	180	97	117	119	92	605	
44.	H. Loser	... Germany ...	15	180	120	180	109	604	
45.	R. Hagel	... Sweden ...	7	138	180	180	96	601	
46.	G. Navarro	... France	100	130	91	142	121	584	
47.	C. Bergamaschi	Italy	117	167	95	108	76	563	
47.	E. Shailor	... United States ...	—	124	79	180	180	563	
49.	R. Das ...	... Netherlands ...	172	79	89	101	115	556	
50.	M. Vondruska	... Czechoslovakia	137	180	69	86	67	539	
51.	H. Nielsen	... Denmark ...	78	115	83	142	113	531	
52.	J. C. Guyot	... France	—	152	180	44	151	527	
53.	J Pouliquen	... France	88	60	180	37	152	517	
54.	A. A. Tennissen	Netherlands ...	171	85	—	127	123	506	
55.	G. Woodworth	Ireland ...	137	56	112	67	88	460	
56.	F. Sussdorf	... Saar	107	—	180	80	83	540	
57.	E. Balasse	... Belgium ...	95	79	180	—	88	442	
58.	R. Molinari	... Monaco ...	141	—	95	107	30	373	
59.	O. Czepa	... Austria	58	164	42	68	34	366	
60.	G. Zigic	... Yugoslavia ...	180	180	—	—	—	360	
61.	H. Waldhauser	... Saar	—	76	—	180	94	350	
62.	G. M. Cornelissen	Netherlands ...	—	72	74	79	115	340	
63	C. de Cosio	... Mexico	150	—	—	79	105	334	
64	M. Libert	... Belgium ...	45	75	41	71	68	300	
65.	J. Verges	... Mexico	—	180	34	42	9	265	
66.	J. Graves (P) Beck	... Canada	108	—	—	51	57	216	
67.	E. Blasche	... Austria	—	92	—	110	—	202	
68.	C. Aubertin	... Monaco ...	137	10	—	—	—	147	
69.	A. King	... Australia ...	69	—	—	—	—	69	
70.	F. Hillcoat	... Argentina ...	43	2	—	—	—	45	
71.	G. Skalla	... Austria	9	—	—	—	—	9	
72.	R. Aubertin	... Monaco ...	8	—	—	—	—	8	
	P. Hakansson	... Sweden ...	—	—	—	—	—	—	
	J. Gerny	... Czechoslovakia	—	—	—	—	—	—	

(P) Indicates proxy flown.

TEAM AGGREGATES FOR THE FRANJO KLUZ TROPHY

1.	Great Britain	2556	8.	Ireland ...	2198	15.	France ...	1716
2.	Italy	2498	9.	Czechoslovakia	2116	16.	Austria ...	1332
3.	Argentina ...	2422	10.	Switzerland ...	2087	17.	Denmark ...	1284
4.	Germany ...	2394	11.	Sweden ...	2018	18.	Finland ...	817
5.	United States	2386	12.	Mexico ...	1977	19.	Saar ...	800
6.	Yugoslavia ...	2267	13.	Netherlands	1889	20.	Australia ...	777
7.	Canada ...	2239	14.	Belgium ...	1792	21.	Monaco ...	520

Nordic flier Ed Wogulus, ponders the mysteries of the sky, Taft, California.

Wakefield flier Bob White launches into a fiery dawn.

Majestic climb-out by the 1937 Michael Roll design reconstructed by Cliff Silva.

Larry Clark's "Albatross", a neat design by Wakefield Champ George Reich.

Everyone's favorite, the Comet Sailplane.

"Brick" Brickner and "Foo-2-U-Too" an Obarski design.

Andy Faykun shows off his Cahill "Clod hopper".

Cliff Silva's Michael Roll ship proudly shows off its classical lines and "Baby Cyclone" up-front.

Dick Gildersleeve and his rendition of the Yugoslavian Wake design by Kmoch.

Carmen Critchlow launches for hubby Bob at Taft.

Superb Chet Lanzo Class E record holder by Vince Costanzo, Los Angeles.

Hardy Broderson fires up his Class D ship at Taft.

Bill Hunter and Satellite.

Class D beauty by R. Simpson.

Jim Parker, El Paso, Texas and "Toker" A2.

Brian Hartill holds Andy Faykun's great unlimited rubber ship.

Late afternoon sky at Taft shows off the structure of Lee Hines' Nordic.

Cardington Air Shed No. 1 during the 1976 Indoor Free Flight World Championships.

The beauty of structure, form and aerodynamics all displayed in this fine example by Eduard Ciapola of Poland.

World Champion of 1974, Ryszard Czechowski of Krakow, Poland.

Dieter Sebenmann of Zurich, Switzerland, developed this long tail moment design from theoretical optimization studies.

Bud Romak, 1976 World Champ, set to launch on winning flight.

Off to Plovdiv, Bulgaria, from Victoria Station.

Channel crossing to Oostende, Belgium.

All aboard the "Orient Express".

Trackside scenery near Linz, Austria.

World Championships Hotel and headquarters, Plovdiv, Bulgaria, 1975.

Eugeny Verbitsky, USSR, considerately displayed the details and inner workings of his elegant F1C model. Second place 1975.

Off to Paris and on to Marigny and the Pierre Trebod, the perpetual beauty of France beckons.

The pleasant village of Gaye, France. Pierre Trebod award reception and banquet held near church at upper center of photo.

Hans Seelig, Munich, checks over his shoulder wing streamliner.

Whoops! Danish model comes to grief.

Tom Prosser of Australia farms wheat and flies fine Wakefields.

Tom Koster, 1977 F1C World Champion.

Judy Barnes holds brother David's Nordic. Sixth place in fly-off, 1975 Pierre Trebod.

Impish Klaus Wetterburg of Denmark lights fuse on Bob White's Wakefield, proxy flown by France's Pierre Chaussebourg at '76 P/T.

Peter Seelig receives his Pierre Trebod cup.

Beautiful cups, in the tradition Criterium Pierre Trebod.

New style F1C model by Roy Collins, Great Britain.

Lining up at Le Gayon bar.

Le Cheval Gris, happy headquarters for many of the Pierre Trebod regulars, country wedding in progress.

Early Wakefield test flight, Roskilde, Denmark.

World Champions Tom Koster, Denmark, F1C, Kim Dong Sik, N. Korea, F1B and Kostadin Abadjiev, Bulgaria, F1A, 1977.

Sergey Sharron, USSR, at Roskilde with sharp new model. Dural skinned wing, very clean cowling, fine attention to details and attractive trim.

The Little Mermaid perches on a rock in Copenhagen harbor, symbol of a gracious people, the Danish hosts of the 1977 Free Flight World Champs.

The Lord Wakefield Trophy. Witness and tribute for fifty years of the grand free flight spirit of international sporting competition.

8 Design Survey

F1A Nordic A2 Towline Glider

	Designer	Country	Source
1	Abadjiev	BG	WFFP
2	Allnutt	CDN	Free Flight News
3	Baguley	GB	Free Flight News
4	Baines	GB	WFFP
5	Banerjee	India	Free Flight News
6	Bauer	USA	Scatter
7	Barnes	GB	Free Flight News
8	Challine	France	WFFP
9	Chancey	USA	Scatter
10	Deubel	BDR	Free Flight News
11	Dvorak	CSSR	Free Flight News
12	Ekhtenkov	USSR	Free Flight News
13	Gronlund	DK	Free Flight News
14	Grunnet	DK	Vol Libre
15	Hacken	HOL	Scatter
16	Hines	USA	Scatter
17	Horejsi	CSSR	Scatter
18	Isaacson	USA	Scatter
19	Lepage	France	Free Flight News
20	Lepp	USSR	Free Flight News
21	Lepp	USSR	Free Flight News
22	Lopez	Spain	Aeromodeller
23	Motsch	BDR	Vol Libre
24	Murray	SA	Free Flight News
25	Parker	USA	Scatter
26	Pontanari	Italy	AVA
27	Seelig	BDR	Scatter
28	Simons	AUS	MAG
29	Spooner (SPL)	GB	WFFP
30	Stewart	GB	Free Flight News
31	Tchop	USSR	WFFP
32	van Eldik	HOL	Free Flight News
33	Vollbrecht	BDR	Free Flight News
34	Walters	USA	Scatter
35	Wilkening	BDR	WFFP

F1B Wakefield

	Designer	Country	Source
1	Barnes	GB	Free Flight News
2	Benedini	RA	Free Flight News
3	Davis	USA	Scatter
4	Dupuis	France	Vol Libre
5	Edwards	AUS	WFFP
6	Ghio	USA	Scatter
7	Gildersleeve	USA	Scatter
8	Gildersleeve	USA	Scatter
9	Gouverne	France	Vol Libre
10	Greaves	GB	Free Flight News
11	Hofsass	BDR	Free Flight News
12	Kongsberg	DK	East Anglian News
13	Kycklingen	S	Free Flight News
14	Loffler	DDR	Free Flight News
15	Mangiarini	Italy	AVA
16	Neumann	BDR	Bartabschneider
17	Pavely	GB	East Anglian News
18	Pollard	GB	Aeromodeller
19	Ruyter	HOL	Free Flight News
20	Samokish	USSR	WFFP
21	Sik	PRK	WFFP
22	Smitz	USA	Scatter
23	Szen	PRK	Free Flight News
24	Thomas	CDN	Model Airplane News
25	Van Leuven	AUS	Free Flight News
26	Wantzenriether	France	Free Flight News
27	Wells	GB	East Anglian News
28	Wells	GB	East Anglian News
29	White	USA	WFFP
30	Woodhouse	GB	Free Flight News
31	Zachalmel	A	Bartabschneider
32	Zetterdahl	S	Modellflygnytt

F1C FAI Power

	Designer	Country	Source
1	Anderson	USA	Scatter
2	Bond	GB	East Anglian News
3	Carroll	USA	WFFP
4	Collins	GB	Scatter
5	Collins	GB	WFFP
6	Faux	GB	Free Flight News
7	Ferrero	France	Modele
8	Gieskieng	USA	Scatter
9	Gieskieng	USA	Scatter
10	Koster	DK	Modelflyve Nyt
11	Lyons	USA	Scatter
12	Nygren	S	Free Flight News
13	Olofsson	S	Modellflygnytt
14	Phair	USA	Scatter
15	Reda	BDR	Free Flight News
16	Rounsaville	USA	Scatter
17	Schaller	CH	WFFP
18	Seelig	BDR	WFFP
19	Sharon	USSR	Krilya Radiny
20	Truppe	A	Vol Libre
21	Verbitsky	USSR	Modellflygnytt
22	Wolff	USA	Scatter

F1D Indoor Microfilm

	Designer	Country	Source
1	Andrews	USA	Aeromodeller
2	Barr	GB	WFFP
3	Blount	GB	Free Flight News
4	Czechowski	POL	Aeromodeller
5	Kowalski	USA	Indoor News & Views
6	Richmond	USA	WFFP
7	Romak	USA	Romak
8	Servaites	USA	Servaites
9	Siebenmann	CH	Siebenmann
10	Thomas	CDN	Free Flight News

F1G Coupe d'Hiver

	Designer	Country	Source
1	Callegari	Italy	AVA
2	Dowsett	GB	Aeromodeller
3	Dupuis	France	Dupuis
4	Linstrum	USA	Linstrum
5	Martegani	Italy	AVA
6	Matherat	France	Matherat
7	Pavely	GB	East Anglian News
8	Savage	CDN	Free Flight News
9	Van Leuven	AUS	MAG
10	White	USA	WFFP

F1A

F1A 1

Etalon - Kostadin Abadjiev, Pazazdjik, Bulgaria

Kostadin is no newcomer to the FAI Free Flight scene although the name was not well known in the West. He has been flying in the World Champs since 1953 and was in fact tenth place in 1971 at Säve, Sweden. He flies Power also and did so at Roskilde. His entry in F1A was a last minute decision.

Etalon represents an excellent example of a strong and serviceable tactical airplane that at the same time gives little ground in efficiency to more extreme examples. An Isaenko-type hook is used, made up from sheet metal and giving circle and zoom release at 2.5 to 3.0 kg unlatch tension. Airfoil is similar to the B 7457 and wing is fully sheeted top and bottom with 1.4 mm balsa. There is no TE member, the sheet is beveled together. Ribs are a 30 mm pitch and have balsa webbing between. A unique aluminum tube wing spar is used that tapers in wall thickness from 1 mm at the root. Bore is 4 mm. Wing attachment is by a single 4 mm steel rod 52 mm long together with a short 2 mm pin nearer the TE. Forward fuselage is an aluminum tube capped with a machined nose that carries an internal adjustable nose weight. Diameter is 25 mm and a Seelig-type timer is semi-submerged within the tube. Tail boom is a balsa tube of two laminations of 1.5 mm. Stab weighs 12 g and an I beam balsa spar of 16 mm top and 10 mm bottom flanges. Planform is rectangular with ribs at 20 mm spacing. Covering is tissue. The wing is covered with silk and has a braided string turbulator on the upper surface 5 mm aft of the LE. Wing weight is 170 g. There are large wheels on the autorudder adjustment screws to allow easy setting without tools. A well-built, tough, efficient Champion.

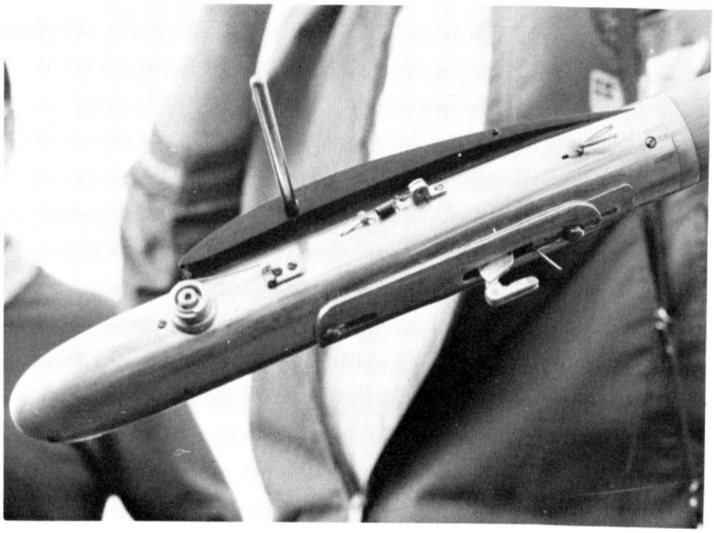

Abadjiev's Nordic used machined dural front end, Isaenko type hook.

Top F1A sportsmen of 1977. Andres Lepp, second place; Kostadin Abadjiev, first place; and Werner Kraus, third.

F1A 2

PA-20 Nordic - Peter Allnutt, Ontario, Canada

Peter has campaigned this model all over the world, and with good effect. Most memorable success was the second place at the Plovdiv World Champs, 1975. The outrigger elastic turbulators used were more popular a few years back, having given way to the less vulnerable surface strips. But Peter feels he gets better results with the outriggers. A number of modifications to the tow system have been tried with the current system being an impulse release working off a swinging arm that controls the circling rudder position. A mild zoom is also possible with careful handling.

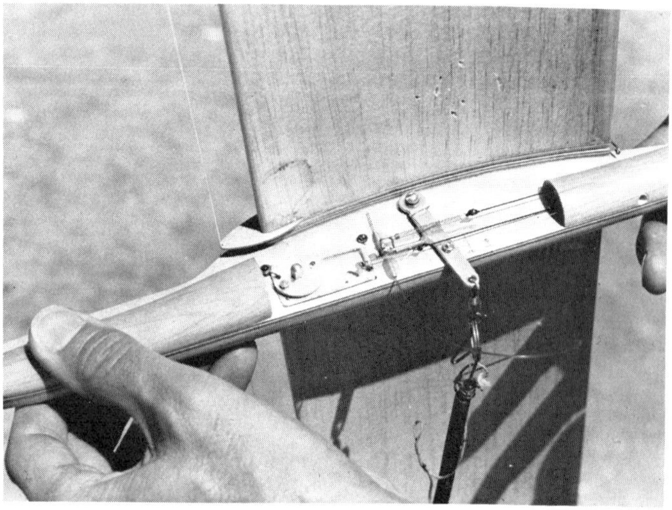

Newer version has sheeted wings. Note impact marks made by recoil spring. Allnut PA-20.

F1A 3

Nordic by Jim Baguely, Worlingham, England

The design was conceived in our caravan at the 1972 Nationals, when, having broken an A2 in the 30 knot winds and an A1 partly consumed by sheep downwind, drawing it seemed the best thing left to do! The idea was an all weather A2. The *Classic* (as in *F.F.N.* and *Aeromodeller*) tail moment was reduced from 28" to 26", tail area increased from 72 sq. inches to 80, wing chord increased from 5¾" to 6" and span reduced from 82" to 78" accordingly. The wing and tail sections were, I must admit, sketched. The first two, one of which I still have, had an all sheet top surface to the wing and tail construction has changed slightly.

They flew well — so well I stopped using the *Classic*, and did as well in contests as my current state of patience after flying free flight for too many years will allow. Even in *dead* air they were full of surprises like maxes at 5 AM in British Nationals which shouldn't happen? The Allnutt empirical/statistical/instinctive formula predicts 145 secs compared with 165 secs for the *Classic*, yet I would favor this one in such conditions. One small quirk was that tow could get out of control in rough weather. Everything — theoretical to desperation was tried but still it was an occasional let down. I can only think now that despite the sheeting it was an unbalanced wing structure causing twist under load.

For the 1976 Trials they were built with flat center section and tapered tips. They were useless! Not that I think anything is wrong with such a layout, only the particular combination I tried, so they were modified with new tips and polyhedral. A little experimentation to find the correct towhook position and the occasional tow problems of the first version are gone, presumably with the very tensionally rigid center sections now used.

In conclusion, as the model now stands I see no need for any changes. It glides and tows as well as anything it is matched against, is tough and reliable and *I know it well* so I can judge when it is in lift on the tow. A maximum or no-maximum is a question of patience and gamesmanship, i.e. how much you want to win and are prepared to put into flying to do so. Only an infallible thermal detector, luck or circle tow (as has been ably demonstrated by continentals for years and our 1977 team more recently) will do better. Dead air performance is not all, but good flying characteristics and reliability are!

Brian Baines of the RAF Model Aircraft Association uses twin timers and fully visible works to the circle towhook.

ing better use of the air than the opposition. The boundary layer control exerted by the multi-spar layout is perhaps one factor. The model looks a bit dated but it still wins. The thick airfoil gives good wing rigidity and no flutter problems. The model zoom launches beautifully. I reckon the gain in height of perhaps 10-50 ft. depending on thermal strength. This is more important than the circle tow facility. The towhook system is fully described in FFN April 1975.

F1A 4

Sunshot 5 by Brian Baines, Lincolnshire, England

I have tried many different airfoils, tapered tips, high aspect ratio but am forced to concede the basic six-inch chord multiple spar Shoaf 4738DF wings with wash-in on inside of turn. They are the best models I have built and flown. Wash-in is very important balanced against rudder offset. Many times the model maxes even when I make a mistake and it has the uncanny knack of working its way back into lift and then mak-

F1A 5

G90 F3 Nordic - Prasanta Banerjee, Calcutta, India

The G90 is a very ambitious looking, potent design. Tests were planned to systematically test the effect of variations in airfoil maximum mean camber while keeping everything else constant. No information yet on those plans but Prasanta writes that performance is in the range of 150 to 160 secs, indicating sub-critical boundary layer flow separation has been avoided, probably by the self-turbulation of the spar/tissue roughness.

Prasanta is editor of the *Calcuttan*, the monthly newsletter of the Calcutta Model Aero Club (16 Surjya Sen Street, Calcutta 12 India). He has been working hard to spread interest in Free Flight in his country and would appreciate any help he could get toward that end. Biggest problems seem to be getting useful materials at affordable prices and also a shortage of knowledgeable instructors.

F1A 6

Zenith Nordic by Ken Bauer, Orange, California

I spent quite a bit of time designing the *Zenith* before I started building. I wanted a glider that would be good in both the early morning rounds and the afternoon thermals. The ship came out 1½ oz. underweight and since the tail was light, the ballast was placed slightly aft of the CG which was set at 55%.

The tow system, I think, is the best part of this airplane. There is no way that the line can come off until the line is tight overhead and is given a "twang". The system is beautiful because it is so simple with no extra gadgets to get fouled up. System idea came from the FFD December '74 which showed the Danish system. I added the lever shown in the drawing so I could get circle tow also. I can also get a zoom launch by just hauling the glider up fast and pulling down while swinging it around into its right glide circle. When the glider is in just the right position, and I am pulling a few pounds, I give the line a "twang" and the model zooms off. I need more practice with this set-up as that is most important.

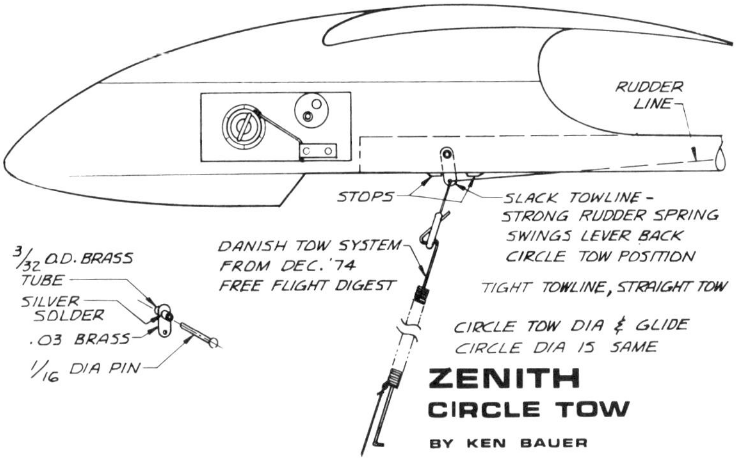

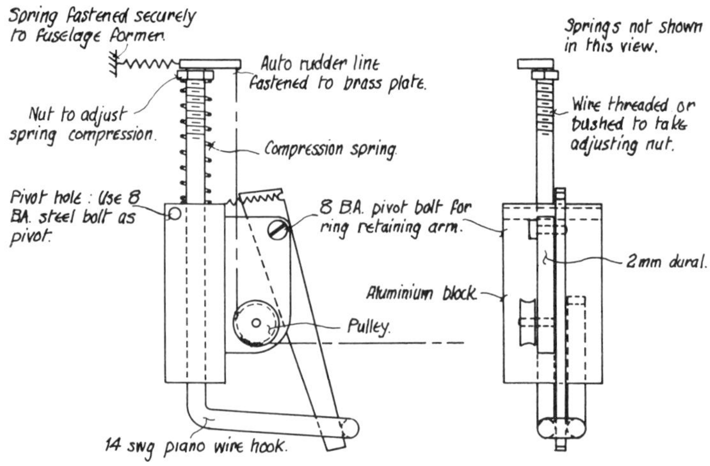

CURVED AIR - Towhook details

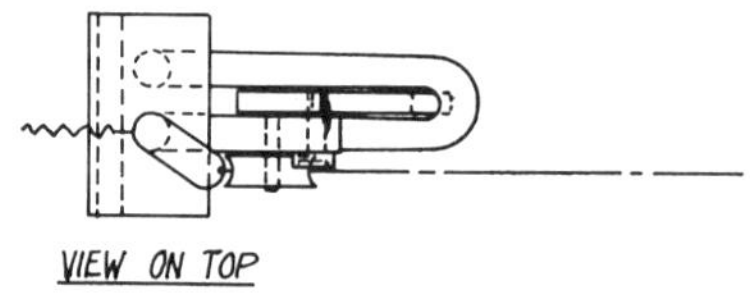

F1A 7

Curved Air - David Barnes, Liverpool, England

Dave started out with the basic outlines of Pavel Dvorak's '71 World Champion to which he added wing structural concepts from Langevin's *Osprey*. Objective was to get a strong and stable model that would withstand the stresses of a fast catapult launch. The tapered wing seems to contribute to stability in transition from circle tow, to zoom, to glide. The model is generally unsophisticated in construction but is effective in light lift and tricky conditions.

David Barnes added some Langevin-type structure to a Dvorak-style design.

F1A 8

Galcha N. 4 Nordic - J. Challine, Longport S/Orge, France

The French have been using side mounted towhooks as demonstrated originally by Jean Marie Berthe. The automatic circling produced can be tricky but the French have proved it manageable. The *Galcha* became the *Champion de France* in 1975 and the same year placed fifth in the World Champs at Plovdiv.

F1A 9

Hyperion Nordic - Don Chancey, Plano, Texas

Don used the Czech-style hook for his initiation into the thrills of circle tow with the *Hyperion* serving him well. The basswood spars are an indication that high zoom loads were anticipated. Generous washout at the tips was built in at initial construction and this gets the wing working more efficiently. An unfortunate line cross at the '75 World Champs went unseen and fouled chances for a good score.

F1A 10

Glider F1A - Arno Deubel, Saarbrucken, W. Germany

This Nordic placed 10th at Wiener Neustadt World Champs in 1973, and a similar model was flown there to 15th place by Wolfgang Schmelter. The solid balsa wing is made in the same manner as the S.P.L. Airfoil is that used by Rudi Lindner in his *Spinne*. Towhook is a springy wire that releases the ring on the line by impulse, but there is no circle tow capability. Arno flew at Plovdiv World Champs, 1975, and placed in the fly-off at 34th.

F1A 11

Saper Nordic - Pavel Dvorak, Prague, Czechoslovakia

Although this model is somewhat dated, being the 1971 World Champion, it is of interest because of its highly-flexible tapered wings which are becoming fashionable for zoom launches. Two rather deep spruce spars are used and these are reinforced with 0.5 mm music wire. Also of special interest is that this model is available in a kit and there are a number of examples now flying in competition in Europe. It is produced in Czechoslovakia but is available in Western Europe and England.

F1A 12

EVG 1172 - Valerij Ekhtenkov, USSR

This is the 1973 World Champion model flown at Wiener Neustadt, Austria. Ekhtenkov nosed out Vladimir Krejcirik of Czechoslovakia in the exciting final fly-off held at dusk (7:15 P.M.). A Very pistol flare started the two with Ekhtenkov towing straight ahead to a spot where he circled. While the Czech released after a light tow-up the Russian whipped his model round and round and zoomed it off with some gain in height. The models both held well and then settled into some 500 spectators who hit the dirt to make room.

Wing structure is more complex than normally seen. More attention is paid to distribution of structure. The sheet box fuselage looks a bit old fashioned compared to FG tubes but it is lighter and stiffer.

Valeriji Ekhtenkov, USSR, uses D tube leading edge inbaord and open truss outboard all laced with diagonals.

Ekhtenkov's 1973 winner gets some attention here at Plovdiv,

F1A 13

Great Gull IV - Sven Gronlund,
Kopenhagen, Denmark

This glider shows a number of features that have appeared in the newer Danish gliders of Grunnet, Koster and others. Dating from the early 70's, the *Great Gull* series have attempted to improve wing moment of inertia, torsional stability and strength. Sven has actually measured moment of inertia of his wings. An earlier model, SG69 had I/S = 1030 gram centimeters. An improvement to I/S = 635 was gained with *Great Gull* IV. Torsional rigidity is greatly enhanced by the enclosed box spar. In spite of all this design effort the *Great Gull* IV wings have been broken three times. One time it crashed into building scaffolding. Balsa used was apparently a poor quality as the breaks across the grain appear to be so sharp that they could have been made with a razor. This is a sign of "wind check damage", caused during growth of the tree. *Great Gull* V will have a rolled balsa tail boom, front fuselage of glass fibre and a nicer flex joint. Impulse towline release is used.

Per Grunnet and Danish-style Nordic.

F1A 14

Jessica by Per Grunnet, Assens, Denmark

Jessica was designed back in 1971 (after returning from the Säve WC) and was intended for some kind of circling device (not the Lepp-type hook) which was finally built-in. I did not foresee the demands to strength that the catapult-launch implies. The wing is definitely too weak in the central part. An extra double-spar with webbing of plywood will do wonders in the inner 20-25 cms. I have actually broken the wing twice: the releasing tension for the hook is now 3 kgs.

The model was finished two months before the Bulgarian World Champs. It flew very well right from the start — the problem was that I didn't! We had our first child in 1972, and my modelling activity was low until the spring of '75.

The wing turned out to be very heavy, 85 grams for each wing-half and that's about 15 grams too much. I think that tapering spars, lighter and thinner spars in the tip and thin jap-tissue will make it possible to reach a weight of at least 75 grams per half, even with a stronger mid-wing. On a new model I have omitted the torsion box in the wing tip but used too much hardwood in the center-wing, so it has not become lighter.

The fuselage construction was first tried out in Denmark by (who else) Thomas Koster. It soon proved to be very good especially when we learned to use silicone-sealant for the glass-fibre front shells (your close-up of the *Jessica* front shows signs of hasty repairs!) and is now considered *Danish standard.* The boom is rolled from 1mm medium-light balsa and covered with thin glass-fibre. Total weight on the *Jessica* is 22 grams (without tailplane). It could easily be made lighter, Svend Gronlund has made one weighing 18 grams. The flex-joint has given me some problems when flying on ploughed fields: when sand gets into the flex-joint, the trim is spoiled.

The trim includes some warping of the wing. I have trimmed the model to make the best possible catapult-launch, and then trimmed the glide and the towing. This rather awkward technique has actually given me better results than my previous trimming where first I trimmed the glide, then the circling towing and last catapult.

You start out with a warp-free wing. Give both tips 2-3 mm wash-out. Now trim the model to tow straight (and make some reasonable narrow turns on the line) and try the catapult. The model will probably make — or try to make — a loop. Then in case of a turn to the right give the right inner wing about 1-2 mm wash-out. Now trim the tow straight, and try the catapult-launch. The model should go into a smooth catapult-launch gaining 4-5 meters height (if you pull hard

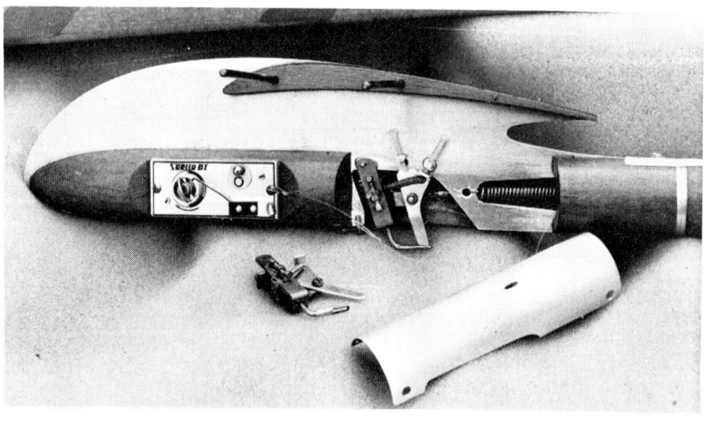

Danish Nordic fuselage details. Tail boom flex-joint tensioned with internal coil spring.

enough). If it stalls and the glide is OK then try a bit more wash-out in the inner wing (the right inner wing) but be careful. Too much wash-out may lead to a spiral-dive.

The rudder position in the catapult launch is, of course, important; too much rudder, and the model will spiral down, or at least get no gain in height. With a release tension of 3 kgs I find the model rather indifferent to the rudder position in the catapult. I hope to increase the release tension on my new model. Five kgs would be around the maximum of tension that you can apply in absolute dead-air conditions, so that will be my aim.

The model is trimmed with tilt in the tailplane to help the circling in the glide (this is not on the plan). The angle of tilt is about five degrees. The CG-position is important — if moved further back, the model will perform badly in turbulence.

I prefer the model to fly 2-3 circles per minute in calm air. In a thermal it will tighten up a bit. The model is not trimmed to the stalling point. I have tried, but it seemed that no increase in dead-air time was achieved, so I trimmed it back, it can fly straight without stalling now.

By the way, I've just built a tailplane with flat underside for the model. I will try it instead of the Hacklinger now used. I hope to get more speed in the catapult-release and thus more height, but I don't believe in it really. Once I tried it for an old model, and the result was a decrease in all respects, stability and flying time.

I may add that, of course, I trim the model whenever there is a possibility, and in spite of my flying a lot (in good weather every day) there is always something to change in the trim.

(Editor's Note:) Klaus Wetterburg, also from Denmark sent along some more information on *Jessica* that may not be apparent on the drawing. Wing joint is made with three pieces of 3 mm piano wire fitted in brass tubes. Ribs in wing and tail are 1.5 mm balsa. The leading edge box is made from 1.0 mm strong balsa (18 g per 10 cm x 150 cm sheet) in the center and from lightwood (10-12 g) in the tips. L.E. and T.E. are from strong, hard balsa in the center and light in the tips. Fin is glued together from two pieces of 1.5 mm quarter-grained balsa with a mylar hinge. Rear fuselage with fin, tailplane holder, etc. (without stab) weighs 25 g. About 60 g ballast is used after location of the CG at 55%.

F1A 15

Pretender Nordic by A. Hacken, Sittard, Netherlands

Our contest season is already ended and I have to say that I didn't make the team. This because of a few reasons. The first reason is that the others were a bit too good. Another reason why I couldn't make the team was, that I lost three models at critical moments this season. And although I build my models fast, a loss of three models is too much. But I am not discouraged by this, because it is something that is in the game. At the moment I am working on two new models, which are specially designed for catapult launching. These models will also have a VIT system which will work as follows: during straight towing the stab will be at glide incidence, during circle tow it will have negative incidence (T.E. up for about 2 mm), when speeding up the model before launching the stab will have the normal incidence again, but after the model is released into its glide turn the stab will again have a negative incidence for about one or two seconds. With this I will try to obtain that the model will lift its nose a bit so that more height can be gained.

Apart from this system I am going to equip all my models with an electronic, FM, recovery aid. With this recovery aid I hope that the models will be more easily found when they land somewhere in woods which are plentiful at our regular flying fields.

At the moment we are also working in our team at epoxy-glass cloth front ends for our gliders. The first results are very good. The front ends are strong and very smooth too. For the tail end we have changed to the Ronytube rods.

I would like to tell you in which way the Dutch team is elected, but I honestly have to say that this method is only possible in a small country like the Netherlands. The year before the Championships (European or World Champs) there are six club contests held on the same place (Rozendaalse Heath near Arnhem). The best score from four out of these six contests are counted together and the three people with the highest score out of these four contests are our team.

The advantage of this system is that one has one year's time to prepare himself for the European or World Champs. The disadvantage of this system is that someone who is 'on form' in the year the Champs are held can't make part of the team.

F1A 16

Ultimax II Nordic - Lee Hines, Costa Mesa, CA

The *Ultimax II* represents a current design trend that is not at all out of the ordinary. It avoids extremes that might prove impractical. The CH407 wing at 5.75 root chord is a good choice. Strength and stiffness of the wing appears adequate, with plenty of diagonals, and spruce at key points. The FG boom represents a concession to the practical, perhaps *"Super" Ultimaxes* will have lighter balsa booms?

Lee usually goes off to the far side of the field and finds his own lift after circling. He is usually followed by a swarm of HLG nuts (or vice-versa?). The W.F.F.A. annual competition at Taft in F1A has twice fallen to Lee's expertise and now he holds the U.S.A. National record in F1A.

Czech.Team for 1977 shown at one of their practice sessions. From left to right: C. Patek, V. Patek, Adlt, Hubert, Dvorak, Horejsi, Kubes, Klima and Rado. Team manager Jiri Kalina takes it easy in the grass.

F1A 17

Bohemia Nordic by Ivan Horejsi, Pilzen, Czech.

This model was built in the '72-'73 winter and though I have built other gliders since, none of them have been better. This ship won the 1974 Czech Champs in very tough competition in a fly-off after 14 consecutive maxes.

Model uses a Seelig timer and the Czech-style cicular hook. Boom is of fiberglass made by a friend, and it is on the heavy side. As for the airfoil, it is the B7457d on top and the Lindner on the bottom. A surface turbulator is used. bottom of the wing is covered with both silk and modelspan. More recently (77WC) I have equipped it with a delayed rudder deflection. The rudder does not deflect to its full glide position at the moment of the zoom but 5-10 seconds later. This system proved to be of great efficiency as it, in fact, removes the great problem of spinning-in instead of gaining extra height.

I discovered this system when flying my R/C soarer some three years ago. I have been using it since. I understand that some of the Russians are also using this system.

F1A 18

Wishbone Nordic by Bob Isaacson
Westminister, CA

The *Wishbone* was built in June 1974. It is my eighth Nordic and first effort at designing (or copying) my own. I wanted a model that would circle-tow, zoom-launch, thermal well, and get better calm air time than the *Ultimate Dragmaster*. So far the only objectives it seems to meet is that it thermals very well (even in very light ground lift) and does between 160 and 170 seconds in calm air. However, my efforts at circle-towing have resulted in several bad crashes.

I have found the cause of some of my problems with circle-towing. If I can find a circle system that is reliable and learn to *use* it I will equip two of my new models with it. Then lots of practice and hope for the best.

(Ed:) Bob has been looking quite good lately with an acquired ease at circle towing, zoom launch, thermal sensing, etc. Bob does gain altitude on his zoom although attempts by your editor to show this on film haven't been too successful. Bob has a new wing structure for his Nordic that is much stiffer and doesn't flutter like his previous ones. He has been using the fibre glass boom and likes it (it is a very light one) but would also like to try a rolled balsa boom. The one bit of training that he would like to get more of is the crowded body-contact launch scramble usually seen at W/C. That is a condition one must be prepared to cope with as you cannot hold back and hide from the maddening crowd.

F1A 19

Pam-R-Veilleux - Remy Lepage,
Bougival, France

The 1973 Pierre Trebod winner shown here, has a rib and spar wing construction with the inboard panels covered with sheet balsa while the outer panels are covered with tissue. The port wing carries one degree more incidence than the starboard. A glass fibre reinforcement is used on the TE. Timer shown on side of fuselage is not mounted on top in the newer version called *P. A. Mystique*. This design also has the rudder mounted aft of the stabilizer.

F1A 20 & 21

AL19 & 29 Nordics - Andres Lepp,
Tallinn, Estonia, USSR

The AL 19 dates to 1971 and earlier, reflecting the earlier Russian design trend using an aluminum spine in the nose for wing attachment. Compared to AL 29, which Andres flew at Roskilde in 1977, it can be seen that there are many similarities in both design and structure. It might also be added that some of the other top Russian fliers deviate much from these general characteristics.

Lepp hooks up while Tchop stands by. Uses slim box fuselage, subrudder.

Similarities between the two models include wing airfoil (B6356-b) and stab, structure of wing and stab and underslung fin. Differences include a slightly smaller stab and larger wing on AL 29 and 125 mm tip dihedral instead of 140 mm. Inner panels on AL 29 have a small amount of dihedral. An auto-aileron was tried on AL 29 to help control the zoom pattern but this was abandoned in favor of a time operated rudder. At Roskilde the aileron was replaced with original structure. The many repairs on this model indicated a vigorous training program. And this no doubt is what makes Lepp such an outstanding flier. At Plovdiv Lepp had the misfortune of losing his best models under the wheels of a car (AL 25) just before the start of the fly-off but still placed 10th. At Roskilde, in a very close, hard fought contest, he placed second.

There has been some speculation in the modelling press as to why Europeans do not use wash-in on the inside wing panel to control thermal spinout, while US models do. First, it is not correct to make such generalizations, continent by continent. Secondly the assumption that European thermals are gentle and small compared to the US is also faulty. Differential wing warps can be a powerful tool in getting a specific flight pattern under a fixed set of conditions. But on the other hand it is speed sensitive and can hamper overhead circling and towing. Lack of torsional rigidity often contributes to thermal spinout as does, of course, unintentional washout on the inner side. If these are attended to, a no-twist wing should work fine.

Lepp's AL25 held here by Verbitsky at the 1975 World Champs. Structure carefully repaired a number of times.

F1A 22

Euro 297:MT - Jesus Lopez, Madrid, Spain

This Nordic has a good win record although we have little news of activities in Spain. Fuselage design dates to that of the Ritz *Continental*. Although strong, it is now felt that light weight at the extremities is more important. Lopez made the fly-off at the Plovdiv W/C and placed 30th.

F1A 23

Muli 77 Nordic - Hermann Motsch, Schwallbach, Germany

Herman won the 1977 Pierre Trebod with this model in a ten man fly-off. The towhook is the Russian style of Andres Lepp, with circle and zoom function, but no rudder delay. Wing airfoil is a modification of the B8356 b3. Mean camber line is unchanged but the upper surface ordinates are 90% original while the lower surface is 110% of the original. Model glides to the right and carries ½ deg of wash-in on the right wing. Separate wing panels balance at 34% of the semi-span. Wings weigh 190 g and have two I beam spars with spruce caps. Additional caps and webbing are carried inboard. Still-air time with catapult start is believed to be 190 sec.

Hermann Motsch, winner of the 1977 Pierre Trebod.

Wing details of Motsch's "Muli."

Underneath details of "Muli."

F1A 24

Faifly 72-1 - Neil Murray, Natal, S. Africa

This model, although somewhat dated, has given good results with a good, bouncy glide. Basic design is an off-shoot of Elton Drew's *Lively Lady*. Biggest difference appears to be the wing structure which is fully sheeted with 1/32 balsa. Triangular turbulators are built into the top sheet near the LE. Airfoil LE radius has been reduced from the *Lively Lady*. Towhook is a home-grown version of the Russian hook and is easier to make. Fuselage forward pod is a neat affair bent-up from sheet aluminum wrapped around a balsa core.

F1A 25

Toker Nordic - Jim Parker, El Paso, Texas

Jim comments that he started out in the 1973 semi-finals with a six-inch chord rectangular-tip ship that he has evolved into the current *Toker*. He tried a lot of different stabs and also a thicker symmetrical rudder. The original fuselage and towline were co-designed with Jim Farmer and have been kept throughout the development period with only minor modification.

The *Toker* is said to be tough and reliable and the wing structure is exceptionally warp resistant. A set of 1/8 and 3/32 inch diameter wing wires were permanently bent 25 degrees in high winds with no damage to the wings. With the PR-7 airfoil it is an all-purpose glider. Another version uses a hybrid airfoil made up from the CH407 for the first 10% and the GF-6 for the aft 90%. The hook shown has two main parts, the main support which mounts the unit to the baseplate and the cage which carries the spring, are 3/8 in. aluminum. The hook is threaded at the top and is locked to the brass spring stop at the upper end of the spring by a #2-56 screw. This screw slides in a slot in the spring cage. The idea is to keep everything oriented properly, and also to limit down travel of the hook.

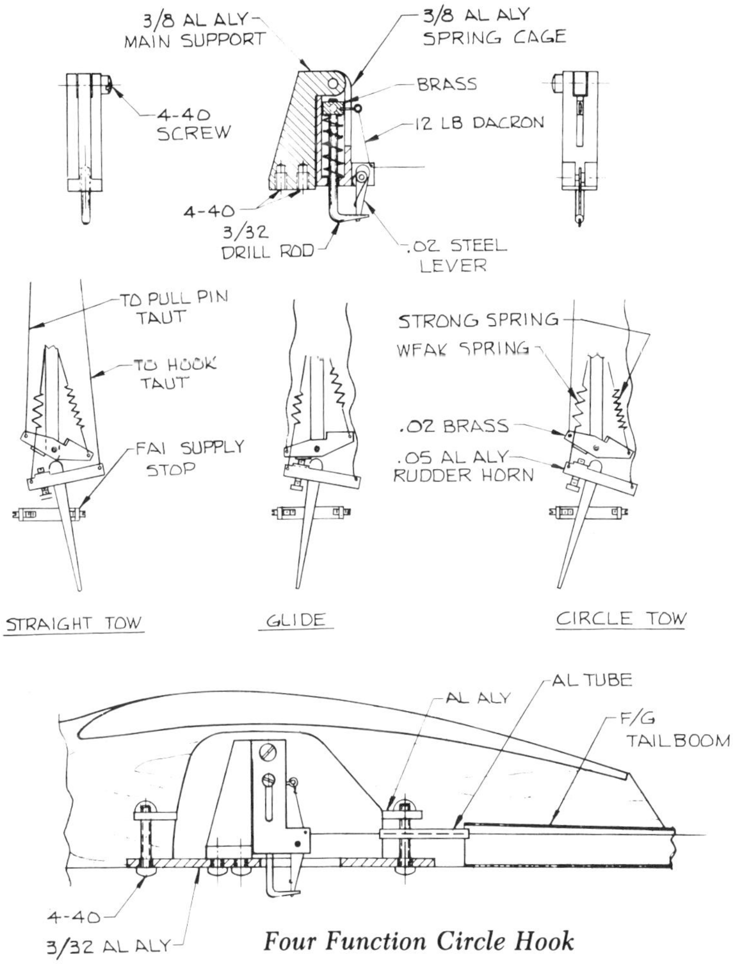

Four Function Circle Hook

Jim Parker, El Paso, Texas

The hook release lever is held into the slot at the rear of the spring cage by a 1/15 in. O.D. brass tube which is slotted and spread a little at each end and to keep it in place. The lever is mounted on the 1/16 in. tube at the rear of the assembly with the torsion spring on one side and a spacer of 3/32 in. O. D. brass tube on the other. See rear view of the assembly.

The circle tow rudder assembly is similar to many we have seen, except that Jim has managed to simplify the works. Both the pivoting idle bar and rudder control horn are made from bent sheet parts. The idle bar has a 1/4 in. long piece of brass tube soldered to it at the pivot point to help the bar rotation stay in a fixed plane. A tang on the idle bar is bent up and the threaded tang on the rudder control horn is bent down, so that the adjustment screw will line up and operate the rudder in the glide mode. This means that the idle bar is to be mounted in the rudder a little lower than the control horn, to allow the screw to line up properly.

Jim now uses .008 in. braided control line for a rudder cable, instead of 12 lb. Dacron shown in our sketch. Make sure the stronger spring is the one which pulls the idle bar to neutral position against the side of the rudder.

F1A 26

A2-16 - Fabio Pontanari, Varese, Italy

A somewhat different approach is used on this Nordic to provide circle tow. As the tow ring slides forward on the springy wire hook it also rocks the assembly and pulls the rudder straight. Less tension and the ring slides aft letting the rudder return to the circle position. A strong pull in the aft position releases the ring and a pull-out pin to allow the rudder to turn to the glide position. Circle tow direction is opposite to the glide direction.

F1A 27

Sigermodell '75 P/T - Peter Seelig, Munich, Germany

Peter's model won the 1975 Pierre Trebod competition in a fly-off with seven others. (Werner Kraus was second). The high aspect ratio fully tapered wing looks very efficient, although the section used, Go 803, has considerable undercamber and does not appear suitable for high speed zoom launches. And in fact Seelig's hook allowed no circle function. The springy wire hook (Hofsass type) releases by impulse. The multi-spar wing also looks to be lacking somewhat in torsional stiffness. No doubt this will be attended to when he adopts the circling style.

Forward fuselage is made up of a neat fiberglass molding to which the F/G tail boom is neatly integrated. Timer, a well cared for Seelig, no doubt.

F1A 28

Illusion - Dave Simons, N. S. W. Australia

Dave sent two models to be proxy flown at the Plovdiv World Champs by Willi Kamp. A certain amount of hard luck ensued, including the loss of the No. 1 ship while testing. Following the W/C experience, the remaining model (*Illusion*) was being returned to Australia by Allan Edwards and Dave Anderson. They stopped off at the Pierre Trebod and among other things arranged to have your editor proxy fly *Illusion* there. Since the editor's Nordic was lost somewhere in England, this all made some sense or at least it seemed to at the time. A rather rapid model/flier orientation program ensued including the taping of a pair of pliers to the nose (of the model). A creditable 15th place resulted and the Australian team carried on far into the night in celebration.

Since that memorable start Dave has built new models, flown in the W/C at Roskilde, at the Pierre Trebod (now on his own) and last heard from was gypsying over Europe with a band of fellow modellers.

F1A 29

The S.P.L. Glider by Brian Spooner, Purley, England

The S.P.L. present design is the culmination of over six years development by the Bavarian trio of Schmidt, Puttner and Lasczlo. Main feature is the high aspect ratio, solid balsa wing and the tee tail. In spite of the large span it is possible, by adjustment of dihedral and twist, to trim the model to fly in large diameter circles which tighten automatically in the presence of lift.

A number of slightly different versions have been tried by the S.P.L. trio, Riener Hofsass and myself. Those of Schmidt and Puttner have a 25mm O/D dural tube front fuselage with the timer (a cut-down Tatone) mounted internally (M.A.N. Dec. 1965); the nose is a push-on fit and is removed to set the timer. Hofsass and I use Seelig timers mounted horizontally in a cut-out forward of the pylon. Puttner flies models with the original section and planform with rounded tips, but Schmidt has changed to equal taper planform, squarish tips and uses the Thomann aerofoil; this he maintains, has increased the still air performance. Hofsass has used both the equal taper and the swept-back wing (again with squarish tips and the Thomann section) while I have retained the swept-back configuration and Thomann section for my two versions. Schmidt and Puttner also have storm models in which the root chord is raised to 160 mm and the span reduced to about 2200 mm.

Hofsass and I used a pylon/tongue laminated from two pieces of 1 mm dural and strengthened with a continuous strip of spring steel. Schmidt and Puttner use 2 mm, the tongue passing through a slot in the pylon and held in place with two riveted and glued-on face plates. The box in the wings is made with a 2 mm dural core and 0.6 mm dural face-plates; the core must be well-lightened otherwise the weight of the completed boxes is excessive.

In the hands of Schmidt and Puttner, this model has recorded an impressive series of competition successes, culminating in 1968 with myself, Puttner and Schmidt taking second, third and fourth places in the German Championships after a four round fly-off. For a "Sunrise" competition early in 1968, I re-trimmed one of my two S.P.L.'s and, although using a curved plate "thermal" tailplane, averaged 176 seconds which is the best time yet achieved with 50 meter line. Hofsass built a special fuselage of increased length for this type of competition and this, together with a high-aspect ratio tailplane, resulted in his first two flights in the same contest being over 200 seconds. Subsequent stalling on the line and in the glide ruined his chances of being the first person to average over 200 seconds.

Many people may be put off building this type of model because of the solid balsa wings; this should not be so, as they are reasonably easy and very satisfying to build. See the *Aeromodeller* for an article I have written on this construction technique. Herbert Schmidt resolutely maintains that he can build this type of wing quicker than a conventional one. I haven't his patience, but I estimate that apart from the wing boxes, I spent about nine hours on the construction of both halves.

F1A 30

Joke Rotary Nordic by Peter Stewart, Aldershot, England

The model has had three hard years of use and is still in reasonable condition. The first year was notable for its lack of success and I came very close to giving up the whole idea of circle tow. There weren't many to copy from in England at that time (1974) but still after a year spent learning the hard way I discovered I'd forgotten the glide trim in my efforts to get the towing right. Some added tailplane elevation and at last the model started to perform. Still air time is around 145 secs and is limited by the wing section. My latest model incorporates a section of less undercamber, other dimensions the same.

F1A 31

1975 World Champ Nordic - Viktor Tchop, Odessa, Ukraine, USSR

Wing structure of forward D box and I beam are Russian "standard", except in minor details. However Viktor uses a fiberglass tail boom (homemade) tapering from 17 mm to 10 mm. Front end is carved from lime. Towhook is built up from aluminum flat stock and can be made with hand tools without soldering. Straight tow, circle and zoom are all accommodated and adjustable. Wing airfoil is the B6356b formally favored by the Russian Nordic fliers. Viktor adds an

0.8 mm surface turbulator to his.

Of most importance to the high level of performance is thorough training and preparation. All adjustments are learned and applied as a function of wind speed in a regimen that is only changed after many check flights. This gives the flier a confidence to concentrate on tactics and maneuver.

Viktor Tchop, 1975 World Champ; model uses F/G tube boom, D box wing full span.

F1A 32

Casper Nordic - Tony van Eldik, Arnhem, Holland

Tony, at the age of 16, flew at the World Champs in Säve, Sweden in 1971. His model was in a nice fat thermal and it looked like he was heading for 11th place. Then the ship spun in due to a warped wing, dropping a disappointed Tony to 32nd place. From that experience he decided to improve his building skill and, incidentally, to study dental surgery. Now with a *Casper* fleet he has done very well, placing 13th at Wiener Neustadt, 5th at the Pierre Trebod and in high places in meets of his home country. He plans to get into Wakefield and to learn, he will start with a Coupe d'Hiver.

F1A 33

A246 Nordic - Martin Volbrecht, West Germany

This model uses an interesting, very thin wing section apparently developed by the designer. The wing tongue is attached at the bottom of the wing surface because of the thin section. The springy hook as used also by Arno Deubel (F1A 10) is fitted. Martin placed 16th at the Plovdiv World Champs.

F1A 34

Go Bird Nordic - Jim Walters, Tacoma, Washington

Jim decided some time ago to go after the Nordic event, hammer and tongs. And in fact when this editor first met him on his home field (Harts Lake Prairie) he had, besides the hammer and tongs, a cross-cut saw, bow and arrow, string, rope, axe and other miscellaneous retrieval gear (and they call it a Prairie?). Which is to say that Jim is a tough competitor who doesn't give up too easily. The *Go Bird* got him on the US Team for Plovdiv where he was doing great until the sixth round. A premature unlatch of his circle-hook got him in trouble, and 34th place. Bouncing back he made the team again in 1977 and flew at Roskilde. This time with a newer design called *Superbird*, Jim fought the difficult conditions to 15th place. An unusual feature of *Go Bird* is that it flies around with its tail in the air. The wing incidence of 8½ deg. and stab at 3 deg. puts the tail boom at about 3 deg. pitch. Jim's idea here was to get the stab up into less wake disturbance from the wing.

F1A 35

Joker - Fritz Wilkening, Trift, W. Germany

Fritz is an architect, 29 years old and has been model-flying since 1964. In 1972 he started flying on the international scene and he won his first competition, the Coupe d'Amsterdam. Here is what he says about his Nordics.

"Basic to the *Joker* was my *Mathi* which is still flown with success by my brother (see results of British Nationals '76 and Marigny '76) and several other friends here in the north of Germany. The *Mathi* is my Hirschel interpretation.

The *Joker* is an easy-going model. I have flown it under different conditions (light weather, strong winds) and I am very satisfied with it. When the wind blows more than 6 m/sec I put the offset hook into the middle and correct the towing curve. Glide is between 155-160 secs. Weights are; wing = 185g, stab = 10g, fuselage = 235g. Please note that the wing tapers linearly from 145 mm to 135 mm and not as shown on the FFN drawing.

Last year I read the essay by D. Siebenmann in the *Schweizer Aero-Revue* where he writes about theoretical and practical F1A studies. The article was so good that it inspired me to design a new model, called *Grunschnabel*. The *Grunschnabel* has the same wing span as the *Joker* and I use the same wing-section. Now I have completed two models of this new type and I hope they will be as successful as my *Joker*.

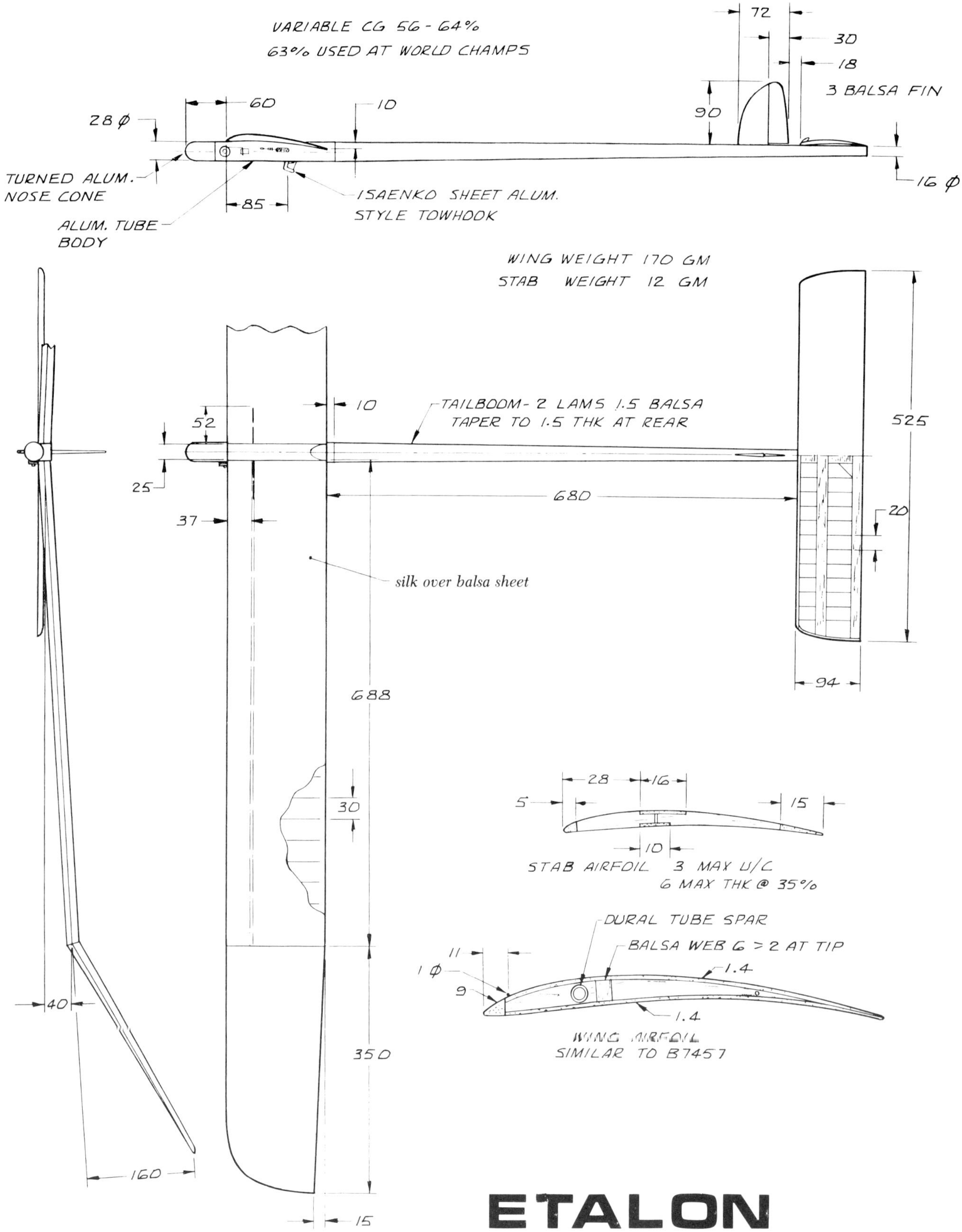

ETALON

1977 World Champion F1A

BY KOSTADIN ABADJIEV, BULGARIA

DIMENSIONS IN MILLIMETERS

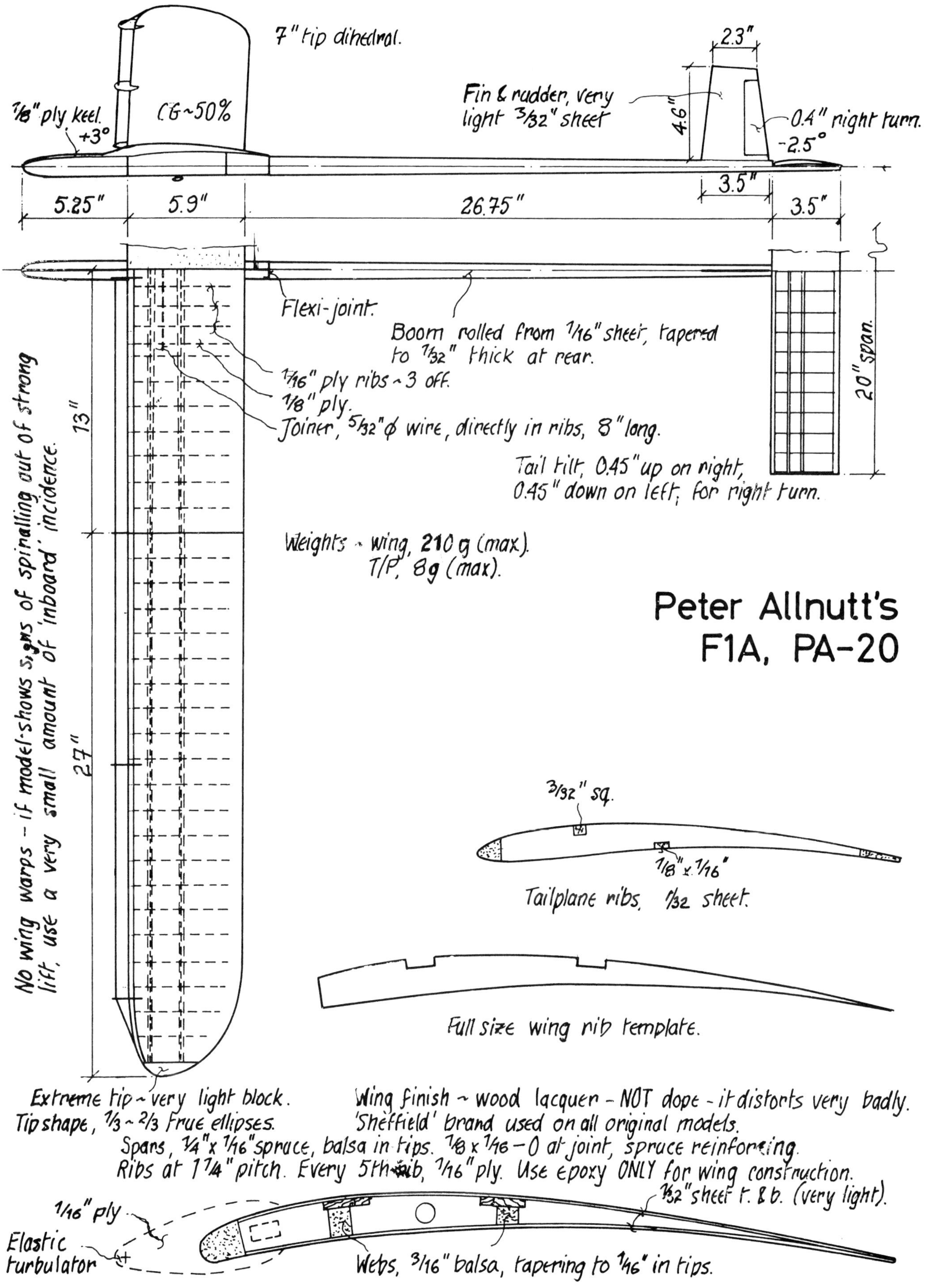

Peter Allnutt's F1A, PA-20

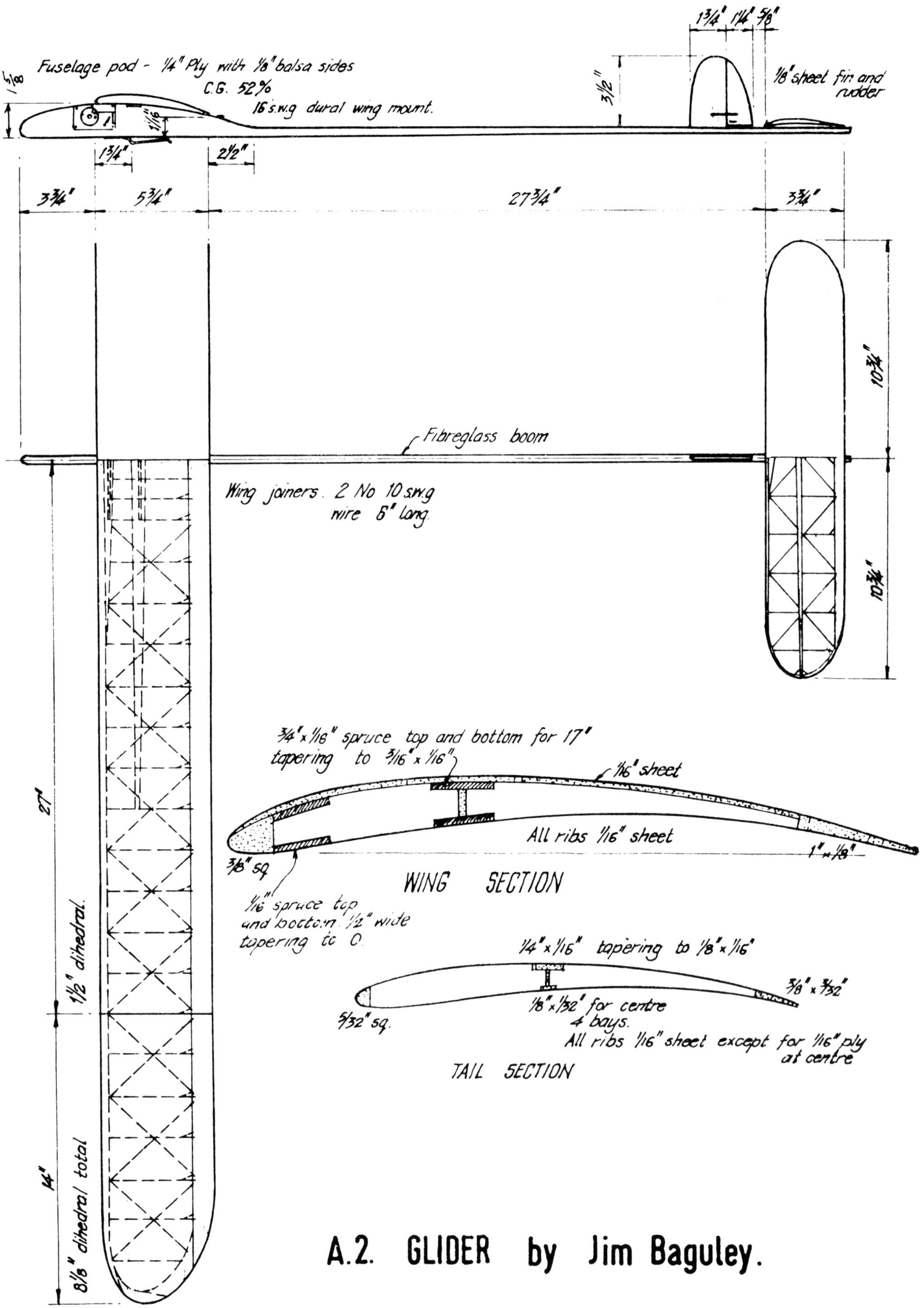

A.2. GLIDER by Jim Baguley.

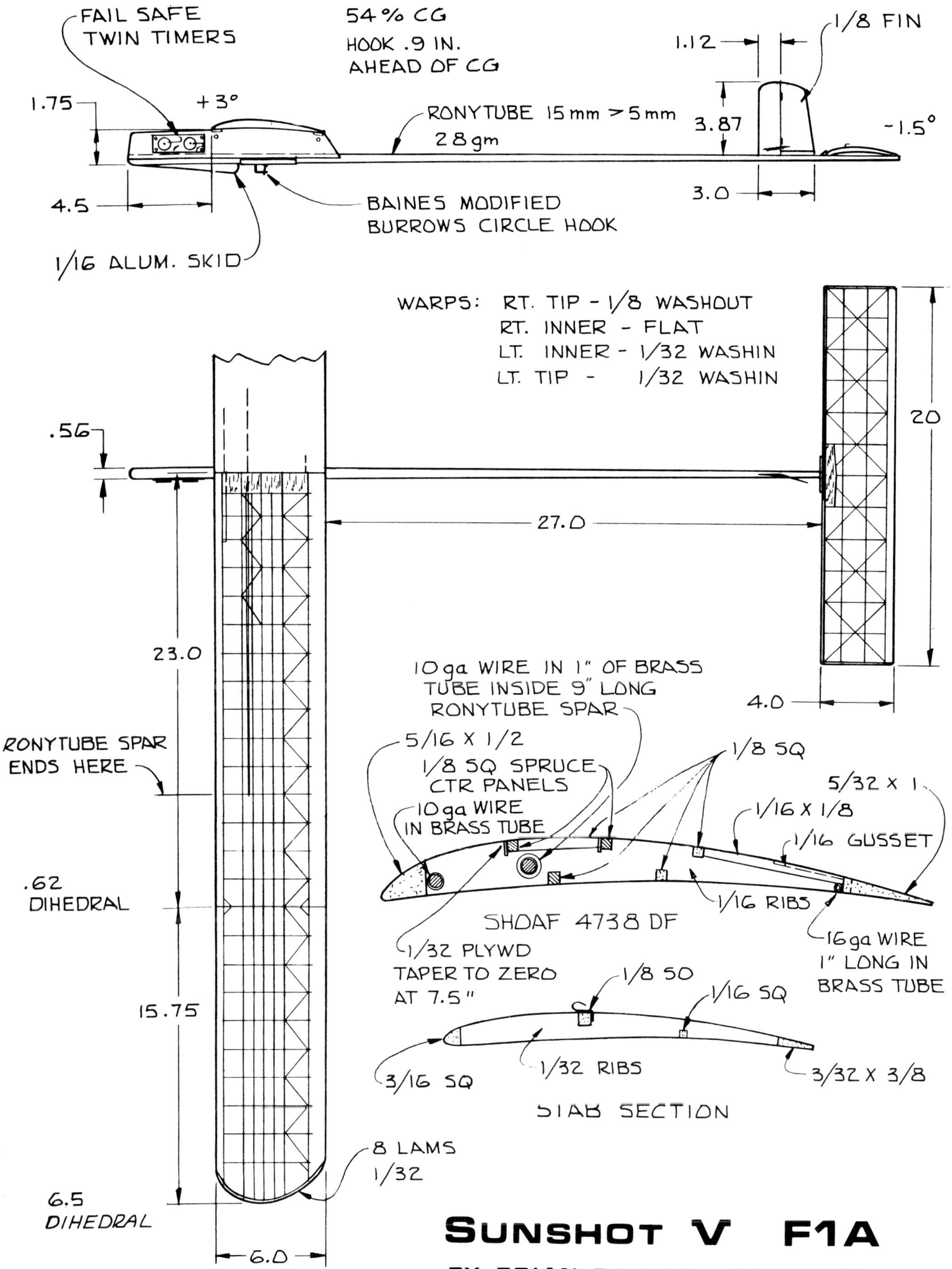

SUNSHOT V F1A

BY BRIAN BAINES — ENGLAND

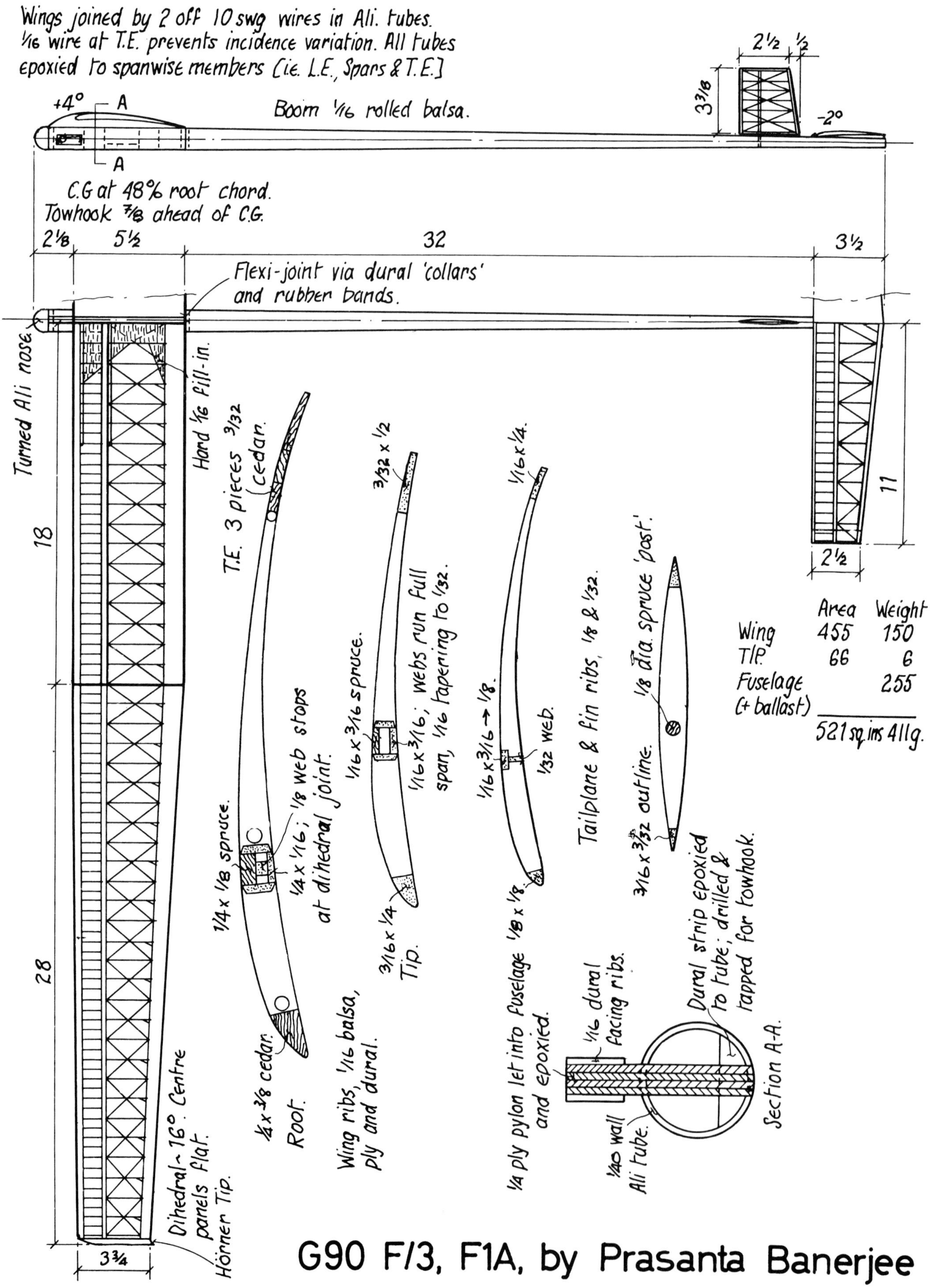

G90 F/3, F1A, by Prasanta Banerjee

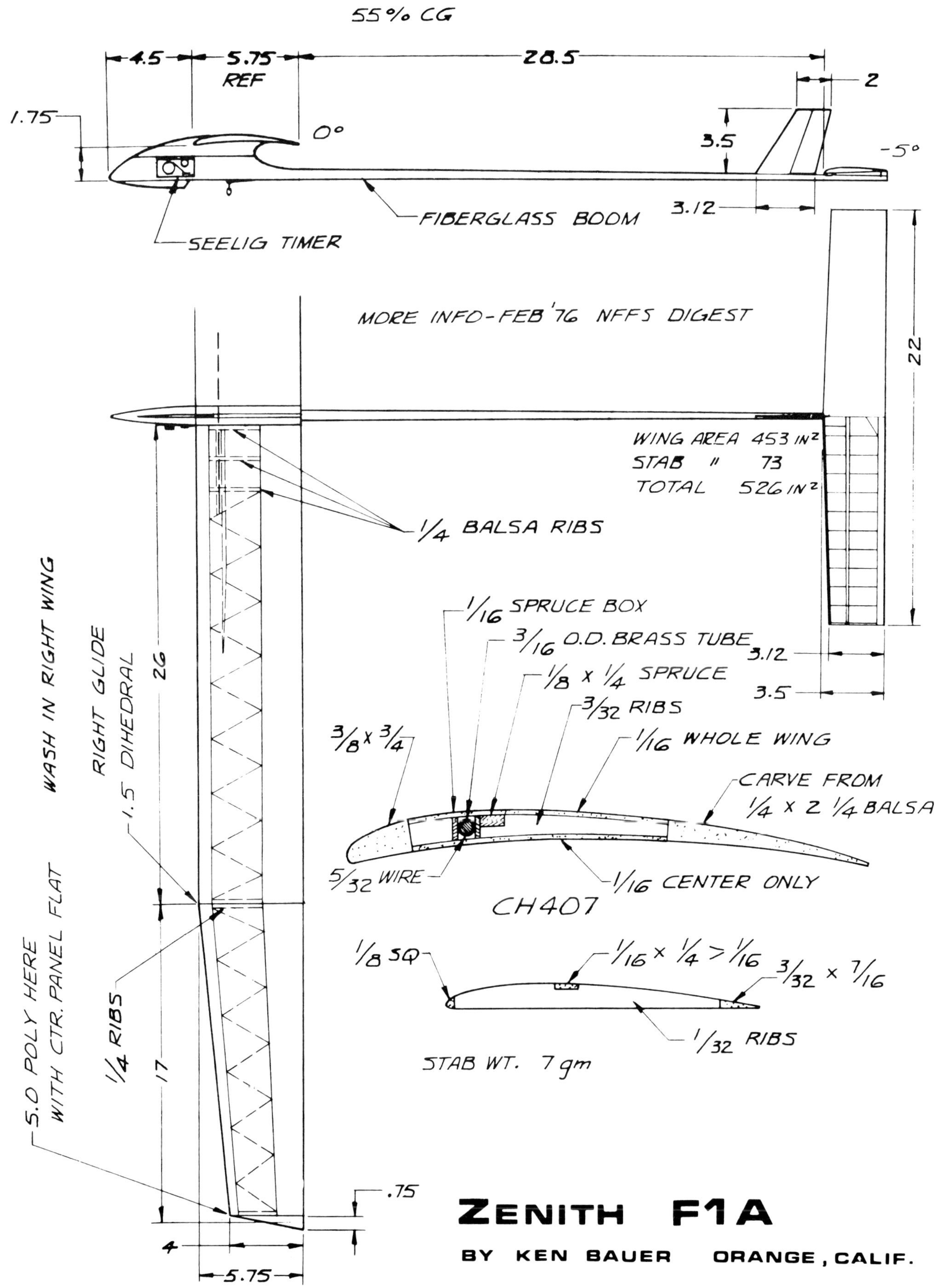

6 Bauer USA Scatter

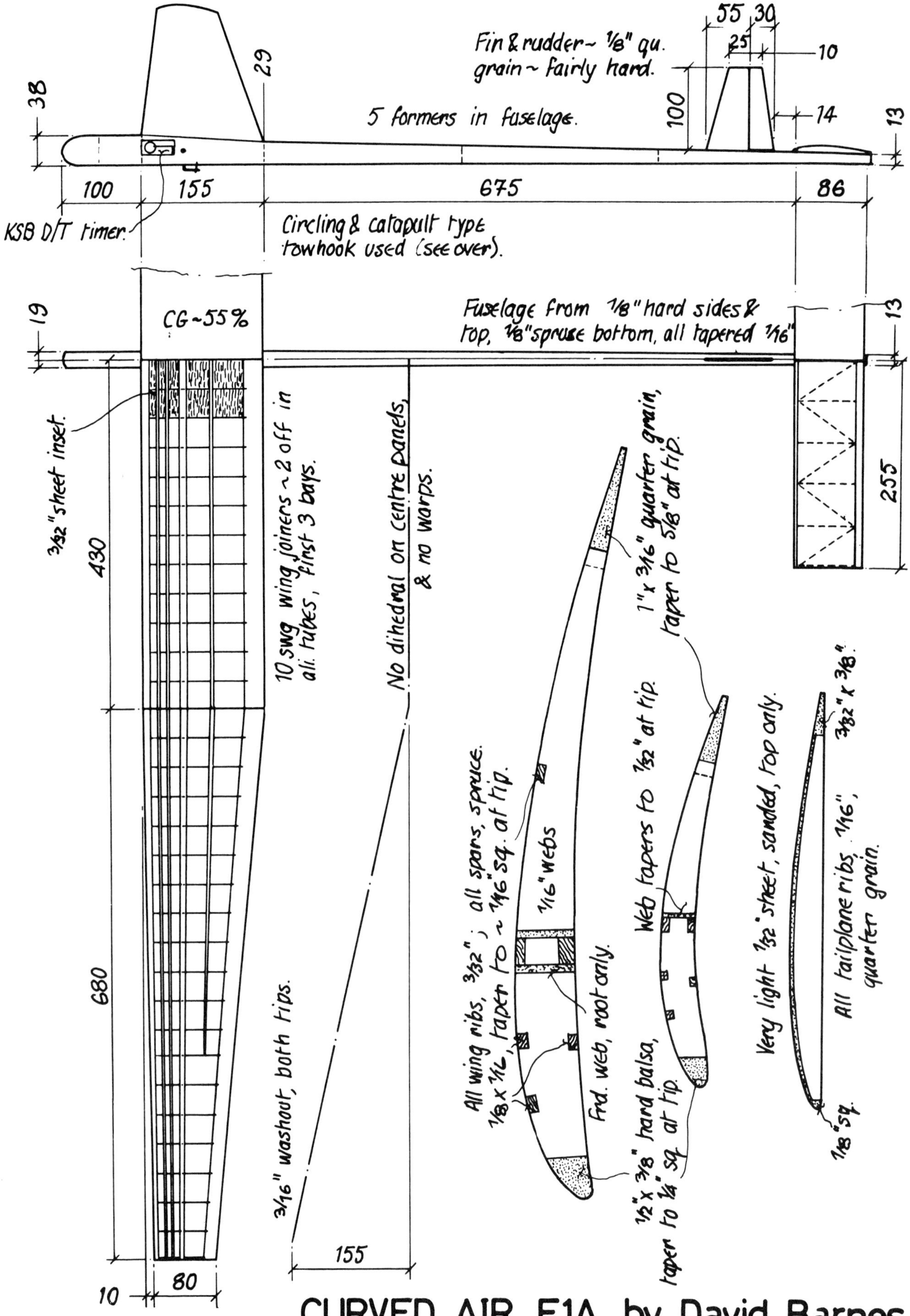

CURVED AIR, F1A, by David Barnes

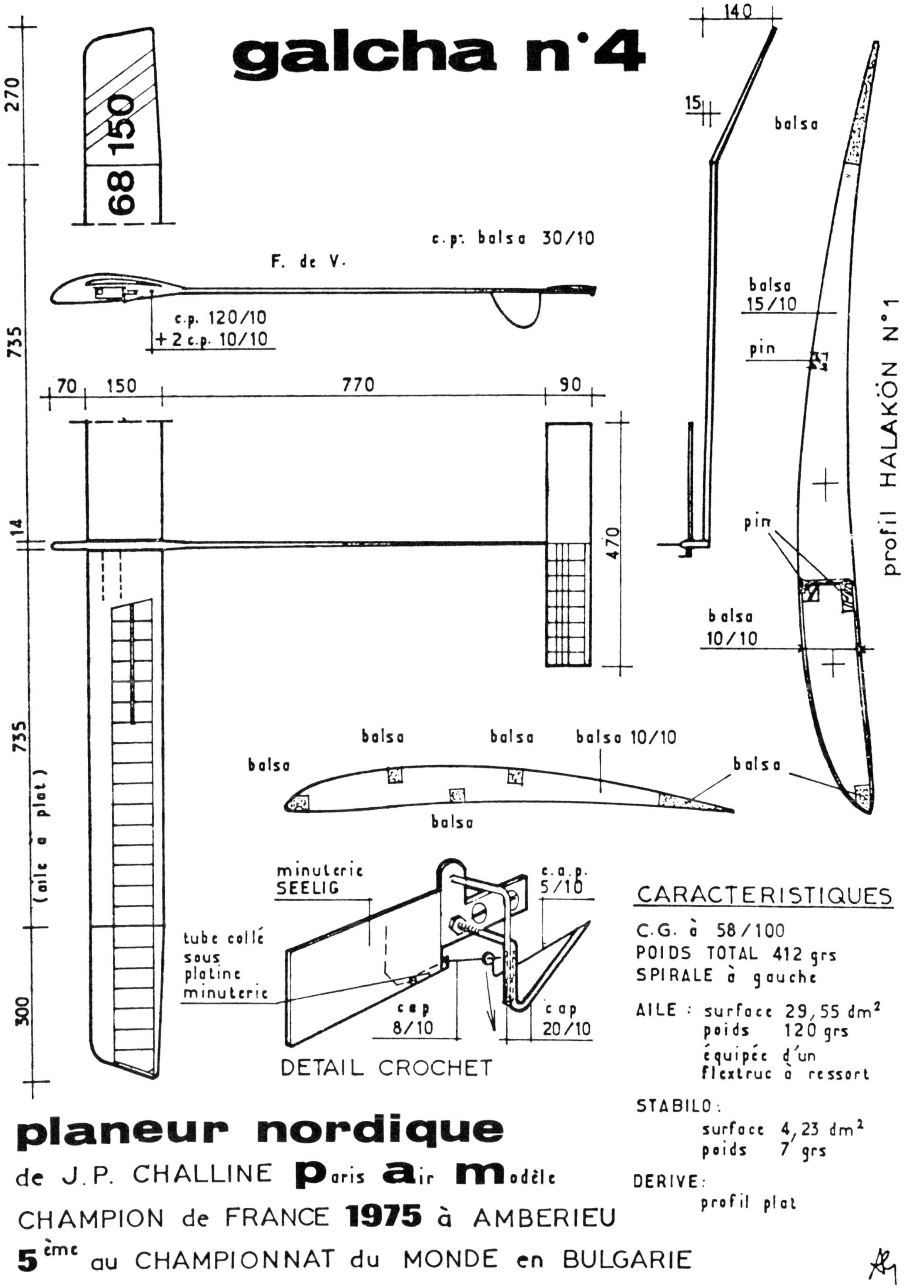

8 Challine France WFFP

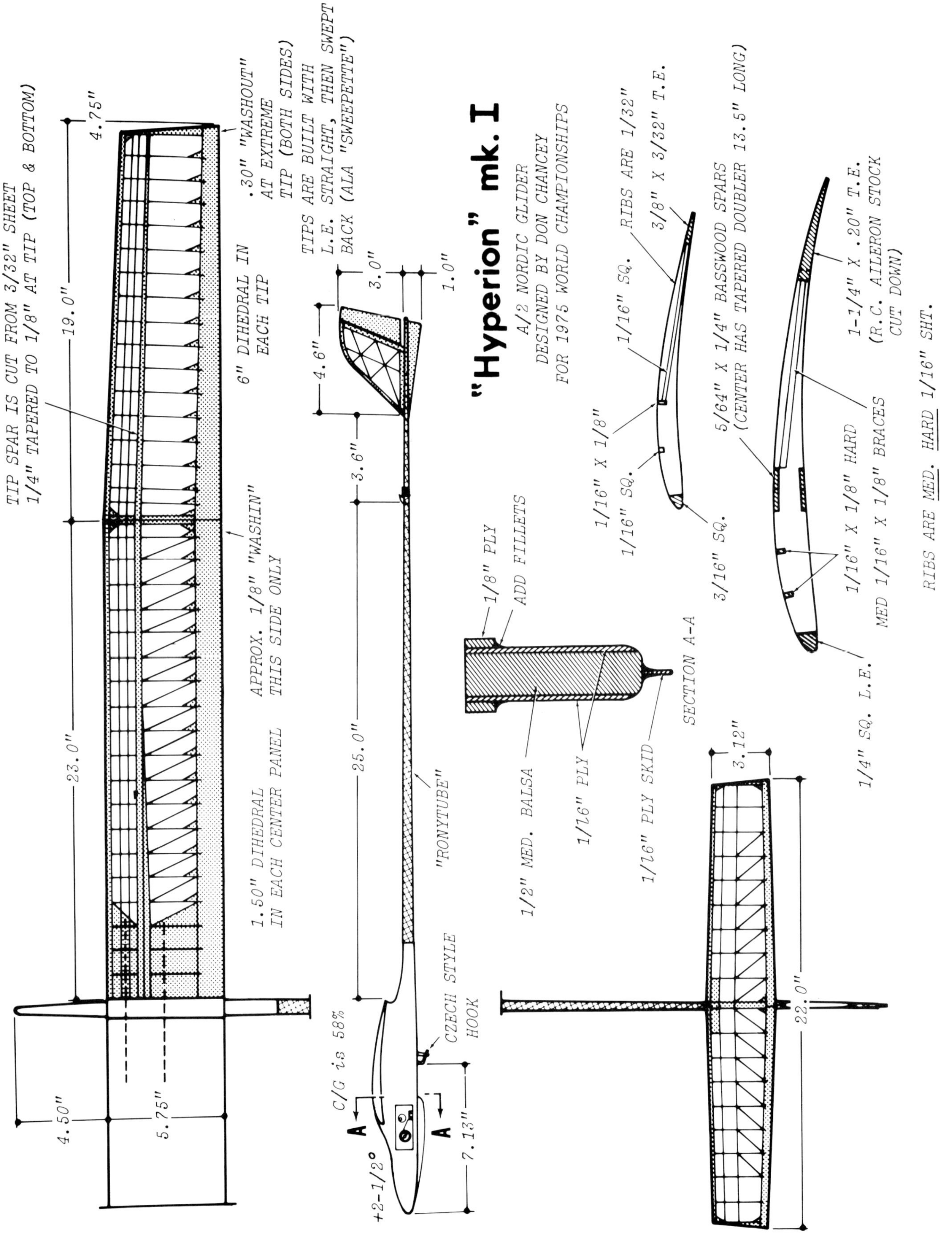

 9 Chancey USA Scatter

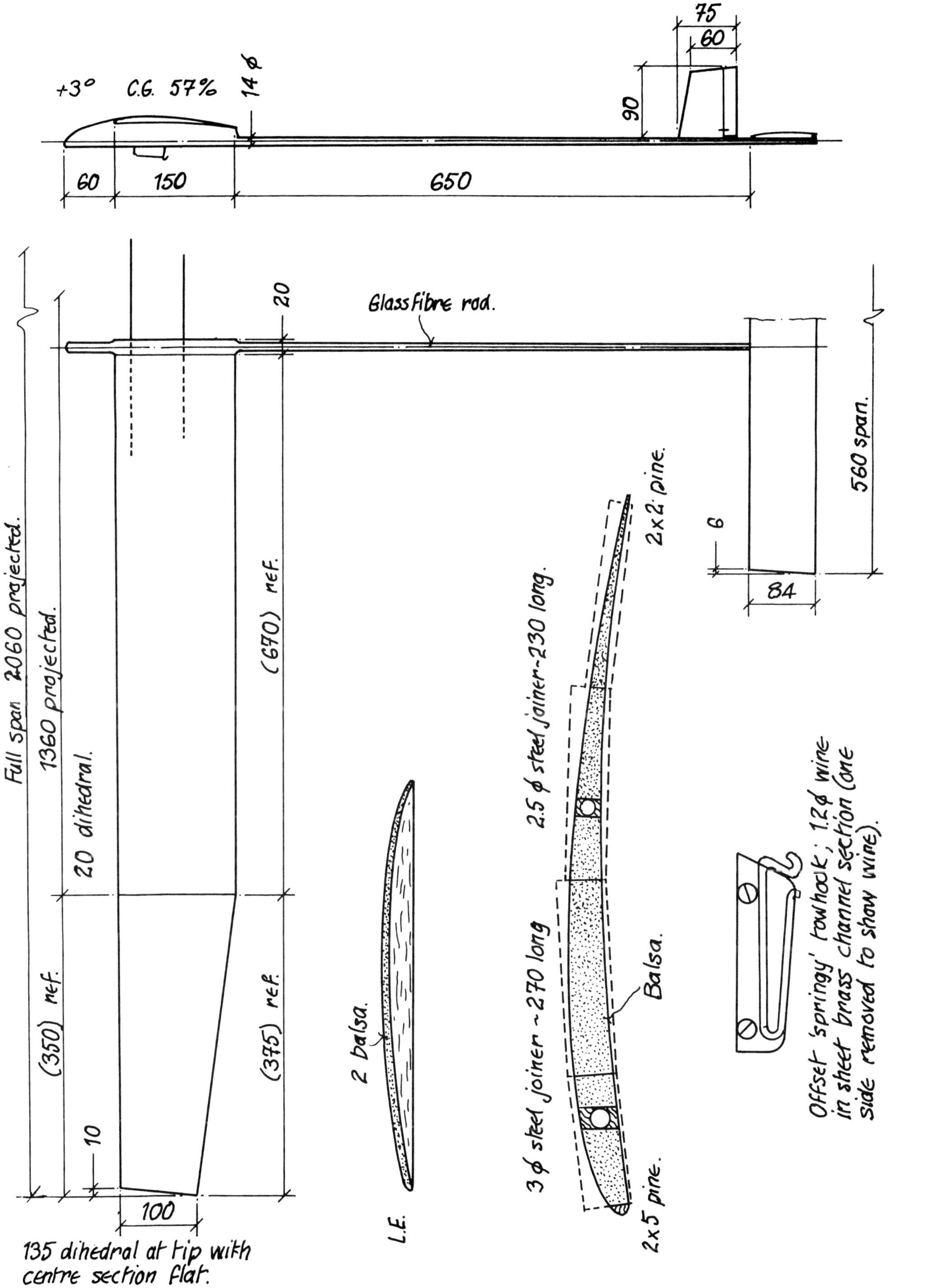

GLIDER, F1A, by Arno Deubel

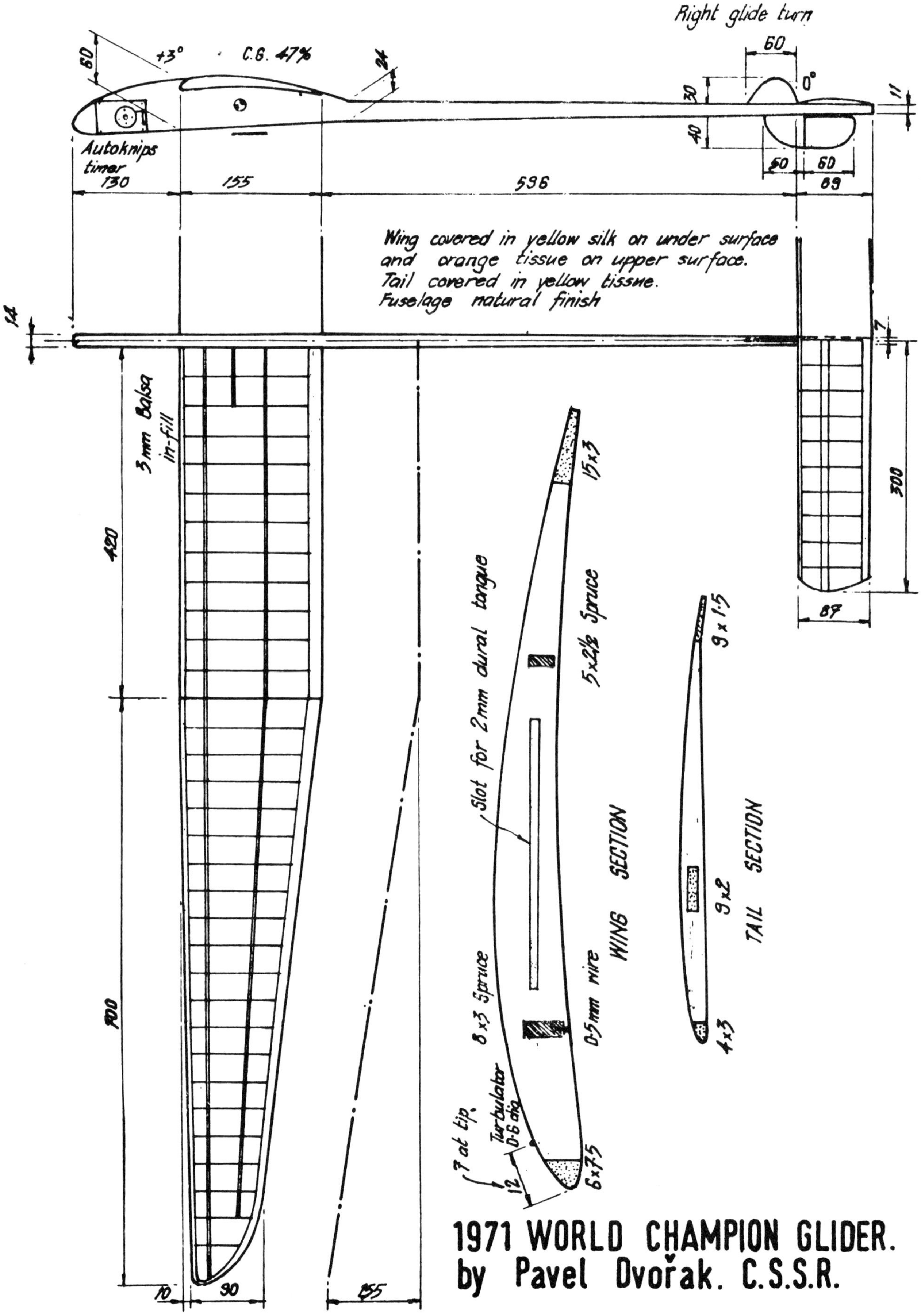

1971 WORLD CHAMPION GLIDER.
by Pavel Dvořak. C.S.S.R.

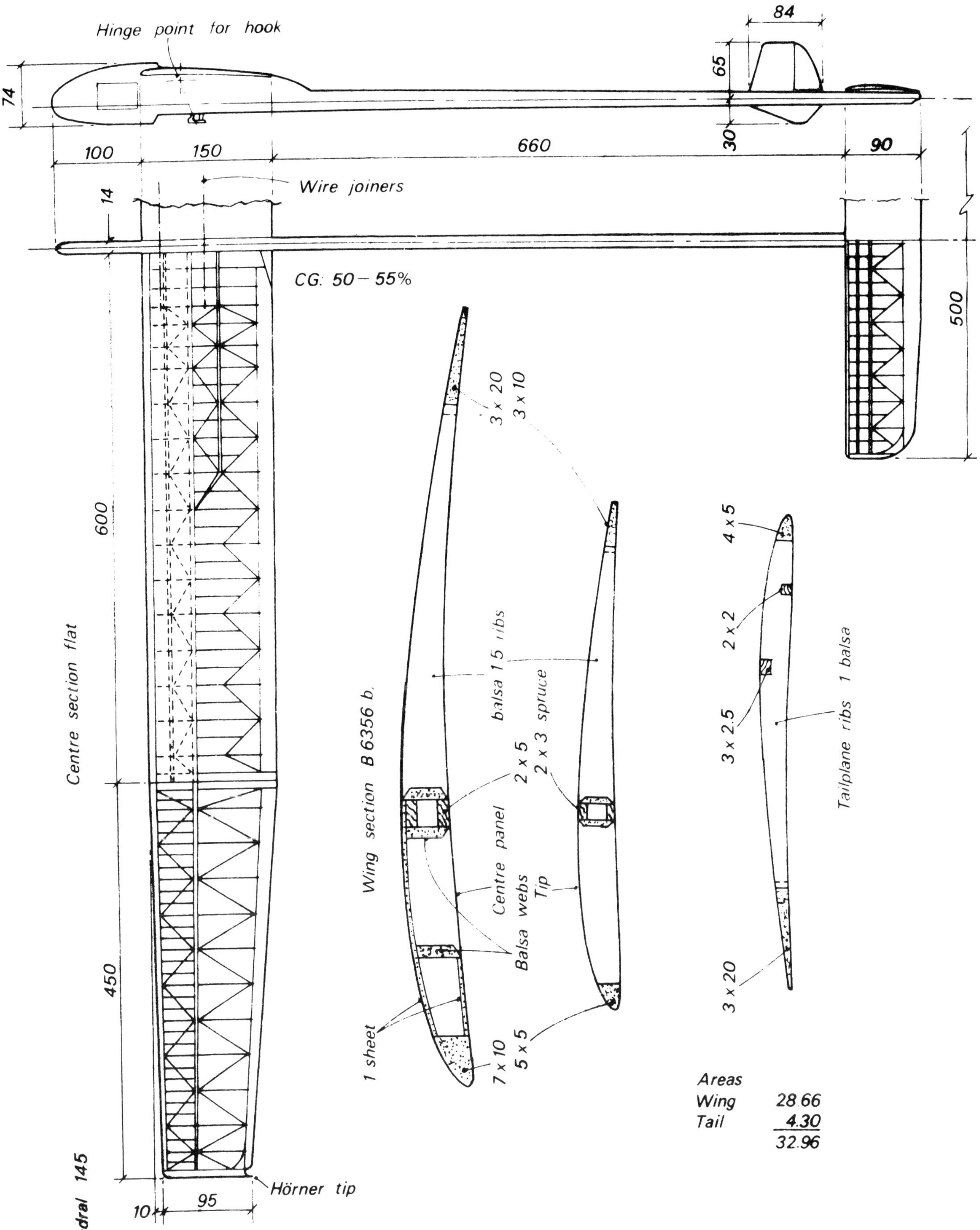

VALERIJ EKHTENKOV's F1A, EVG 1172

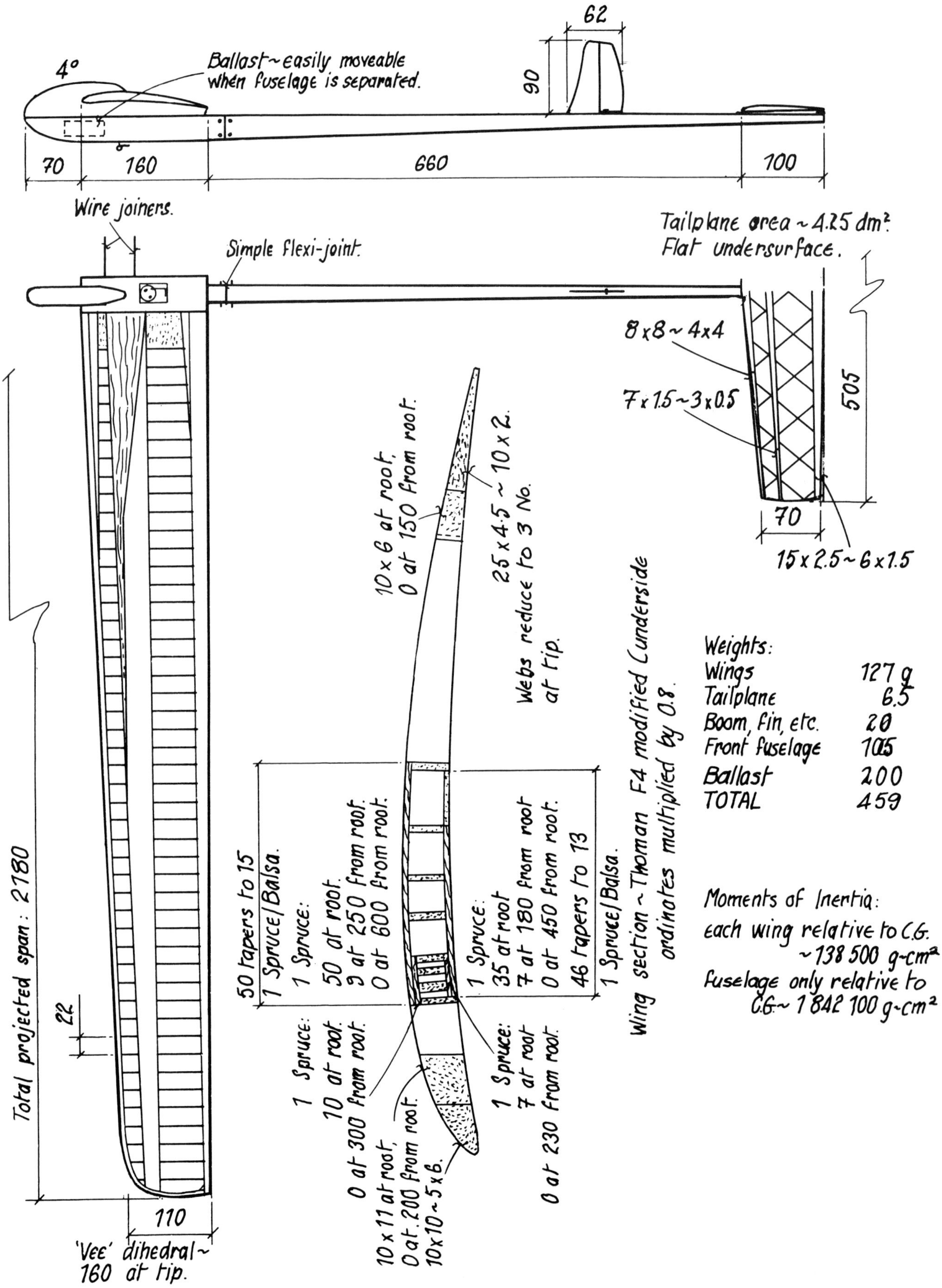

GREAT GULL IV, F1A, by Svend Grønlund

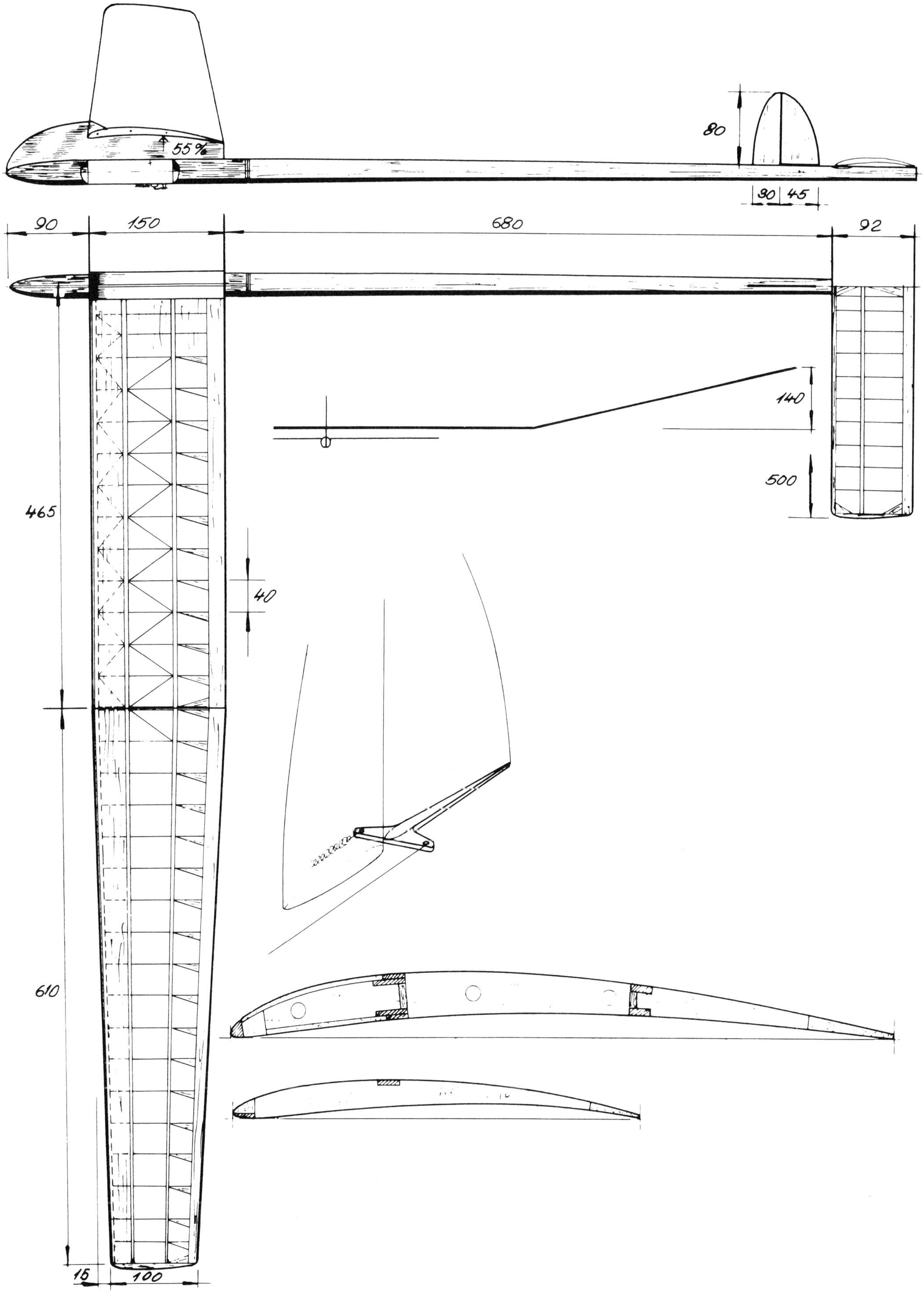
55%
80
30
45
90
150
680
92
140
500
465
40
610
15
100

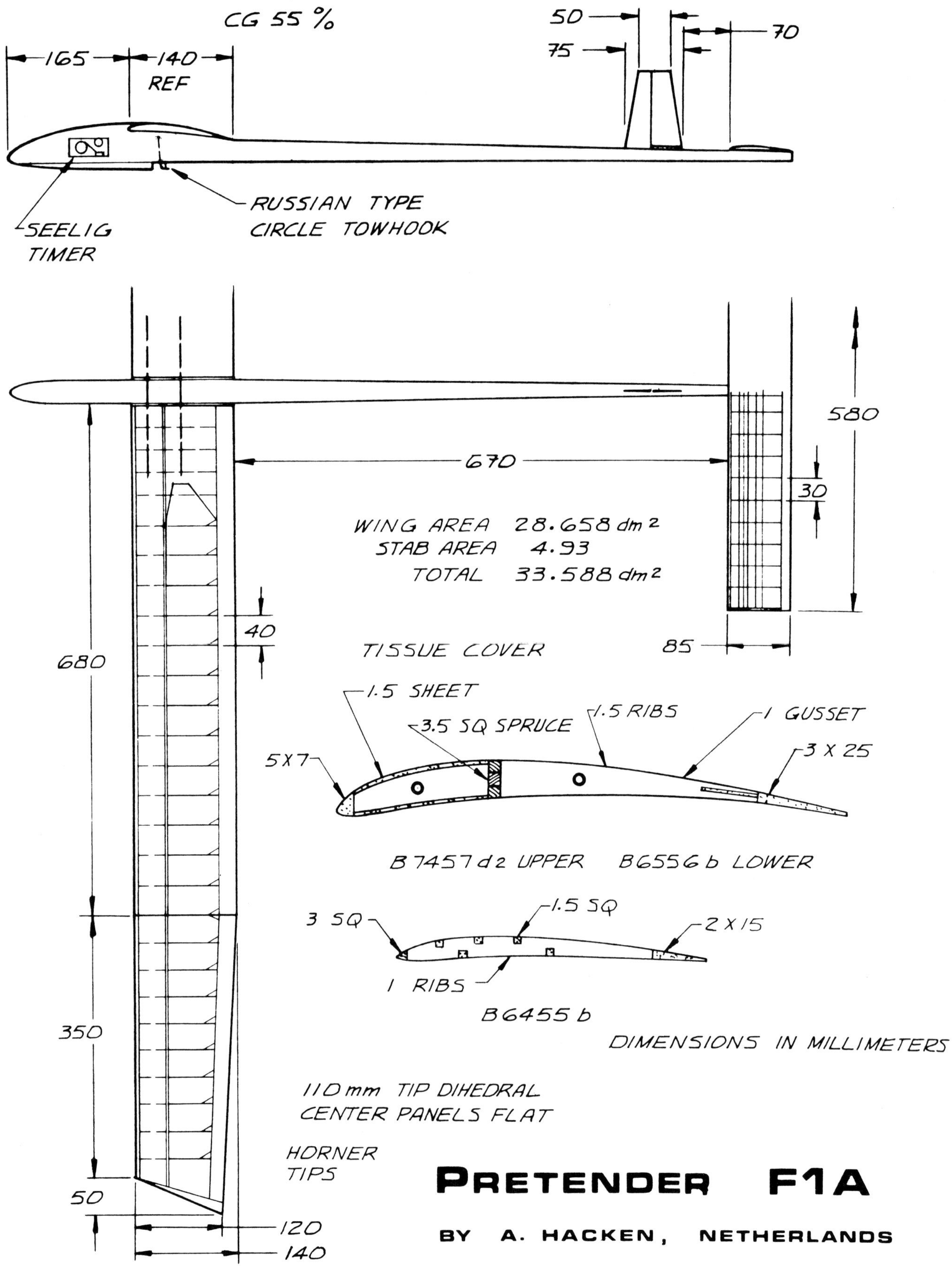

PRETENDER F1A

BY A. HACKEN, NETHERLANDS

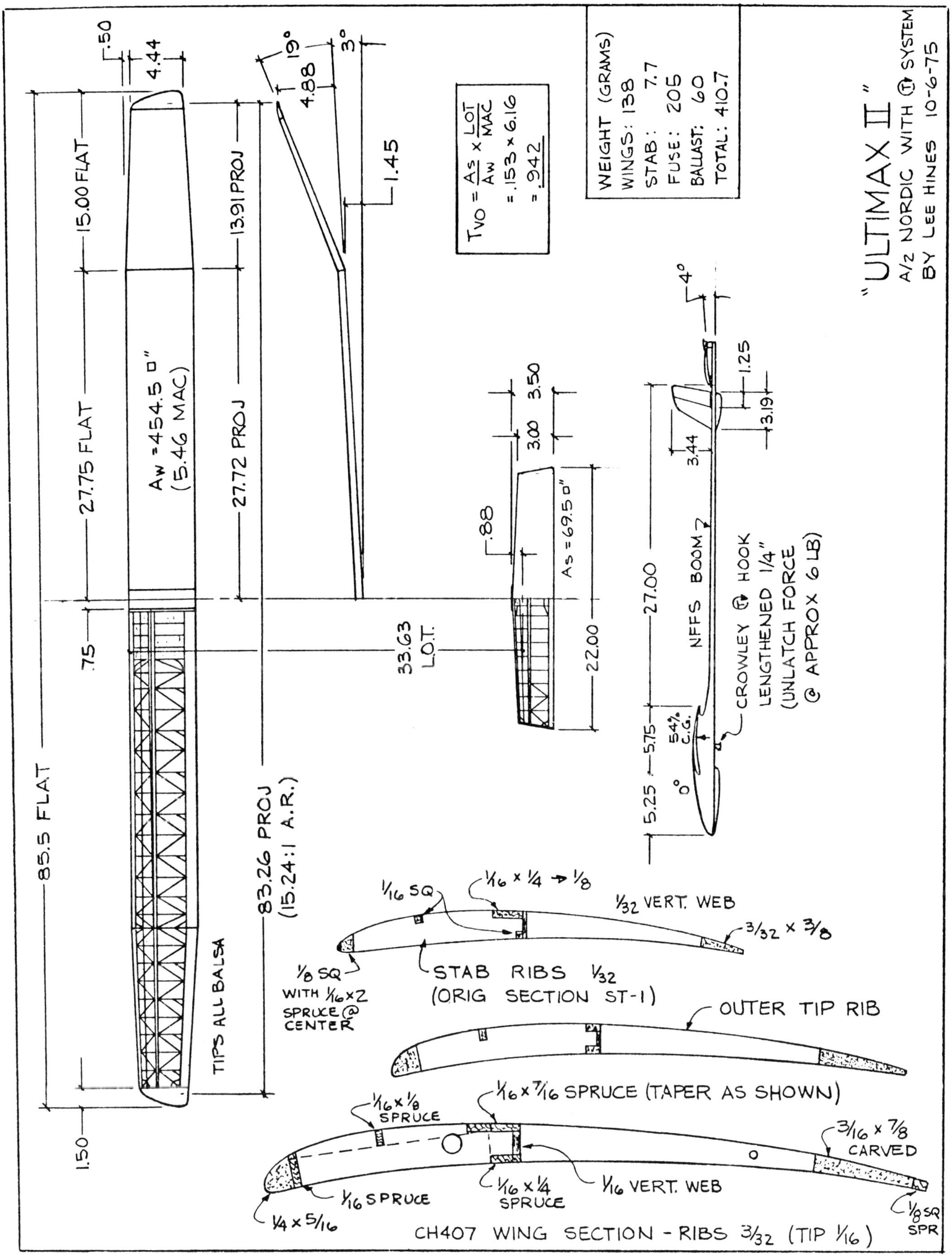

16 Hines USA Scatter

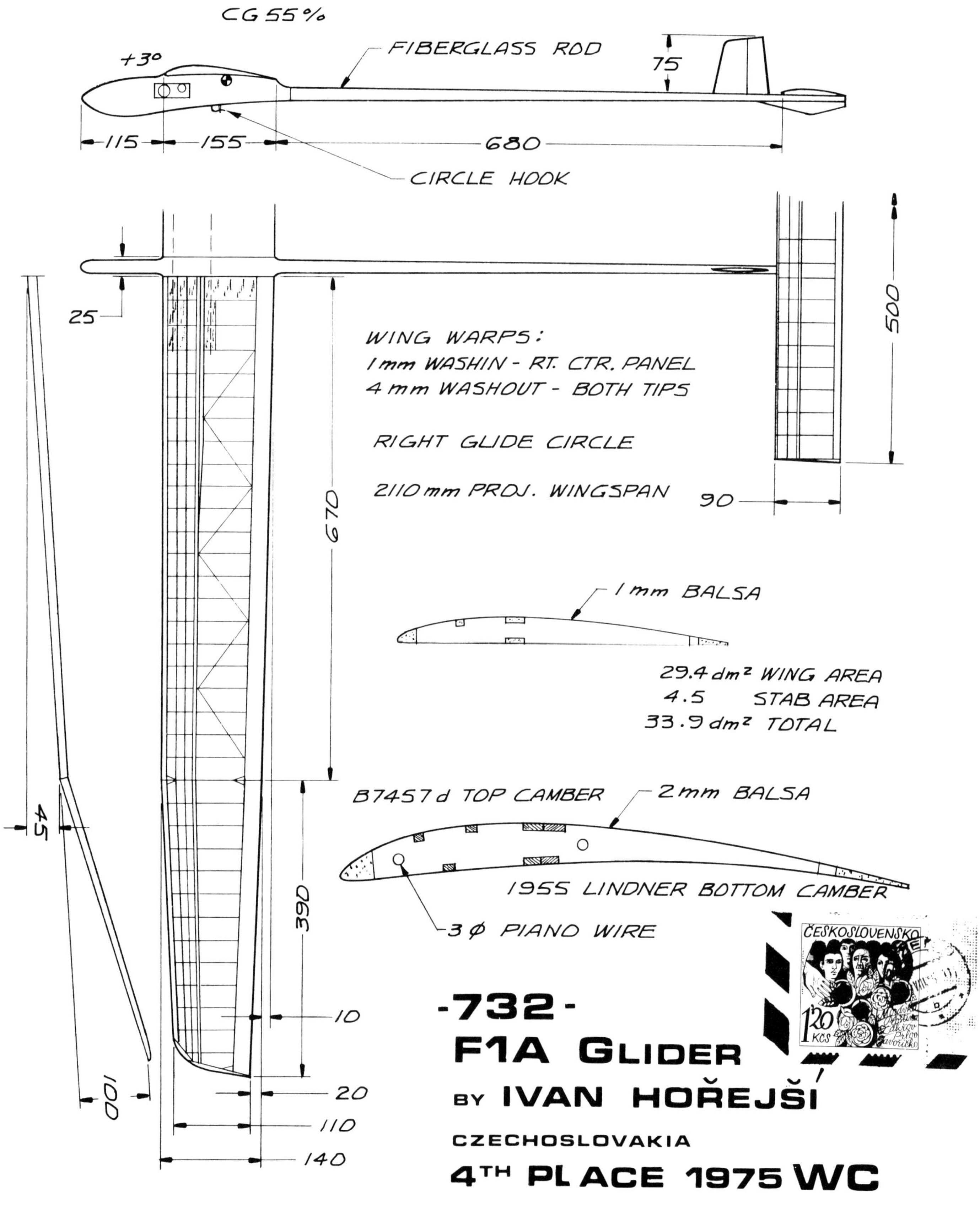

-732-

F1A Glider

by IVAN HOŘEJŠI

CZECHOSLOVAKIA

4TH PLACE 1975 WC

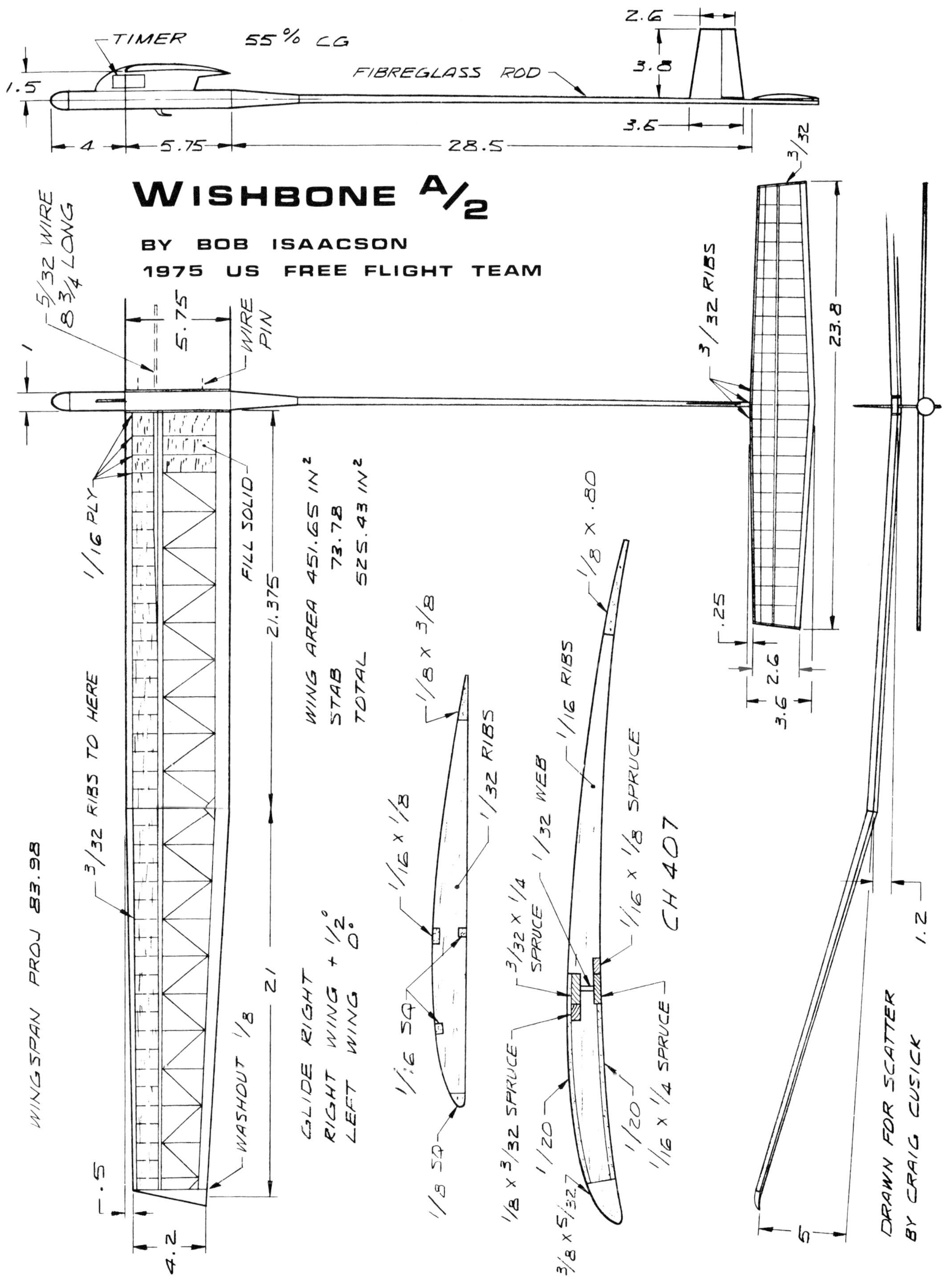

18 Isaacson USA Scatter

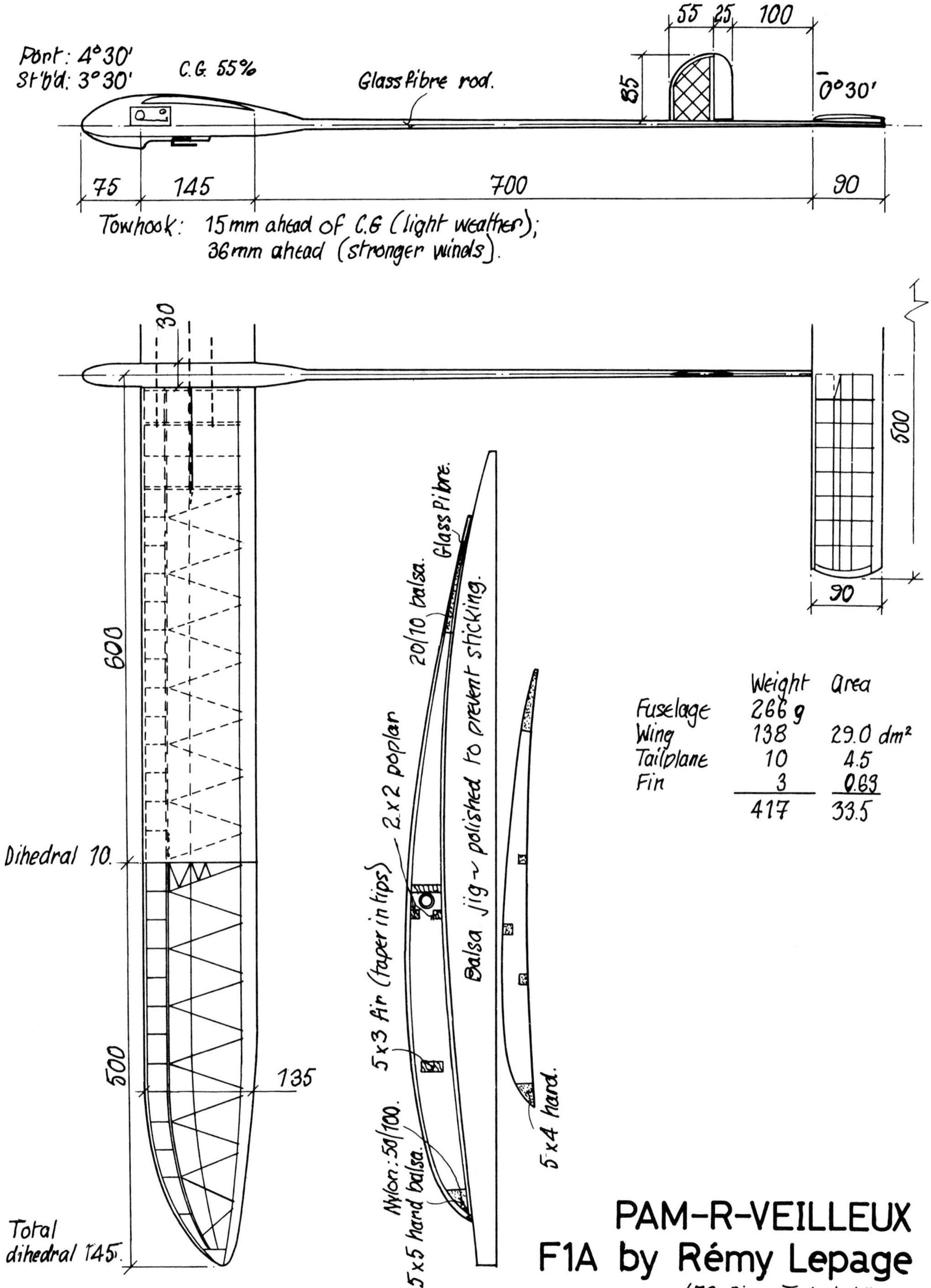

Port: 4°30'
St'b'd: 3°30'
C.G. 55%
Glass fibre rod.
55
25
100
85
0°30'
75
145
700
90
Towhook: 15mm ahead of C.G (light weather);
36mm ahead (stronger winds).
30
500
90
600
Glass Fibre.
20/10 balsa.
Balsa jig ~ polished to prevent sticking.
2 x 2 poplar
5 x 3 fir (taper in tips)
Nylon: 50/100.
5 x 5 hard balsa.
5 x 4 hard.
Dihedral 10.
500
135
Total
dihedral 145.
Weight Area
Fuselage 266 g
Wing 138 29.0 dm²
Tailplane 10 4.5
Fin 3 0.63
417 33.5
PAM-R-VEILLEUX
F1A by Rémy Lepage
'73 Pierre Trebod Winner

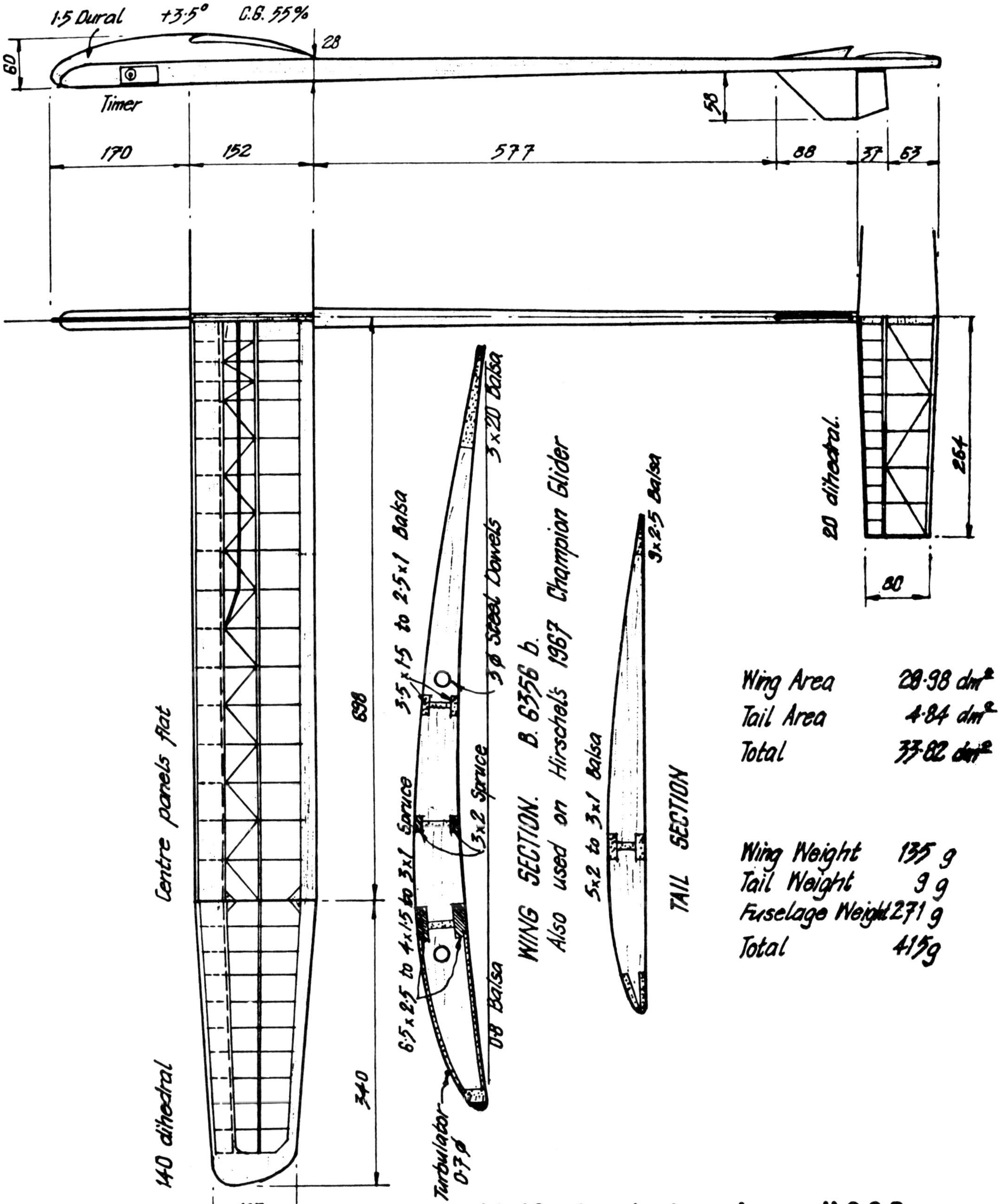

AL-19 by Andres Lepp U.S.S.R.

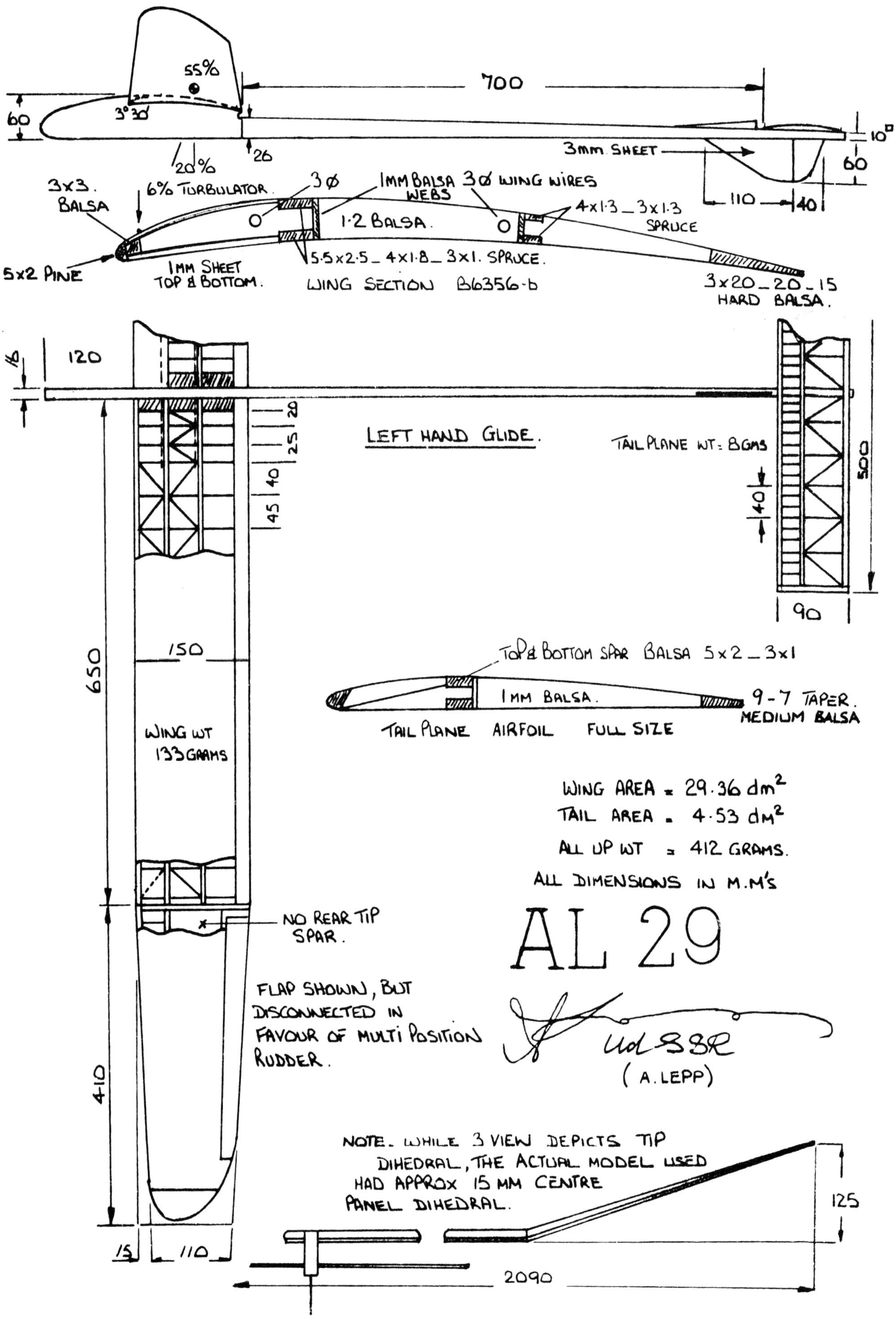
55%
700
60
3°30'
26
20%
10°
3mm SHEET
60
110
40
3x3 BALSA
6% TURBULATOR
3Ø
1MM BALSA WEBS
3Ø WING WIRES
4x1·3 _ 3x1·3 SPRUCE
1·2 BALSA
5x2 PINE
1MM SHEET TOP & BOTTOM
5·5x2·5 _ 4x1·8 _ 3x1. SPRUCE.
WING SECTION B6356-b
3x20 _ 20 _ 15 HARD BALSA.
16
120
20
25
40
45
LEFT HAND GLIDE.
TAIL PLANE WT: 8GMS
500
40
90
650
150
TOP & BOTTOM SPAR BALSA 5x2 _ 3x1
1MM BALSA.
9-7 TAPER. MEDIUM BALSA
TAIL PLANE AIRFOIL FULL SIZE
WING WT 133 GRAMS
WING AREA = 29·36 dm²
TAIL AREA = 4·53 dm²
ALL UP WT = 412 GRAMS.
ALL DIMENSIONS IN M.M'S
NO REAR TIP SPAR.
AL 29
FLAP SHOWN, BUT DISCONNECTED IN FAVOUR OF MULTI POSITION RUDDER.
(A. LEPP)
410
NOTE. WHILE 3 VIEW DEPICTS TIP DIHEDRAL, THE ACTUAL MODEL USED HAD APPROX 15 MM CENTRE PANEL DIHEDRAL.
125
15
110
2090

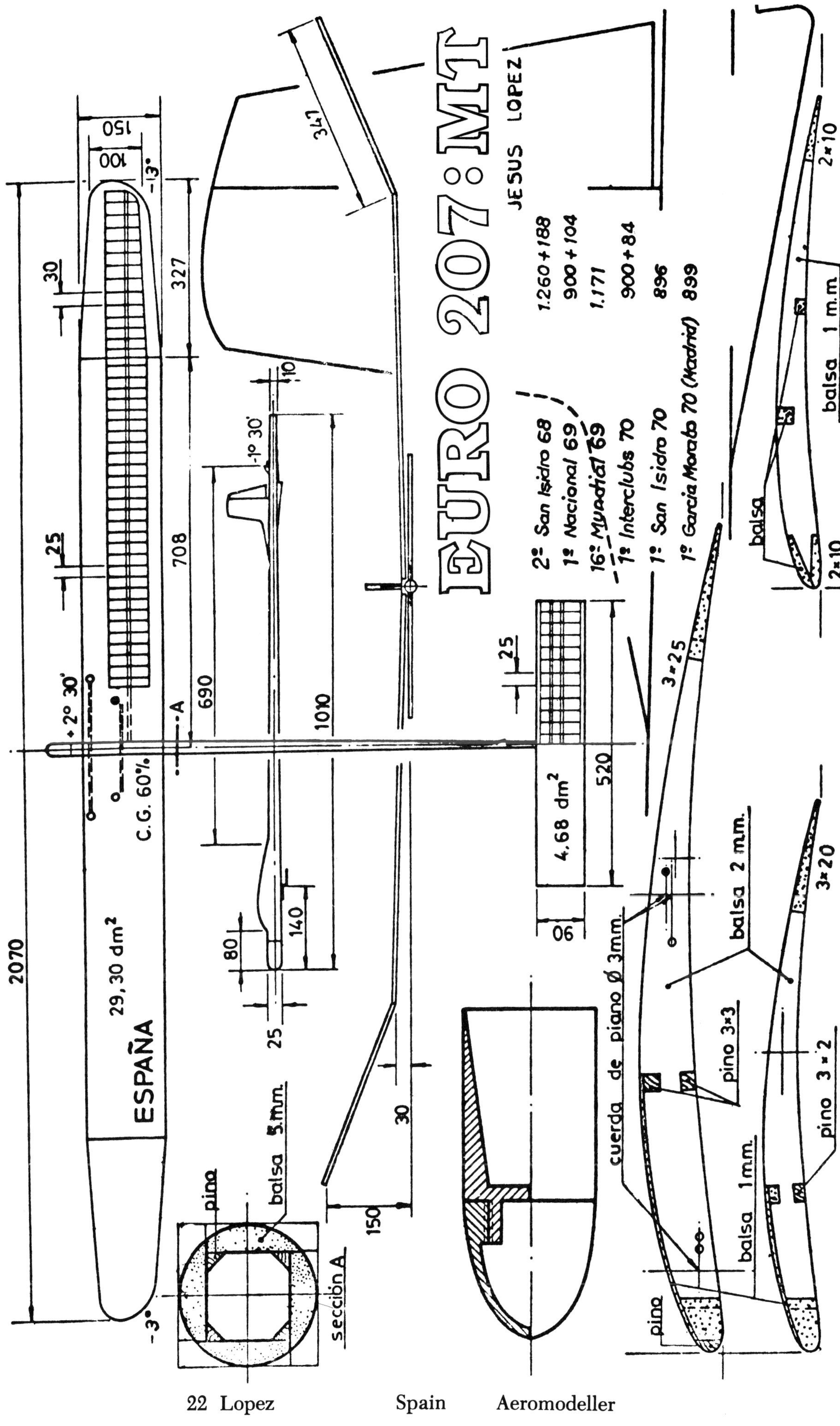
EURO 207:MT
JESUS LOPEZ
2º San Isidro 68 1.260+188
1º Nacional 69 900+104
16º Mundial 69 1.171
1º Interclubs 70 900+84
1º San Isidro 70 896
1º Garcia Morato 70 (Madrid) 899
ESPAÑA
29,30 dm²
C.G. 60%
4,68 dm²
2070
cuerda de piano Ø 3mm.
balsa 2 m.m.
balsa 1 m.m.
pino 3×3
pino 3×2
sección A

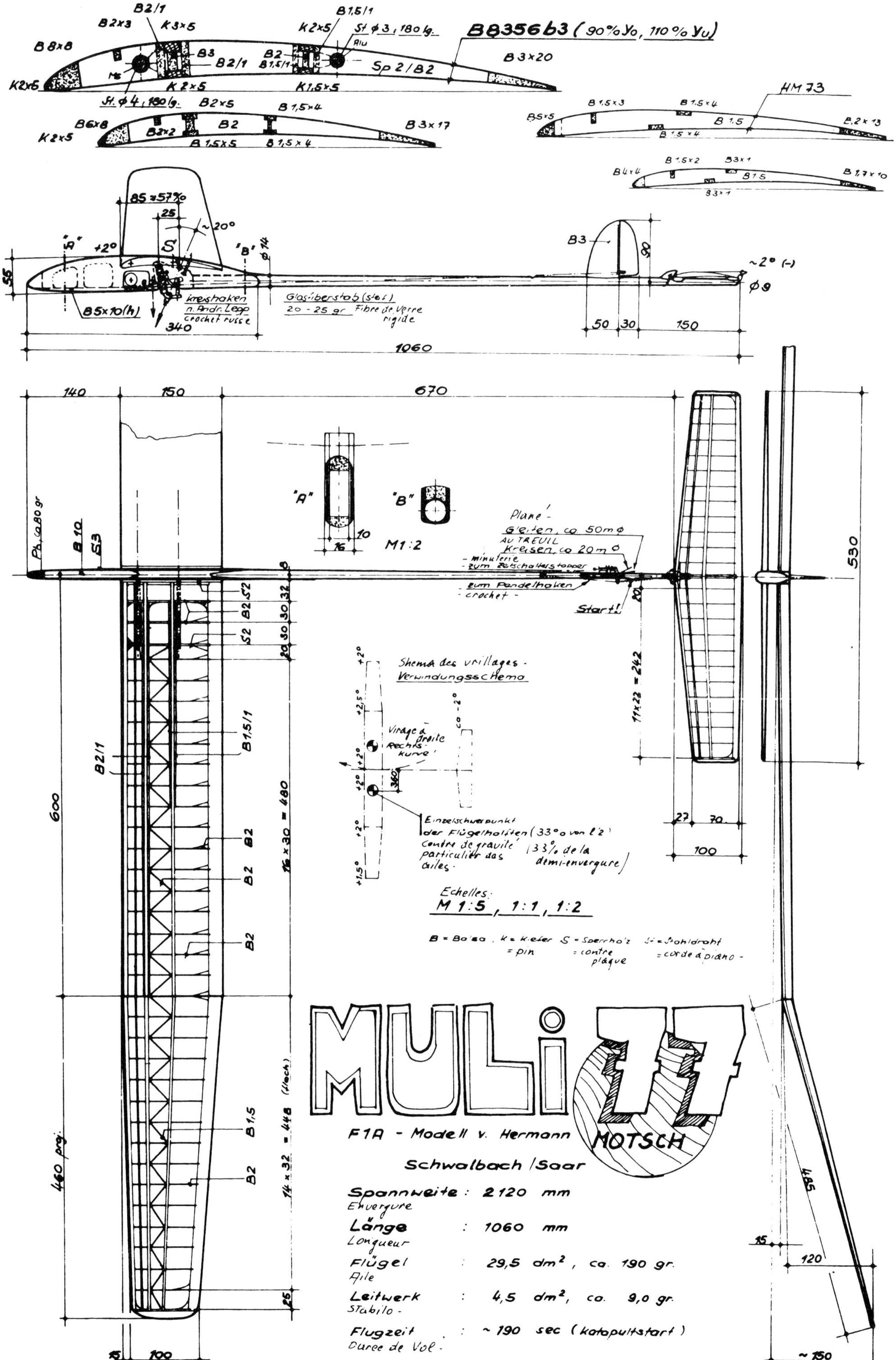
B 8356 b3 (90% Yo, 110% Yu)
HM 73
Glasfiberstab (steif)
20 - 25 gr Fibre de verre rigide
Kreishaken n. Andr. Lepp
crochet russe
1060
"A"
"B"
M 1:2
Plane - Gleiten, ca. 50 m Ø
Au TREUIL - Kreisen, ca. 20 m Ø
- minuterie - zum Zeitschalterstopper
zum Pendelhaken - crochet -
Start!
Shema des vrillages - Verwindungsschema
Virage à droite
Rechtskurve
Einzelschwerpunkt der Flügelhälften (33% von l/2)
Centre de gravité particulier des ailes (33% de la demi-envergure)
Echelles: M 1:5, 1:1, 1:2
B = Balsa, K = Kiefer = pin, S = Sperrholz = contreplaque, St = Stahldraht = corde à piano
MULi 77
MOTSCH
F1A - Modell v. Hermann
Schwalbach / Saar
Spannweite : 2120 mm
Envergure
Länge : 1060 mm
Longueur
Flügel : 29,5 dm², ca. 190 gr.
Aile
Leitwerk : 4,5 dm², ca. 9,0 gr.
Stabilo
Flugzeit : ~ 190 sec (Katapultstart)
Durée de vol

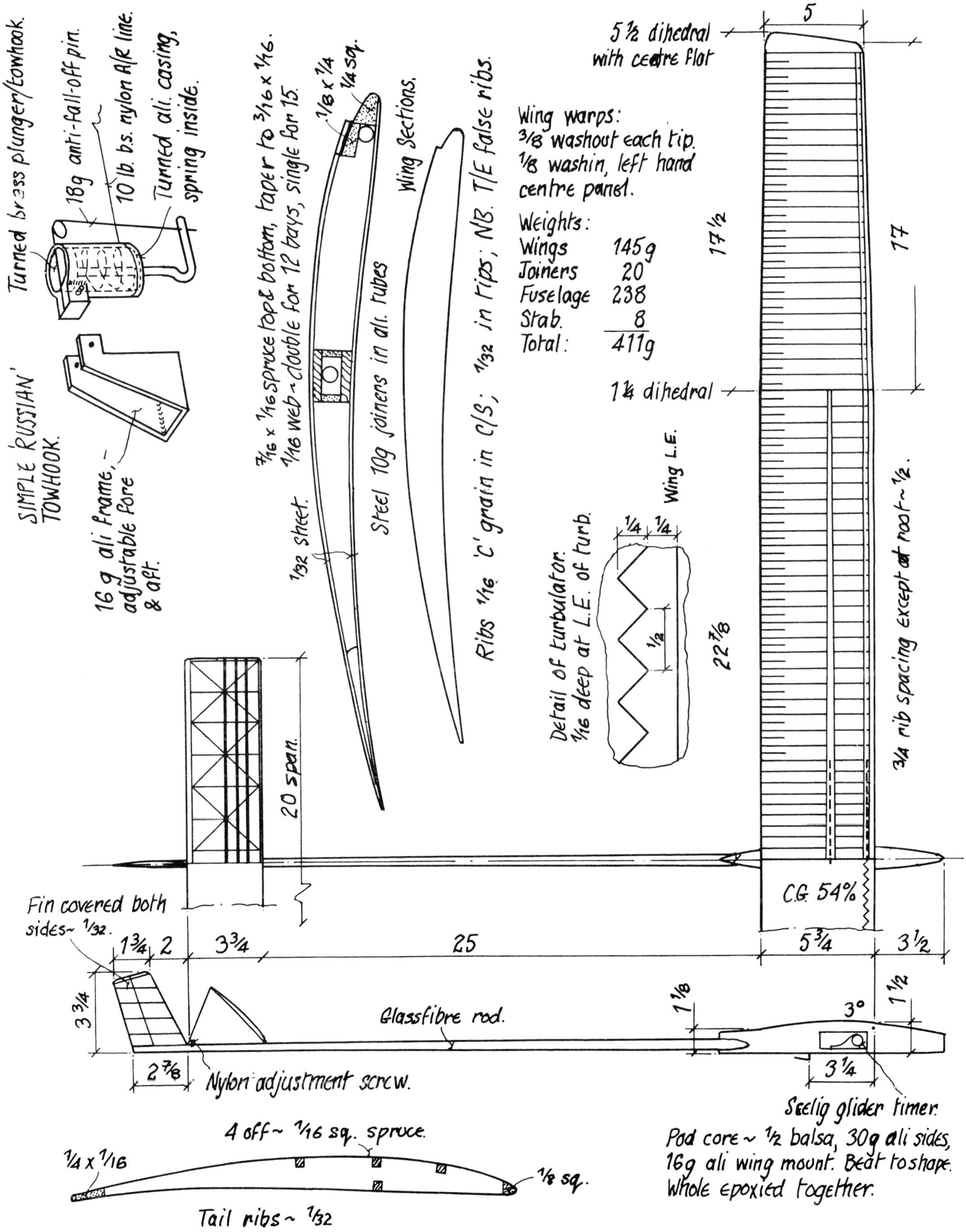

FAIFLY 72-1, F1A, by Neil Murray

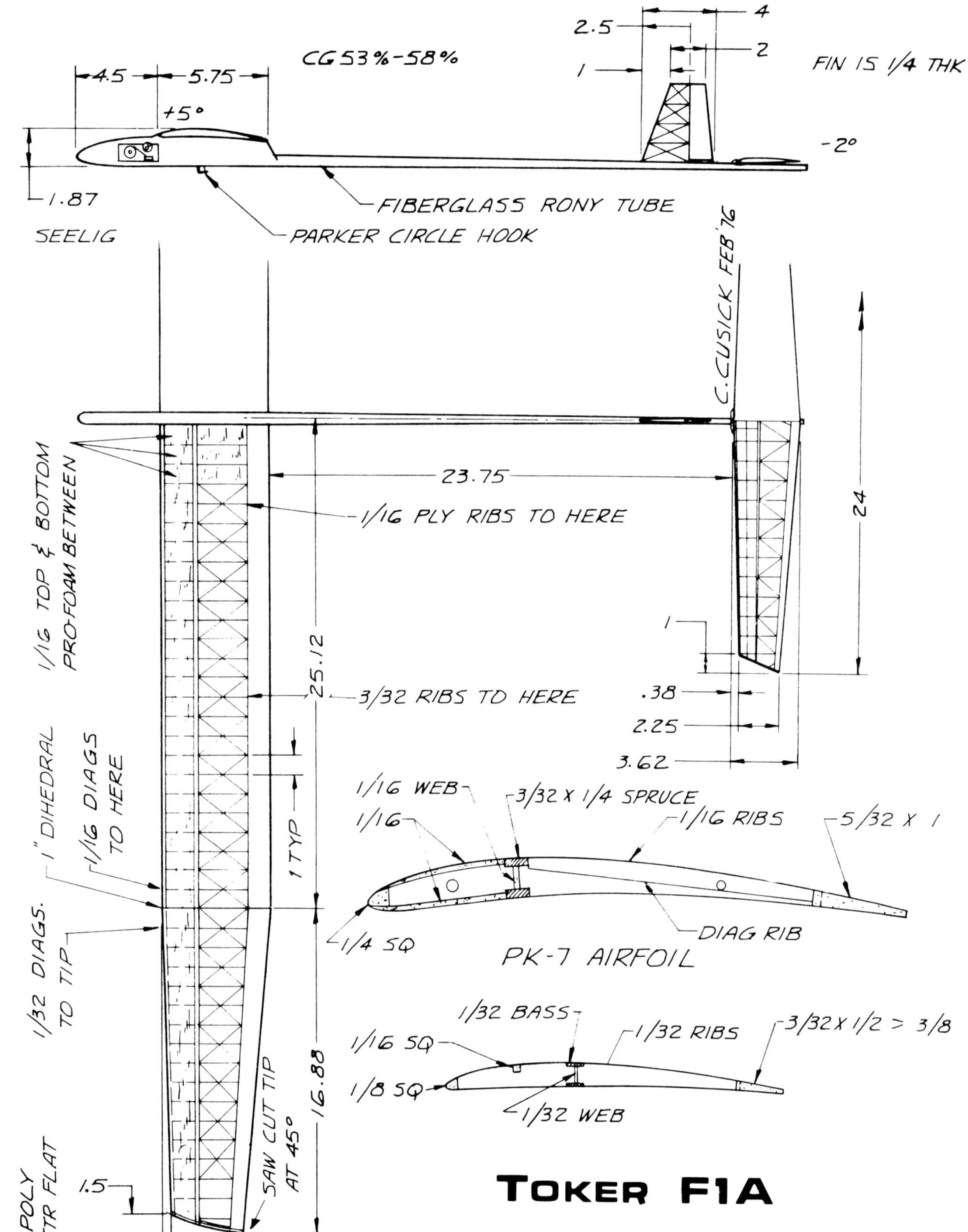

TOKER F1A

BY JIM PARKER EL PASO, TEXAS

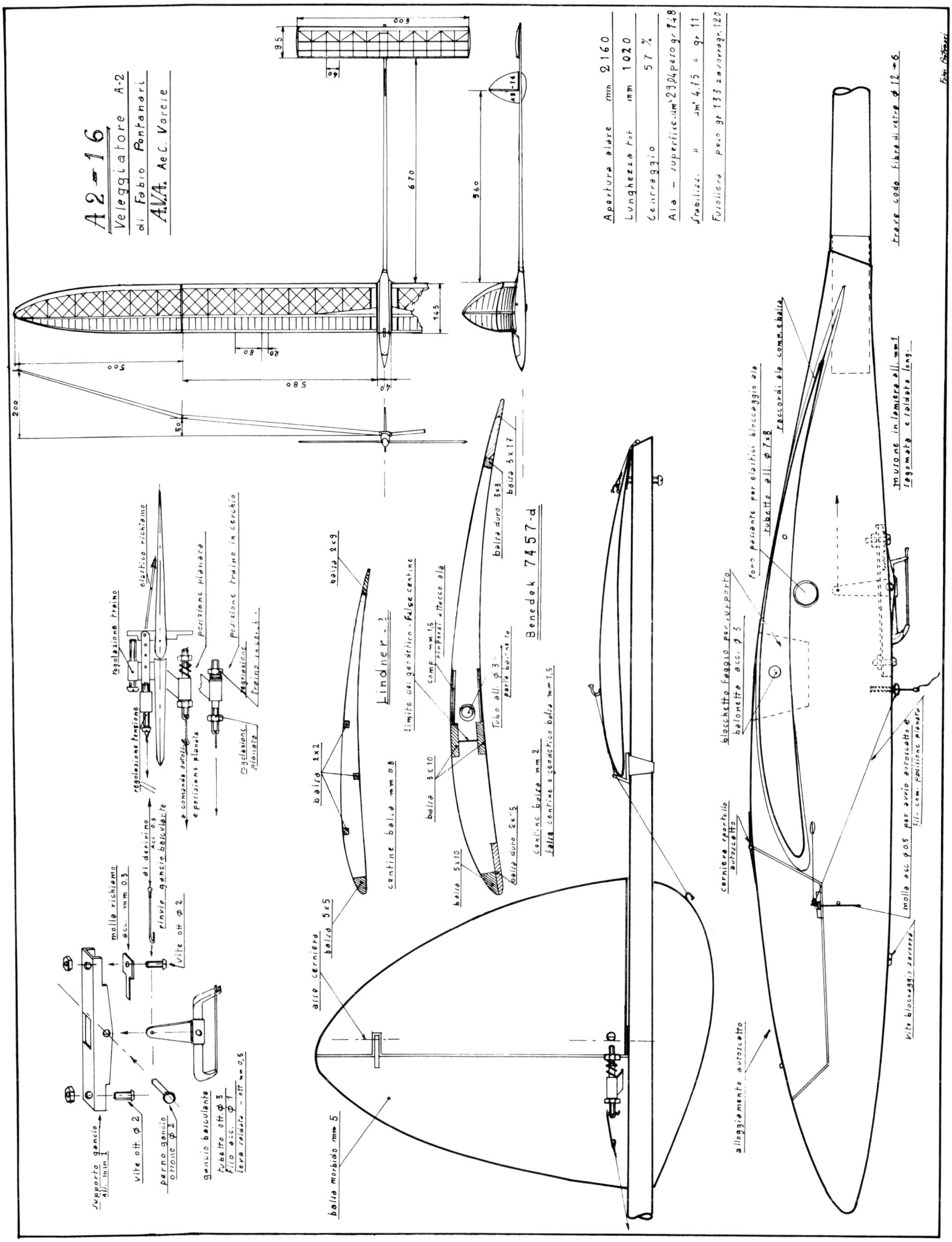
A2-16
Veleggiatore A-2
di Fabio Pontanari
AVA Ae.C. Varese
Apertura alare mm 2160
Lunghezza tot mm 1020
Centraggio 57 %
Lindner
Benedek 7457-d

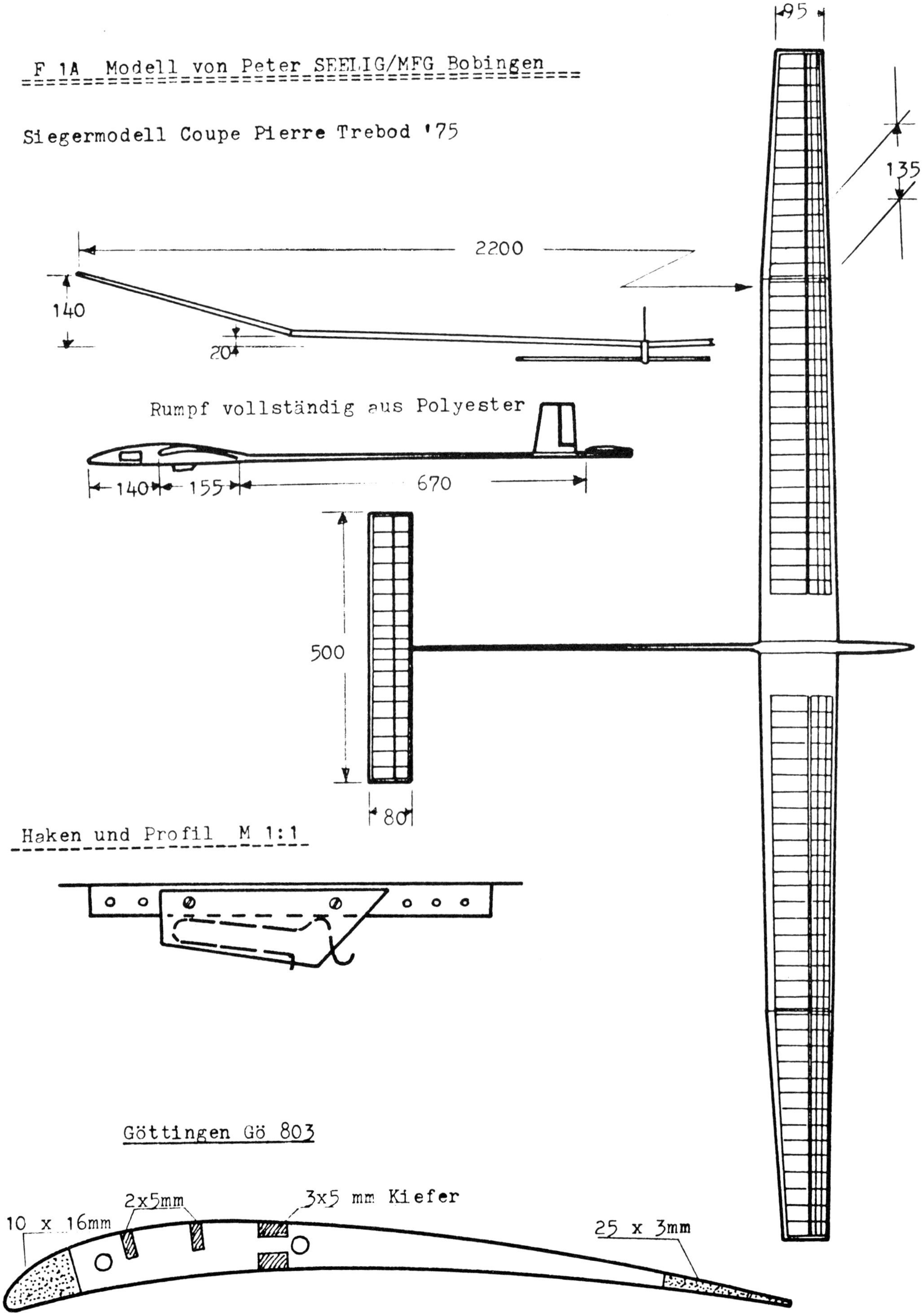

 27 Seelig BDR Scatter

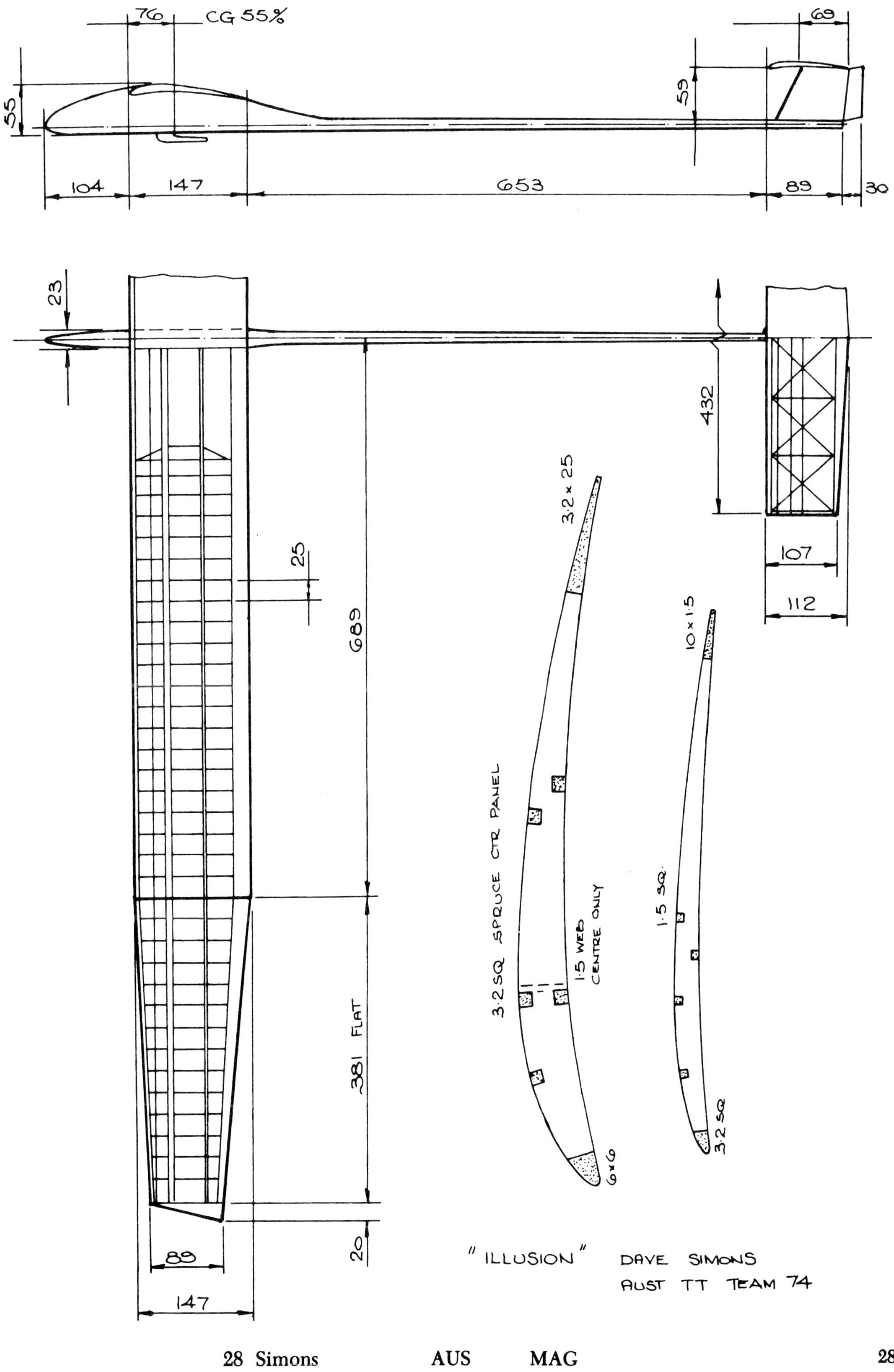
76
CG 55%
69
55
59
104
147
653
89
30
23
432
3·2 × 25
107
112
25
689
10 × 1·5
3·2SQ SPRUCE CTR PANEL
1·5 WEB CENTRE ONLY
1·5 SQ.
381 FLAT
6×6
3·2 SQ
20
89
147
"ILLUSION" DAVE SIMONS
AUST TT TEAM 74

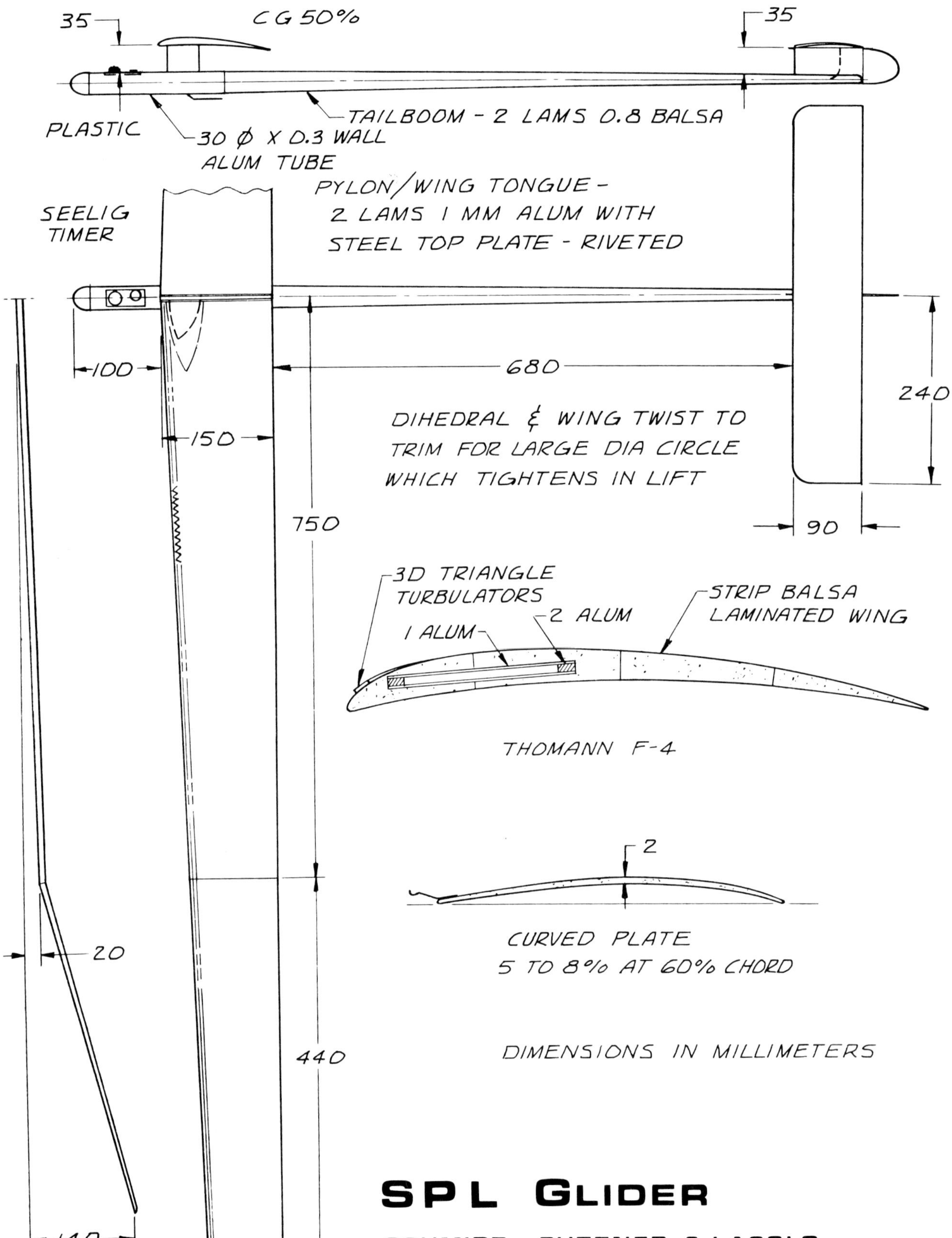

SPL GLIDER

SCHMIDT, PUTTNER & LASCLO

BRYAN SPOONER VERSION

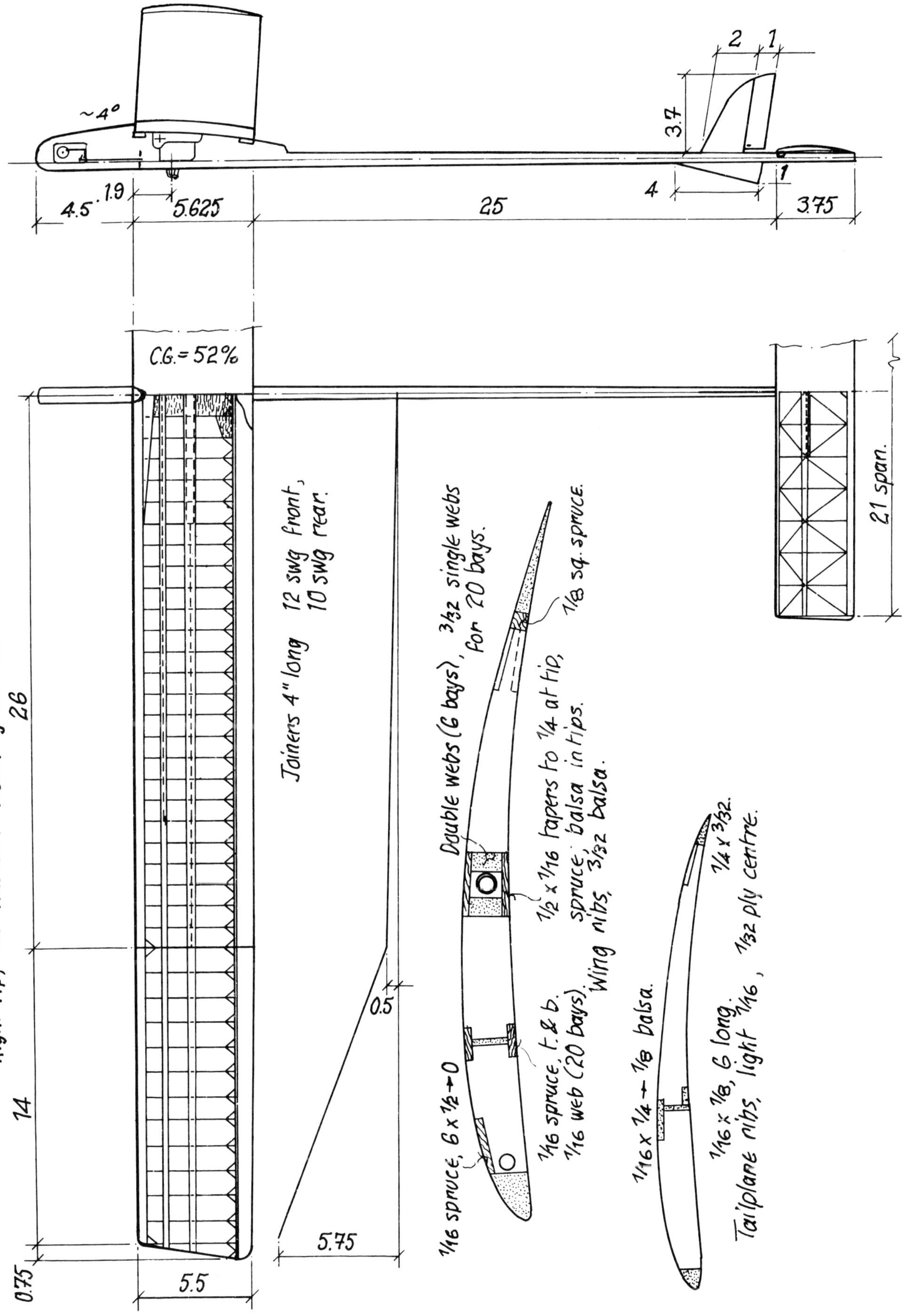

Peter Stewart's JOKE ROTARY A/2

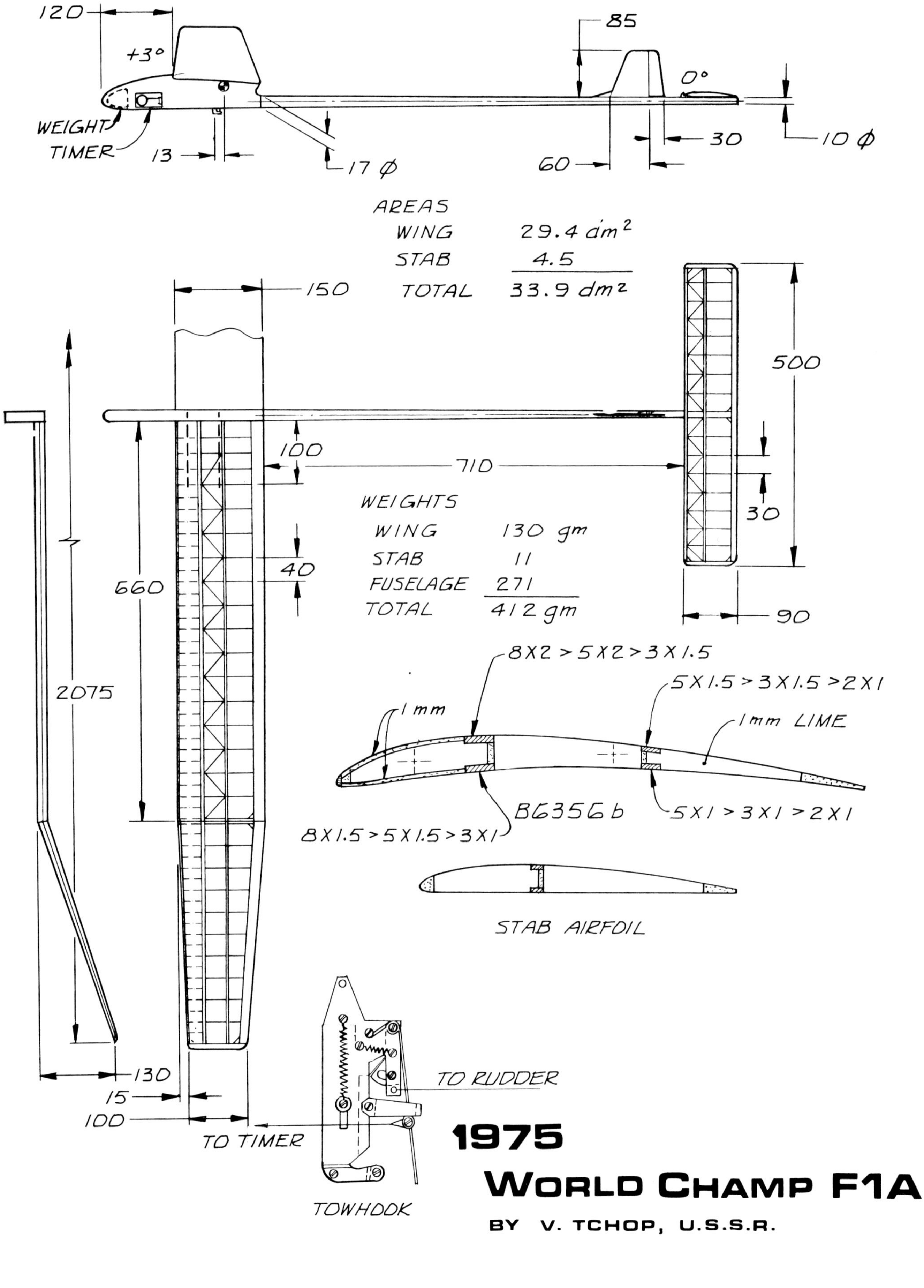

120
+3°
85
0°
WEIGHT
TIMER
13
17 ϕ
60
30
10 ϕ
AREAS
WING 29.4 dm²
STAB 4.5
TOTAL 33.9 dm²
150
500
100
710
WEIGHTS
WING 130 gm
STAB 11
FUSELAGE 271
TOTAL 412 gm
30
40
660
90
2075
8X2 > 5X2 > 3X1.5
5X1.5 > 3X1.5 > 2X1
1 mm
1 mm LIME
B6356 b
5X1 > 3X1 > 2X1
8X1.5 > 5X1.5 > 3X1
STAB AIRFOIL
130
15
100
TO RUDDER
TO TIMER
TOWHOOK
1975
WORLD CHAMP F1A
BY V. TCHOP, U.S.S.R.

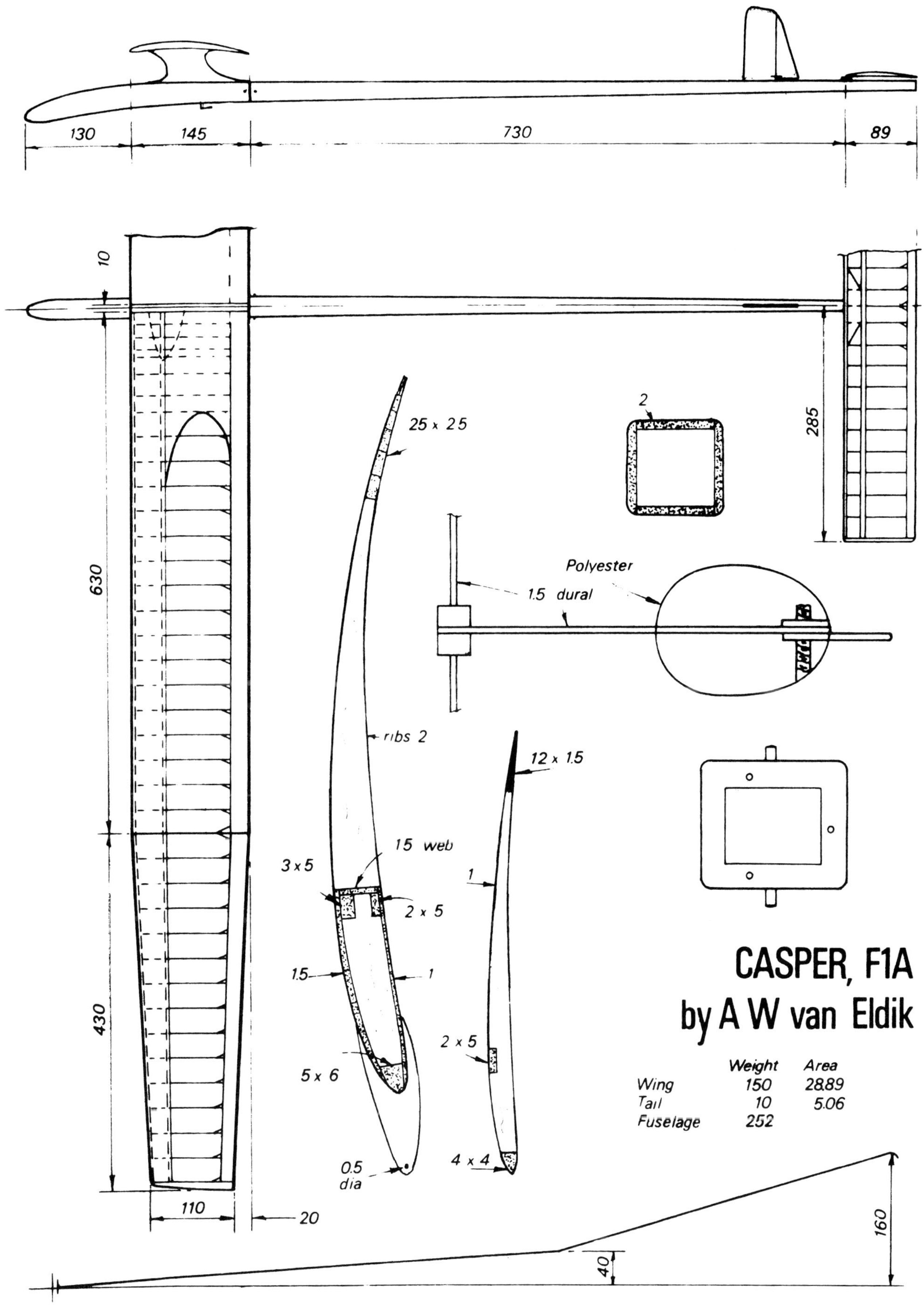

CASPER, F1A
by A W van Eldik

	Weight	Area
Wing	150	28.89
Tail	10	5.06
Fuselage	252	

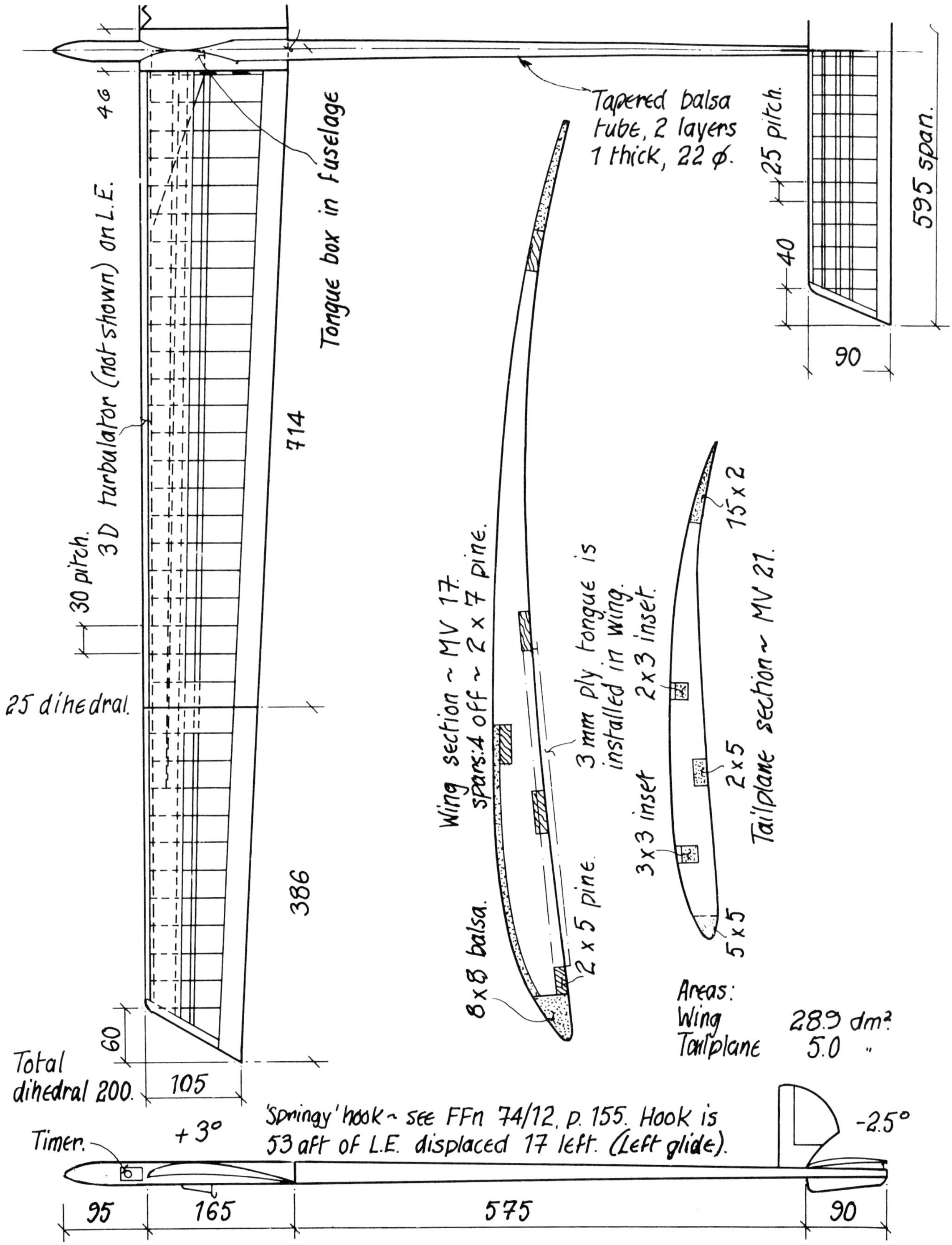

A-246, F1A, by M Vollbrecht

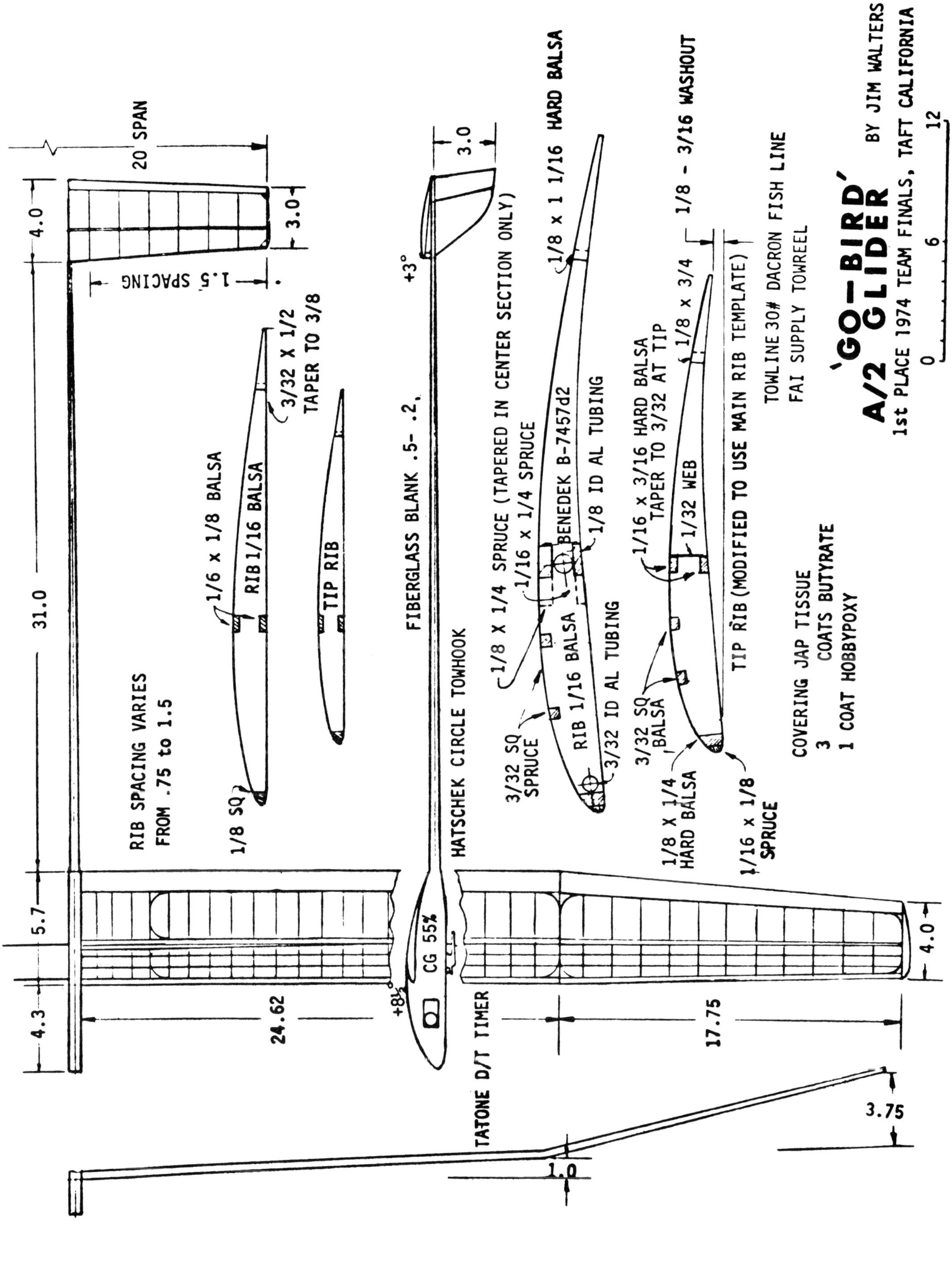
'GO-BIRD'
A/2 GLIDER
BY JIM WALTERS
1st PLACE 1974 TEAM FINALS, TAFT CALIFORNIA
0
6
12
TOWLINE 30# DACRON FISH LINE
FAI SUPPLY TOWREEL
COVERING JAP TISSUE
3 COATS BUTYRATE
1 COAT HOBBYPOXY
20 SPAN
4.0
3.0
1.5 SPACING
31.0
5.7
4.3
24.62
17.75
4.0
3.75
1.0
RIB SPACING VARIES FROM .75 to 1.5
1/6 x 1/8 BALSA
RIB 1/16 BALSA
1/8 SQ
3/32 X 1/2 TAPER TO 3/8
TIP RIB
FIBERGLASS BLANK .5- .2
+3°
+8°
CG 55%
HATSCHEK CIRCLE TOWHOOK
TATONE D/T TIMER
1/8 X 1/4 SPRUCE (TAPERED IN CENTER SECTION ONLY)
1/16 x 1/4 SPRUCE
BENEDEK B-7457d2
1/8 ID AL TUBING
1/8 x 1 1/16 HARD BALSA
RIB 1/16 BALSA
3/32 ID AL TUBING
3/32 SQ SPRUCE
1/16 x 3/16 HARD BALSA TAPER TO 3/32 AT TIP
1/32 WEB
1/8 x 3/4
1/8 - 3/16 WASHOUT
3/32 SQ BALSA
1/8 X 1/4 HARD BALSA
1/16 x 1/8 SPRUCE
TIP RIB (MODIFIED TO USE MAIN RIB TEMPLATE)

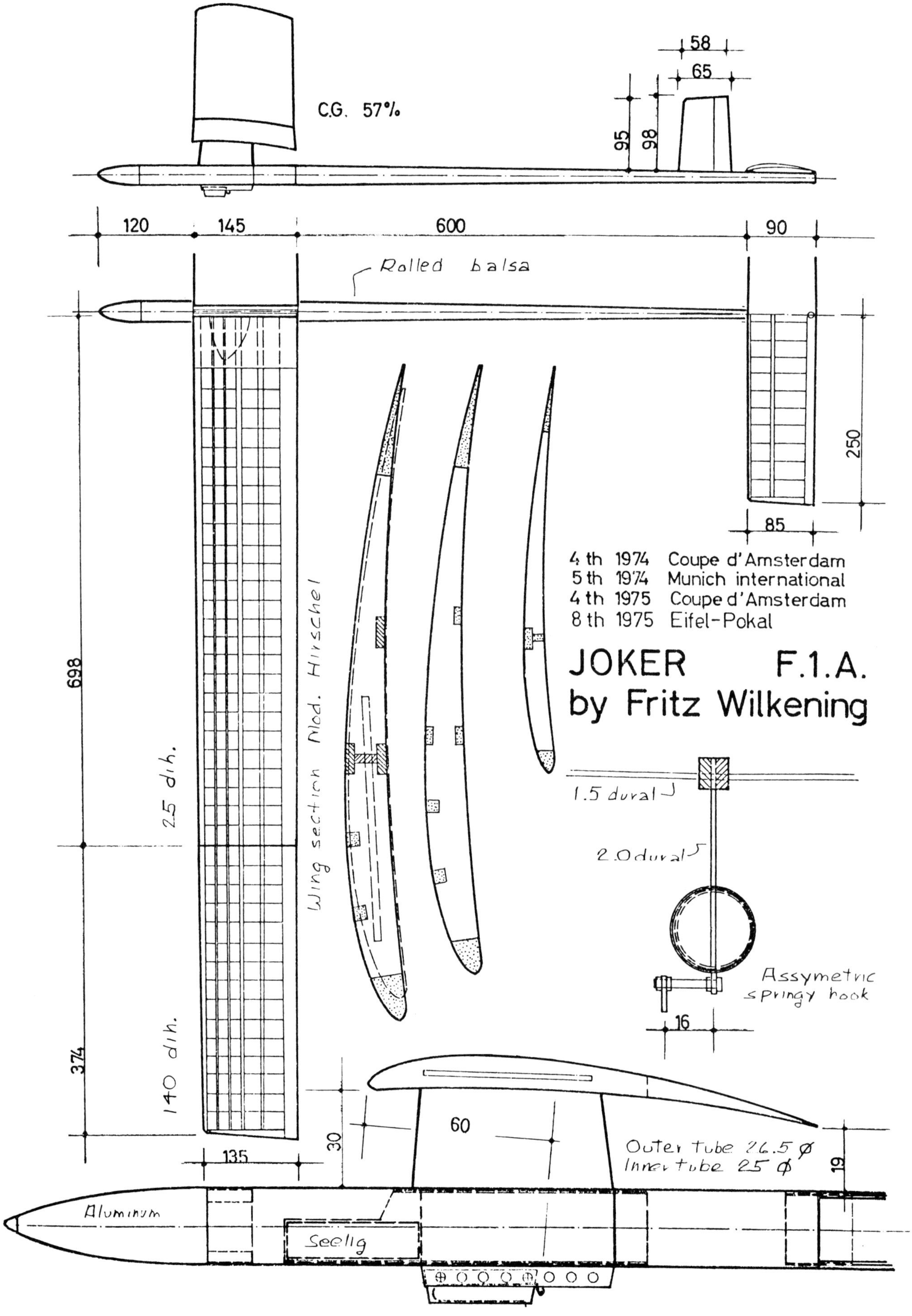
C.G. 57%
58
65
95
98
120
145
600
90
Rolled balsa
250
85
698
25 dih.
374
140 dih.
135
Wing section Mod. Hirschel
4 th 1974 Coupe d'Amsterdam
5 th 1974 Munich international
4 th 1975 Coupe d'Amsterdam
8 th 1975 Eifel-Pokal
JOKER F.1.A.
by Fritz Wilkening
1.5 dural
2.0 dural
Assymetric springy hook
16
60
30
19
Outer tube 26.5 Ø
Inner tube 25 Ø
Aluminum
Seelig

F1 B

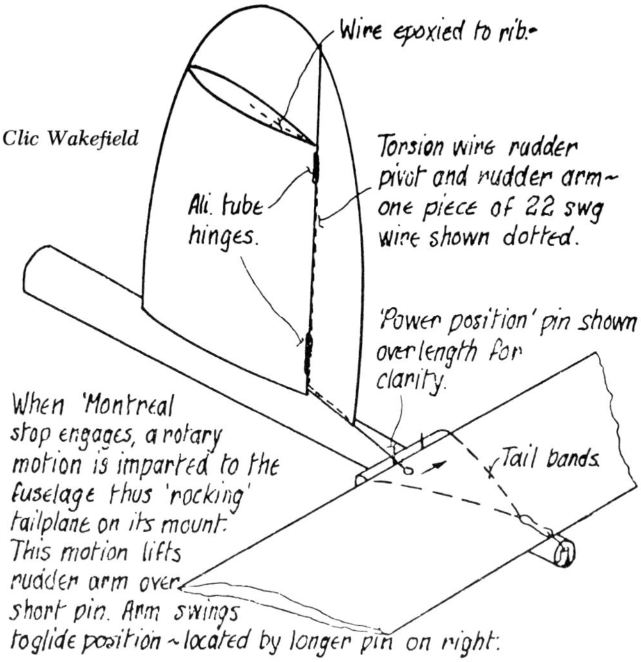

Clic Wakefield

F1B 1

Clic Wakefield - Joe Barnes, Liverpool, England

This model uses the rather daring technique of harnessing a floppy tailplane to activate the auto-rudder when the prop hits its stop. The idea was first tried by Carrol Allen of the USA and surprisingly it does work. There is no need to attach strings, timers or releases here and there. Impetus for this device is the desire to fly R-R for better centering in a thermal. An auto-rudder makes this a lot easier to do but it needs to be reliable. The Clic system does appear to be reliable. Joe says that if you recognize anything of yours on this model please take it as a compliment as he has tried to learn from others.

F1B 2

Veronica 73 - Hugo Benedini, Buenos Aires, Argentina

In 1975 Hugo placed this model 8th in the 14 man fly-off in the World Champs at Plovdiv, and then went on to place sixth at the Pierre Trebod that same year. Fuselage is a silver anodized aluminum tube and balsa tail cone. Wire prop hub rides on a 3mm shaft in two ball races. Stop is by tension pin. Wing is fully sheeted with 0.8 mm balsa. Section is the Saaman/Bousard, of thin, high undercamber and large nose radius. A turbulator is used. Fourteen strands of 6x1 mm Pirelli give 35 sec run off 440 turns.

F1B 3

Groovy Tuna by Jon Davis,
Albuquerque, New Mexico

At the World Champs in 1973 in Wiener Neustadt I used *Groovy Tuna* with the small (34 sq. in.) stab for rounds 1 & 2. For rounds 3, 4, 5, & 6 I used an old standard *Maxine* model with different airfoils and wing construction. For round seven I used my spare *Groovy Tuna* with the large stab (46 sq. in.). I had intended to use two models for the contest but my 7th round dork caused me to resort to my 3rd backup.

The model shown in the drawing is a good still air performer. Bob White and some others can still outdo me in dead air but I think it is mostly a matter of trim for those last 12-15 seconds. For an all-around model the large stab is the way to go. I tried the tiny stab in response to Bill Bogart's Symposium article on stab size and lo and behold it seems to work fine. However I am a little scared of using it in turbulence. I included some alternate airfoils because I think that a slightly different model would result. Perhaps a better all around model.

Jon Davis, Sarah and a rather wildly decorated "Groovy Tuna".

F1B 4

Wakefield - Louis Dupuis, Chatellerault, France

Louis flew this model to second place at the 1976 Pierre Trebod competition. Fuselage is molded balsa with very neat plug-in wing fixing. Wings are fully sheeted while the stab is full geodetic with tissue covering. Flight pattern is R-R with auto-rudder working off the prop stop. A prop run of 26 to 30 secs is obtained from 16 strands of 6x1 mm Pirelli.

F1B 5

1975 Wakefields - Allan Edwards, Hillview, Australia

Allan has long been a champion of advanced design concepts and continues to turn out very technically interesting models. A few seconds lost in the Plovdiv W/C dropped him to 18th place.

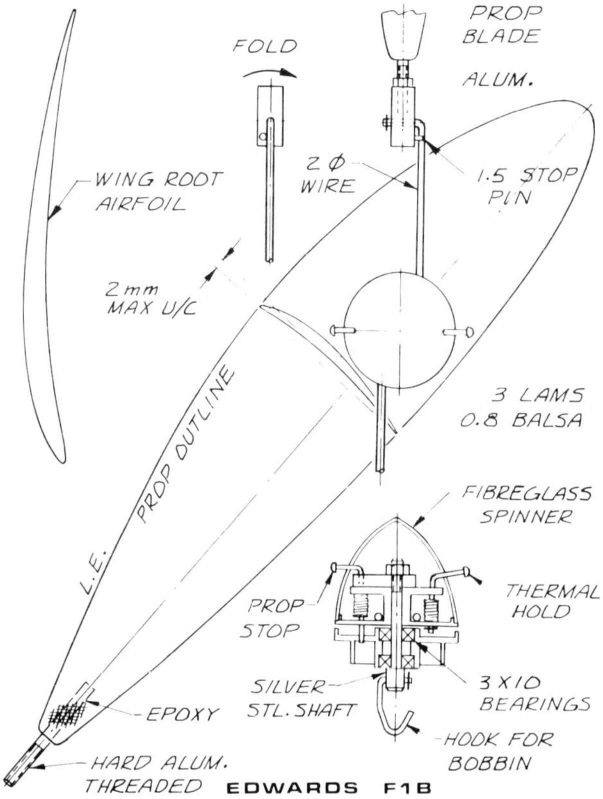

F1B 6

Flux Wakefield - Walt Ghio, Stockton, California

Walt's Wakefield uses an aluminum motor tube combined with a diamond section tail boom. This is a combination also in use by some of the very successful N. Koreans. In Roskilde, Walt had trouble with the damp conditions, as did most everyone. With the mediocre rubber available there was little margin to account for spoiled trim and cool turbulent air.

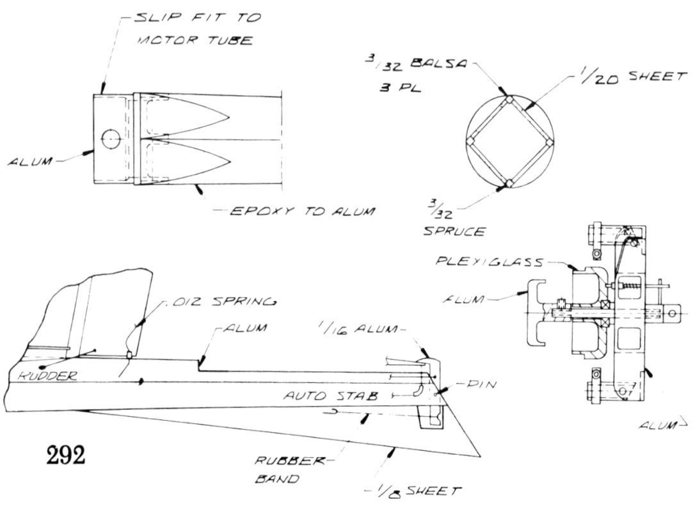

Allan Edwards uses solid wing, placed 18th at Plovdiv,

F1B 7

73-74 Wake -Dick Gildersleeve, Tujunga, California

This is Dick's version of Jim Patterson's design, with a few modifications including a shorter tail moment and a tissue covered stab. A Crocket winding hook is added to the Patterson-type hub. Gott. sectioned wing is skinned with 1/32 sheet balsa at constant chord of 3-15/16 inches. Schwartzbach prop blades were put on a Patterson type hub. Pattern is R-L with no auto surfaces.

F1B 8

French Connection by Dick Gildersleeve, Tujunga, California

I have been flying Vilim Kmoch's (Yugoslavia) Wakefield design for several years. I wanted to try the nearly flat bottom and straight topped section that Kmoch had done so well with in the W/C at Sweden (1971) where he placed second. I couldn't get much from the glide unless the model was placed in a fat thermal. Even with Bob White style turbulators, the glide wasn't up to standard. Haven't figured this one out yet.

When news came that Pierre Chaussebourg and his wife Josette would visit us in July '76, and a contest would he held at Taft so they could fly with us. I decided to make some changes to this model in an effort to place well at this meet. I lengthened the tail boom and made a new wing with the section from Lofler's Wake (B7406f). We had a great time enjoying our friend's visit and as it turned out the model changes did some good as I won the meet.

The original Kmoch construction gave some difficulty with cracking of the full depth spar on DT. Next model will have different spar location and straight rib construction.

F1B 9

Mini Ostrogoth - Emil Gouverne, Nancy, France

The *Mini Ostrogoth* is a rather wild looking machine by most standards but is somewhat conservative from the talented designers' standpoint. The wing sits directly on a long motor tube and there is a little "dog house" near the rear peg to house the KSB timer that operated the all-moving fin and the DT. The stab sits high up on wire stilts that give a fragile look to the affair. There is a thumb operated prop release to assure all motor turns go to the climb. The wing is made from solid balsa in a manner similar to the S.P.L. Nordic. This involves cutting sections out of a balsa plank and then regluing them together so that the grain is oriented with the camber shape for maximum strength. Emil does his wing sanding with a stationary motorized long drum that has sandpaper attached to it. The wing panel is held manually against the rotating cylinder. A vacuum cleaner hose picks up the dust.

To convince you that this is a conservative Gouverne design consider this. Emil has built solid wing Wakefields with aspect ratio as high as 47:1 (span = 111 inches).

F1B 10

Greave's Wake - David Greaves, Leamington Spa, England

Dave, with this model, won the 1975 Criterium Pierre Trebod. The design of the model never proceeded from an overall plan but rather came about through a series of unrelated events. As things were found to be unnecessary, such as an auto-stab, they were just left off, as a contribution to reliability. The John Gard airfoil, with turbulators at 7% and 23%, gives a very good flat, slow glide. Power is 16 strands 6x1 mm Pirelli. Triple fins appear to make the R-R pattern forgiving of too much turn and is a feature David will retain.

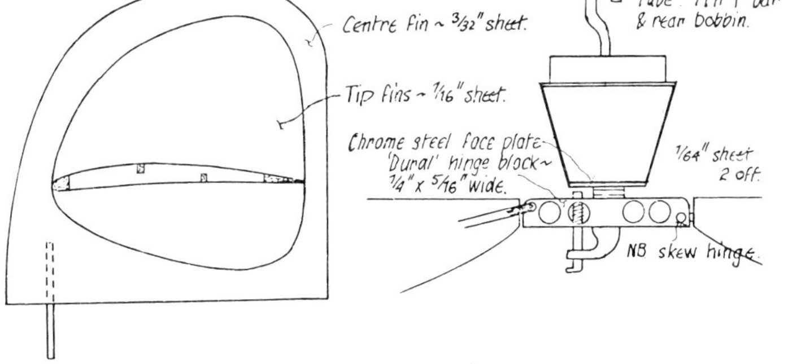

This is Vilim Kmoch's original machine that Dick Gildersleeve (USA) used as inspiration for his "French Connection."

Dave Greaves of Birmingham, England, won the 1975 Pierre Trebod contest with this triple finned ship.

F1B 11

Espada - Reiner Hofsass, Stuttgart, Germany

This model is rather dated, having been flown to 4th place at the 1971 World Champs and 10th place in 1975. However it has a number of interesting features that have been incorporated to improve the ultimate still air performance. The wing is solid balsa of 1972 mm span. Airfoil is very thin and well cambered. A triangular three-dimensional turbulator pattern is built into the wing surface. An all-moving fin and auto-stab are operated from a pylon mounted Seelig timer. Prop has clean low-drag spinner and outriggers. Some of Reiner's personal comments follows. "The drawing of *Espada* in FFN was quite correct. The airfoil is the result of an investigation in aerodynamics and structures. It is thick in percent in the center section due to the higher loads and higher Reynolds number there. While sanding it is necessary to have a good light to see "waves" on the surface (light must come from only one direction). I sand with my right hand and hold the wing in my left hand using a kitchen glove.

Wing is made from one piece of wood but first it is

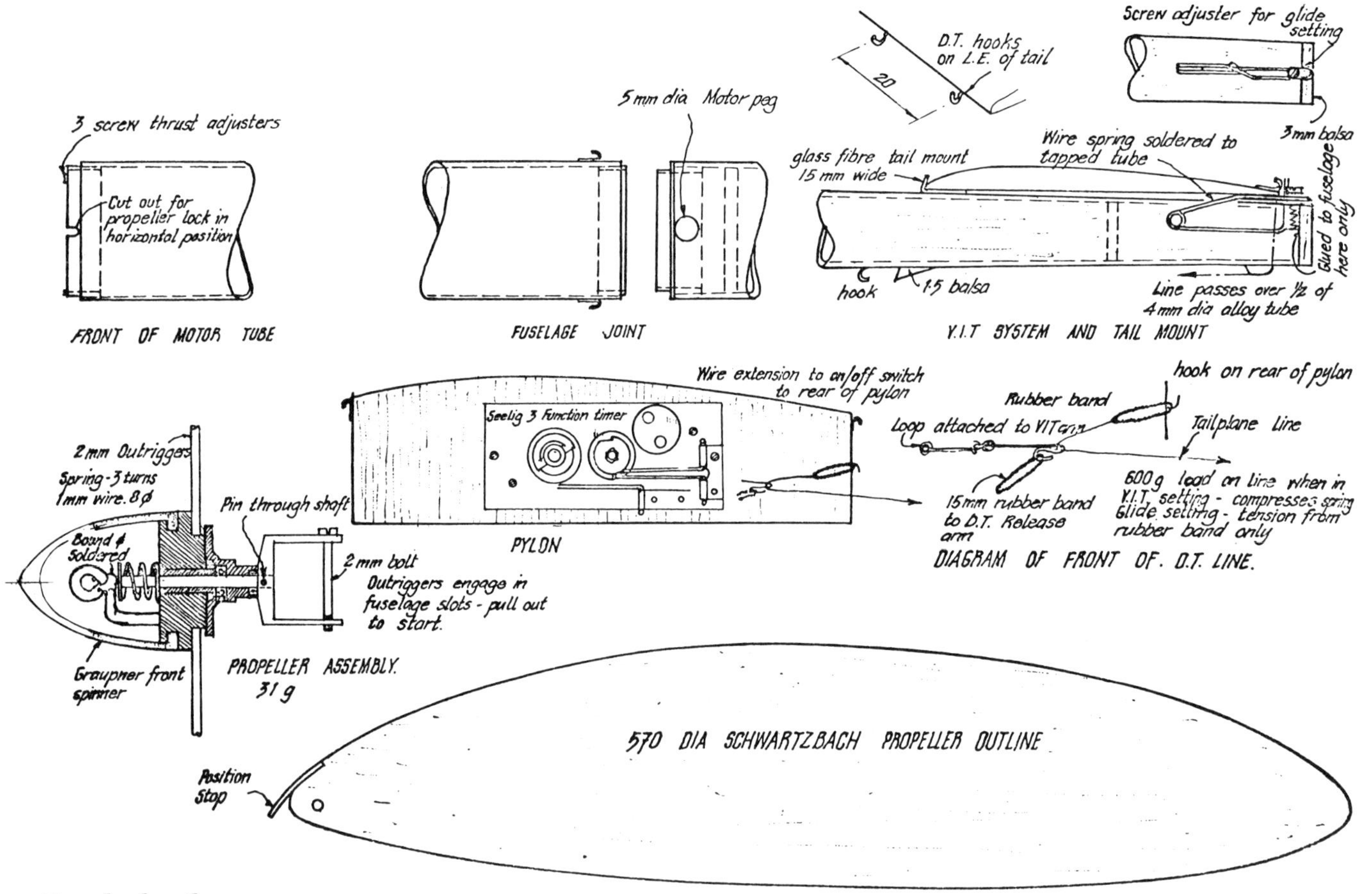

Espada details

sawn into spanwise strips and glued (casein glue) back together so that the quarter grain follows the camber line (very important). Wood density is 4.6 to 4.9 lb/c. ft. and modulus of elasticity is 159 Kg/mm^2. Weight of wing and tissue and dope is 73 g.

I have found that the dihedral is a function of humidity because of spruce LE and TE (see notes on drawing). The tailplane is trimmed with flat L.H. panel and with 3 mm washout in R.H. panel. The R.H. side is tilted down 10 mm. The CG is at 75% chord and displaced to the right 6 mm (right wing heavy). I built the Espada in 1969 but it took me a year to get used to it. In 1970 I managed to beat Loffler in an international contest in winter time. In 1971 it flew very well and in 1972 it was stolen by kids at the meet in Zell em See, Austria, when I had five maxes only together with Dohne, of DDR on a rainy, windy day. Afterwards I built models with single vee dihedral but could not get a good glide. Then I built some smaller models but I can handle the big ones better. This time I missed the team but will holiday in Denmark in '77. See you there.

F1B 12

No. 7401 - Kjeld Kongsberg, Albertslund, Denmark

Kjeld placed this Wakefield seventh in the 1971 World Champs. A 560x720 Schwartzbach type prop is used in a R-R pattern.

F1B 13

Flygande Kycklingen - Flying Chicken Wakefield Committee, Malmo, Sweden

The *Flying Chicken* was designed by a group of well-known Wakefield fliers who in 1974 gathered at Anders Hakansson's chicken ranch in Malmo to design the best Wakefield in the world. Three years prior this group also met there and designed the then world's best Wakefield, *Peas and Punsch*, which promptly became a big flop. Undaunted, with new resolve, the problem was again attacked, along with much liquid refreshments, and chicken.

The main thing established was the general disagreement about what would make the best design. So thankfully your ideas are still safe from rejection.

F1B 14

Wake Winner - Joachim Loffler, E. Germany

The Loffler Wakefield has had a good success topped by the 1973 World Championship win. A number of others have incorporated many of the Loffler design characteristics. Configuration is on the conservative side with nothing extreme, an approach that appears to be statistically desirable feature for those desiring to win. Most interesting feature is the torque operated variable pitch propeller. Prop assembly shown in the sketch has since been modified to include a Montreal-style stop.

F1B 15

Elastico FAI - Angelo Margiarini, Varese, Italy

Airfoil of this Italian Wakefield is not specified on the drawing but appears to be similar to the Gott.342. Deep multi-spars are used which add turbulation but rib strength is probably critical. Ply ribs at the root especially would be desirable. The drawing of the outrigger style prop hub does not show a prop stop. The spring on the 3 mm shaft indicates it is a tension stop and so there is probably a stop dog attached to the bent leg of the bobbin holder that contacts a screw in the back of the nose plug.

F1B 16

Plotschn 15 Wakefield–Ernst Neumann, Munich, Germany

Ernst writes that this last winter has been a busy time for him. He skiis in the Austrian, Italian and French Alps, plays basketball and coaches a girl's team. And of course he has a job (insurance-mathematics).

"Now to the *Plotschn 15*; it was my 18th Wakefield and I was trying to get an F1C type climb pattern. After toying around with VIT and short motor runs (20-25 secs) and 18 strands I turned to variable pitch props. I found the Loffler method to be the best, as it is simple to make and with a good prop works OK. I am still using the Bob White prop which gave the best results on my ships. Sixteen strands 6x1 mm rubber gives 35-37 sec run, quite impressive climb. But I always had trouble with the glide. I tried several sections as for instance the old one I used in Marigny and of which you took a picture of the VP prop. On that model I used a B7406f modified. Upperside (Y_o) were the normal coordinates of a 120 mm chord but undercamber (Y_u) of a 140 mm chord B7406f. Actual wing chord was 120 mm. The climb wasn't very fast, but the glide was very good, maybe I'll try it on a Nordic.

Plotschn 15 had a so-called fast wing section, and the climb was very good but the glide was rather tricky. Maybe the dihedral was too flat, the stab too thick or a second turbulator might have cured the problem, but now it is lost in a corn field.

I am very content with my latest model although it doesn't have contest experience yet, but it shows a very good climb and the glide is good enough to give a flight time of over 200 sec. In contrary to the drawing I now use a modified version of the Loffler method. It is enclosed in a spinner to keep out dirt. Wings, stab, tail boom and rudder are Mylar covered and trim the same as *Plotschn 15*. Inner wing tip washout one deg. (right wing) and outer wing washout three degrees. In my sparetime I am designing a new Wakefield with torque controlled flaps.

F1B 17

Wakefield - Ray Pavely, Ramsden Heath, England

A clever torque controlled prop is used by Ray on this Wakefield. The stop pin also functions as a lock pin for holding. Flight readiness is gained by pulling the prop hub forward. An auto-rudder operated from a timer comes in after four secs for a R-R pattern.

F1B 18

Vitar 2 - Ron Pollard, Newcastle on Tyne, England

Ron has been working on the Wakefield design problem for over ten years and has consistently turned out well-engineered and handsomely constructed examples. The *Vitar 2* represents an important step in that striving for excellence. Ron was quick to see the need for auto systems and so uses auto-stab to control the first five secs or so of burst. Then after about 27 secs uses auto rudder to kick into a right glide turn. Sixteen strands and the Schwartzbach prop helped the climb and of course led to the need for the auto-systems. Use of Koster type wing section helped improve the glide and this basic design formula has been backed up with a foolproof structure. Recognizing that the ribs are the key to a warp free wing, Ron uses very hard 1.5 mm balsa together with TE gussets and cap strips on their upper surface. Wear points are especially reinforced and hardened to eliminate vagaries in trim requirement. Ronytubes, a tapered F/G tube, (sold by Ron) are used as reinforcing spars in the wing roots.

Ron Pollard and his "Vitar" at Roskilde, Denmark.

F1B 19

Dutch Threat by Pim Ruyter, De Meern, Holland

My original models, MVG 87-88 were lost in early 1975 and this was the reason I flew so poor during the 1975 W/C. My replacement models were not as good. These models had an asymetrical wing and caused a so-called hooking effect during the initial burst. A second reason was the rubber which became sluggish in the Plovdiv heat. The year 1975 was also the year of qualification for the European Championships. I was not able to make it for the same reasons (I was second in 1974). The selection for the 1977 W/C took place in 1976 during the flying season. I was unable to attend due to a five month business trip to the USA (Mass.)

Details and differences of the 1976 model from that shown on the drawing are:

Fuselage - Two parts (front/aft); wing mount, glass cloth epoxy covered; rudder, 3 mm sheet actuated by Montreal stop; tail incidence operated by Seelig two function timer.

Wing - Center panel, full-sheeted 1 mm with spruce 1x8 integrated main spar (previous construction caused flutter); section modified B6456F with sharp LE; center panel span reduced to 390 mm, turbulator used on wing.

Stabilizer - Not geodetic, rib spacing 20 mm, same spar structure.

Propellor - Glass cloth covered, very soft balsa. weight 12 g per pair.

F1B 20

VM2-77 Wakefield - Sergey Samokish, USSR

The second place Wakefield winner at Roskilde had a number of interesting features which attracted attention. Most apparent was the location of the reworked camera timer set into a streamlined blister atop the alloy motor tube forward of the wing. It operated the auto stab at five secs and auto rudder at 27 secs. pattern was R-R with 3 deg washout in the left panel. Airfoil was the popular B6356b with an 0.6 mm turbulator mounted 4 mm aft of the LE. The prop hub had a clever, and complex, torque operated prop stop and holding pin mechanism all inside the noseplug and spinner. Tail boom was diamond shaped section, with reed longerons. Rib material (balsa) tapered in thickness from 0.2 mm at LE to 1.2 at TE. Prop was 590 mm dia by 650 pitch and had a run of 38 secs with 520 turns of 16 strands of 6x1, '72/'74 Pirelli. In anticipation of turbulent conditions at Roskilde, Sergey trimmed the chord of his prop blades from 25 mm to 15 mm at the root. This speeded up the prop run to 32/30 secs, which proved helpful in the actual conditions.

Pym Ruyter (rt.), Ernst Reitteren (lt.) and Joachim Loffler (ctr.) placed second, third and first, respectively at the 1974 European Championships.

Details of the second place 1977 W/C Wakefield of Samokish. Design and construction is very carefully thought out and executed. No development stalemate here.

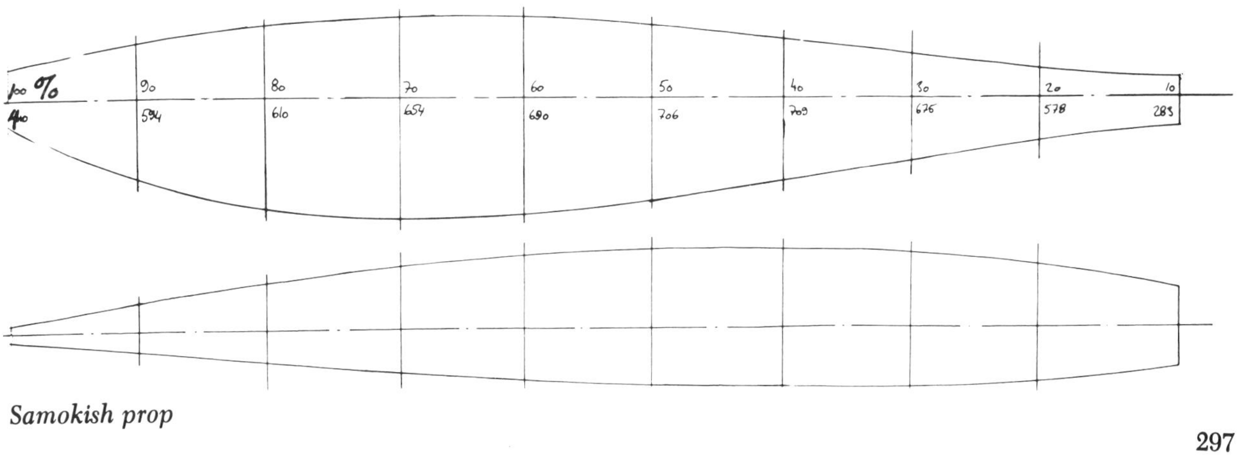

Samokish prop

F1B21

1977 World Champ - Kim Dong Sik, N. Korea

All the N. Korean Wakefields that have been seen in the West (starting from the 1971 World Champs) have appeared to have the same general characteristics. These seem to reflect the Russian design influence which came about through instruction and transfer of information and materials. That the N. Koreans have learned their lessons well is quite obvious and in fact have surpassed their mentors through concentration and hard work.

The design is based on moderate proportions that are suitable for a broad range of weather conditions. It won at Plovdiv under ideal weather. It won again at Roskilde in depressingly gusty drizzle. Rubber used was Filati and it must have been good. Some suspected it was outstanding but it is unlikely that their success hinged directly on the rubber. Thermal detection by one of their managers played an important and possibly decisive role as he seemed to have that sloppy air figured out quite well.

Model design and construction features are described on the drawing. The following comments can fill in other data. Prop uses an S-shaped hook, short outrigger blade hangers and a tension operated stop. Winding is done without protective tube or prop shield. DT is by mechanical timer which is started by a launch grip release. Since the auto-rudder operates off the timer, consistent timer start is needed. There is a lot of spruce and pine (or equivalent) used in the construction. This may not be by necessity but by preference, as there seems to be no shortage of the best modelling material in their equippe.

F1B22

US Team Wakefield by Willard Smitz, Kenosha, Wisconsin

This model design was influenced by John Gard's 1969 5th place World Champs ship. I made a few changes to suit my tastes. In the calm early rounds it flies very well. For the other rounds I fly a shorter ship which is a part Joe Bilgri and John Allen design.

The 1975 World Champs was quite an experience. Had a good time at the Pierre Trebod also. Many thanks for rescuing Frank Wolff and me at Marigny. It did show that to survive in a strange country be sure to have some of their money!

Leaving Marigny Frank and I took the rental car back to Chateau Thierey and got some French Francs! Went to Paris for an hour of walking around the block from the train station, then time to get back on board for the train to Frankfurt. Frank left for home and I went on to Wurzburgh and Ulm on some company business. Arrived home for Labor Day pretty well worn out. See you next time!

World Champion (1977) Wakefield of Kim Dong Sik, North Korea. Dedicated team effort has paid off.

F1B23

Ko-Bu-Ki - Paik Chang Sun, N. Korea

This model is the 1975 World Champion, flown at Plovdiv, Bulgaria. Under ideal weather conditions 14 advanced to the fly-off. Half of these advanced to the six minute fly-off where *Ko-Bu-Ki* scored 303 seconds to edge out Bob White of the USA by 22 seconds. The N. Korean presence in FAI Free Flight was thus firmly established with this individual and team wins. Paik Chang Sun is the instructor of the N. Korean modelling institute. The model is quite similar to the one of Kim Dong Sik who placed 2nd in 1973 and 1st in 1977. Model uses a prop-stop actuated auto-rudder and pattern is R-R. One degree of left thrust is used. Prop, rudder combination appeared to be very nicely matched as the ship climbed in a good pattern. Quite a bit of non-balsa parts were used but this seemed to be by preference and not necessity.

F1B24

Hesitator - Mike Thomas, Ontario, Canada

The unusual feature of this model is the delayed release prop mechanism. Mike holds the fully wound model with his hand on a time trigger. He gives the model a tremendous heave after galloping across the field (a cricket bowl launch). The timer advances and in about one second a mechanism releases the prop stop. The blades, heretofore trailing back along the fuselage, bestir themselves and start flailing around to pull the model onward and rescue it from its

Mike Thomas, Canada, shows off his timer-delayed prop release launch technique.

precarious situation. Mike claims he does get more height with this rig. At Plovdiv he made it to the fly-off only to break a motor leaving him little time to get ready and launch. The flight was 171 secs, good enough for 13th place, and a team bronze medal for Canada. A complete building article for this model was published in the May 1974 Model Airplane News.

F1B 25

Santa Catalina - Paul Van Leuven,
Queensland, Australia

The version illustrated in the plan was designed and built some time ago. Since then Paul has become somewhat dissatisfied with performance of single fin models (that is without AR and VIT) and now uses a twin fin layout. He has found that twin fin models fly well, as if on rails, particularly for simple designs. Looking back over the years and the records both Alan King and Bond Baker won the World Champs with twin fin models. Bob White placed on the U.S.A. team three times in a row, and placed second in 1975.

F1B 26

Flemmard XIV - Jean Wantzenreither,
Farebersviller, France

This model became the Champion of France in 1973 and has a number of interesting and somewhat unusual features. The stabilizer sits atop a wire and aluminum saddle and stilt arrangement. Motor tube is 1.5 mm balsa with heavy paper inside and out. Plastic ring reinforcements are placed at nose and rear peg. Wing is fully sheeted with no spars. Triangular turbulators are built into the sheet. Tailplane is multi-spar and tissue covered. Prop is Schwartzbach type 560/900 with an outrigger hub.

F1B 27 & 28

Wakefields by Bob Wells, Hornchurch, England

The 1968-76 Wakefield (27) is of course an old one, being built for the 1969 World champs when I was lucky enough to be on the team. I have of course built several since then including the 1975 Wakefield (28) which was built for the 1975 Champs in Plovdiv. I am pleased to say that this model is my best Wakefield yet but I would not stick my neck out and put a time on its performance.

My thoughts on F/F at the moment are mixed. I still think that free flight contest models are the only type worth flying and I have tried radio glider, team racing etc., but have not been impressed with them. Open rubber is my favorite with Wakefield a close second. I feel FAI contests have degenerated to a tactical rat-race and the quality of the model now means nothing. If all this sounds gloomy it is not the way I feel. I am looking forward to building and flying free flight models and even winning a contest now and again, but it would be more fun with fewer design restrictions.

F1B 29

Vol Libre Wakefield - Bob White, Monrovia, California

Bob has established an exceptional win record with this good looking model.

1971	3rd	place	World	Champs,	Sweden
1973	5th	place	World	Champs,	Austria
	1st	place	Pierre	Trebod,	France
1975	2nd	place	World	Champs,	Bulgaria
	3rd	place	Pierre	Trebod,	France

And of course he has won and placed highly in many other meets in the U.S.A. The R-L pattern he uses makes for simpler adjustment and doesn't require an auto rudder. Every item in the model in fact is arranged to avoid complexities. This does not mean however, that the structure is simplified. Bob puts a lot of detailed work in developing a structure of great precision. Then he concentrates on a step by step trimming procedure, which is then followed by a tactical approach in competition to develop his thermal sensing ability. Here are some of Bob's comments:

"I carve prop blades quite thin for controlled flaring during the first part of the power run. I use a turbulator on the prop blades as I think it helps during the cruise portion of the flight. The flight adjustments

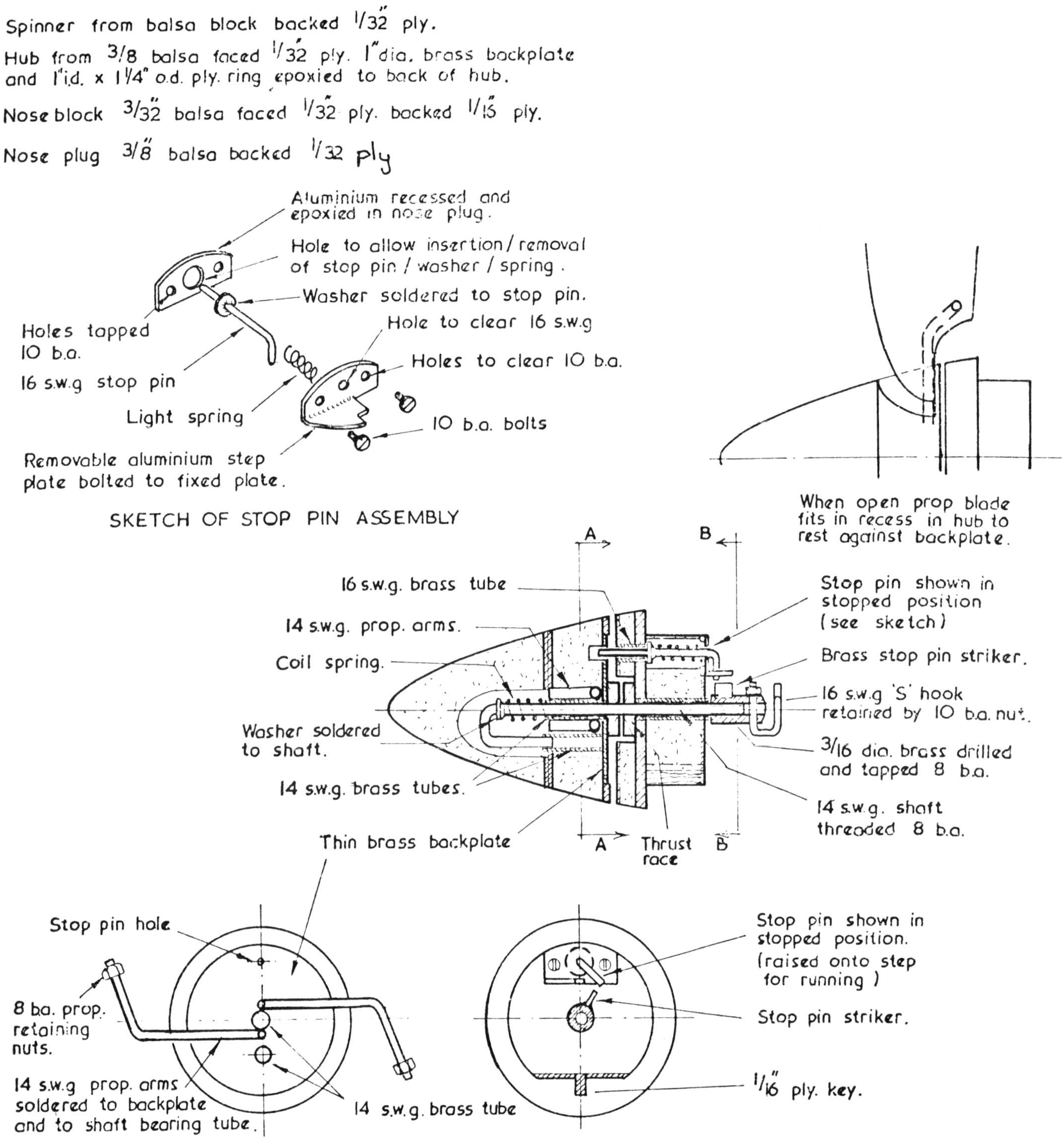

Bob Wells prop details

are conventional, except for a few important points. I fly the model in a R-L pattern. I use about 4 deg right thrust and 3 deg down thrust. The rudders are cambered for a left turn. I use an aft CG position. The CG of each model is determined by tests in the following way: I tape lead ballast to the bottom of the model, then adjust the tail incidence for a proper glide. Then I move the weight back, and remove some of the negative incidence from the stab to again get a good glide. I continue doing this until the model is on the verge of being longitudinally unstable. That is, it continues in a straight line from whatever direction you throw it, with no tendency to settle into a normal glide. Then I move the weight forward again until the model seems stable. This is done with each new plane, as they are all different even when I try to make them the same.

With this adjustment procedure I try to get a straight-out climb at first then a gradual turn to the right as the power decreases. In still air the glide is to

Bob White and second place Wakefield at 1975 World Championships.

the left in a wide open circle. I tighten the turn for turbulent air. I use a thermister type thermal detector, a 30-ft. mylar ribbon on a 20-ft. pole, and cat-tail fluffies to test the air. The best Wakefield will not max in bad air. I try to launch into a thermal, or at least avoid "down-air". Also, I like to choose my own air. Watching other planes for lift is fine sometimes, but often there are none to watch.

F1B 30

Whiskas - Mike Woodhouse, Norwich, England

Mike flew this Wakefield in the 1975 World Champs at Plovdiv. Deceptive soap bubbles seemed to be his undoing early on and he ended up a rather dismal 42nd place. The model used a Burrows glass fiber motor tube and tailboom, and had a R-R pattern with the help of a prop stop actuated auto-rudder. Sixteen strands of Pirelli drove a Schwartzbach pattern prop 22 in. dia. by 28 in. pitch. Some of Mike's comments are here excerpted from FFN.

"I stuck to a low aspect ratio wing, 50 in. x 5 in. for my basic model to maintain good stability. I like the B6405b wing section but I have modified it to have a thicker TE. The geodetic structure keeps the weight down but retains rigidity and avoids the need for turbulators. In 1967 in correspondence with Paul Lagan, we agreed that we needed a machined prop hub with a 'Montreal' stop. It was designed by post. Paul made them and we furthered the development. Murray Stringer designed the mass produced version from it, then FFN took over. On my version the nose block is solid 'Delrin' to replace the nylon block housing used in the original unit. The stop pin is 0.080 in. dia. (as is now standard) but with a coil spring, and the wire hangers are wrapped with fuse wire and soldered."

F1B 31

Wakefield Modell - Hans Zachalmel, St. Polten, Austria

This model placed 3rd in the 1975 World Champs at Plovdiv, placing one second behind Bob White and 23 secs behind the winner, Paik Chang Sun. Hans has generally performed well in the international meets including 14th, 1973 W/C and 16th, 1977 W/C. The wing is two piece and fully tapered, root to tip.

F1B 32

Alba Constellation - Jan Zetterdahl, Spanga, Sweden

Jan has yet to reach near the top of the list at the World Champs but his models have always shown a technical excellence and good workmanship. *Alba* has a super short nose to get the mass near the CG and reduce longitudinal moment of inertia. Prop blade tips actually end up aft of the wing TE on fold. A somewhat complex prop hub and nose plug account for some weight up forward and a prop-stop auto-rudder mechanism is used. Pylon has a clamp arrangement around the alloy motor tube. It is adjustable for pylon position and prevents those embarrassing moments when a shock like a blown motor knocks the pylon off.

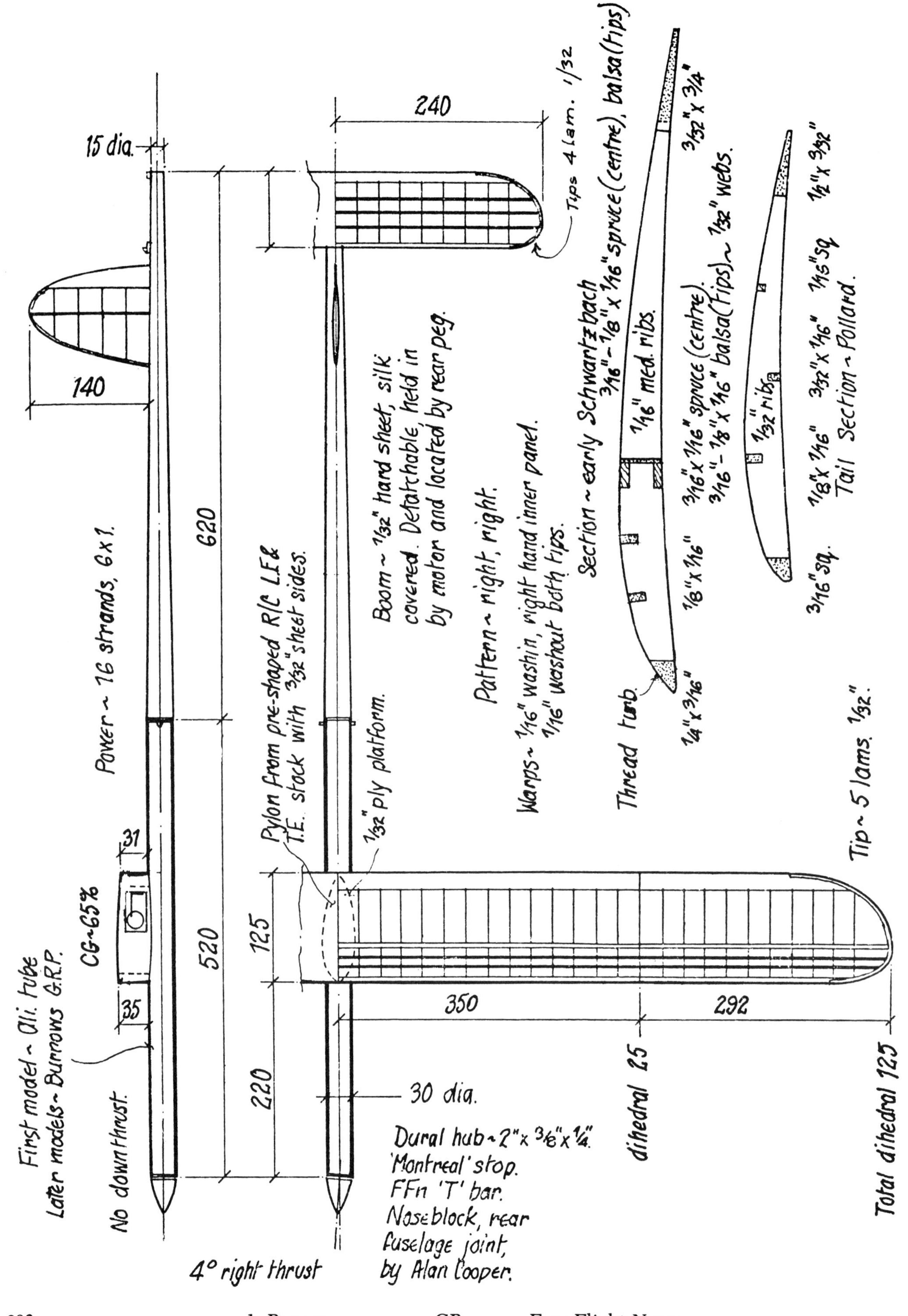
15 dia.
140
620
520
240
125
350
292
220
31
35
CG~65%
30 dia.
Tips 4 lam. 1/32
First model ~ Ali. tube
Later models ~ Burrows G.R.P.
No downthrust.
4° right thrust
Power ~ 16 strands, 6 x 1.
Pylon from pre-shaped R/C L.E. & T.E. stock with 3/32" sheet sides.
1/32" ply platform.
Boom ~ 1/32" hard sheet, silk covered. Detatchable, held in by motor and located by rear peg.
Pattern ~ right, right.
Warps ~ 1/16" washin, right hand inner panel. 1/16" washout both tips.
Section ~ early Schwartzbach
3/16" - 1/8" x 1/16" spruce (centre), balsa (tips)
3/32" x 3/4"
1/16" med. ribs.
3/16" x 1/16" spruce (centre)
3/16" - 1/8" x 1/16" balsa (tips) ~ 1/32" webs.
1/8" x 1/16"
1/4" x 3/16"
Thread turb.
1/32" ribs
1/2" x 3/32"
1/15" sq
3/32" x 1/16"
1/8" x 1/16"
3/16" sq.
Tail Section ~ Pollard.
Tip ~ 5 lams. 1/32".
Dural hub ~ 2" x 3/8" x 1/4".
'Montreal' stop.
FFn 'T' bar.
Noseblock, rear fuselage joint, by Alan Cooper.
dihedral 25
Total dihedral 125

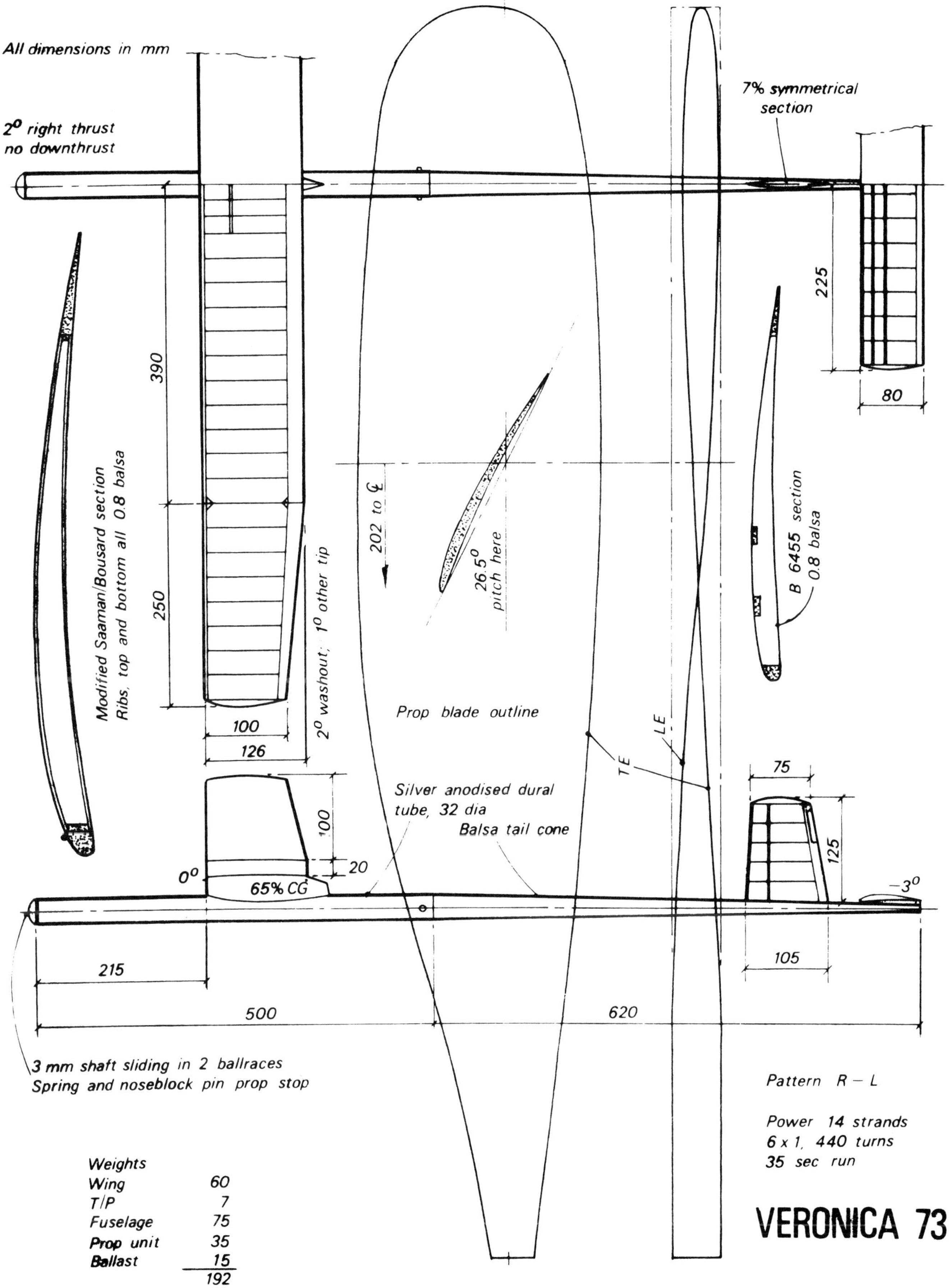
All dimensions in mm
2° right thrust
no downthrust
7% symmetrical section
225
80
390
250
100
126
Modified Saaman/Bousard section
Ribs, top and bottom all 0.8 balsa
2° washout; 1° other tip
202 to ℄
26.5°
pitch here
B 6455 section
0.8 balsa
Prop blade outline
TE
LE
75
Silver anodised dural tube, 32 dia
Balsa tail cone
100
20
0°
65% CG
125
−3°
105
215
500
620
3 mm shaft sliding in 2 ballraces
Spring and noseblock pin prop stop
Pattern R – L
Power 14 strands
6 x 1, 440 turns
35 sec run
Weights
Wing 60
T/P 7
Fuselage 75
Prop unit 35
Ballast 15
192
VERONICA 73

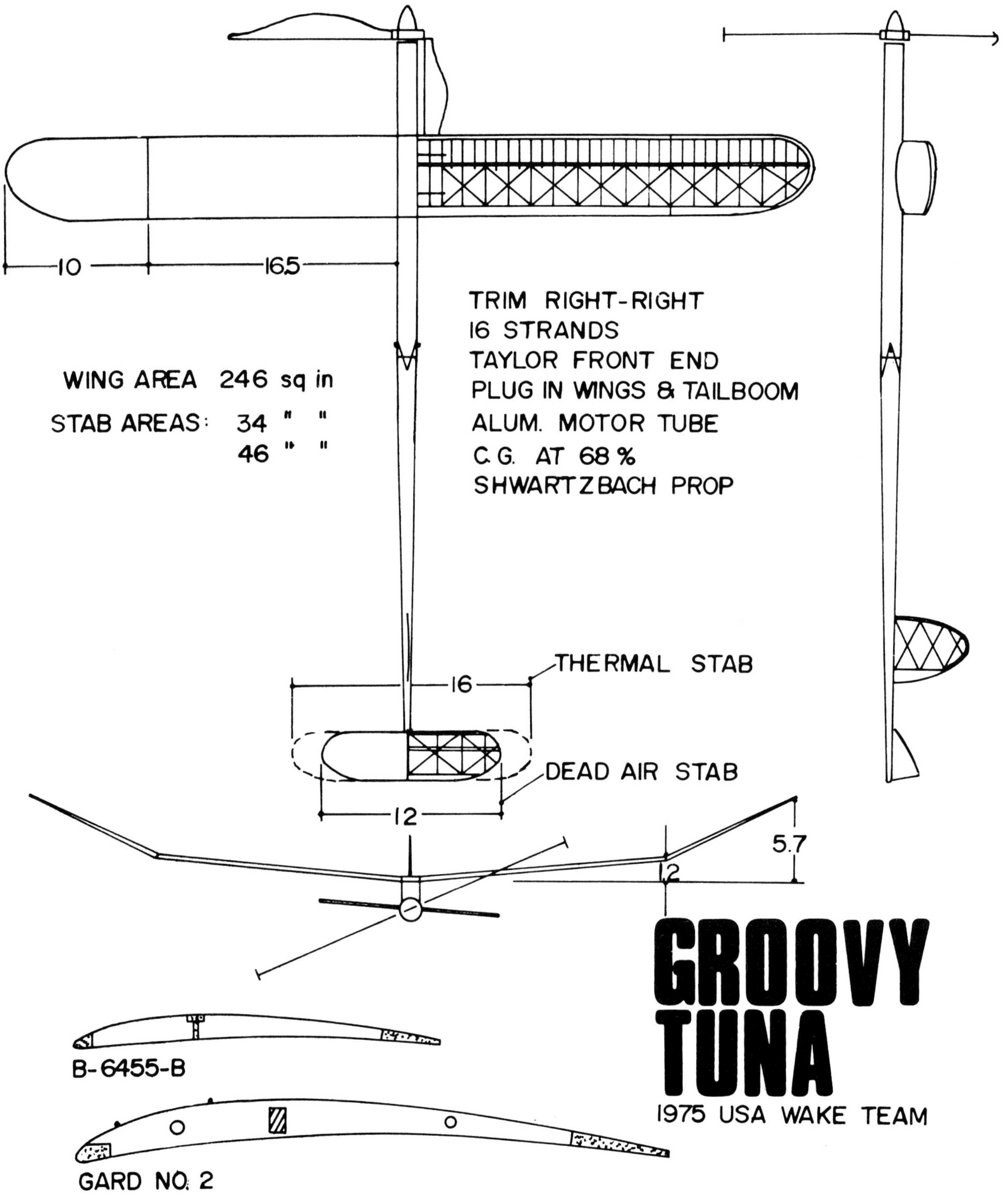

 3 Davis USA Scatter

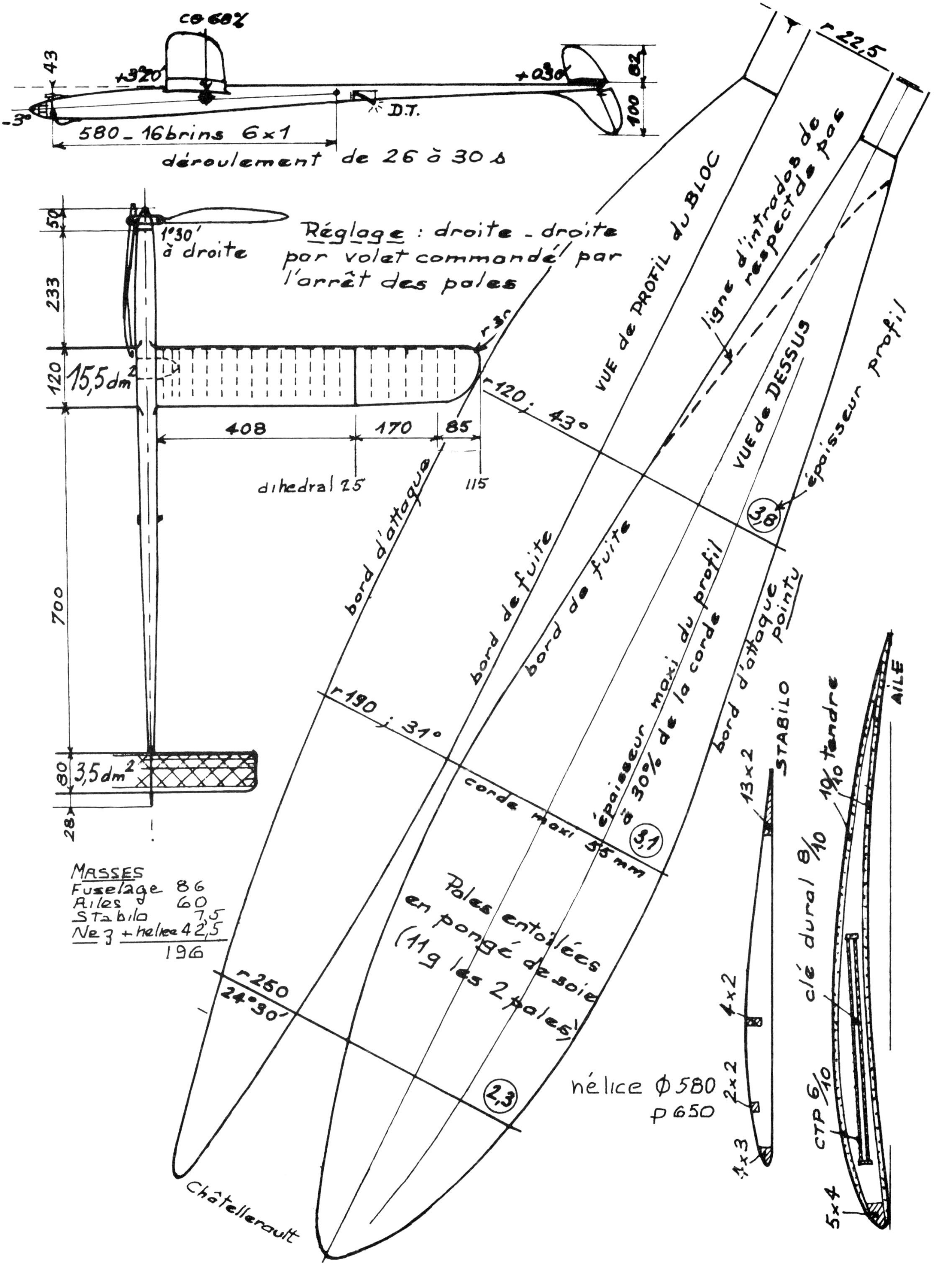

4 Dupuis France Vol Libre

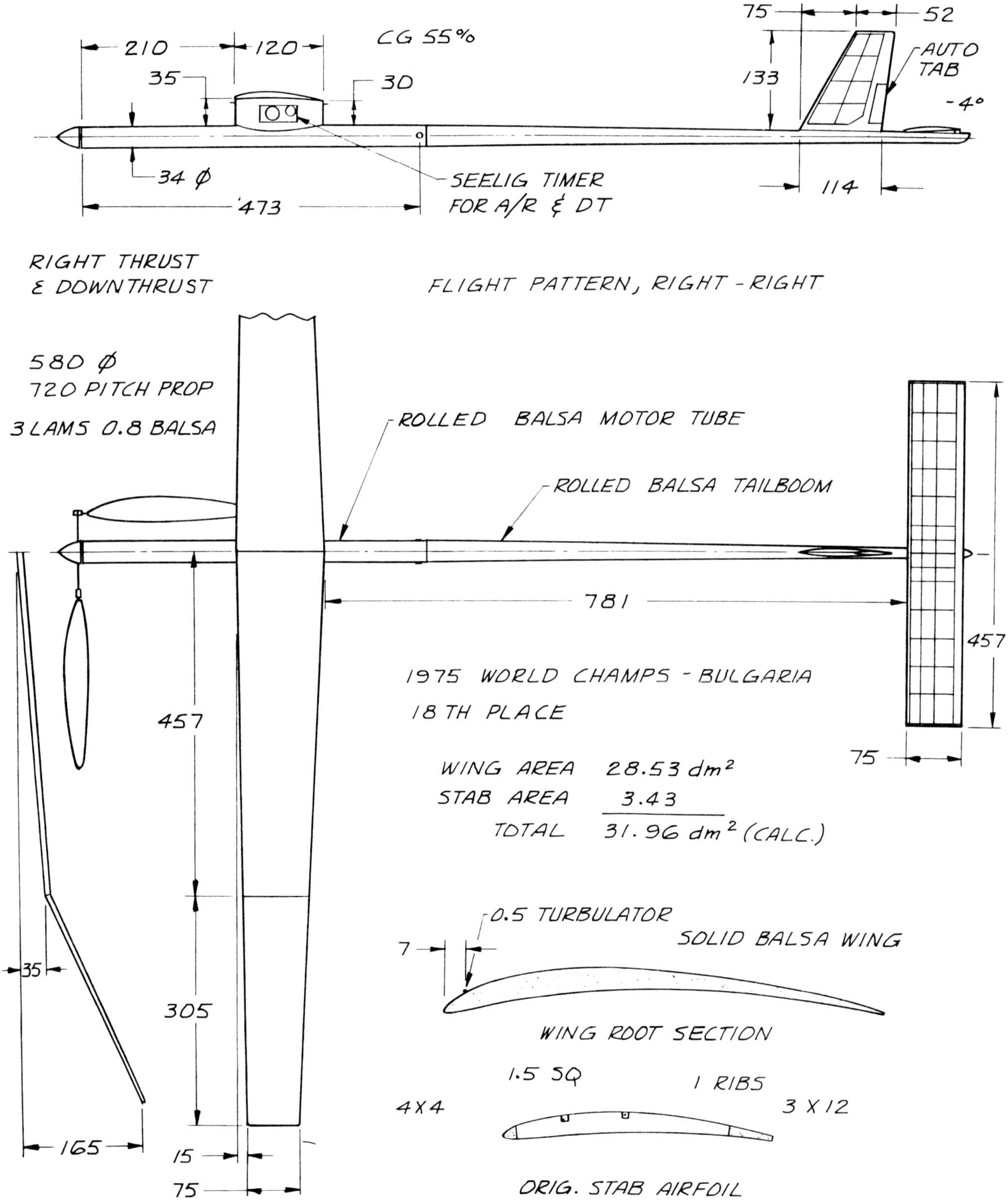

DIMENSIONS IN MILLIMETERS

1975 No. 2 & 3 F1B

BY ALAN EDWARDS, AUSTRALIA

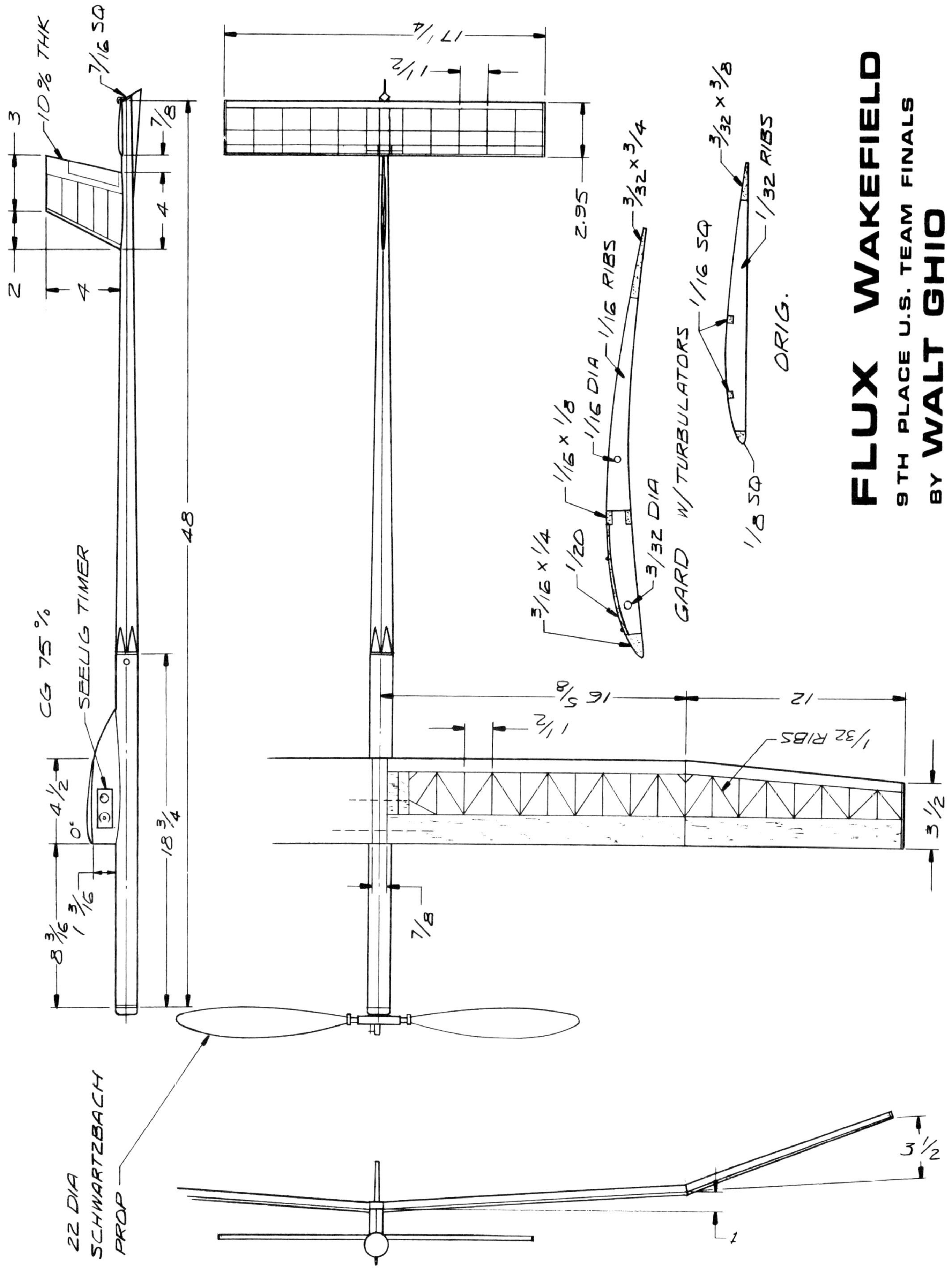
FLUX WAKEFIELD
9 TH PLACE U.S. TEAM FINALS
BY WALT GHIO
10% THK
7/16 SQ
7/8
3
4
2
4
17 1/4
1 1/2
2.95
3/32 x 3/4
1/16 RIBS
1/16 DIA
1/16 x 1/8
3/32 DIA
1/20
3/16 x 1/4
GARD W/TURBULATORS
3/32 x 3/8
1/32 RIBS
1/16 SQ
1/8 SQ
ORIG.
48
CG 75%
SEELIG TIMER
4 1/2
0°
18 3/4
8 3/16
1 3/16
16 5/8
12
1 1/2
1/32 RIBS
3 1/2
7/8
22 DIA
SCHWARTZBACH
PROP
3 1/2
1

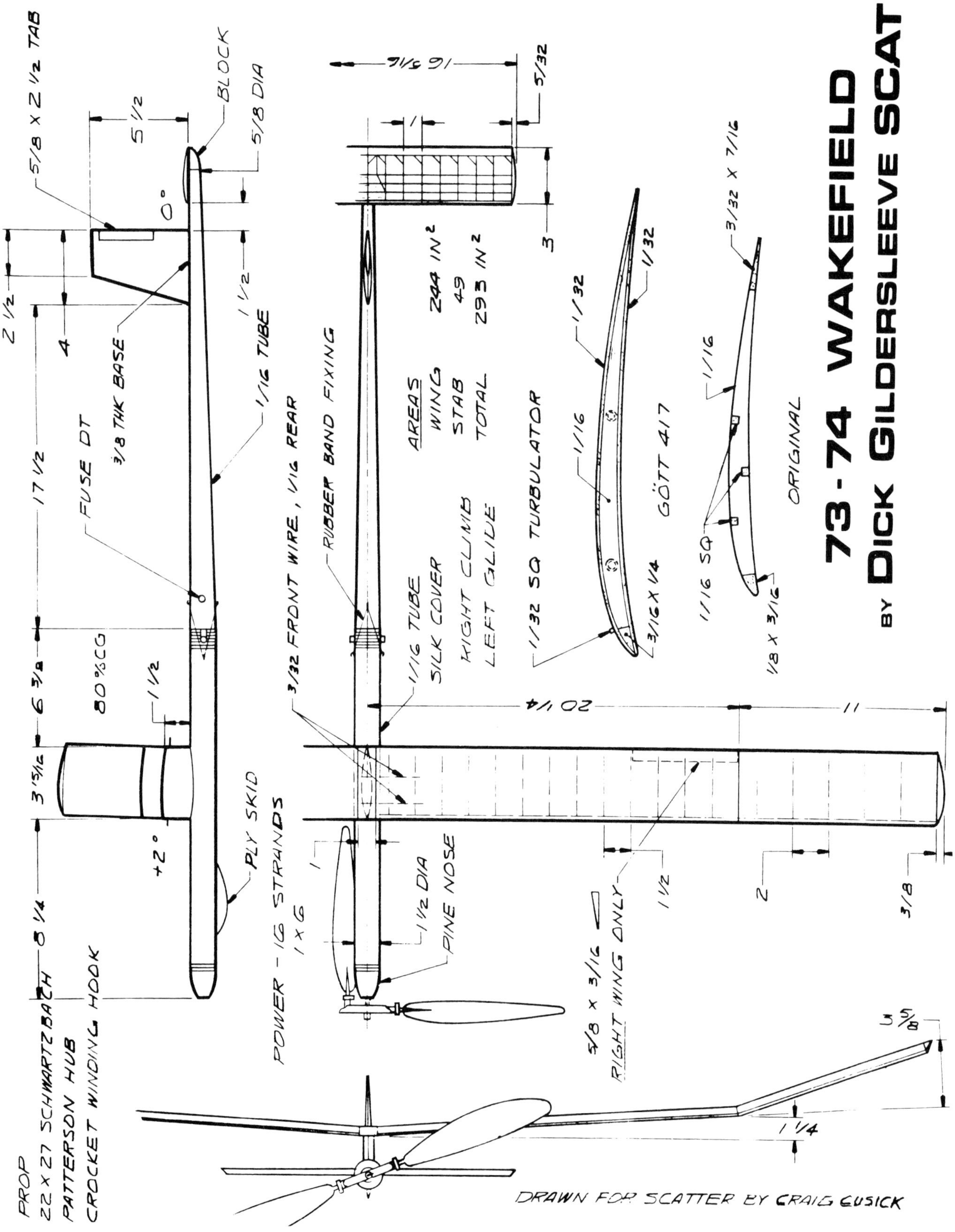

7 Gildersleeve USA Scatter

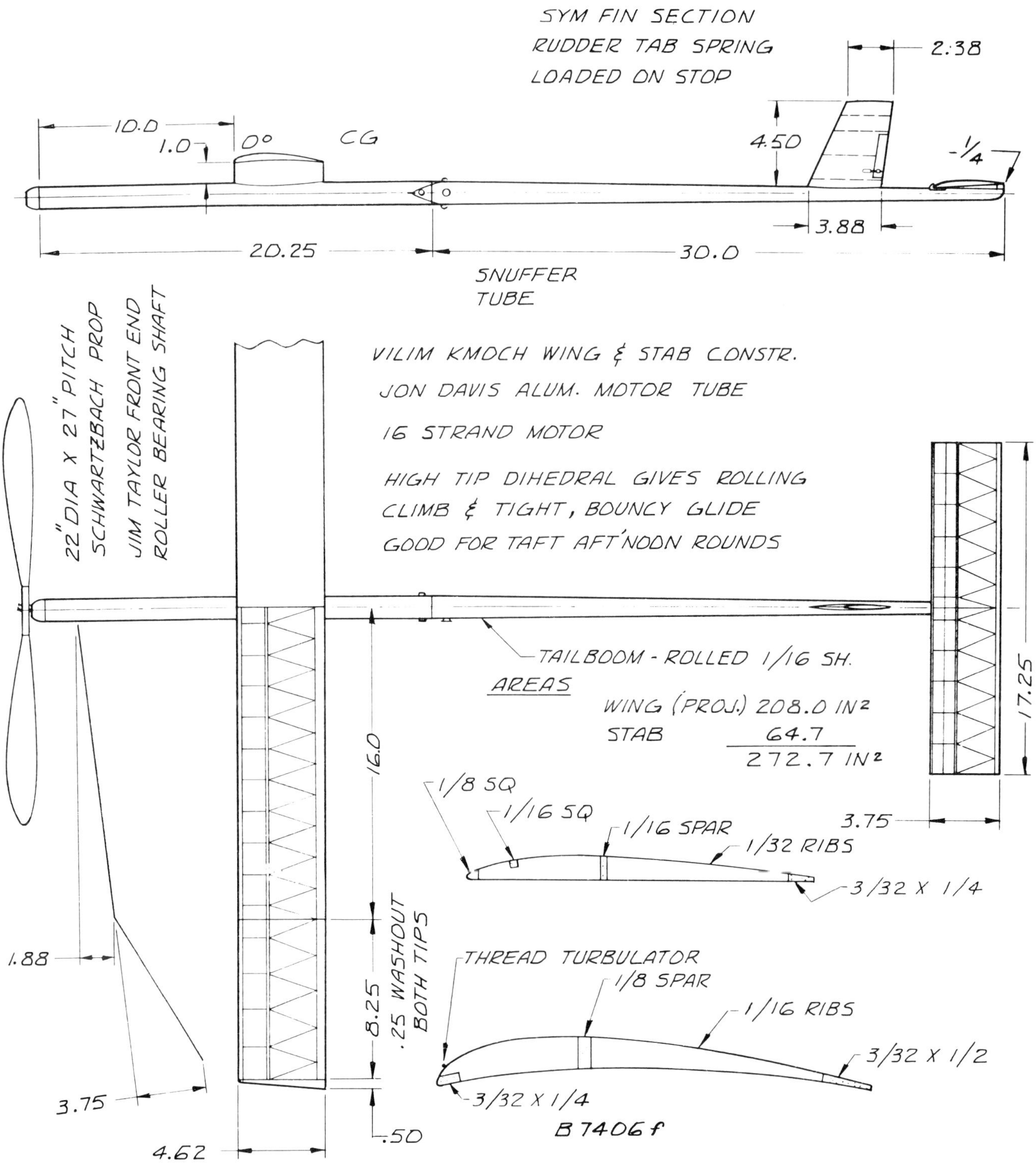

WEIGHTS

WING	33.4 gm
STAB	6.7
FUSELAGE	100.9
PROP	45.1
BALLAST	5.0 → 191.0 gm

FRENCH CONNECTION F1B

BY DICK GILDERSLEEVE S.C.A.T.

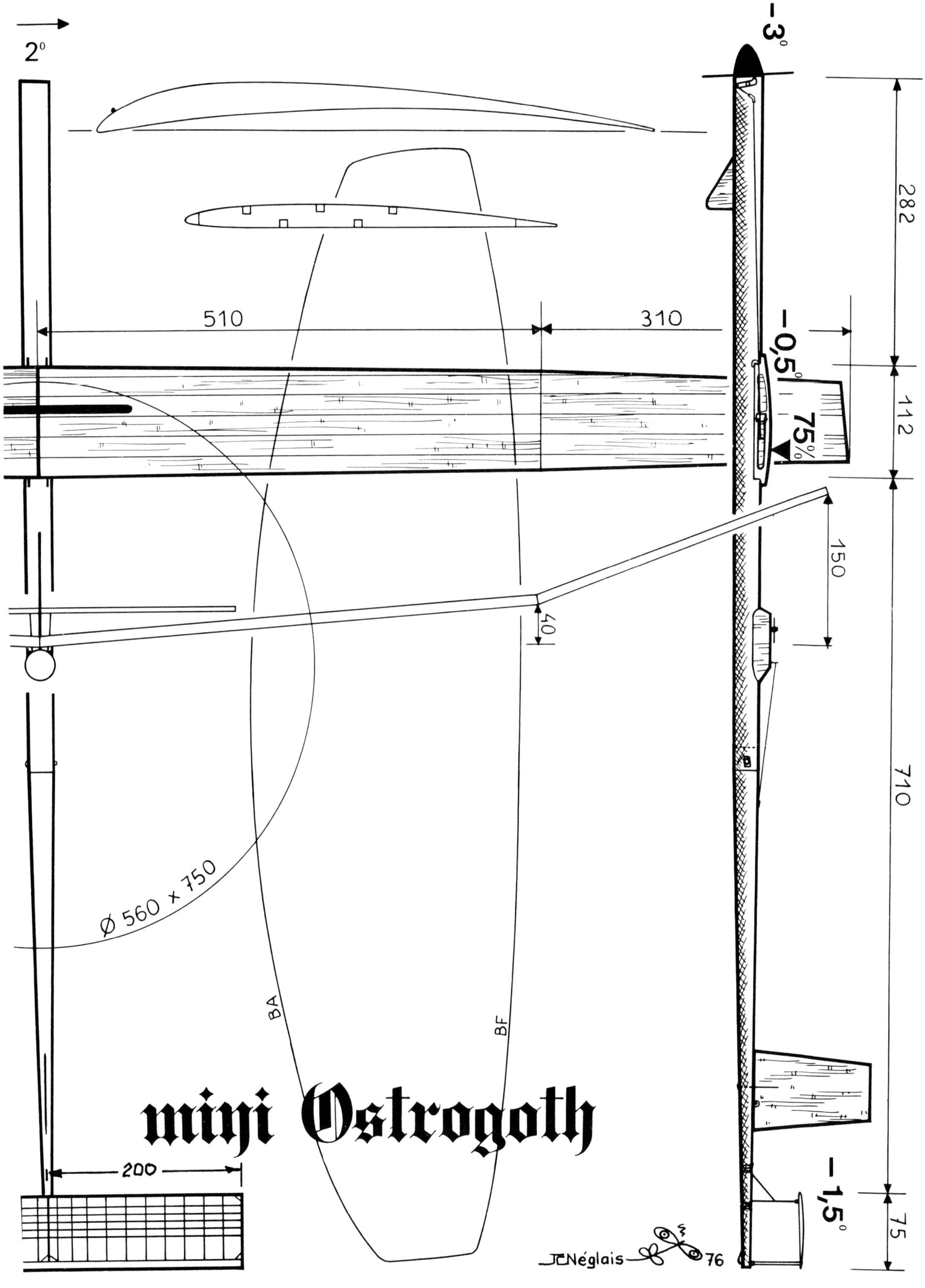

9 Gouverne France Vol Libre

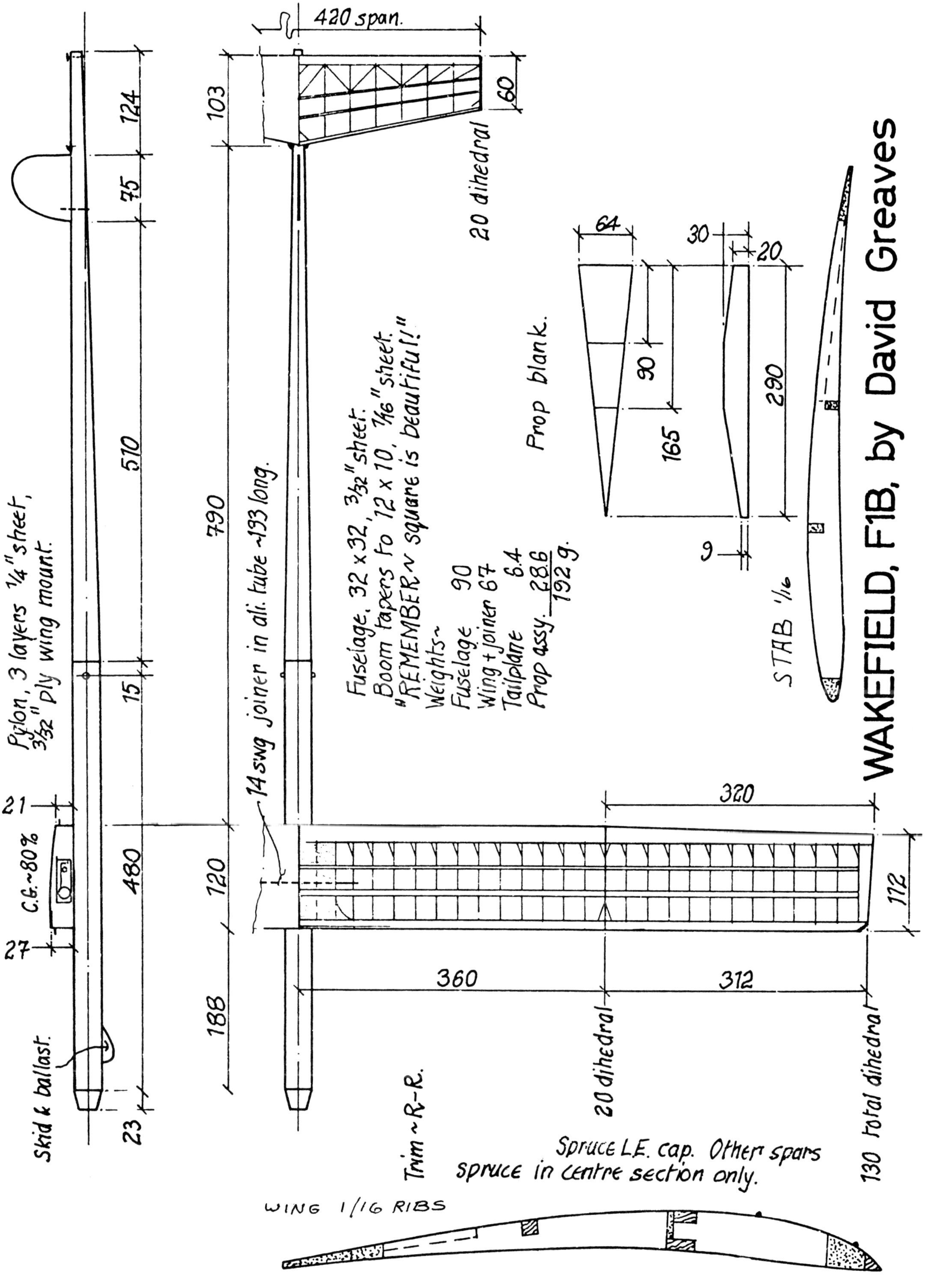
WAKEFIELD, F1B, by David Greaves
420 span.
103
60
124
75
510
15
480
21
27
23
790
120
188
320
172
360
312
20 dihedral
Pylon, 3 layers 1/4" sheet, 3/32" ply wing mount.
C.G.~80%
Skid & ballast.
14 swg joiner in ali. tube ~133 long.
Fuselage, 32 x 32, 3/32" sheet.
Boom tapers to 12 x 10, 1/16" sheet.
"REMEMBER~ square is beautiful!"
Weights~
Fuselage 90
Wing + joiner 67
Tailplane 6.4
Prop assy. 28.6
192 g.
Prop blank.
64
90
165
30
20
290
9
STAB 1/16
20 dihedral
130 total dihedral
Trim ~R-R.
Spruce L.E. cap. Other spars spruce in centre section only.
WING 1/16 RIBS

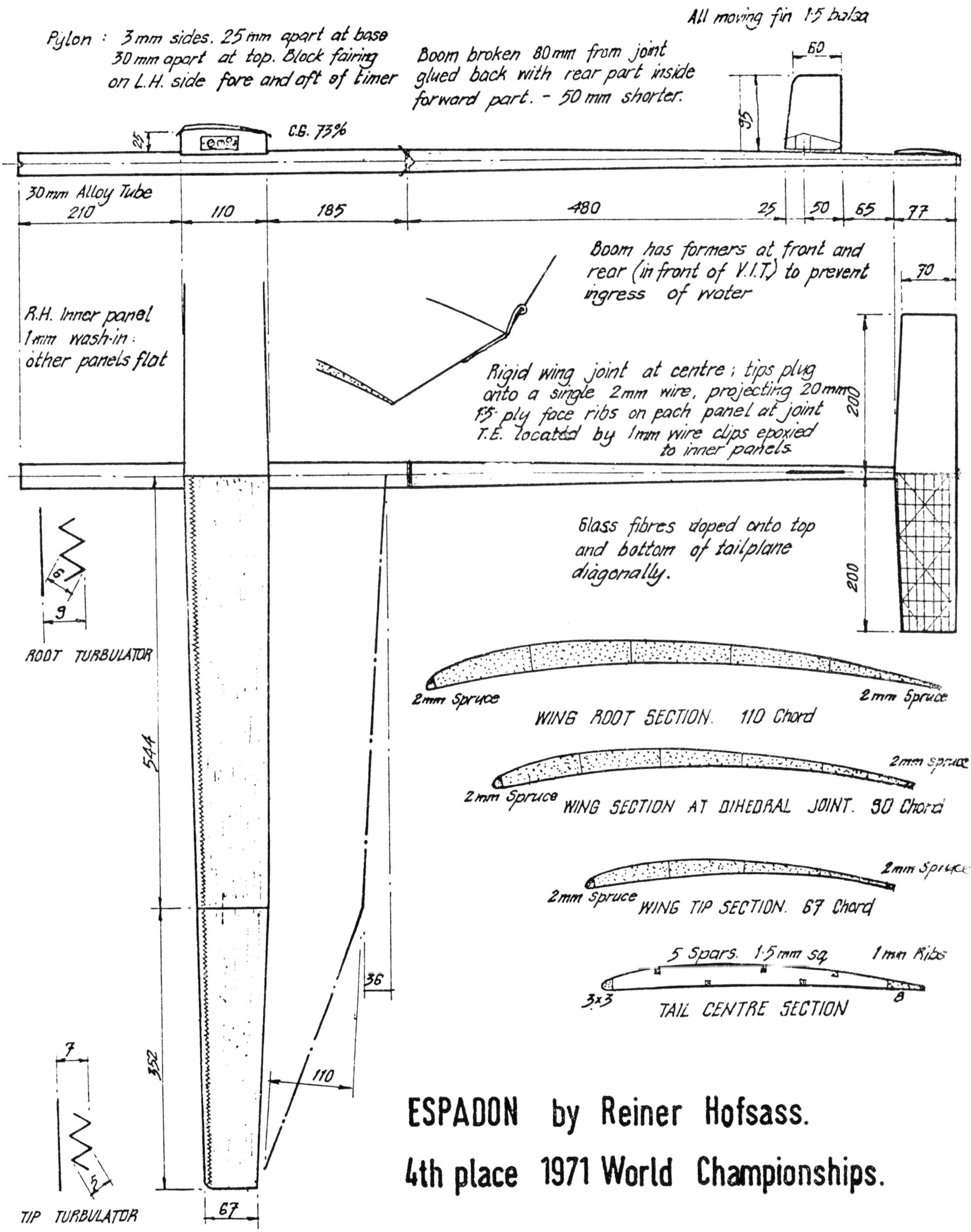

ESPADON by Reiner Hofsass.

4th place 1971 World Championships.

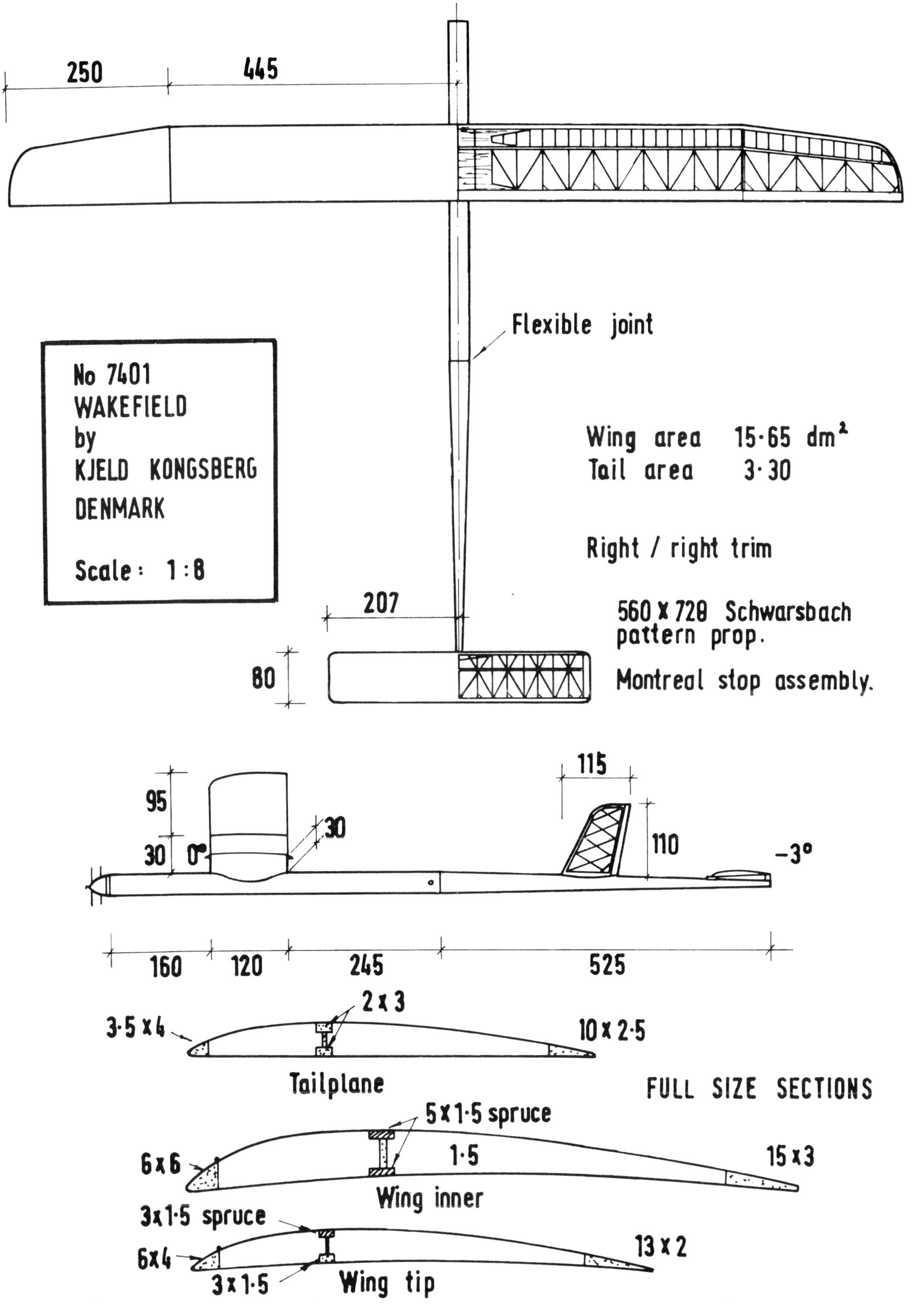

12 Kongsberg DK East Anglian News

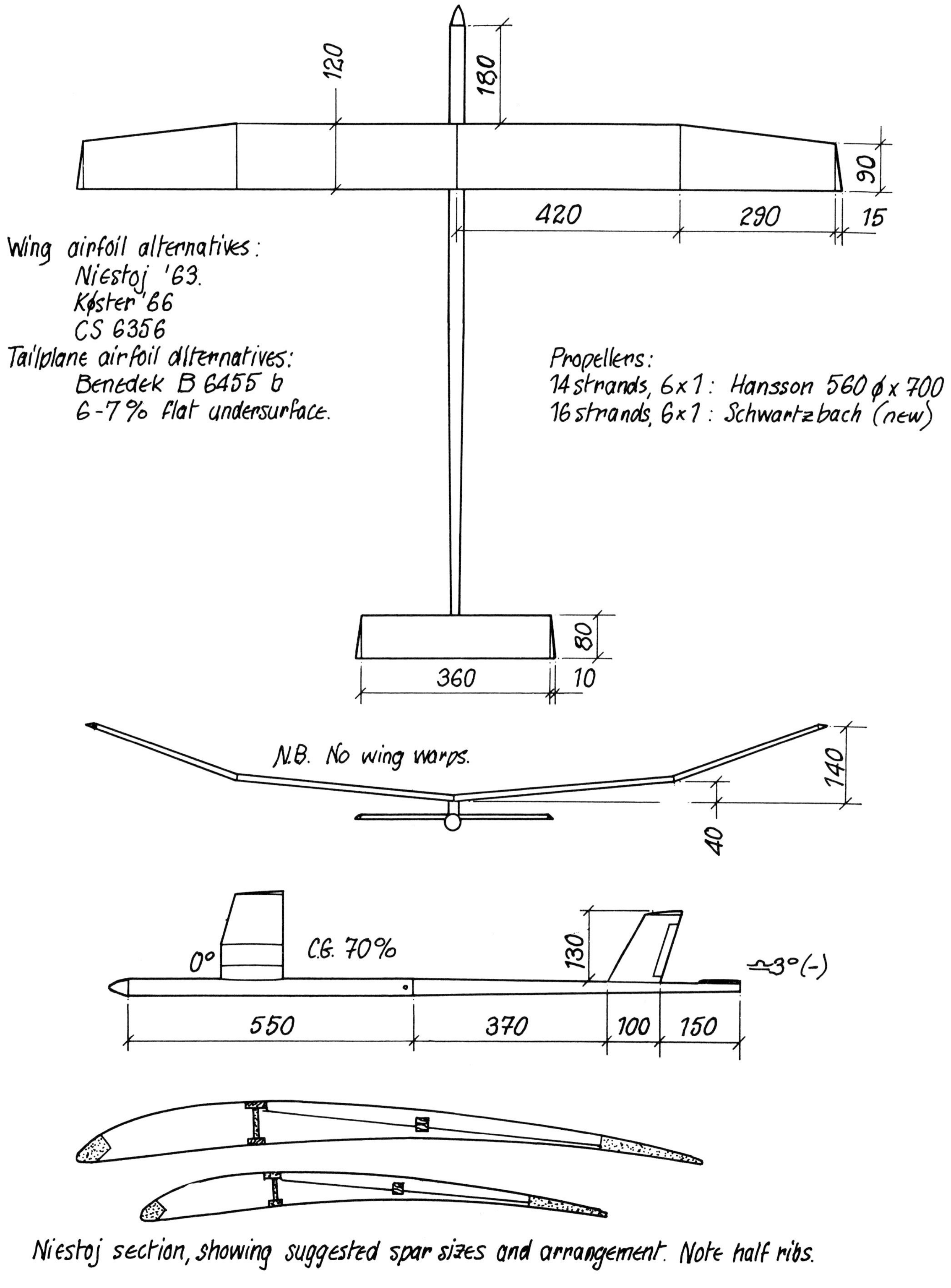

Niestoj section, showing suggested spar sizes and arrangement. Note half ribs.

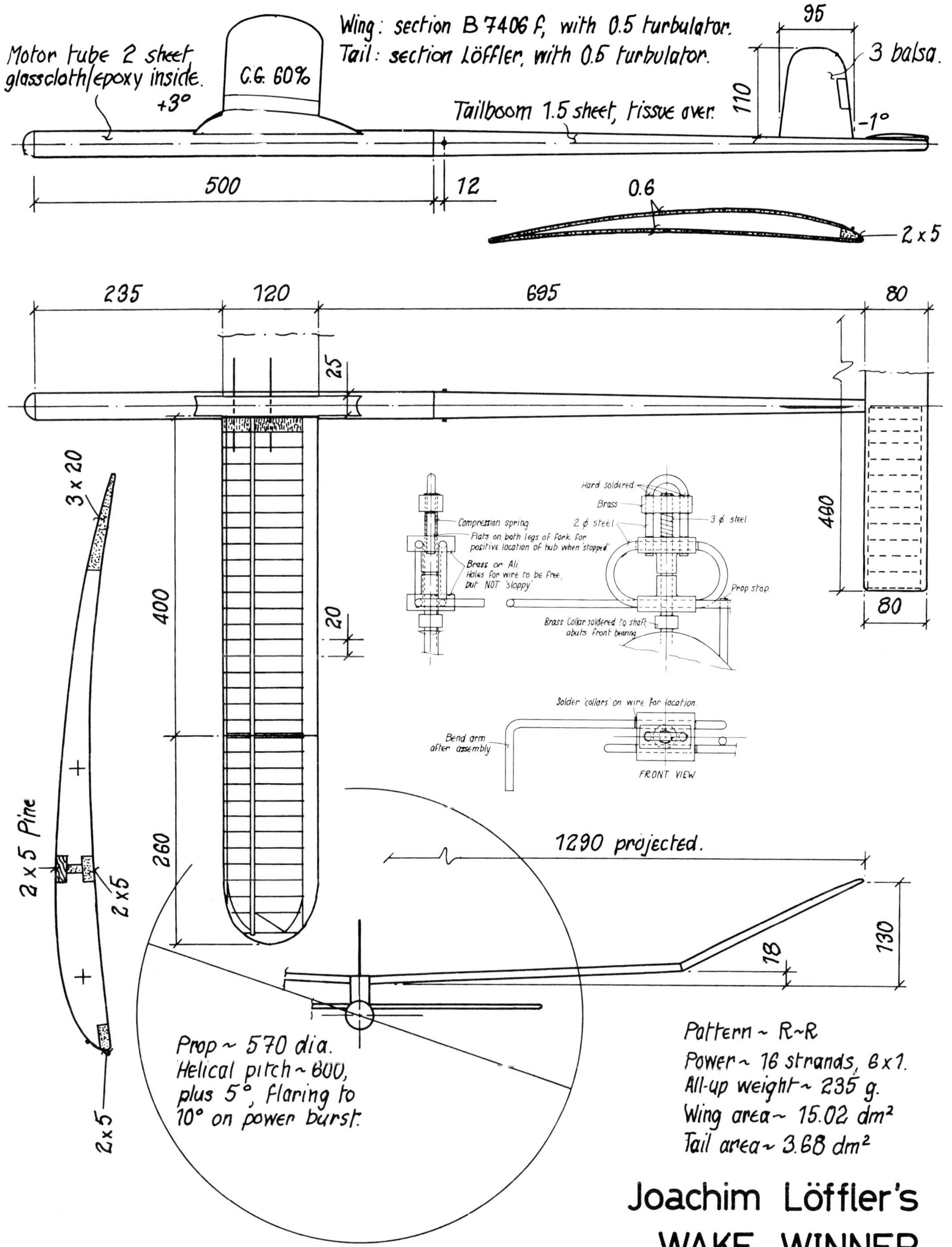

Joachim Löffler's WAKE WINNER

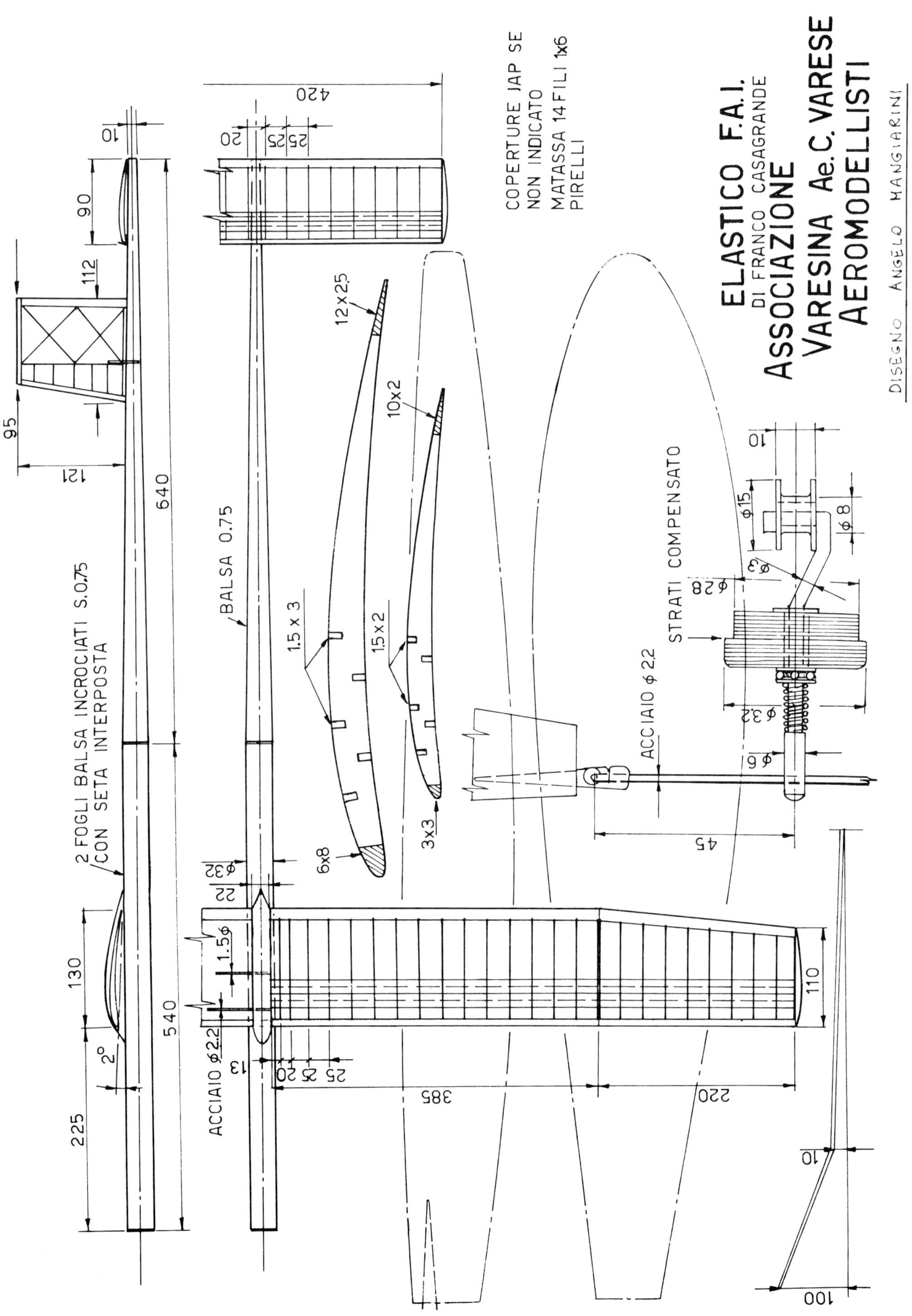

 15 Mangiarini Italy AVA

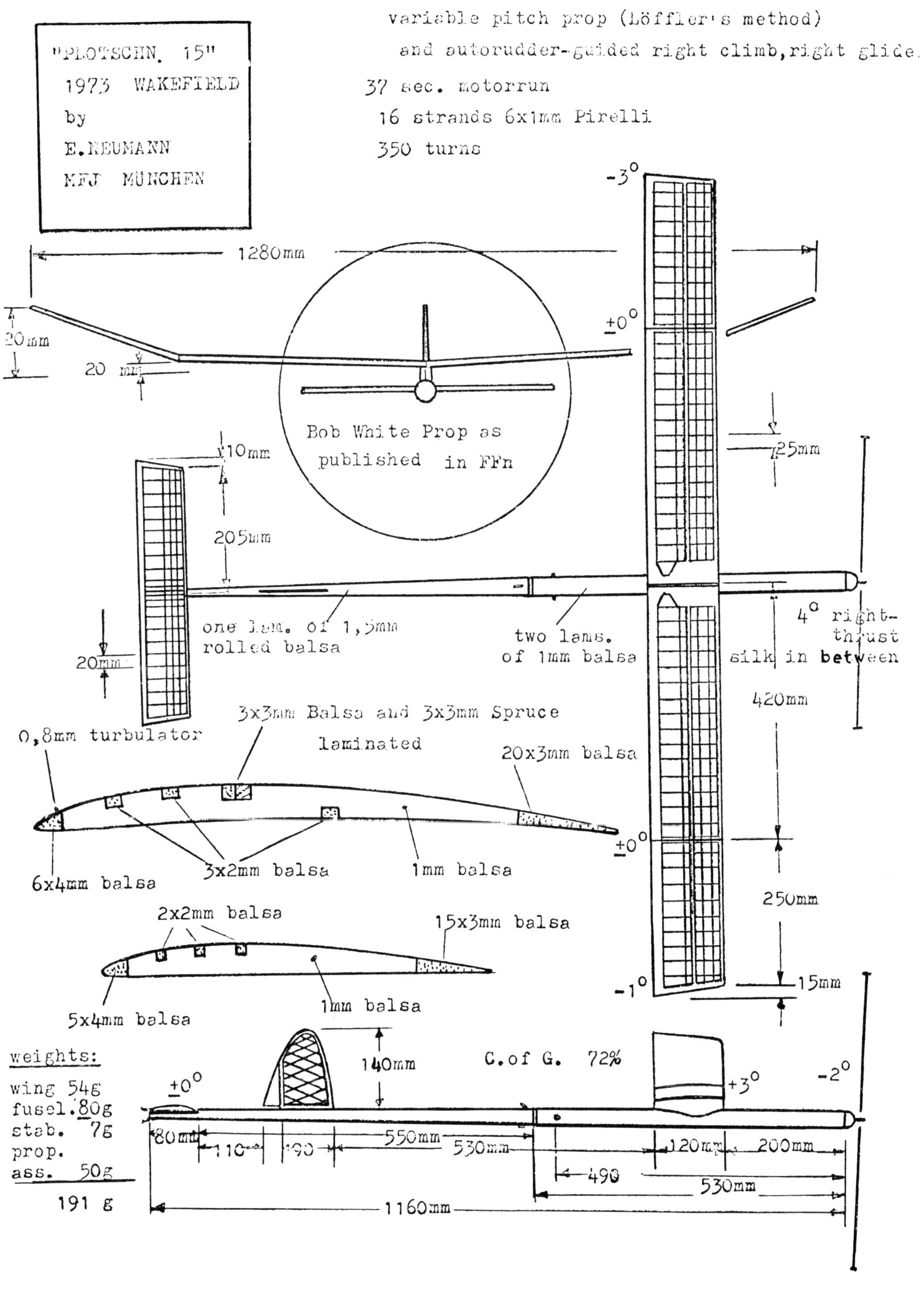
"PLOTSCHN. 15"
1973 WAKEFIELD
by
E.NEUMANN
MFJ MÜNCHEN
variable pitch prop (Löffler's method)
and autorudder-guided right climb, right glide.
37 sec. motorrun
16 strands 6x1mm Pirelli
350 turns
1280mm
20mm
20 mm
-3°
±0°
Bob White Prop as
published in FFn
10mm
205mm
25mm
one lam. of 1,5mm
rolled balsa
two lams.
of 1mm balsa
4° right-
thrust
silk in between
20mm
420mm
0,8mm turbulator
3x3mm Balsa and 3x3mm Spruce
laminated
20x3mm balsa
±0°
6x4mm balsa
3x2mm balsa
1mm balsa
250mm
2x2mm balsa
15x3mm balsa
-1°
15mm
5x4mm balsa
1mm balsa
weights:
wing 54g
fusel. 80g
stab. 7g
prop.
ass. 50g
191 g
±0°
140mm
C.of G. 72%
+3°
-2°
80mm
110
190
550mm
530mm
120mm
200mm
490
530mm
1160mm

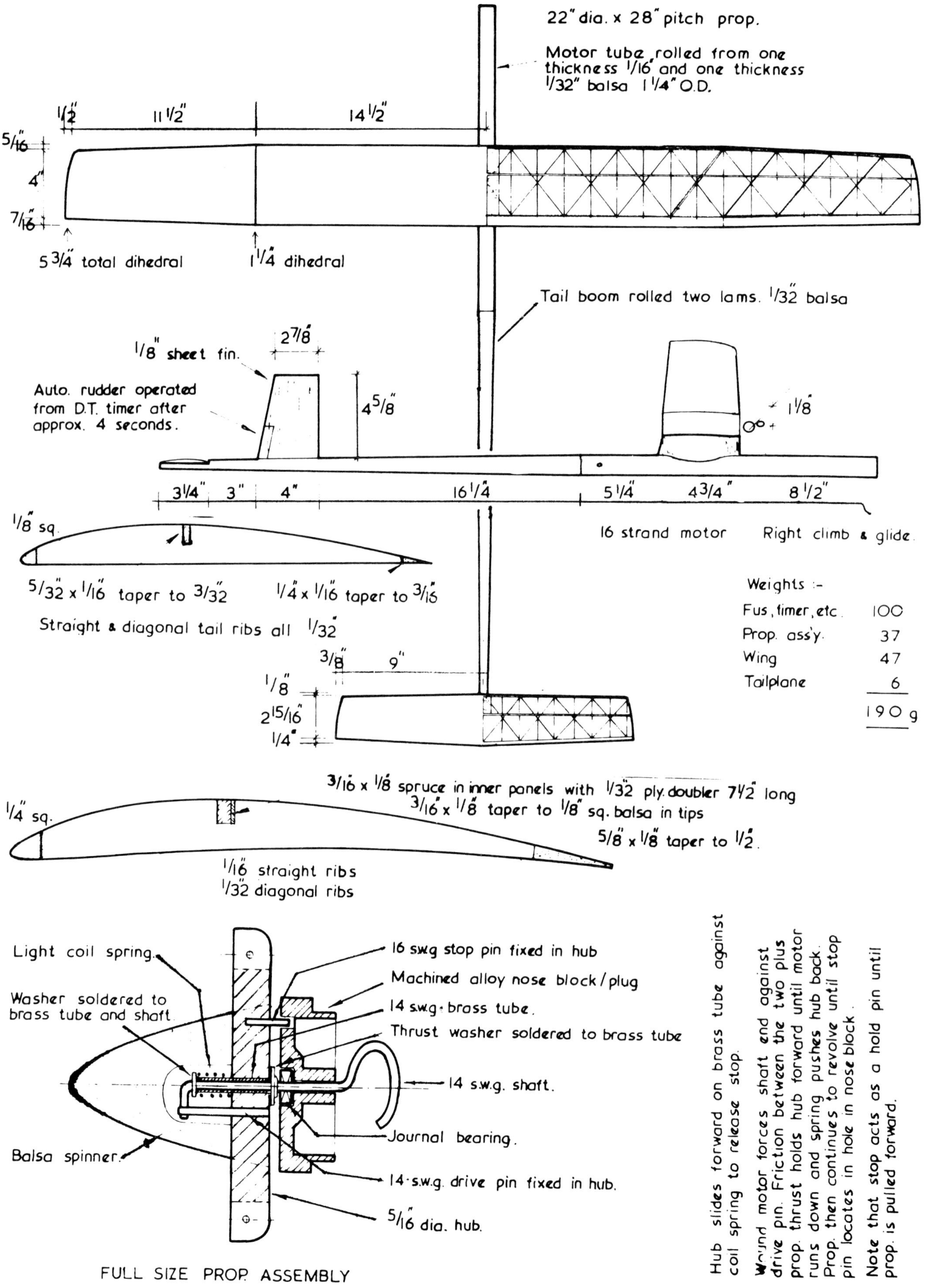
22" dia. x 28" pitch prop.
Motor tube, rolled from one thickness 1/16" and one thickness 1/32" balsa 1 1/4" O.D.
1/2"
11 1/2"
14 1/2"
5/16"
4"
7/16"
5 3/4" total dihedral
1 1/4" dihedral
Tail boom rolled two lams. 1/32" balsa
1/8" sheet fin.
2 7/8"
Auto. rudder operated from D.T. timer after approx. 4 seconds.
4 5/8"
1 1/8"
3 1/4"
3"
4"
16 1/4"
5 1/4"
4 3/4"
8 1/2"
1/8" sq.
16 strand motor
Right climb & glide.
5/32" x 1/16" taper to 3/32"
1/4" x 1/16" taper to 3/16"
Straight & diagonal tail ribs all 1/32"
Weights :-
Fus, timer, etc. 100
Prop. ass'y. 37
Wing 47
Tailplane 6
190 g
3/8"
9"
1/8"
2 15/16"
1/4"
3/16" x 1/8" spruce in inner panels with 1/32" ply doubler 7 1/2" long
3/16" x 1/8" taper to 1/8" sq. balsa in tips
1/4" sq.
5/8" x 1/8" taper to 1/2".
1/16" straight ribs
1/32" diagonal ribs
Light coil spring.
Washer soldered to brass tube and shaft
16 swg stop pin fixed in hub
Machined alloy nose block / plug
14 swg brass tube.
Thrust washer soldered to brass tube
14 s.w.g. shaft.
Journal bearing.
Balsa spinner.
14 s.w.g. drive pin fixed in hub.
5/16" dia. hub.
FULL SIZE PROP. ASSEMBLY
Hub slides forward on brass tube against coil spring to release stop.
Wound motor forces shaft end against drive pin. Friction between the two plus prop. thrust holds hub forward until motor runs down and spring pushes hub back. Prop. then continues to revolve until stop pin locates in hole in nose block.
Note that stop acts as a hold pin until prop. is pulled forward.

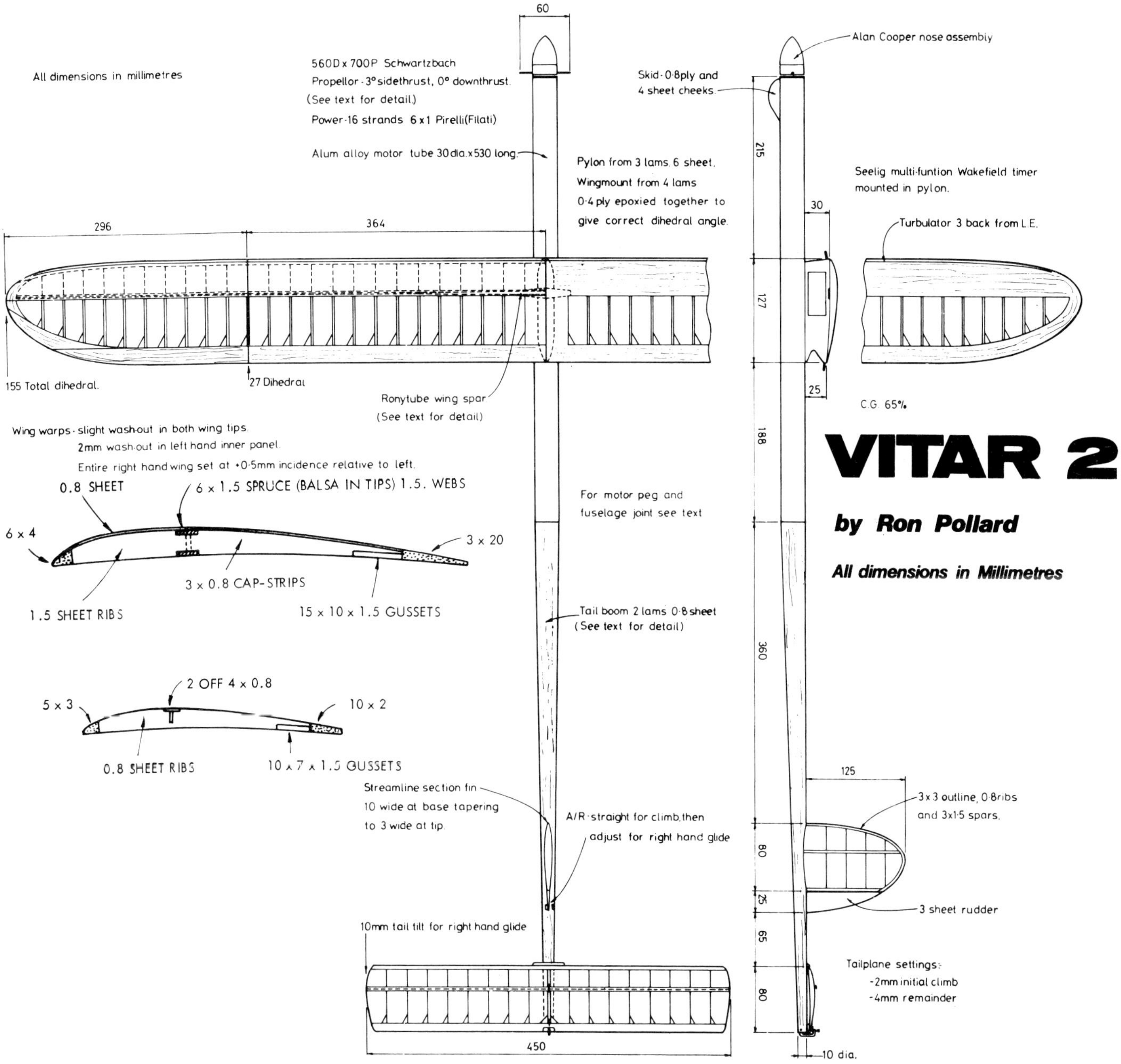

VITAR 2

by Ron Pollard

All dimensions in Millimetres

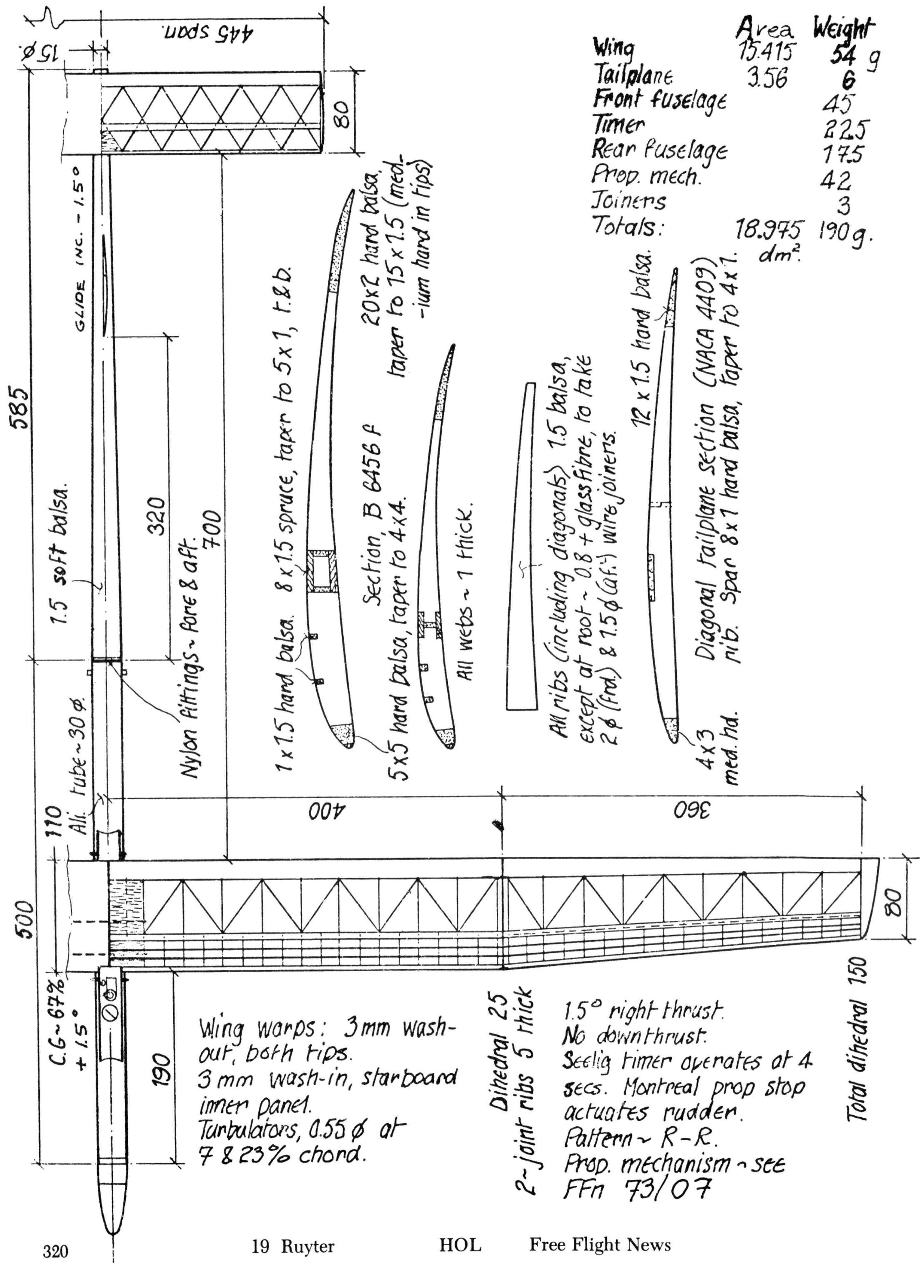

 19 Ruyter HOL Free Flight News

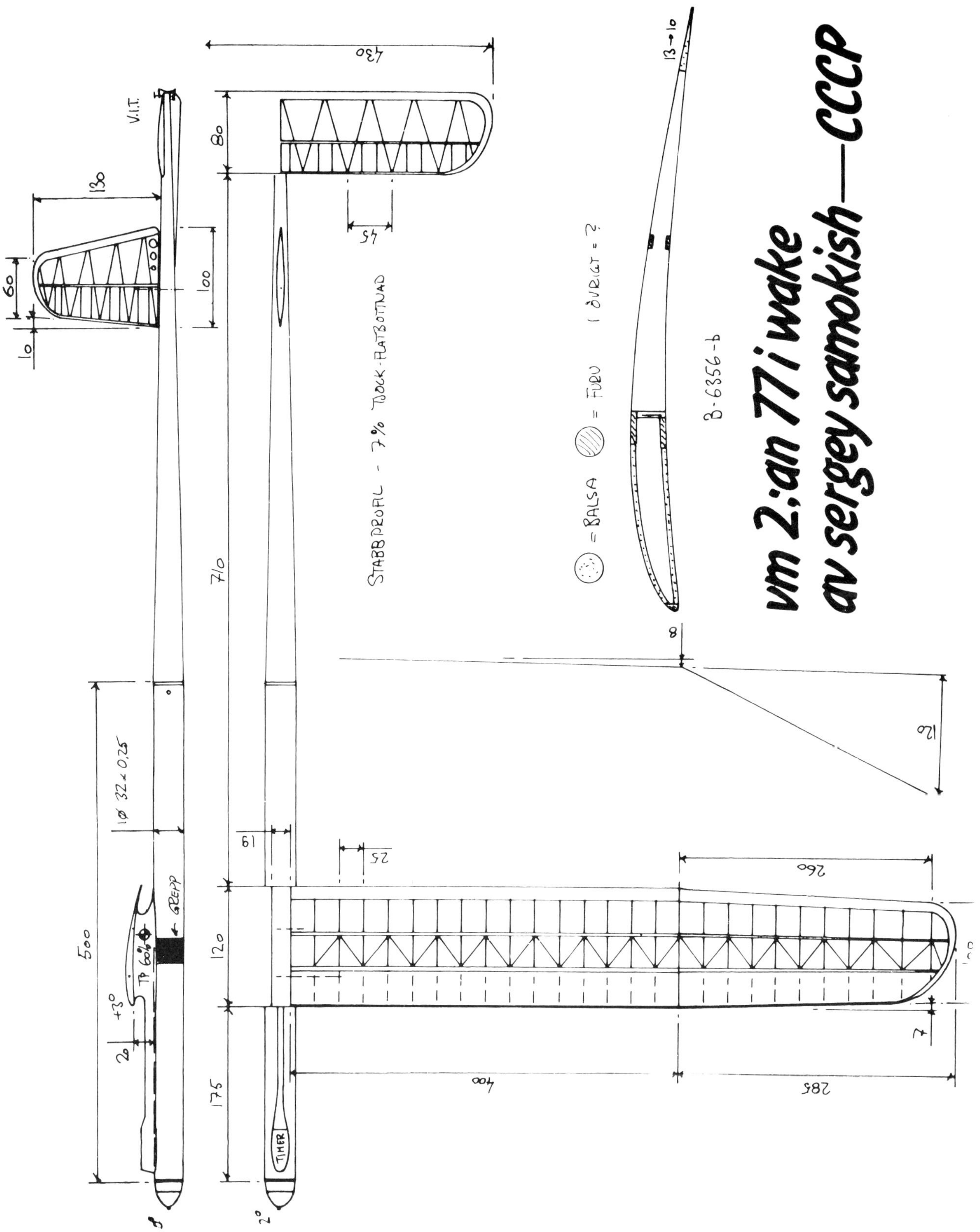

20 Samokish USSR WFFP

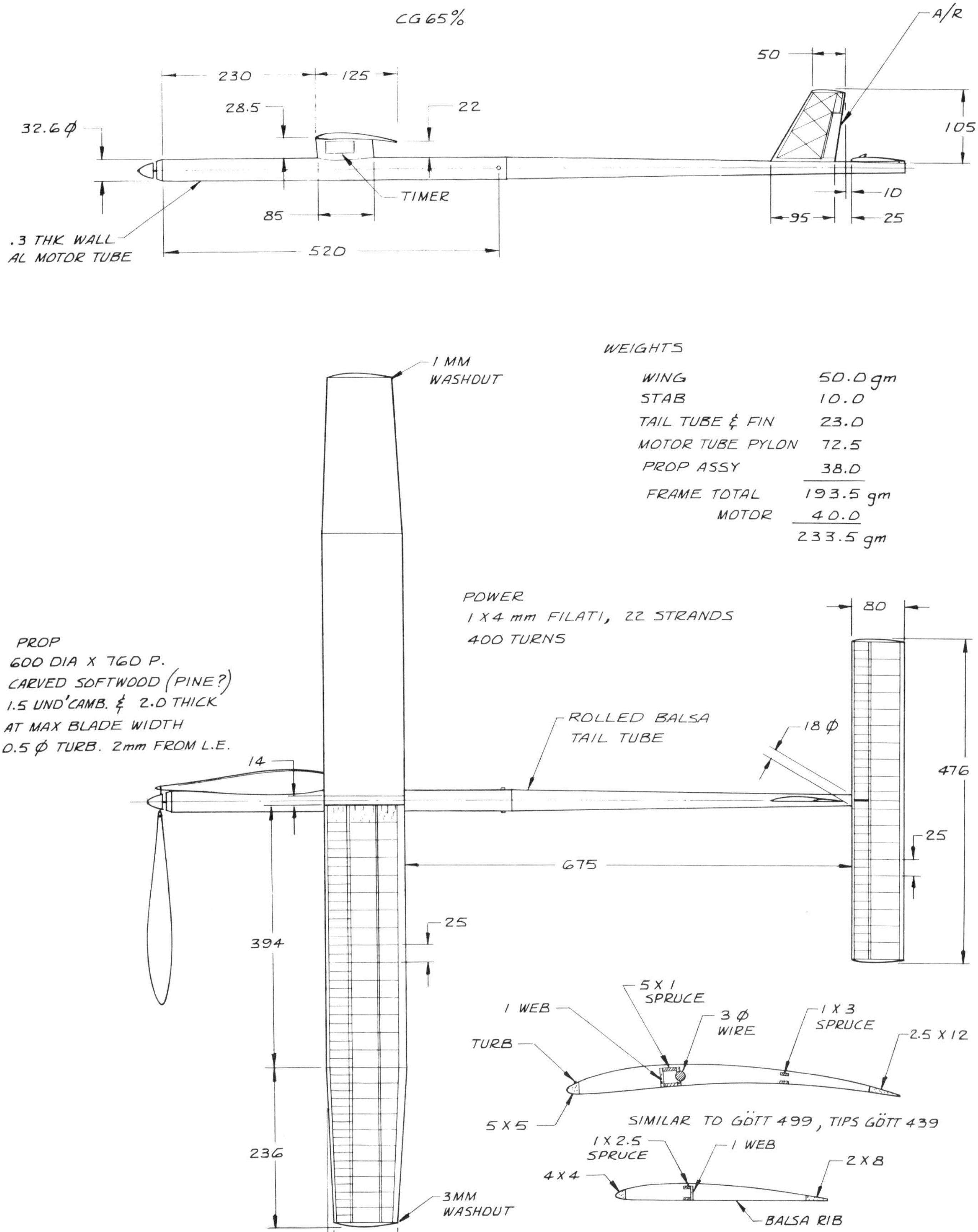

1977 World Champion F1B

BY KIM DONG SIK, NORTH KOREA

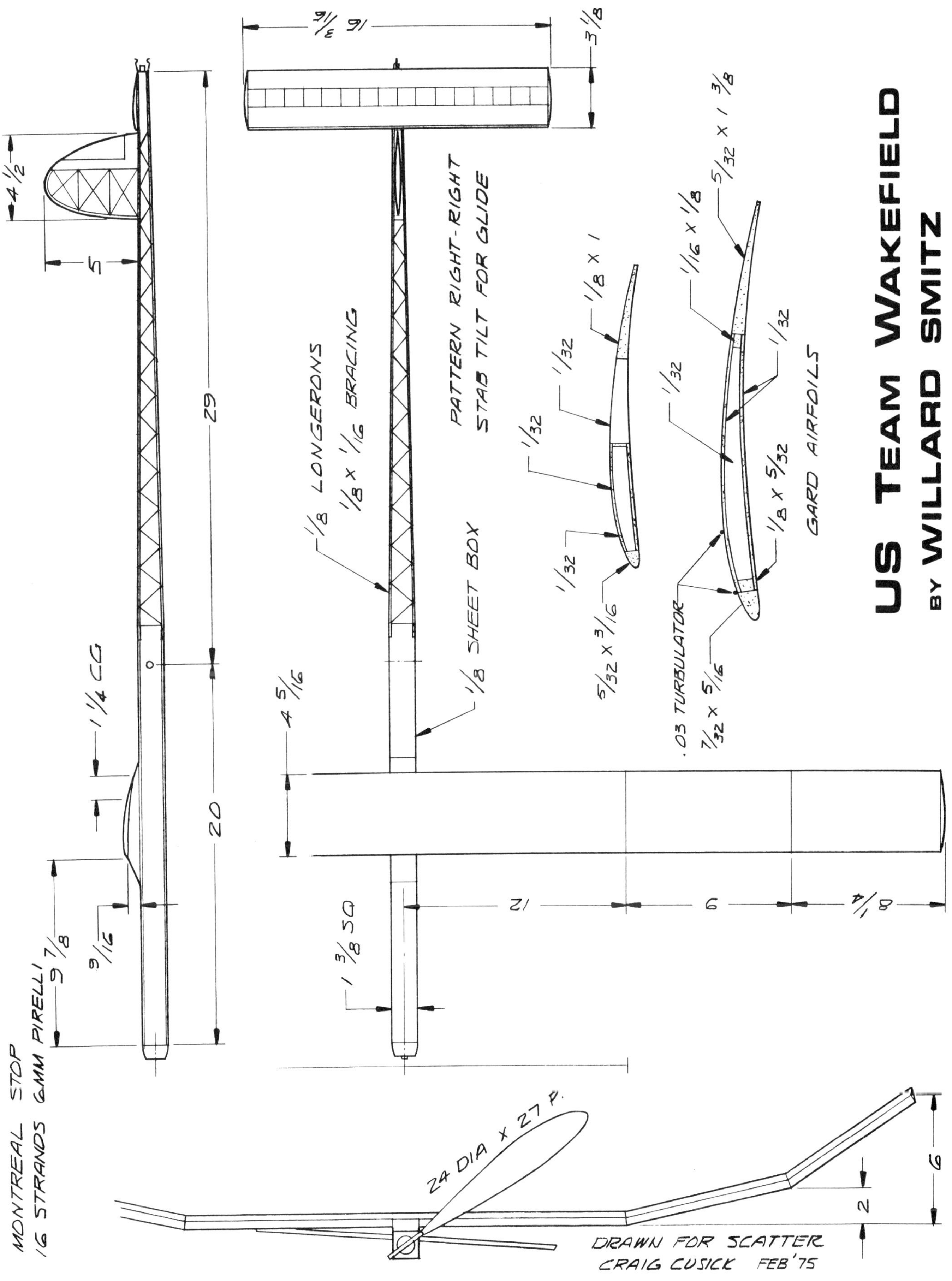
US TEAM WAKEFIELD
BY WILLARD SMITZ
MONTREAL STOP
16 STRANDS 6MM PIRELLI
1/8 LONGERONS
1/8 x 1/16 BRACING
PATTERN RIGHT-RIGHT
STAB TILT FOR GLIDE
1/8 SHEET BOX
1 1/4 CG
1 3/8 SQ
GARD AIRFOILS
.03 TURBULATOR
24 DIA x 27 P.
DRAWN FOR SCATTER
CRAIG CUSICK FEB '75

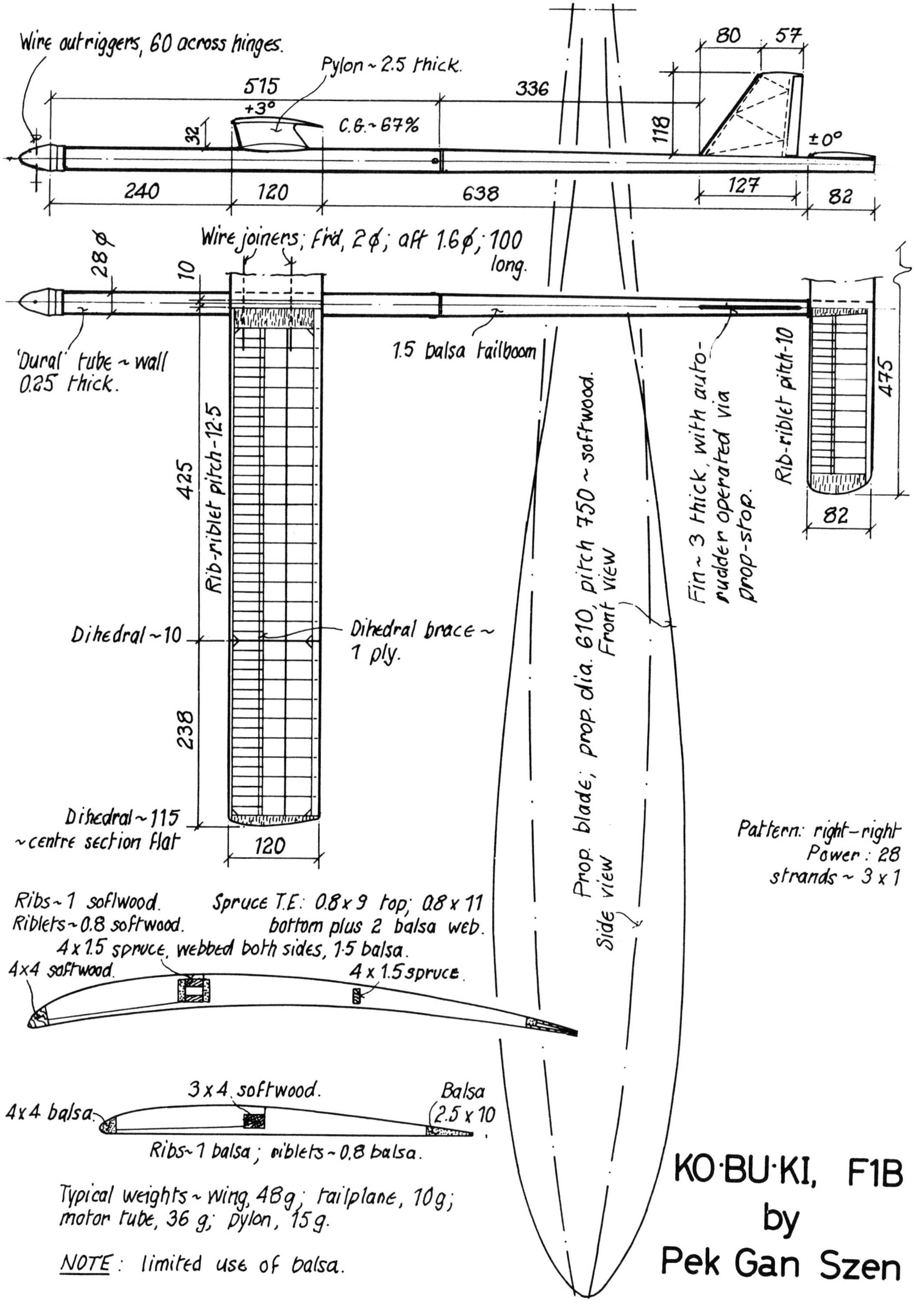

KO·BU·KI, F1B
by
Pek Gan Szen

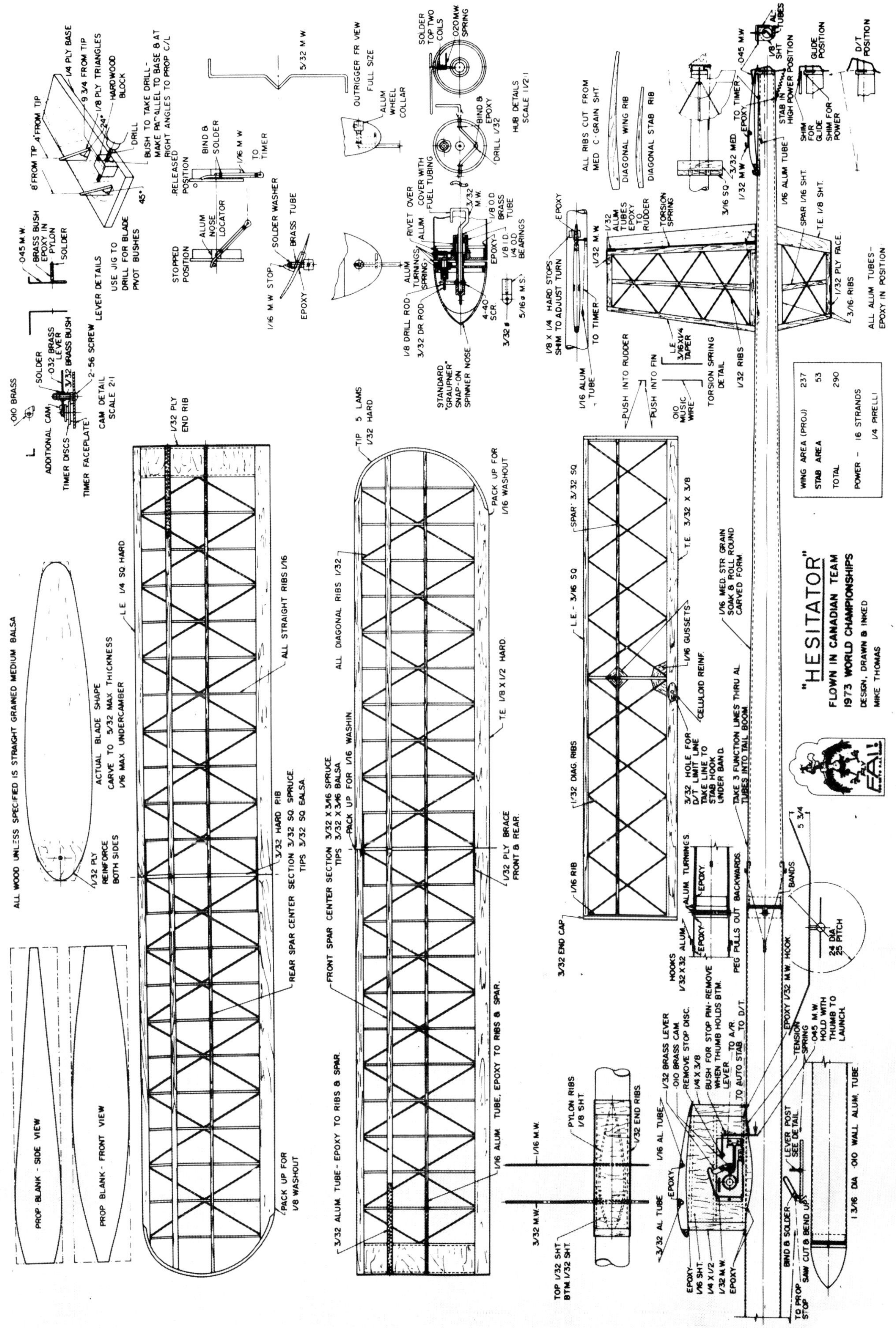

"HESITATOR"
FLOWN IN CANADIAN TEAM
1973 WORLD CHAMPIONSHIPS
DESIGN, DRAWN & INKED
MIKE THOMAS
WING AREA (PROJ) 237
STAB AREA 53
TOTAL 290
POWER - 16 STRANDS
1/4 PIRELLI

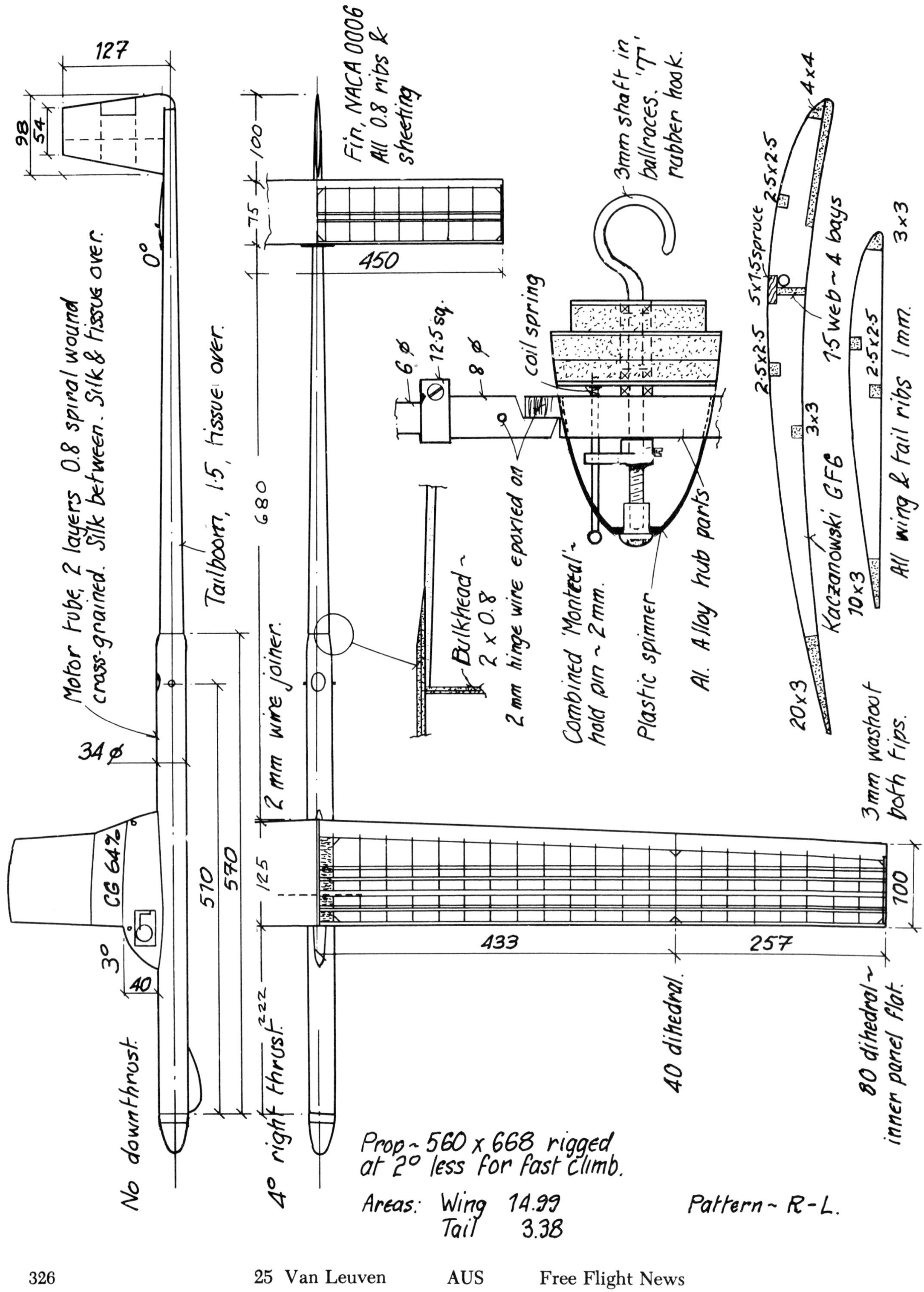

127
98
54
0°
Motor tube, 2 layers 0.8 spiral wound cross-grained. Silk between. Silk & tissue over.
Tailboom, 1.5, tissue over.
34 ⌀
CG 64%
3°
40
570
570
No downthrust.
4° right thrust
222
680
125
2 mm wire joiner.
100
75
450
Fin, NACA 0006 All 0.8 ribs & sheeting
Bulkhead ~ 2 x 0.8
6 ⌀
12.5 sq.
8 ⌀
coil spring
2 mm hinge wire epoxied on
Combined 'Monteal' ~ hold pin ~ 2 mm.
Plastic spinner
Al. Alloy hub parts
3mm shaft in ballraces. 'T' rubber hook.
4x4
2.5x2.5
5x1.5 spruce
2.5x2.5
1.5 web ~ 4 bays
3x3
Kaczanowski GF6
20x3
3x3
2.5x2.5
10x3
All wing & tail ribs 1 mm.
3 mm washout both tips.
433
257
100
40 dihedral.
80 dihedral ~ inner panel flat.
Prop ~ 560 x 668 rigged at 2° less for fast climb.
Areas: Wing 14.99
Tail 3.38
Pattern ~ R - L.

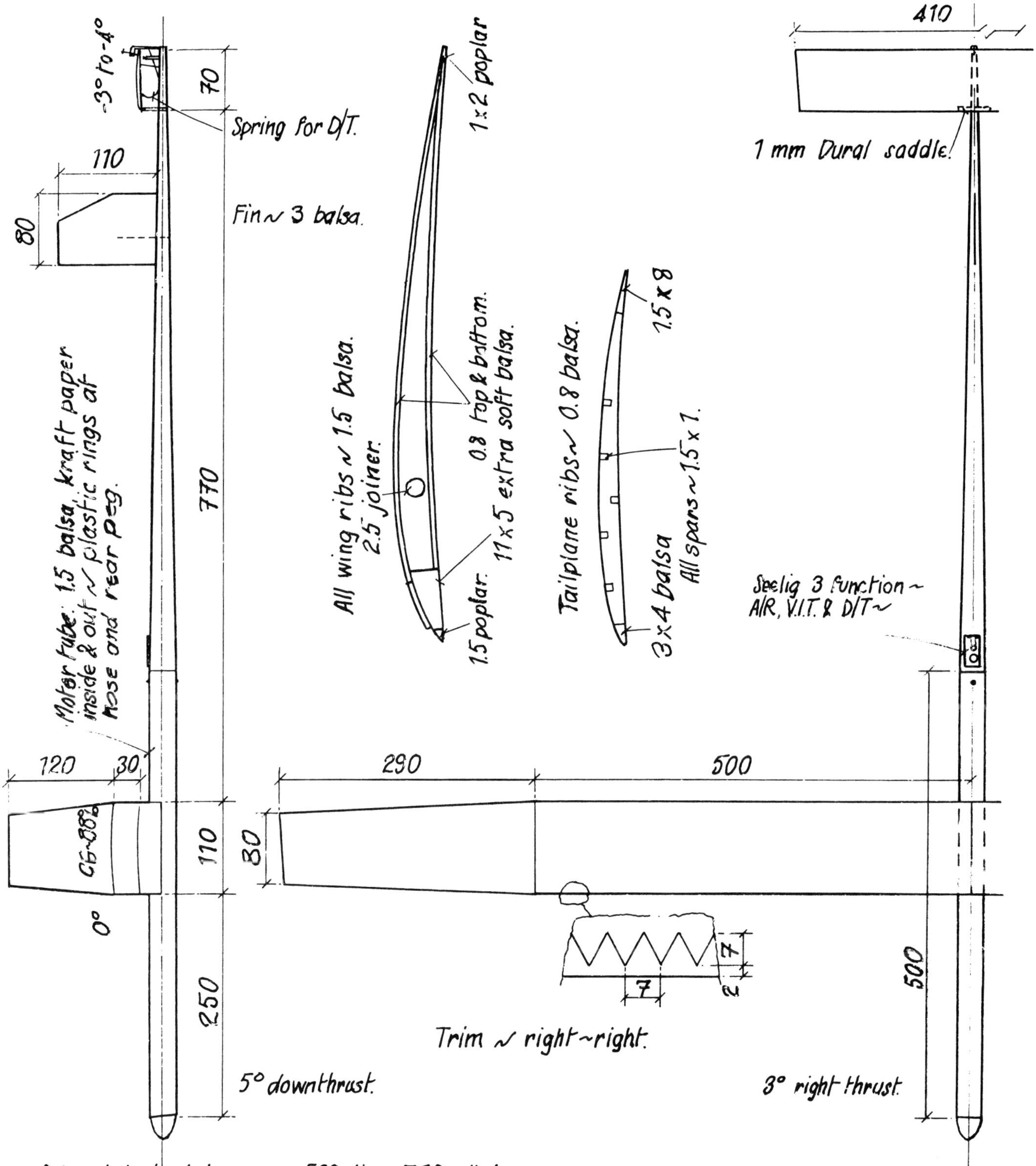

Schwartzbach-style prop ~ 560 dia. x 720 pitch.
Power ~ 18 strands, 6 x 1 ≃ 25 seconds run.

Weights:

Motor tube & pylon	38	Wing	66
Tailboom & fin	27	Tailplane	7
Prop.	37	Timer	20

FLEMMARD XIV, F1B
by J. Wantzenriether

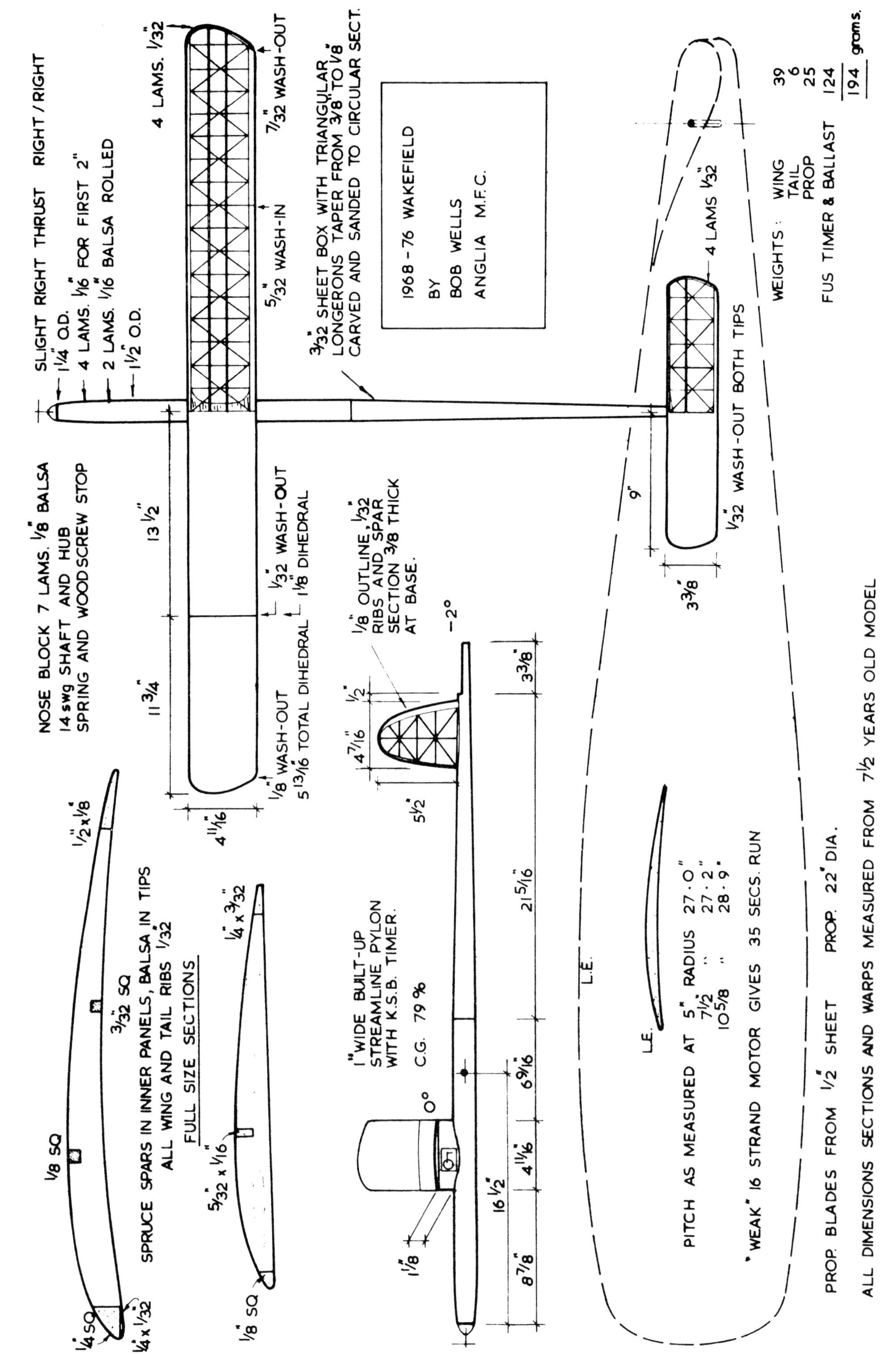
1968 - 76 WAKEFIELD
BY
BOB WELLS
ANGLIA M.F.C.
SLIGHT RIGHT THRUST RIGHT / RIGHT
1¼" O.D.
4 LAMS. 1/16" FOR FIRST 2"
2 LAMS. 1/16" BALSA ROLLED
1½" O.D.
4 LAMS. 1/32"
7/32" WASH-OUT
5/32" WASH-IN
3/32" SHEET BOX WITH TRIANGULAR LONGERONS TAPER FROM 3/8" TO 1/8" CARVED AND SANDED TO CIRCULAR SECT.
NOSE BLOCK 7 LAMS. 1/8" BALSA
14 swg SHAFT AND HUB
SPRING AND WOODSCREW STOP
13½"
11¾"
1/32" WASH-OUT
1⅛" DIHEDRAL
1/8" WASH-OUT
5 13/16" TOTAL DIHEDRAL
4 11/16
4 LAMS 1/32"
9"
1/32" WASH-OUT BOTH TIPS
3⅜"
1/8" OUTLINE, 1/32" RIBS AND SPAR SECTION 3/8 THICK AT BASE.
−2°
½"
4 7/16"
5½"
3⅜"
21 5/16"
1" WIDE BUILT-UP STREAMLINE PYLON WITH K.S.B. TIMER.
C.G. 79 %
0°
6 9/16
4 11/16"
16½"
1⅛"
8⅞"
½" x ⅛"
3/32" SQ
1/8 SQ
¼" SQ
¼" x 1/32"
SPRUCE SPARS IN INNER PANELS, BALSA IN TIPS
ALL WING AND TAIL RIBS 1/32"
FULL SIZE SECTIONS
¼" x 3/32"
5/32" x 1/16"
1/8" SQ
L.E.
L.E.
WEIGHTS : WING 39
TAIL 6
PROP 25
FUS TIMER & BALLAST 124
194 grams.
PITCH AS MEASURED AT 5" RADIUS 27·0"
7½" " 27·2"
10⅝" " 28·9"
"WEAK" 16 STRAND MOTOR GIVES 35 SECS. RUN
PROP. BLADES FROM ½" SHEET PROP. 22" DIA.
ALL DIMENSIONS SECTIONS AND WARPS MEASURED FROM 7½ YEARS OLD MODEL

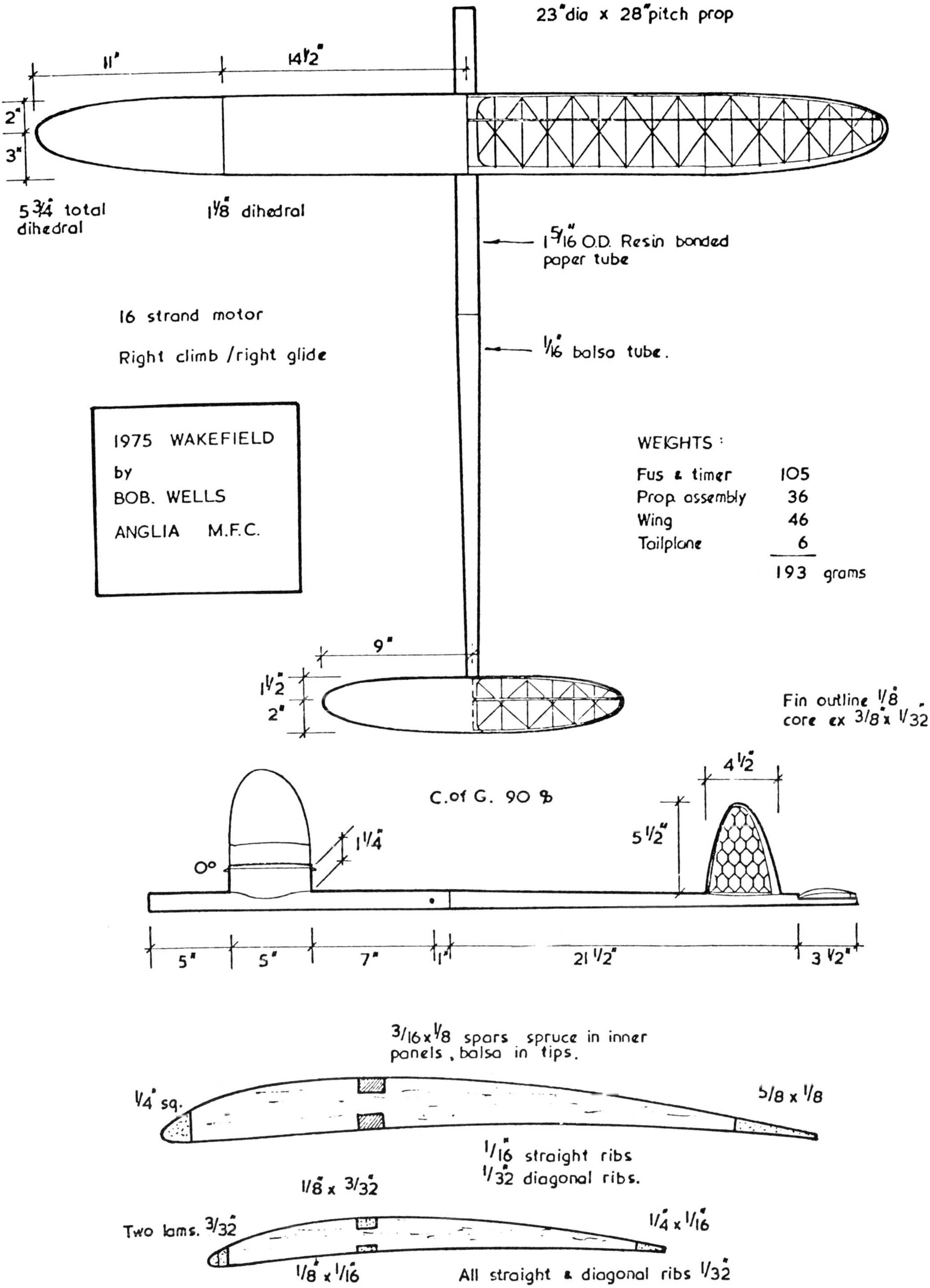

28 Wells GB East Anglian News

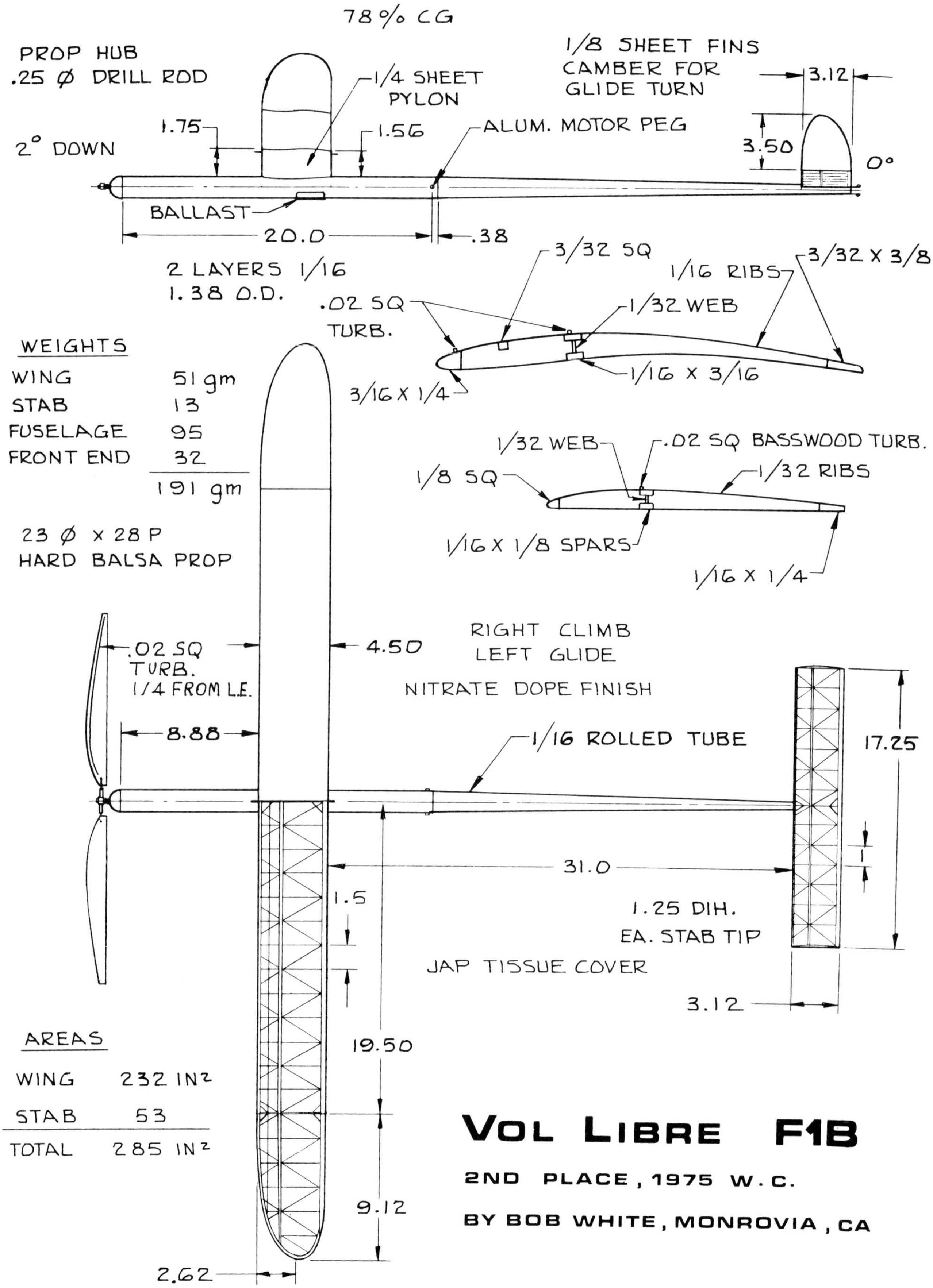
78% CG
PROP HUB
.25 Ø DRILL ROD
1/4 SHEET
PYLON
1/8 SHEET FINS
CAMBER FOR
GLIDE TURN
3.12
1.75
1.56
ALUM. MOTOR PEG
2° DOWN
3.50
0°
BALLAST
20.0
.38
2 LAYERS 1/16
1.38 O.D.
3/32 SQ
1/16 RIBS
3/32 X 3/8
.02 SQ
TURB.
1/32 WEB
3/16 X 1/4
1/16 X 3/16
WEIGHTS
WING 51 gm
STAB 13
FUSELAGE 95
FRONT END 32
191 gm
1/32 WEB
.02 SQ BASSWOOD TURB.
1/8 SQ
1/32 RIBS
1/16 X 1/8 SPARS
1/16 X 1/4
23 Ø X 28 P
HARD BALSA PROP
.02 SQ
TURB.
1/4 FROM L.E.
4.50
RIGHT CLIMB
LEFT GLIDE
NITRATE DOPE FINISH
8.88
1/16 ROLLED TUBE
17.25
31.0
1
1.5
1.25 DIH.
EA. STAB TIP
JAP TISSUE COVER
3.12
19.50
AREAS
WING 232 IN²
STAB 53
TOTAL 285 IN²
9.12
2.62
VOL LIBRE F1B
2ND PLACE, 1975 W.C.
BY BOB WHITE, MONROVIA, CA

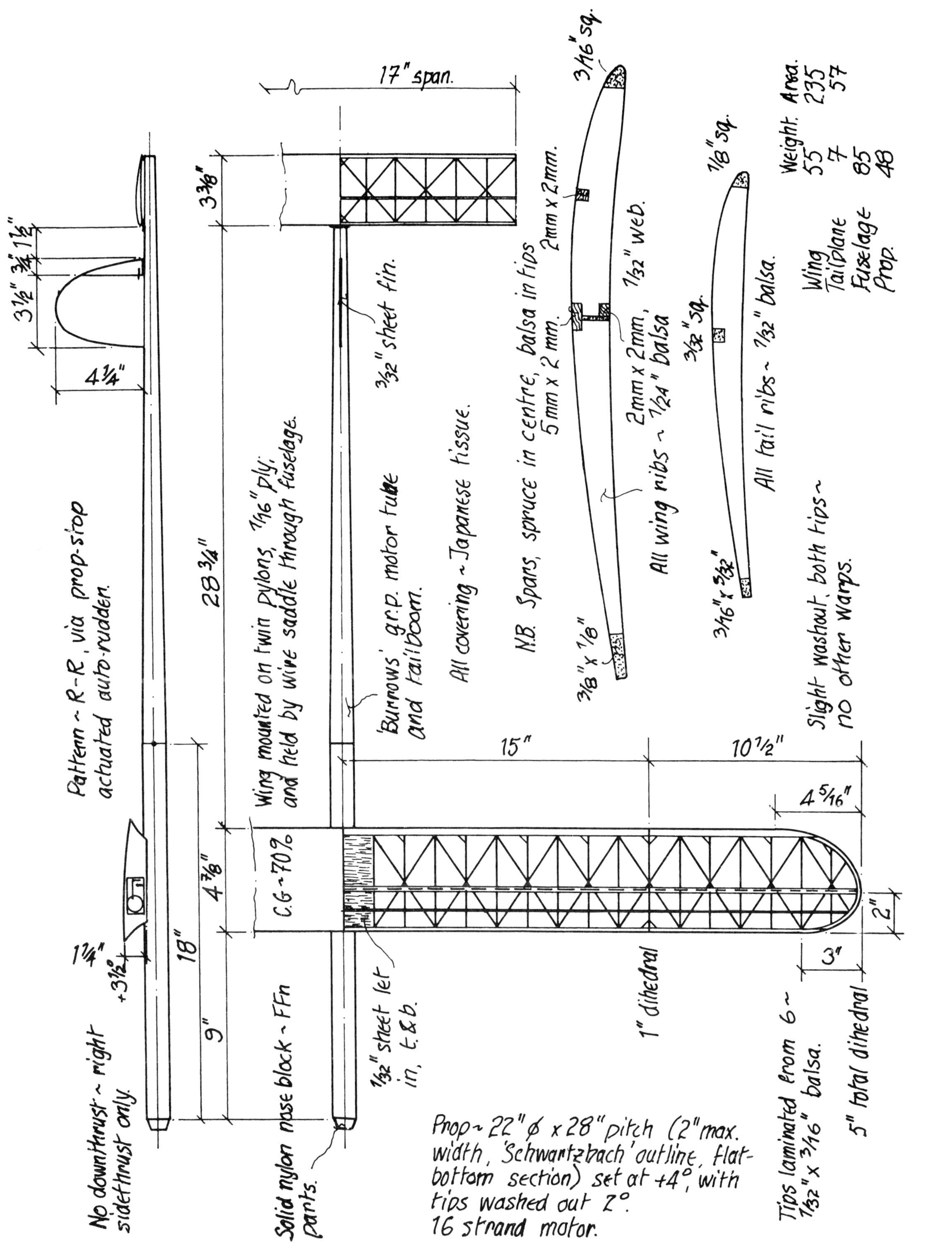

17" span.
3/16" sq.
1/8" sq.
Weight. Area.
Wing 55 235
Tailplane 7 57
Fuselage 85
Prop. 48
3 3/8"
2mm x 2mm.
1/32" web.
3 1/2"
3/4"
1 1/2"
4 1/4"
3/32" sheet fin.
N.B. Spars, spruce in centre, balsa in tips
5 mm x 2 mm.
2mm x 2mm, 1/24" balsa
3/32" sq.
1/32" balsa.
All wing ribs ~ 1/24" balsa
All tail ribs ~ 1/32" balsa.
28 3/4"
Pattern ~ R-R, via prop-stop actuated auto-rudder.
Wing mounted on twin pylons, 1/16" ply; and held by wire saddle through fuselage.
'Burrows' g.r.p. motor tube and tailboom.
All covering ~ Japanese tissue.
3/8" x 1/8"
3/16" x 3/32"
Slight washout, both tips ~ no other warps.
15"
10 1/2"
4 5/16"
4 7/8"
C.G ~ 70%
18"
1 1/4"
+3 1/2°
2"
3"
9"
1/32" sheet let in, t.&b.
1" dihedral
5" total dihedral
No downthrust ~ right sidethrust only.
Solid nylon nose block ~ FFn parts.
Prop ~ 22" ø x 28" pitch (2" max. width, 'Schwartzbach' outline, flat-bottom section) set at +4°, with tips washed out 2°.
16 strand motor.
Tips laminated from 6 ~ 1/32" x 3/16" balsa.

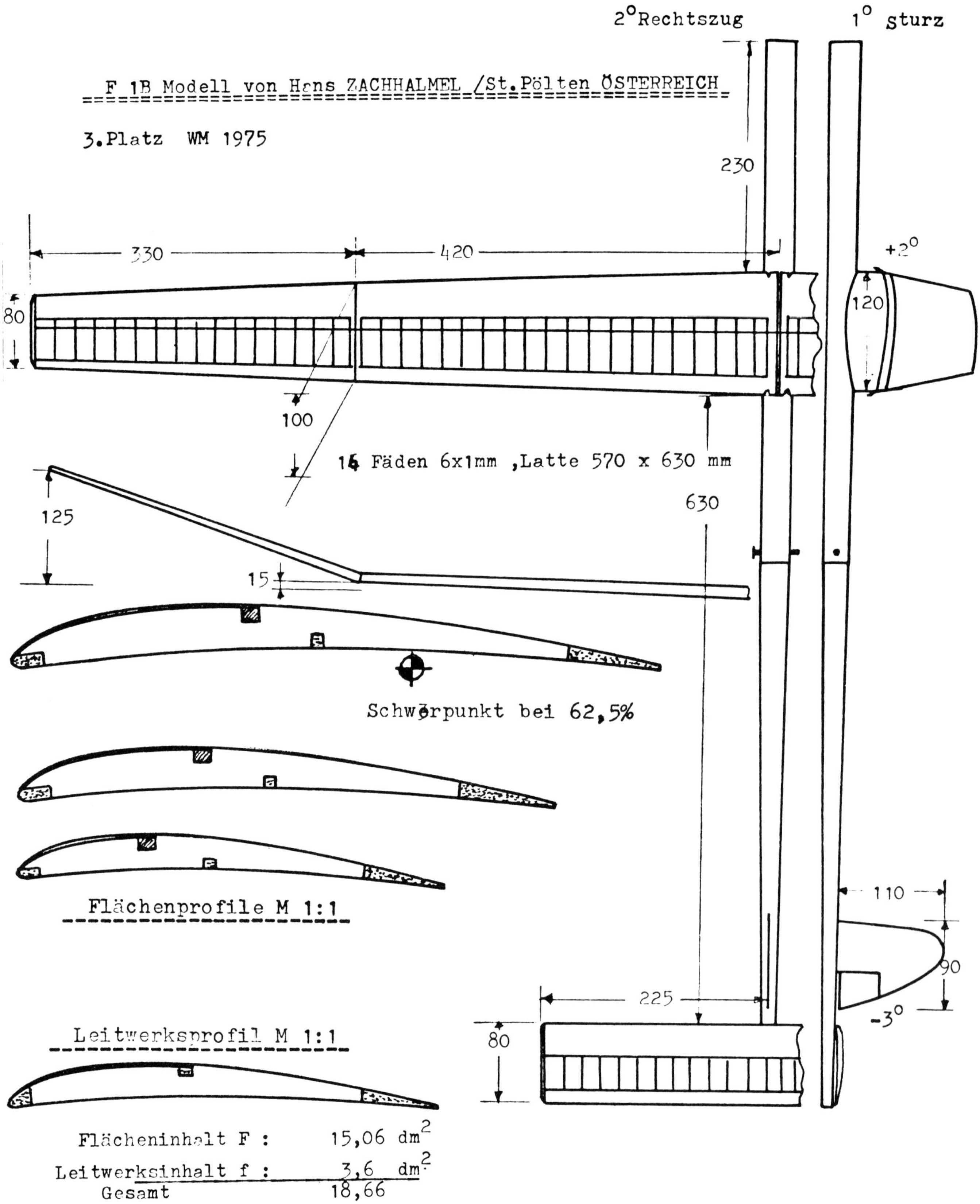
F 1B Modell von Hans ZACHHALMEL /St.Pölten ÖSTERREICH
3.Platz WM 1975
2°Rechtszug
1° sturz
230
330
420
+2°
80
120
100
14 Fäden 6x1mm ,Latte 570 x 630 mm
630
125
15
Schwerpunkt bei 62,5%
Flächenprofile M 1:1
110
90
-3°
225
80
Leitwerksprofil M 1:1
Flächeninhalt F : 15,06 dm²
Leitwerksinhalt f : 3,6 dm²
Gesamt 18,66

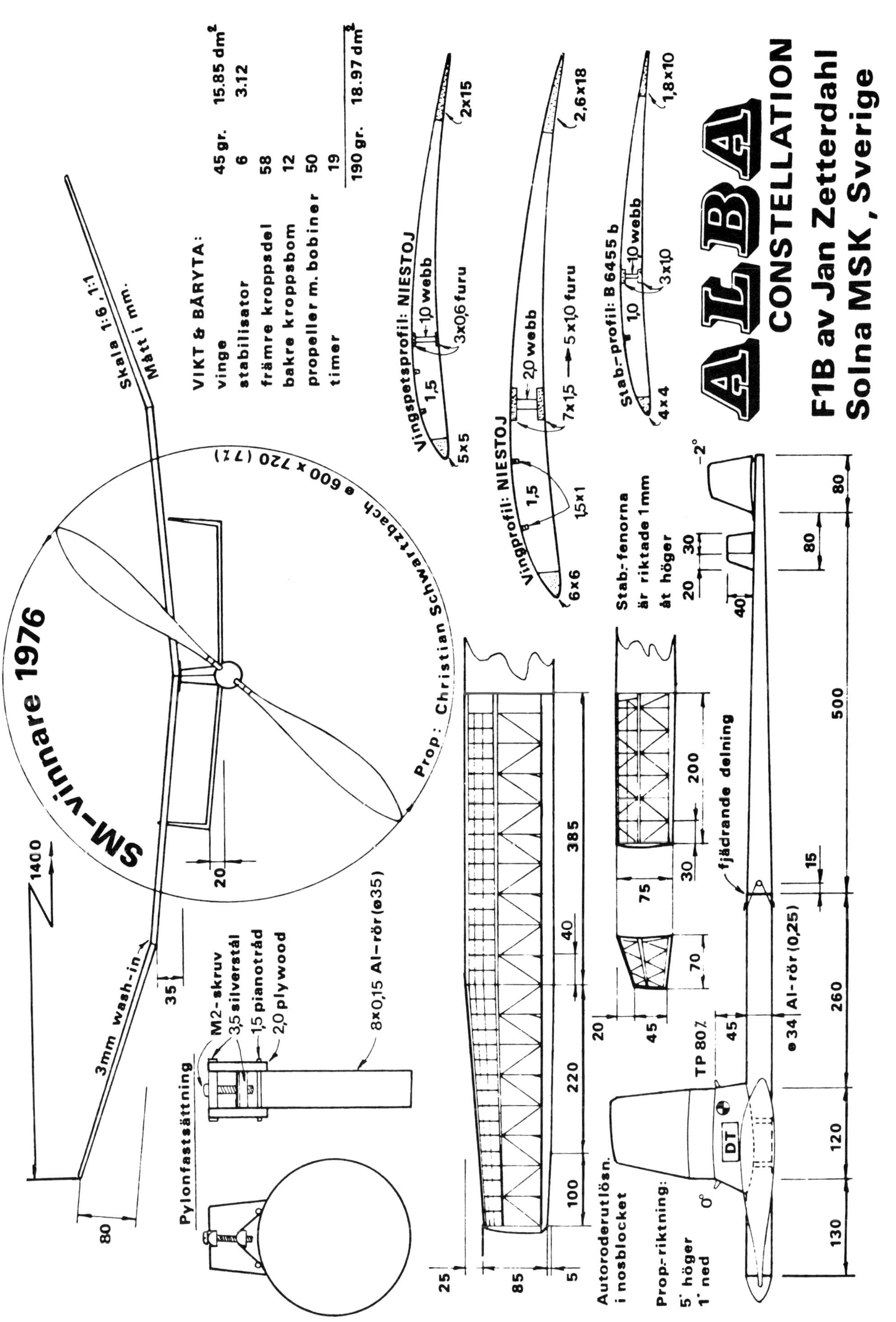
SM-vinnare 1976
ALBA CONSTELLATION
F1B av Jan Zetterdahl
Solna MSK, Sverige
Skala 1:6, 1:1
Mått i mm.
VIKT & BÅRYTA:
vinge 45 gr. 15.85 dm²
stabilisator 6 3.12
främre kroppsdel 58
bakre kroppsbom 12
propeller m. bobiner 50
timer 19
190 gr. 18.97 dm²
Vingspetsprofil: NIESTOJ
Vingprofil: NIESTOJ
Stab.-profil: B 6455 b
Prop: Christian Schwartzbach ø 600 x 720 (72)
3mm wash-in
Pylonfastsättning
M2-skruv
3,5 silverstål
1,5 pianotråd
2,0 plywood
8x0,15 Al-rör (ø35)
Autoroderutlösn. i nosblocket
Prop.-riktning: 5° höger 1° ned
Stab.-fenorna är riktade 1 mm åt höger
fjädrande delning
Al-rör (0,25)
TP 80%
DT

F1 C

F1C 1

Atlas FAI Power by Rol Anderson, Toledo, Ohio

Thanks for the inquiry on the *Atlas*. Sorry to say I don't have any three views or even full size plans. Since the advent of all sheet wings and stabs, I have gotten in the habit of laying everything out on a sheet as I go.

Wing rib spacing is 2¼ in. with ¼ in. ribs at panel ends. The sections are joined like hand launch glider wings; cut to angle and butt glued with epoxy. Glass tape three inches wide goes on the center joint. Tissue and dope covered wings average 180-190 grams. Latest wings, three coats K & B epoxy with red bottom, white tips on top average 170-180 grams. Stab rib spacing is 1-3/8 in. with several at ½ in. in center area. Stabs finished with tissue and dope average 28 grams. Latest stabs with three coats of K & B clear epoxy, about 22 grams.

George Versaw uses about 3/16 in. wash-in, right center panel... My latest ships are using 3/32 in. wash-in, right panel; 3/32 in. wash-out, left panel.

Doc's engine set-up, our props and pattern have all been developed together. As George says, we attempt to achieve as near a vertical climb as possible. The ships vary in pattern somewhat, but basically they are climbing near vertical, doing a left roll due to the washed-in wing. About ½ to one full turn in a ten second run. We use just enough right rudder on power to keep the nose to the right of vertical.

The Rossis are ported and timed to develop peak horsepower at about 26,500 RPM. Our latest experiments indicate the best altitude with our seven in. dia. by 2½ in. pitch prop. Many people believe that the props quoted at 7x3½ (Cox copies) have 3½ pitch, and think we are extremely low. Not true because the Cox copies are not true pitch. Most will average out at the equivalent of a true pitch of three in. or even 2¾ in. pitch. We are using wood props right now, to cut the run down. The weight is half that of fiberglass and thus stop in half the time. When the pattern we fly is right, our run down is only .3-.4 sec. However, we are working on a brake for the next year.

ATLAS AUTO STAB

ROL ANDERSON

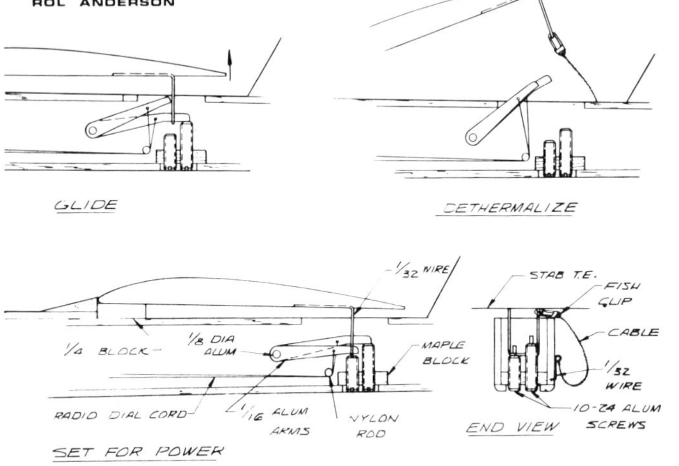

F1C 2

Anglian Power - Paul Bond, Anglia, England

Paul's model has an interesting trigger arrangement at the auto-stab mechanism. Latching and unlatching takes place right at the stab TE. Stab is sheeted with 1/16 balsa and also has a spruce LE making it a little on the heavy side at 36 g. Wing is fully tapered and in two pieces for easy transport. Full depth 1/8 spruce spar in inner panels gives a lot of strength.

F1C 3

Moonraker Power - Ed Carroll, Riverside, California

Ed worked out the general approach of this design while flying and talking with Doug Galbreath in Davis, California. Since then the original *Eros* concept went through an evolutionary change. Aspect ratio went up to 10.5/1, stab area came down to 22%, tail moment increased and wing section thickness decreased from 9% to 8.5%. All these things helped up the performance and the ship has won a goodly share of contests. Construction is straight-forward and the ship is rugged. A full description is given in the March 1975 Model Airplane News.

F1C 4

FAI Power by Roy Collins, Hornchurch, England

The wing is made of balsa weighing .5 to .55 oz. per 36x4x1/16 in. sheet. Nine inches of the port and starboard wing are covered in one oz. glass cloth, I always make the wings in halves, I feel it is easier to keep them on jigs therefore keeping settings. Though the wings are over 6 ft. this construction will take DT load on runways.

The fuselage tube has 1/8 in. med. balsa sheet fitted vertically along the whole length of the balsa tube to the rear of the top fin, this is to take knocks such as when the model DT's onto fences. This design has been very easy to trim, in fact the latest version has 1/32 in. packing on the rudder power trim and flew straight away into a contest, I no longer look on high aspect ratio as only a calm air model;it is just as good in a wind and cold as in calm and hot.

The total weight is 27½ oz.; wings 7½ oz., stab 1 oz., fuselage 19 oz.

Roy Collins contemplates his Rossi powered F1C model. This is earlier version without cowl. Wings are two piece, rubber banded to pylon.

Front end of Ken Faux's new model.

Rear end of Ken Faux's new model; foam rubber block helps push stab up to glide setting.

Collins' front end of new ship. Steel spar fits in slot of wing root. Coil spring goes through hole to hook wings together. Timer start switch aft of timer. All molded in fiberglass.

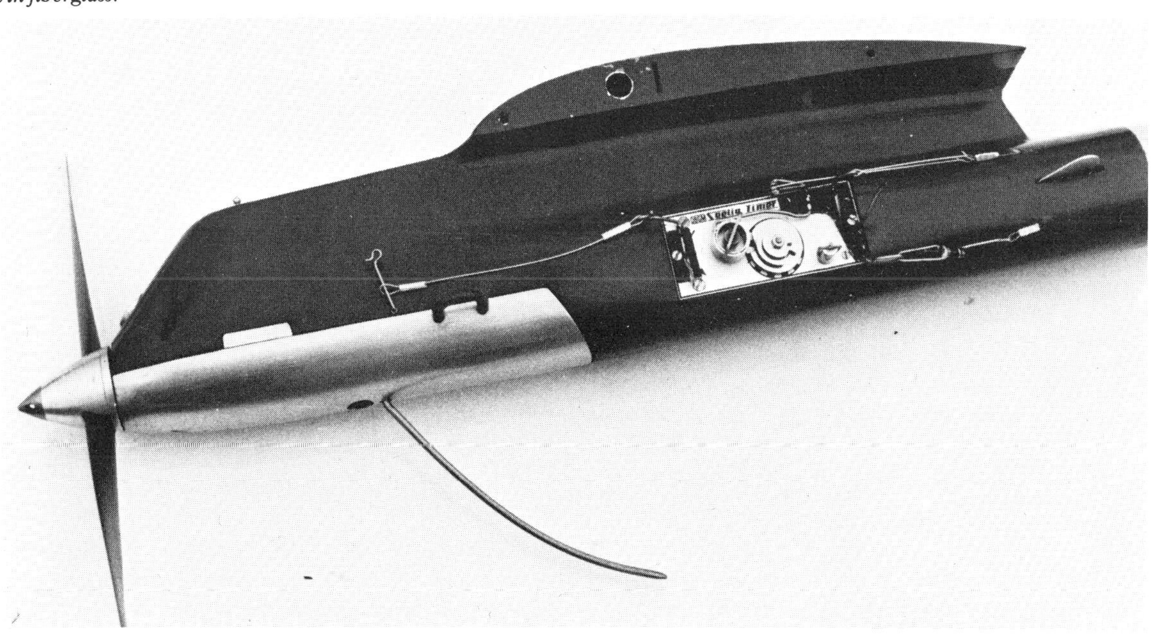

F1C 5

1977 F1C - Roy Collins, Hornchurch, England

Roy, together with Ken Faux, designed a new series of power models with the "modern" look of fully cowled engine, fiber glassed front end, "speed" pan, and plug-in wings. Ken's version is similar to Roy's but has triple fins. Heart of the model is the complex F/G moulded front end that forms the engine cowl, wing root fixing, tail boom attachment and surface for sticking on the engine pan. Roy makes these mouldings and also the alloy engine pans. The pan is held in with one screw. Round tailboom is rolled up from two layers of 1.5 mm balsa and covered with glass cloth. Wings have brass boxes (from Graupner) for fitting a 10 x 1 mm steel spar joiner. Wings are kept snugged in by a spring that is hooked from the right panel to the left panel, passing through an alloy tube in the center section. A special tool must be used for the hook-up. Wing section is on the thickish side and thinner ones will be tried next. Stab section is a Verbitsky type, high nose entry and reflexed TE.

F1C 6

1974 F1C - Ken Faux, Hornchurch, England

Ken placed this model into the fly-off at the 1975 World Champs. And then with his chances looking quite good the high revving prop blew apart on launch. So much for sure things!

Wing section is the well-known B8353 b/2 which is fairly blunt in the nose and with a high entry. Ken uses a Burrows glass fiber tube for the fuselage boom. Stabilizer and triple fins are Danish style. Two-piece wing is joined with an 8 S.W.G. steel wire in brass tubes. Collins pan is rotated 45 deg to the right so that the exhaust stays clear of the timer and pylon.

F1C 7

Bijou - Denis Ferrero, Varennes, France

The elliptical dihedral of this power model, if not new, nonetheless is a clever idea. Since most wings are built on jigs anyway these days, why not a curved jig. With shallow curvature there is little difficulty coping with the compound curves resulting. Rest of the model is conventional and conservative in layout with large stabilizer, short tail moment and low aspect ratio wing. Denis wraps the Rossi cylinder with an aluminum heat shield, and he gets impressive smooth runs. This ship won the 1976 Criterium Pierre Trebod. Details - wing spars are 5x2 mm spruce, section NACA 4409, weight 190 g, LE & TE are laminated balsa. Stabilizer uses two spruce spars, 3x2 mm and three balsa spars 2x2 mm. TE is 15x3 mm balsa, LE is 4x3 mm balsa, weight is 35 g. Fuselage is a molded tube using kevlar and carbon fiber. Diameter goes from 37 mm to 15 mm. Engine pan machined from dural, engine an ABC Rossi turns 30,000 RPM. Prop is a Ferrero special from Kevlar fiber. Fuselage weight 530 g.

F1C 8

Sour Cream - Bill Gieskieng, Denver, Colorado

The Bill Gieskieng "Flapper Story" is described in detail in the 1975 July/Aug. and Sept./Oct. issues of Scatter. This description by Bill will just cover the highlights.

"This ship is really a mixed up airplane. The wing was meant to test a fiberglass fuselage idea but since the glass fuselage dragged on, and on, a special quickie body was built. I wanted to keep it simple and conventional appearing. The wing was designed around some nice pieces of wood meant for R. C. ailerons. The solid flaps were easy to fabricate but lack torsional strength. A very bad feature of the "five sectioner" is the flap coupling so close in to the body. There is a certain amount of sliding allowed by flap couplings and being in so close to the center the flap loads are not well handled. In Taft's soupy air the ship became uncontrollable. It would dodge to the right and then to the left. I just lost touch with it. I think the basic design is good for a utility flapper concept but without the "V" stab and the extra dihedral break. Solid flaps may be all right but they should be thicker than the *Sour Cream's,* also to add some more structural rigidity they should be somewhat wider. The glide on this ship is not in the same class as the *Siren-Dipity* but then it was never expected to be such with its lower camber. Something like the *Sour Cream* would make a good beginner-in-flappers model. I would be tempted to try a flat center section with just tip dihedral of 24°; a slightly larger stab, flat, with a rectangular shape. The flaps could be built-up or solid but wider, such as 33%.

F1C 9

Siren-Dipity by Bill Gieskieng, Denver, Colorado

The *Siren-Dipity* was designed '71-'72 as a fair weather fly-off ship. I was getting tired of shoulder wing designs and decided to hoist the wing up into the air for a change. I wanted something stable and forgiving but yet capable of high performance. Originally the airfoil was meant to be the Gott.81, but a section that I had drawn up for Don McGhee's first flapper looked so good that I decided a switch was in order. The thin section with a relatively blunt, and well-rounded nose was derived from the Gott.400. The 400 is one of the very few sections tested with a flap at various angles. The thing that drew my interest was that the drag with the flap "up" was very low compared to other under-cambered sections. In fact I had never seen another undercambered airfoil that had its lowest drag point near the zero-lift angle. The *Siren-Dipity* version was thickened slightly by decreasing the undercamber. The flap hinging position was changed from 40% to 35% for structural reasons.

The flap angle used on the model for glide is much

SOUR CREAM AUTO STAB

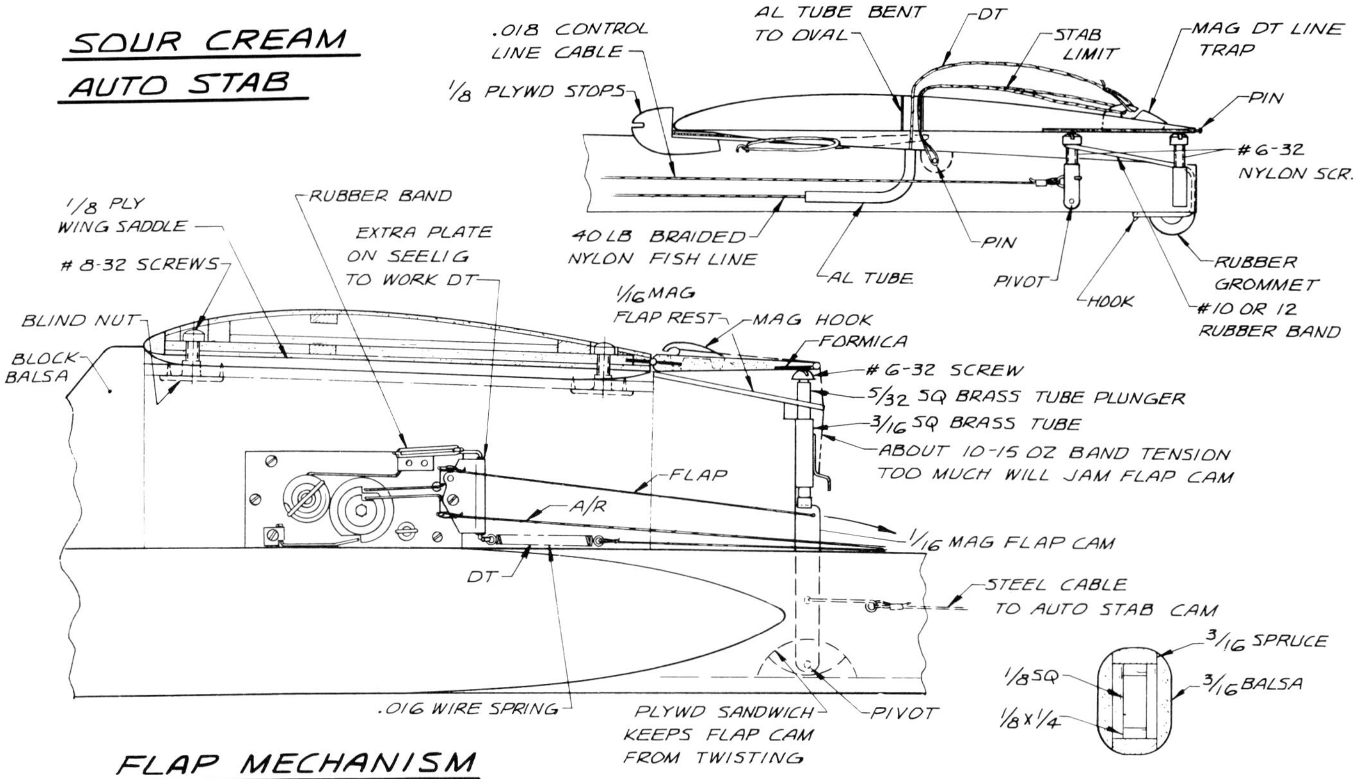

FLAP MECHANISM

FUSELAGE SECTION

SIREN-DIPITY AUTO STAB

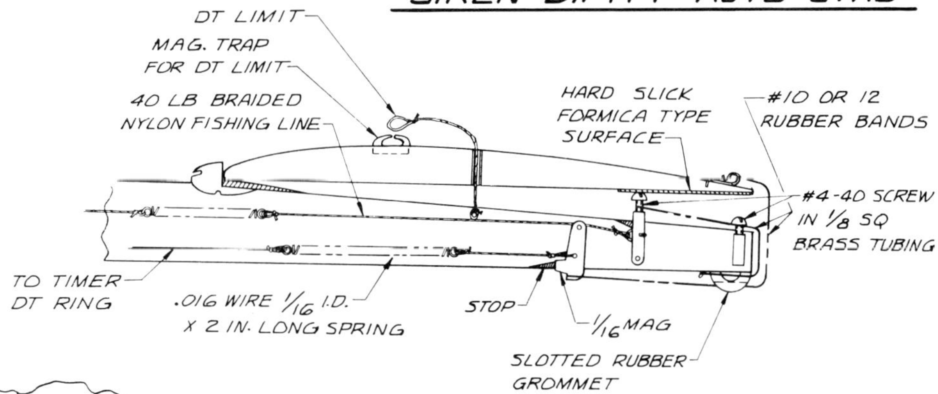

FLAP JAMMER FOR WASH-IN

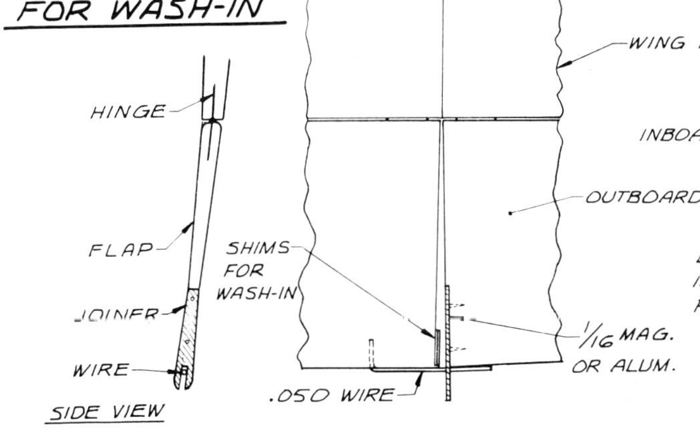

FLAP JOINER

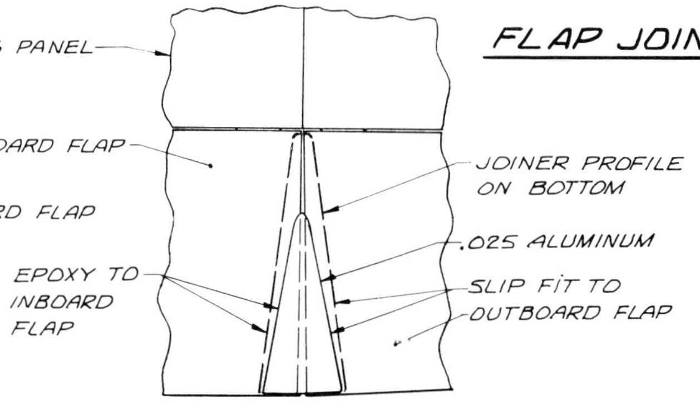

more than that found on the basic Gott.400 and corresponds to one of the flap angles tested. In the N.A.C.A. chart showing relative merits, the flapped Gott.400 comes out "average" in gliding ability compared to similar "Nordic like" sections. It is far inferior to the Gott.81 but then the '81 is impossible for the power phase without double flaps. The under-camber in the main wing makes for difficult construction and I would prefer a flat section for that reason.

The ship was originally designed without fins on the stab. From calculations it seemed feasible to use the "V" stab in making up the needed area. But it didn't work out that way. On an early test flight the ship made a sudden divergent move and flipped over on its back. Fins went on pretty quick! I'm not certain but that it couldn't use more fin area than it now has.

The long body is a pain in the neck. It is a built-up boom and it is very tedious to build. Because of the great length the stab had to be very light and fragile and was constantly getting cracked spars and changing the trim. Recently I removed the light mylar covering and besides sticking in some spruce spars, covered it with synthetic Japanese tissue. It is now quite strong but c.g. has gone back. The ship is more erratic now in gusty winds and although there is sufficient longitudinal stability the more rearward c.g. may call for more fin. Originally the ship was fitted with a pacifier tank underneath a bubble canopy. Pacifiers soon went on my crap list since they are a mess to handle and were suspected of blowing pistons and rods due to their unlimited flood-off action. A hard tank was fitted and it worked out much better except for the fact that all the needed tubing hanging out makes the nose of the ship look like it is being strangled by a squid.

Test Flying: The big rule with flappers is to go slow and check everything with extra care. You will know that your mind is safety conditioned when you find yourself repeatedly trying to put the flaps up on your conventional ships.

I start with very short runs of about 1.5 secs. I tie off the autorudder line so there is no chance of it kicking in. I put the DT line, the flaps and autostab all on the same trigger. There is about a second delay on the DT. I run the engine on the rich side and launch as nearly straight up as I can. I launch steeply because I figure that up is the safest direction. After I grow more comfortable with the ship I jump up the engine time second by second. At this point I'm delaying the DT more but still not attempting a glide although I may drop the flaps and stab to help slow it down.

Once I get the power phase sorted out for a full engine run, I start thinking about the glide. I don't always do it, but it is best to start out with a very short DT of around ten or twenty seconds. From then on it's just like trimming a conventional ship. Just remember to adjust the right screws!

What Happens When You Make A Mistake: All of us are subject to human error as we are constantly reminded. The best way to avoid making a mistake with a flapper is to be very, very familiar with it. This means many sessions at home re-cycling the gimmicks and drilling into your mind the principles you are triggering. For the power phase you have to know without thinking: flaps up, stab up. I check my planes three times. Once when I set the timer and hook up the gizmos. A second time just before starting the engine, or at least just before getting into my thermal seeking stance. A third time I give a cursory glance at the flaps, stab, autorudder prior to launch. Because, if you try to fly with your flaps down and the stab up you will immediately find a Rossi spinner digging into your buttocks. If you launch with the flaps down and the stab down it will zoom all over the sky in fantastic maneuvers. If the power phase is all right but something happens to the glide settings, you can have the following: the stab comes down without the flaps, meaning either a vicious nega-dive or a series of outside loops and rolls. If the flaps come down but the stab doesn't it will do a series of hammerhead stalls. If neither the flaps or stab come down but remain in the power setting, anything can happen depending on the decalage. Some flappers will glide, if rather steeply. Others particularly if they miss the transition, will dive straight in. If one of the flaps comes down but the other doesn't you will have to use your imagination as it never happened to me.

F1C 10

Speed Cream - Thomas Koster, Skaevinge, Denmark

The *Speed Cream* is similar to the *Square Cream* shown in the drawing, with the major exception regarding wing section. It is 6.8% thick and flat-bottomed. An intricate fiber glass moulded fuselage front end provides the engine cowl, shoulder wing mount, tail-boom socket and mounting surface for the engine pan. Tail-boom is rolled from balsa and is glass cloth covered. It attaches to the forward fuselage by means of a threaded fitting. Surfaces are all sheet covered and tapered. Wing joining to the root is via 1x 7 mm brass boxes and steel spar. Not shown on the drawing is the mechanism that pulls the climb stab stop down to give a nose down bunt at the top of the straight-up climb. Bill Gieskieng had some interesting comments on Koster's transition.

"You wouldn't believe some of the transitions the Speed Cream made from impossible positions! Sometimes the ship would get off to the left and start to roll. The thing would be cranked left and upside down when the dropped stab and autorudder pulled it upright in time to go into the glide smoothly. The proper pattern was straight up or with a slight hedge to the right, but the main thing was it had a great latitude for coming out smelling like a rose. The action was pretty subtle at times and some may not have even realized what he was doing.

Tom's props were produced in Tom's stove, trimmed by Dave Rounsaville and balanced on my balancing device; carbon fiber by Verbitsky. The engines were reworked by a friend in Germany (an international effort!).

I thought Tom got higher in some practice sessions than he did at the W/C. He was using the plain glass and epoxy prop. The thing sounded scary on the ground because the tips were fluttering (I kept looking around for the nearest bomb shelter) but immediately on launch it settled down into a ferocious clean scream. The carbon versions with the stiffer blades ran nice on the ground and screamed like mad on release. But I thought the ship was slower (Tom didn't think so . . . felt I was just getting used to it). A possible explanation, if I was right, would be that the softer blades opened up on pitch."

(Ed. note: At 29,000 rpm, the tips of a 7-inch dia. prop are traveling at 0.8 Mach no. Most airfoils start to get into trouble with transonic flow separations and drag rise at this speed unless they are carefully contoured. A thin section has less of a problem than a thicker one and lower angle of attack also reduces the problem. Going from static, zero forward velocity to climb speed could reduce the effective blade angle of attack enough to back off from the critical Mach no., and this would be more noticeable with the thicker blade.)

F1C 11

Strutter 7C-Dick Lyons, Libertyville, Illinois

The *Strutter 7C* is the third modification of the 7th design - "Mark VII, Mod C"as the British would say! I won't go into details of the design rationale — but from a practical standpoint I feel that simplicity, structural integrity, and *reliability* are more important in competition than the ultimate in aerodynamic design and streamlining, etc. Most of the details are on the sketch with the most "radical" feature being the foam wing. I don't know if many fliers on the West Coast are familiar with "Pro-Foam". It is a closed cell rigid polyurethane foam — weighs about 1.8 to 2 pounds/cu. ft. and can be worked very easily and will take any kind of glue or dope. I use mostly white glue. The wing is really easy to build and can be knocked out in just a few evenings once you are familiar with it, no hot-wire type cutting is necessary, it is too easy to work with; all I do is cut a piece to size, glue on leading and trailing edges, form airfoil with long sanding block to templates on each end, put in grooves for spars while panel joints are merely "hand-launch glider type" butt joints using epoxy with glass cloth reinforcement on top and bottom and then given a coat of dope and covered first with "00" silkspan and as many coats of dope as you want.

Double covering is required for FAI power models as a single covering does not give enough torsional rigidity — Monokote can be used — it goes over the foam very nicely but is not rigid enough for high speed FAI models.

(Editor: Dick showed us some Nordic wings he made using this technique. They were impressive. if you try this idea make sure you get the right foam. Some types are too heavy and require special glue.)

F1C 12

Reaktans - Urban Nygren, Spanga, Sweden

Power fuselages needn't be round, or all sheet construction for that matter, as proven by Nygren. This very long tail moment design uses a built-up silk covered boom with 3 x 3 mm spruce longerons. Wing uses an interesting box spar of 1.5 mm balsa. This approach eliminates sheet where it doesn't contribute much to the strength thus increasing the strength to weight ratio.

F1C 13

Uncle Remus/Miss Muffett - Lars-G Olofsson, Stockholm, Sweden

These are the World Championship models of 1975, Plovdiv, Bulgaria. Lars flew Miss Muffett in rounds 1-5 and the 6 sec fly-off and Uncle Remus in rounds 6, 7 and the 4-sec fly-off. Both models are of rather simple, straight forward design, of all sheet construction. Triple fins, a Nordic country trend, are used with short nose and long tail moment. Lars makes his own props and the ones used were 177 mm dia by 95 mm pitch. This is considered a relatively slow revving prop (25,000) but Lars claims higher thrust in the climb, especially on the short runs. In the critical fly-off rounds Lars made every tenth of a second of engine run count and then compounded that by trimming to a very good transition.

F1C 14

Oblivion Express by Ken Phair, Denver, Colorado

The flap mechanism has worked flawlessly and is also quite light. The DT arm and its spring loaded lever to push the stab down to glide position has been marginal. I'm working on a revised system, to eliminate any problems.

I am flying the model like a conventional power ship, a right spiral climb with a delay to autostab and flap activation which occur simultaneously. I owe many hours and many destroyed airplanes to Bill Gieskieng. He has offered me the results of his years of experimenting with flappers. My discussions with Bill helped me avoid many mistakes that would have required a long development program to eliminate.

Bill's stab mechanism is basically simpler than mine,

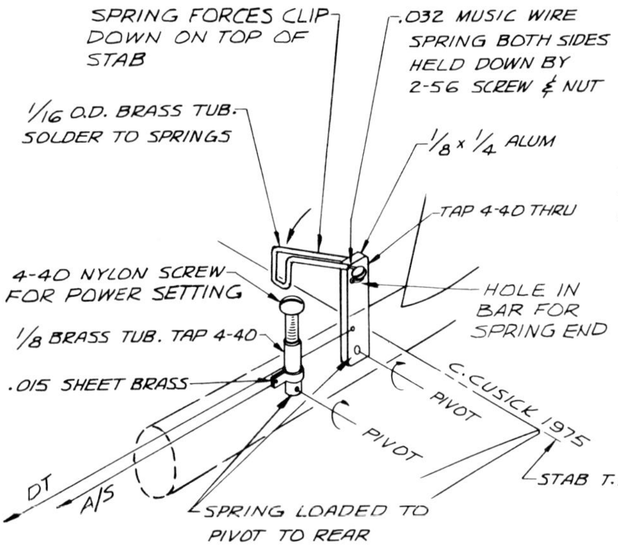

ORIGINAL AUTO STAB

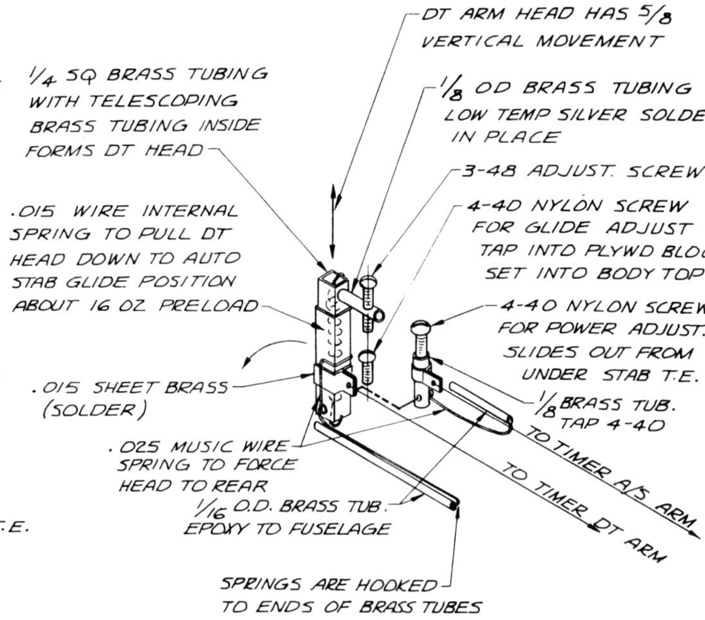

IMPROVED AUTO STAB

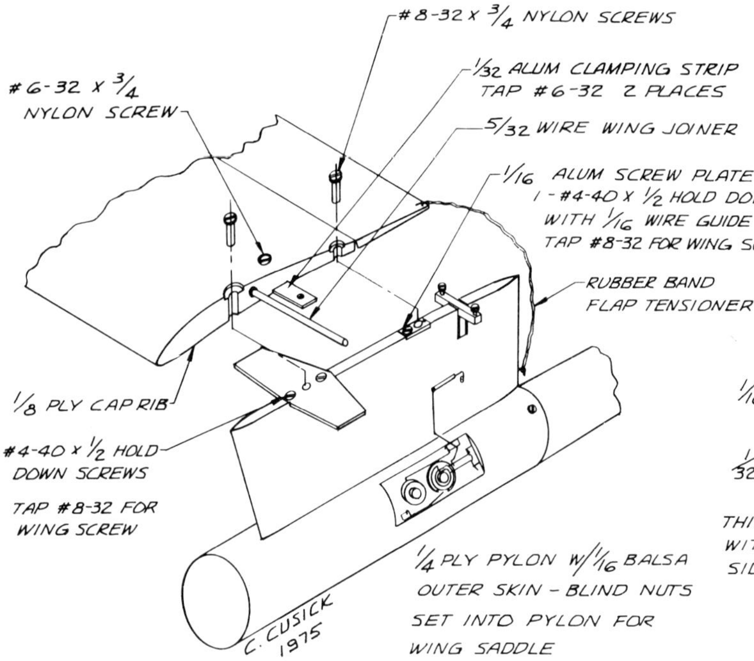

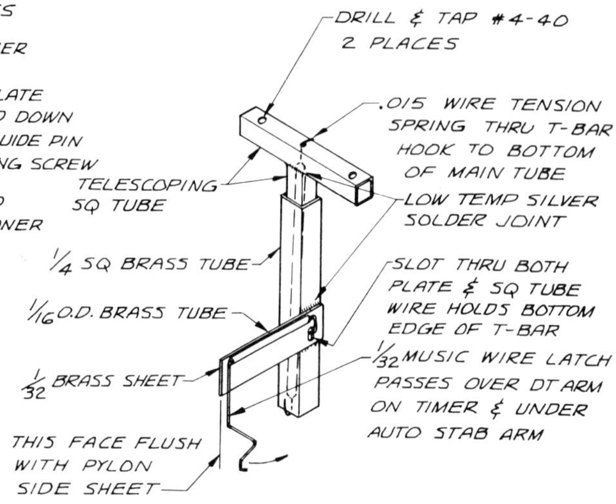

WING FLAP ACTUATOR

OBLIVION EXPRESS
WING MOUNT DETAIL

but mine was designed to be close to the operation of my conventional ship auto-garbage system as possible for two reasons. One, to minimize fabrication and development problems and two, to minimize flying field errors due to different set-ups between flappers and conventional models.

My wing mount is the same system I use on the conventional model but it needs revision to provide greater action. A flapper wing is tougher to key than a conventional wing because of the shorter root chord and twisting of the wing about the model's longitudinal axis affects the flap angles and hence the power pattern.

The wing mount uses a total of four nylon screws; two 6-32's in conjunction with an aluminum binding strip to hold the wing halves together (dihedral is by a 5/32 in. dia. music wire joiner) and two 8-32's thru the joint centerline into aluminum plates secured to the ¼ in. ply pylon. The front plate is also the wing saddle. From a strength standpoint the system seems adequate being neither too strong or too weak. Its only drawback is its lack of 100% positive location of the wing.

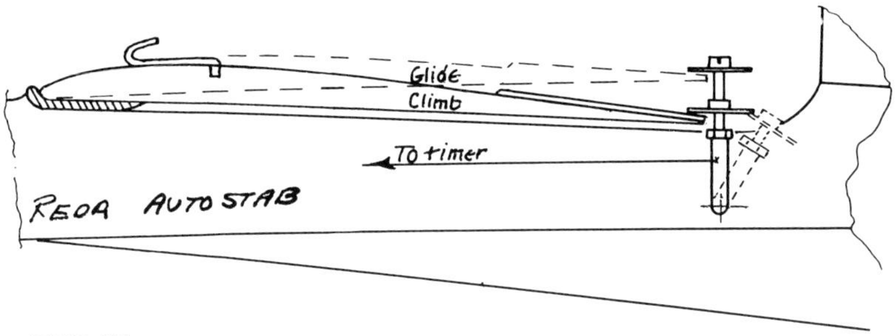

F1C 15

SR 104-12 - Sigfried Reda, Munich, Germany

These models are the result of a serious concentrated effort to produce a performing machine more or less stripped of frills and cosmetic nuances. In the process Reda developed a foam core structure with balsa skin for the flying surfaces which is also coated with glass cloth and resin and reinforced with carbon fiber. Construction is aided with a vacuum wrapping device that uses a pump from a refrigerator.

Reiner Hofsass has this to say about his friend Sigfried. "The most remarkable thing of Reda's models is how you can repair them after a crash. He crashed one during trimming in Plovdiv, epoxied it together and flew with just this model the whole contest to place 4th. To glue the fuselage he uses two aluminum angles as splints to line up the pieces while the epoxy set."

The foam core used is about 1 lb./cu. ft. density. Balsa skins are glued with ½ gram epoxy resin per sq. dm. Glass cloth at 27 g per sq dm. is then attached with the same resin. A plastic sheet pulled down by vacuum holds everything together while the resin sets. Resin: "Epolyth Farblos" (Lackfabrik Bader KG, 7300 Esslingen/Neckar, Schliestrasse 25, GFR) Cloth: Glass-Silk fabric, quality 90001, finish 1, (Interglass Textil GMBH, 7900 Ulm/Donau, Postfach 619, GFR).

F1C 16

Excelsior FAI Power by Dave Rounsaville,
New Milford, New Jersey

I have finally decided on the new models I want to build this winter. Both are similar to my standard *Excelsior* but with higher A/R wings; one at 10/1, the other at 13/1. both use plug-in wings, fiberglass cowled front ends with speed pan engine mounts. One thing that is very important with 7 sec engine runs is weight! The airplane should be right-on or no more than 5 g over. This really makes the ship so much easier to handle and of course it climbs higher.

F1C 17

So Long For Now - Urs Schaller, Florence, Italy

Urs has had a very successful time with this design and the striking lines have been admired everywhere. Urs won both the Italian Championships and the Swiss Champs plus some other comps. in '75. He placed 7th at the '76 Pierre Trebod.

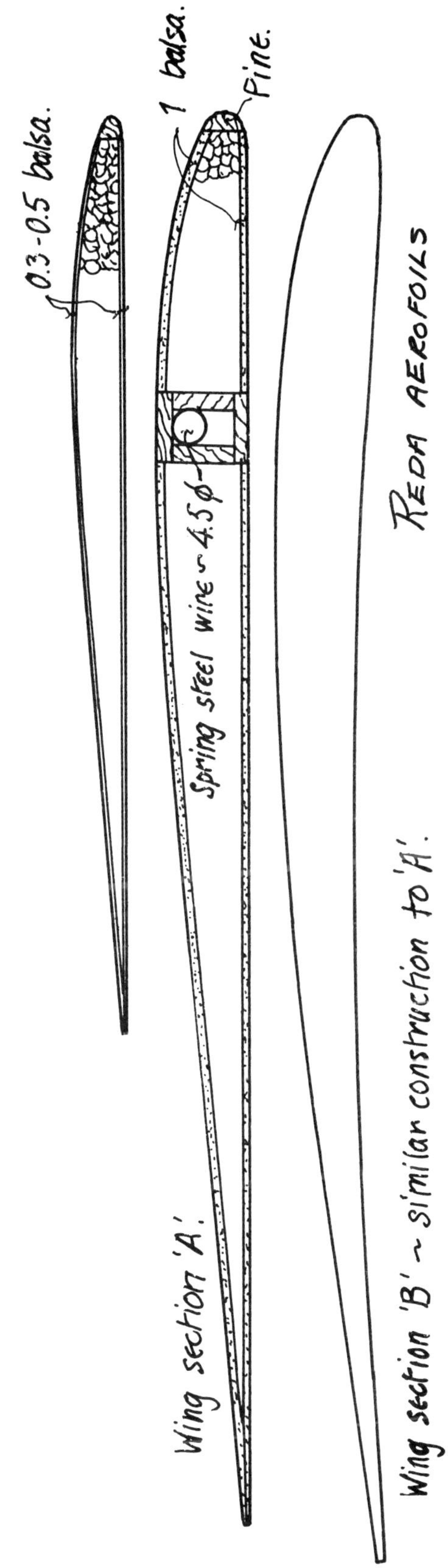

Urs Schaller

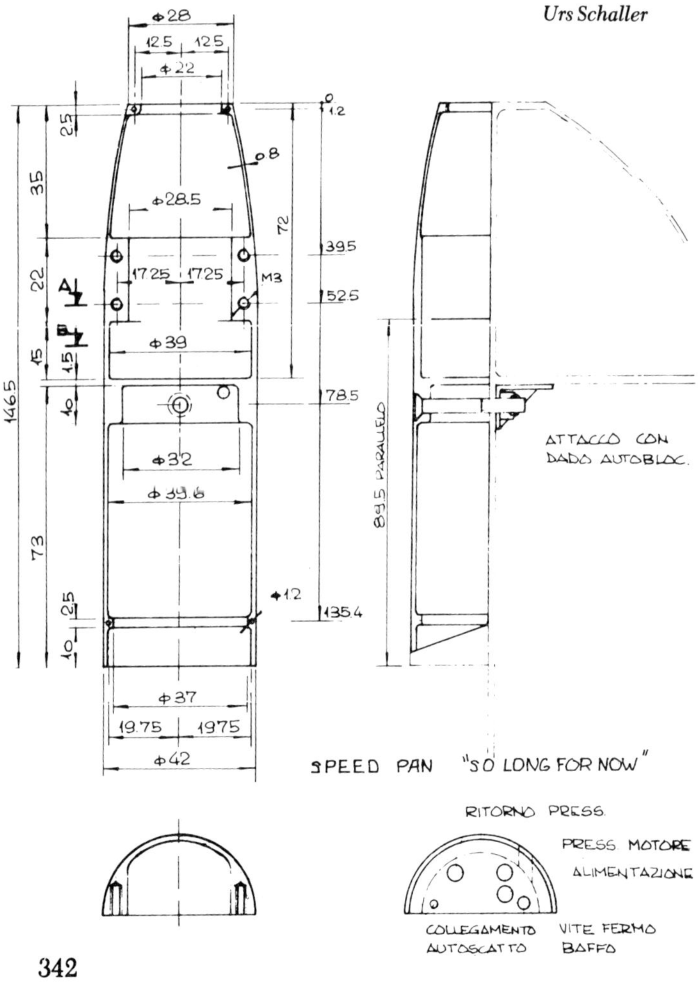

F1C 18

Gambrinus 75 - Hans Seelig, Munich, Germany

Hans Seelig's power model abounds with many clever ideas and sound engineering practice. It looks good too! Forward fuselage and engine cowl is a moulded fiber glass shape to which a bottom-mounted engine pan is attached with one screw. The wings plug into shoulder sockets, and an elliptical section tail boom is grafted on aft of the wing. Engine exhaust is piped internally and dumped out the bottom. Autostab is gripped by swinging claws that are hidden in the fuselage under the stab. Wing is constructed with sheet over a foam core in a manner similar to that used by Reda. An earlier version used the MVA 101 wing section, while a later version, as shown on sheet 2 has a higher aspect ratio and the Wortmann M2 airfoil.

F1C 19

WC-FIA Power - Sergey Sharon, USSR

This model of Sharon's appears to be the last version to retain the strut braced wing. The modern fuselage construction is apparent and an indication of the new concept to come. Sergey flew a later version at Plovdiv, a design with forward fin. Things didn't go his way in the fly-off, when the engine wouldn't start. Switching to a spare, it quit cold on launch. Two years later at Roskilde, Sergey had a new fleet of really beautiful models with alloy sheet covering on inboard wing panels and neat air-brushed trim

F1C 20

Moto 300 - Reinhard Truppe, Klagenfurt, Austria

Reinhard Truppe tied with Siegfried Reda for 4th place in the 1975 World Champs (Plovdiv) flying this tapered-wing all-sheet power model. Two-piece wings are held in place with a 7x1 mm steel spar and brass box sections in the wing roots. A rather high pylon is used and the wing section is flat-bottomed and 8.9% thick. All surface are coated with glass cloth (27g/dm^2) and resin.

F1C 21

BE 34 - Eugeny Verbitsky, USSR

Eugeny Verbitsky needs little introduction in the world of international aeromodelling competitions as he places consistently high, produces technically excellent designs, and has a friendly word and helping hand for everyone. His recent F1C World Champ score reads; 1971, 45th; 1973, 16th; 1975, 2nd; and 1977 with models proxy flown by Ziljberg, he placed 3rd. That proxy situation came about due to his having been bit by a poisonous snake while competing in a contest in the Ukraine. A number of months were needed for his recuperation.

BE 34, the model shown here, was flown at Plovdiv. Its most interesting feature is the intricately machined

Details of Verbitsky's BE34; second place at 1975 World Championships.

alloy forward fuselage and engine mount. The engine pan is held to the metal tube fuselage by means of a cylindrical section that slips inside the tube. This cylinder has two synthetic rubber "O" rings that are clamped by discs. The clamping action squeezes the rubber rings against the tube, holding the engine mount in place only by their friction. The fuel tank fits inside this cylinder and is shaped like a doughnut to make room for the single screw used for tightening the "O" rings. A neat fiberglass cowl slips over the engine and is held down by a tab under a cylinder head screw. Tail boom is a carbon-fiber reinforced-balsa tube which is attached through machined alloy fittings. Wings are skinned with balsa and glass cloth and plug into a machined round stub spar attached to a medium/low pylon. Airfoil is flat-bottomed and about 8.5% thick. A tapered I beam spar is used on the inboard panels. Stabilizer is also sheeted and uses a neat stab platform that blends into the LE. Stabilizer airfoil has a high LE entry and a reflexed TE.

Workmanship on all parts is of a very high order and there appeared to be no short-cuts taken to avoid work. It was obvious that this model was engineered and built by one that perceived and appreciated the beauty of fine machinery.

Not content with this fine design, new models were prepared for 1977. These feature even cleaner front ends and aluminum alloy sheeted wings. The alloy is .03 mm, heat treated (from helicopter blades). Balsa sheet is epoxied to the alloy and then wrapped around the leading edge and ribs. An alloy tube is used for back-up on the LE. Apparently weight can be controlled within limits. All three Russian Team Members in 1977 had models with this type wing structure. Mozyrsky managed to DT one on to power lines. The retrieving crew was treated to an impressive view of sparking and flashing as the metal wings shorted out part of the Danish power grid.

Verbitsky launches a little to the right.

F1C 22

Square V - Frank Wolff, New York, New York

Frank's model is fairly conventional, all-sheet and well constructed. Stab adds some distinction with dihedral. The prop brake used was the shot-pin type developed by Bob Hatschek and fellow club members of the New York Skyscrapers. It really worked fast and stopped the prop horizontally every time. Frank's misfortunate at the '75 W/C stemmed from a crazy dust-devil that damaged his model on the ground between rounds.

Neat stab key on BE34.

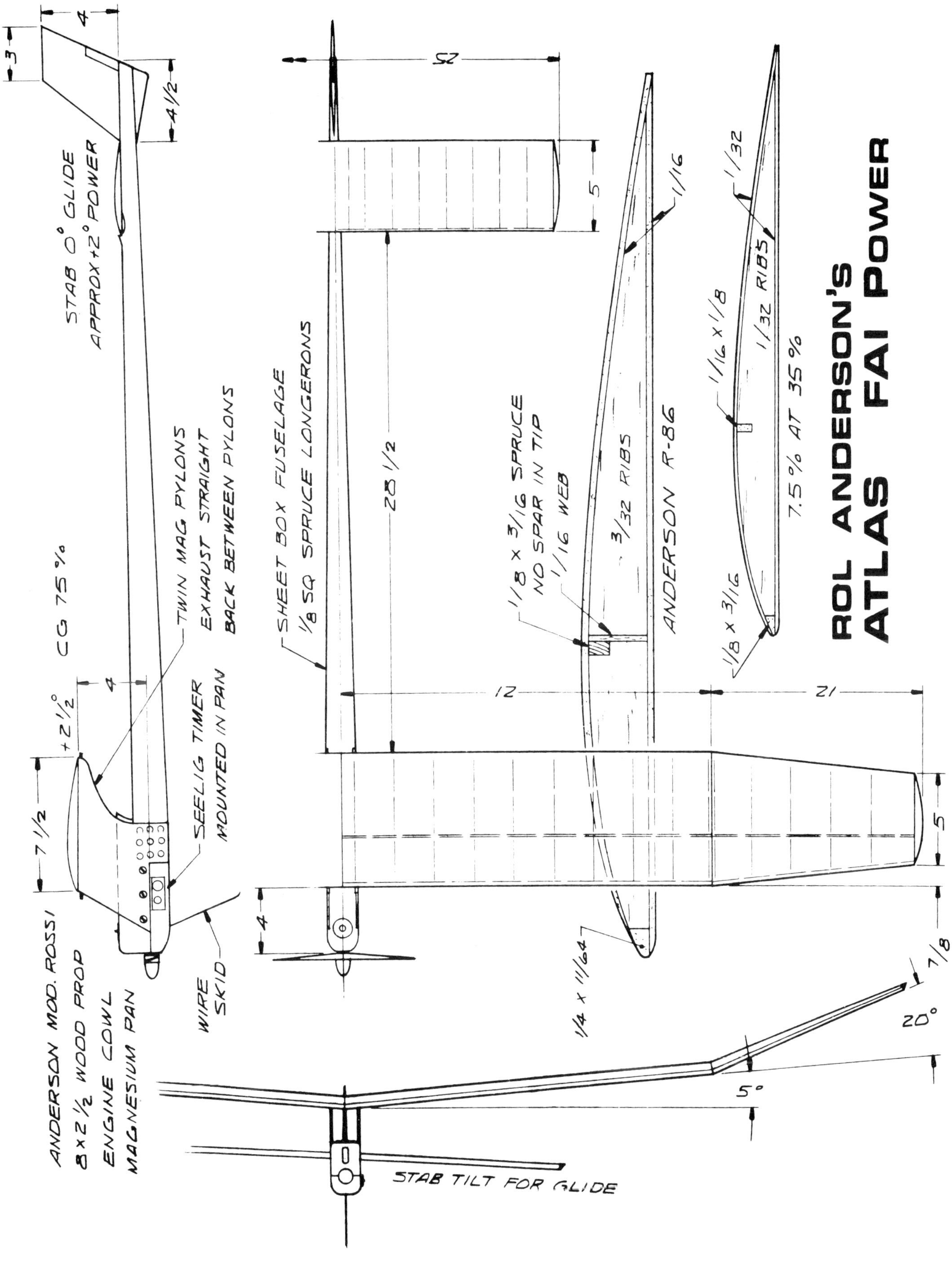
ROL ANDERSON'S
ATLAS FAI POWER
STAB 0° GLIDE
APPROX +2° POWER
CG 75%
TWIN MAG PYLONS
EXHAUST STRAIGHT
BACK BETWEEN PYLONS
SHEET BOX FUSELAGE
1/8 SQ SPRUCE LONGERONS
SEELIG TIMER
MOUNTED IN PAN
ANDERSON MOD. ROSSI
8×2 1/2 WOOD PROP
ENGINE COWL
MAGNESIUM PAN
WIRE
SKID
1/8 × 3/16 SPRUCE
NO SPAR IN TIP
1/16 WEB
3/32 RIBS
1/16
ANDERSON R-86
1/32
1/16 × 1/8
1/32 RIBS
1/8 × 3/16
7.5% AT 35%
1/4 × 11/64
STAB TILT FOR GLIDE
20°
5°
1/8
25
5
28 1/2
12
21
4
3
4 1/2
+2 1/2°
7 1/2

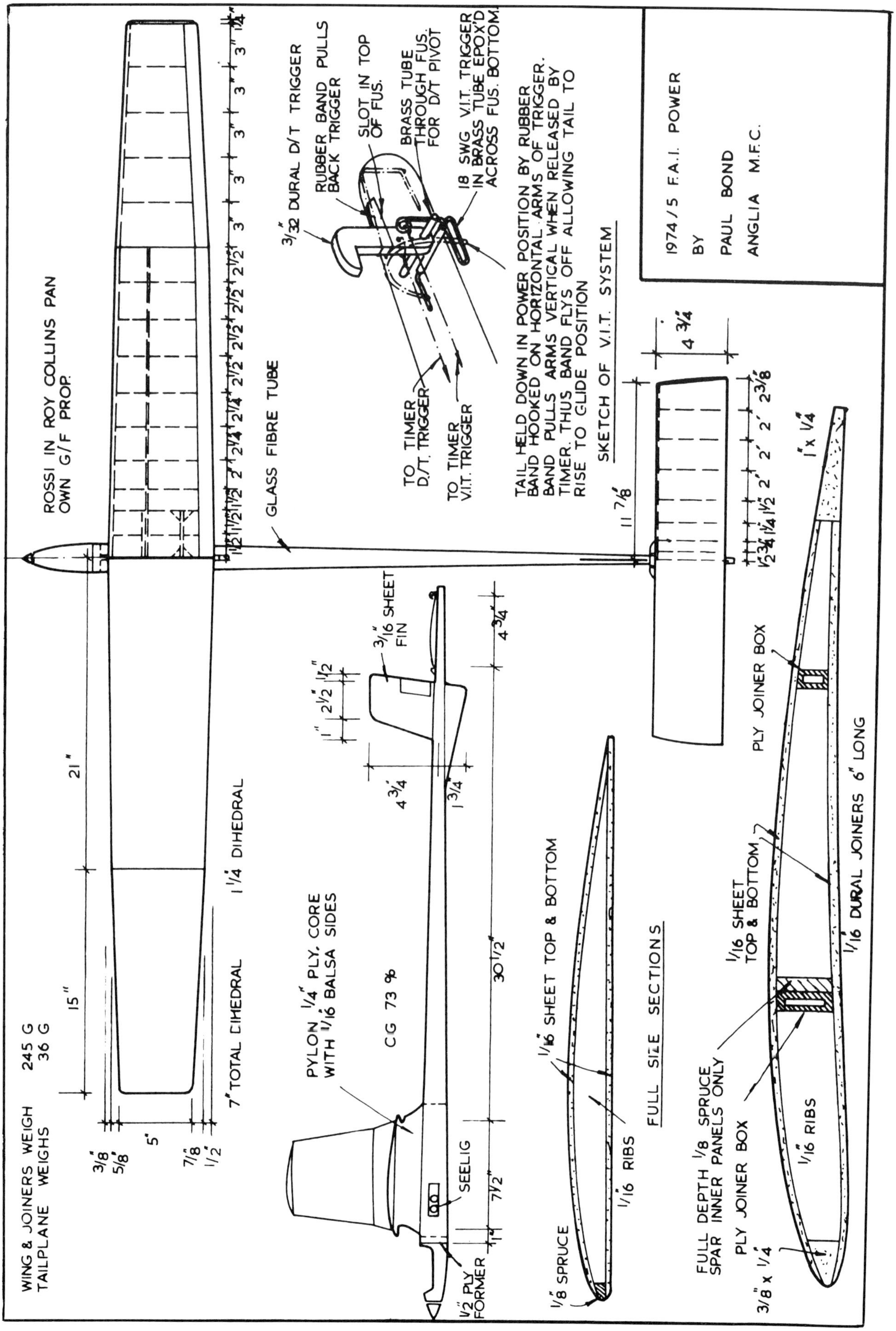
1974/5 F.A.I. POWER
BY
PAUL BOND
ANGLIA M.F.C.
ROSSI IN ROY COLLINS PAN
OWN G/F PROP.
GLASS FIBRE TUBE
3/32" DURAL D/T TRIGGER
RUBBER BAND PULLS BACK TRIGGER
SLOT IN TOP OF FUS.
BRASS TUBE THROUGH FUS. FOR D/T PIVOT
18 SWG V.I.T. TRIGGER IN BRASS TUBE EPOX'D ACROSS FUS. BOTTOM.
TO TIMER D/T. TRIGGER
TO TIMER V.I.T. TRIGGER
TAIL HELD DOWN IN POWER POSITION BY RUBBER BAND HOOKED ON HORIZONTAL ARMS OF TRIGGER. BAND PULLS ARMS VERTICAL WHEN RELEASED BY TIMER. THUS BAND FLYS OFF ALLOWING TAIL TO RISE TO GLIDE POSITION
SKETCH OF V.I.T. SYSTEM
WING & JOINERS WEIGH 245 G
TAILPLANE WEIGHS 36 G
21"
15"
7" TOTAL DIHEDRAL
1 1/4" DIHEDRAL
PYLON 1/4" PLY, CORE WITH 1/16 BALSA SIDES
CG 73 %
3/16" SHEET FIN
30 1/2"
7 1/2"
SEELIG
1/2" PLY FORMER
11 7/8"
4 3/4
1/16" SHEET TOP & BOTTOM
1/16" RIBS
1/8" SPRUCE
FULL SIZE SECTIONS
PLY JOINER BOX
1/16 DURAL JOINERS 6" LONG
1/16 SHEET TOP & BOTTOM
FULL DEPTH 1/8 SPRUCE SPAR INNER PANELS ONLY
PLY JOINER BOX
1/16" RIBS
3/8" x 1/4
1" x 1/4

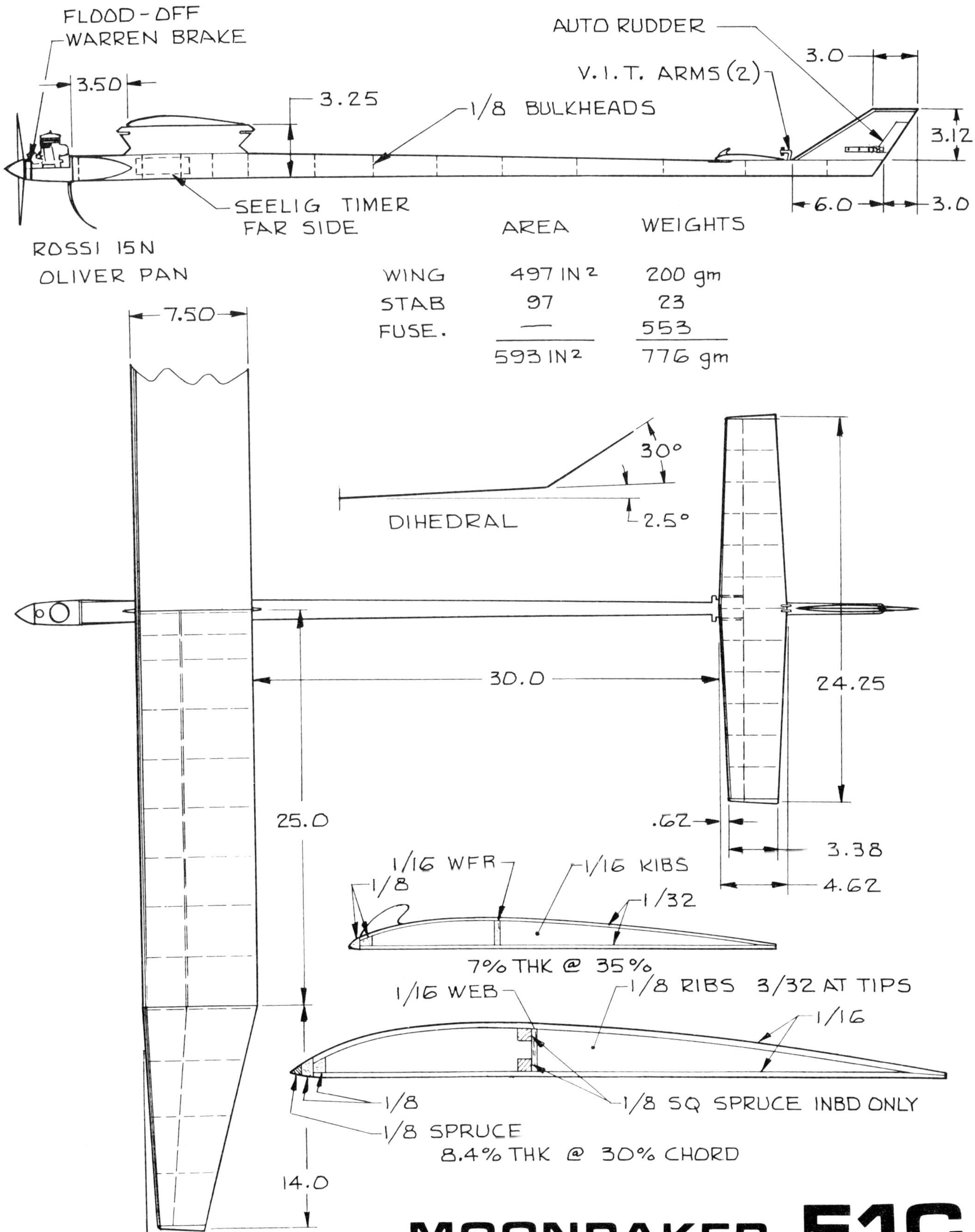

MOONRAKER F1C

BY ED CARROLL, RIVERSIDE, CALIF.

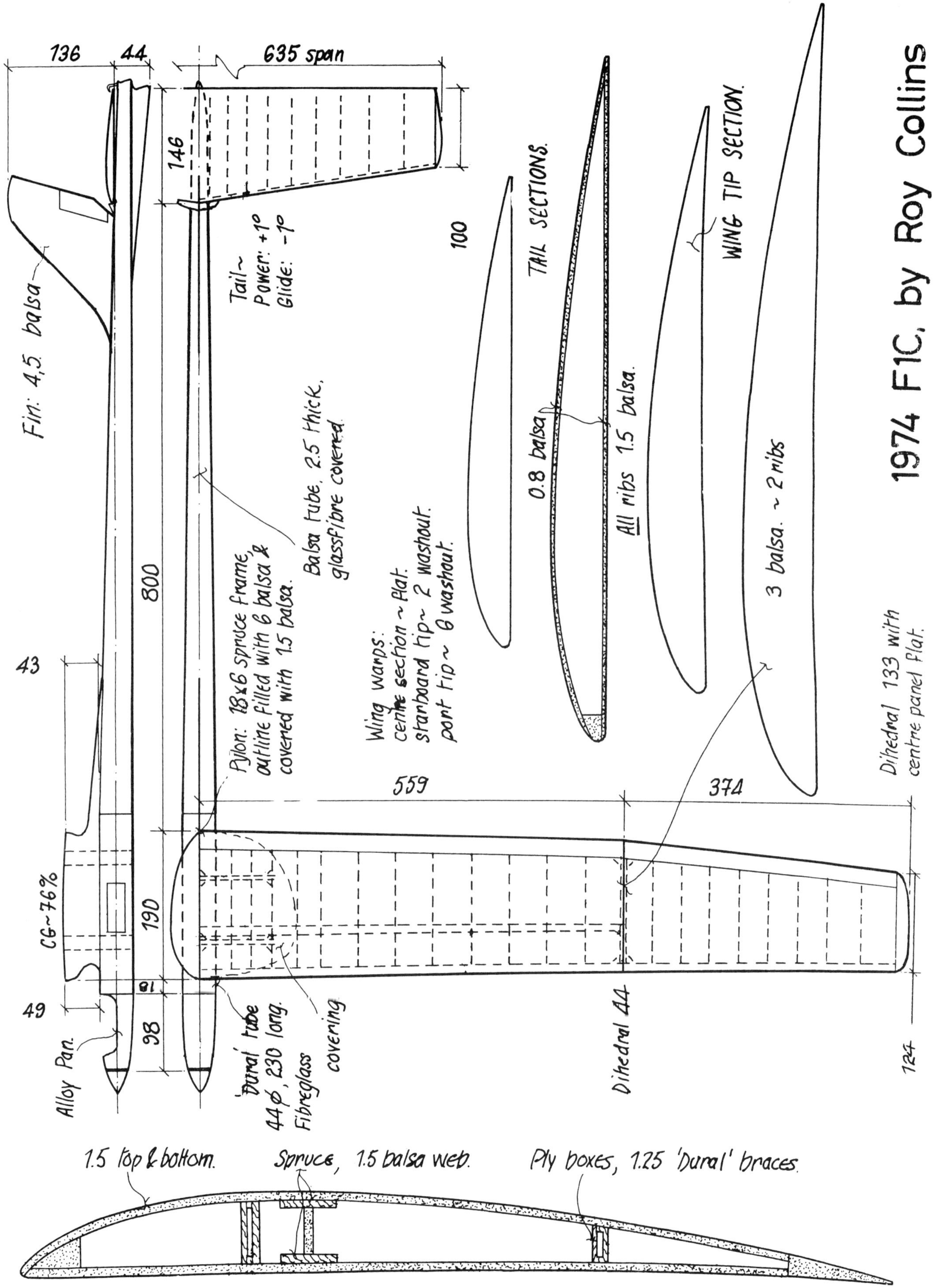

4 Collins GB Scatter

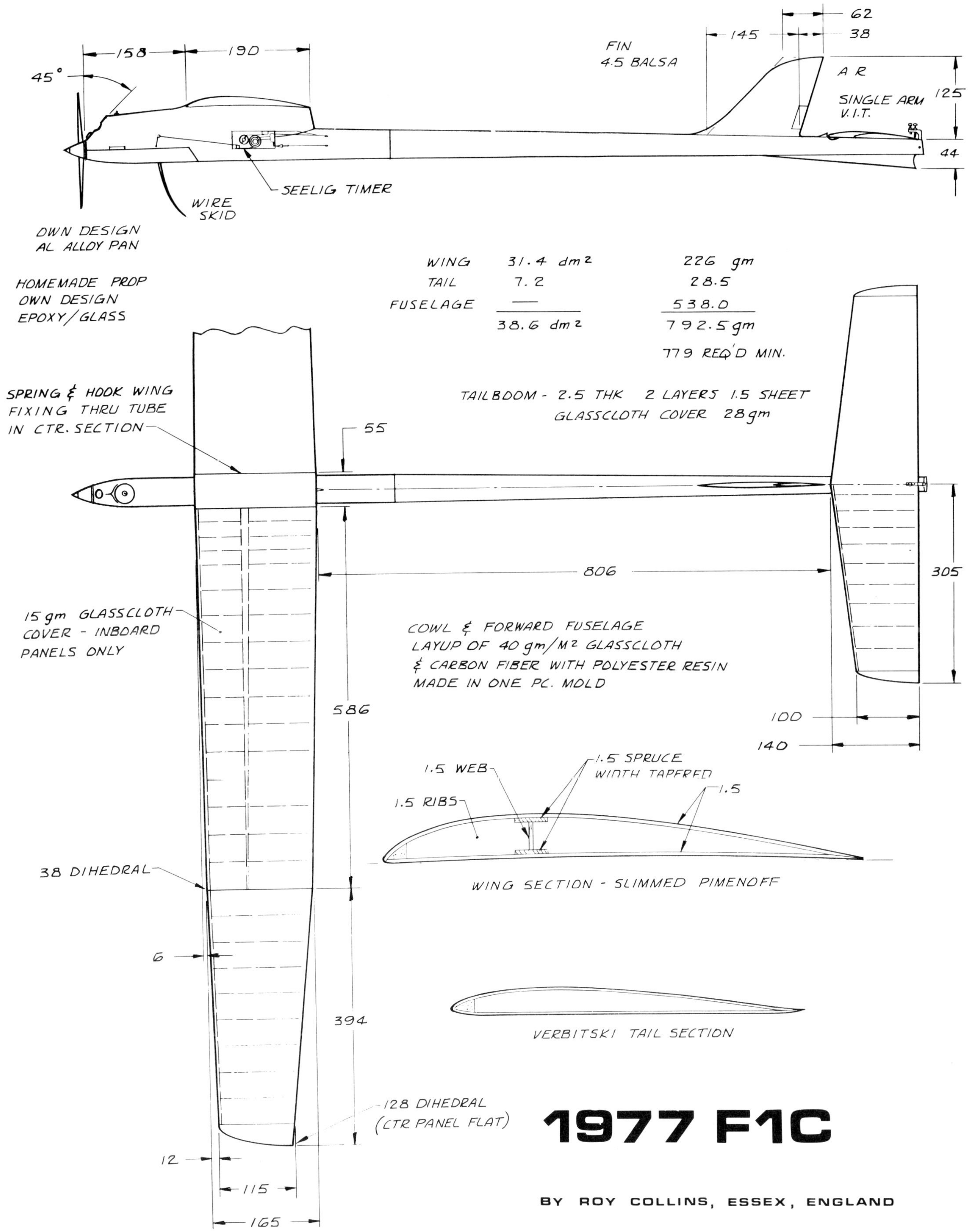

5 Collins GB WFFP

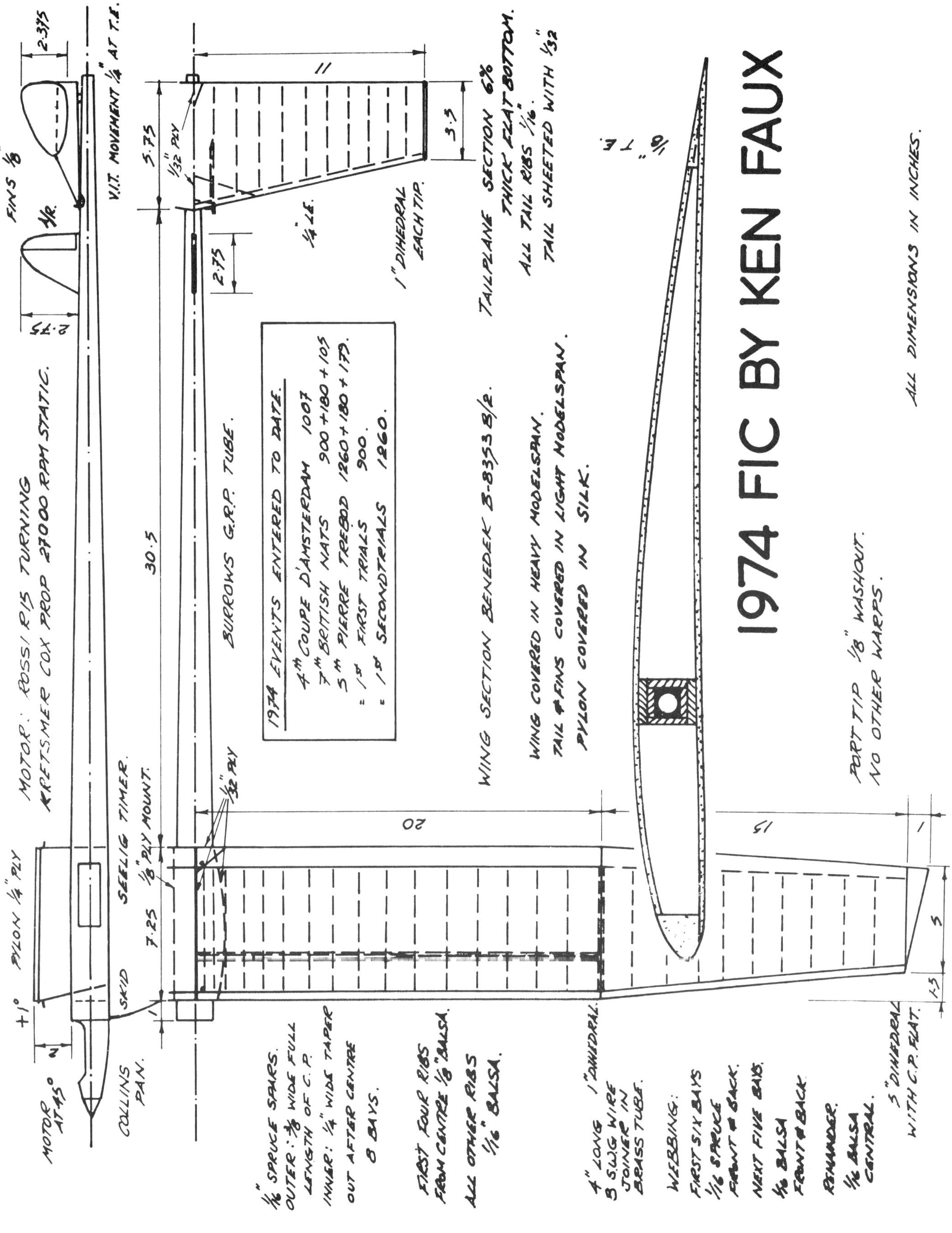
1974 FIC BY KEN FAUX
MOTOR: ROSSI R15 TURNING KRETSMER COX PROP 27000 RPM STATIC.
BURROWS G.R.P. TUBE.
1974 EVENTS ENTERED TO DATE.
4th COUPE D'AMSTERDAM 1007
7th BRITISH NATS 900+180+105
5th PIERRE TREBOD 1260+180+179.
1st FIRST TRIALS 900.
1st SECONDTRIALS 1260.
WING SECTION BENEDEK B-8353 B/2.
WING COVERED IN HEAVY MODELSPAN.
TAIL & FINS COVERED IN LIGHT MODELSPAN.
PYLON COVERED IN SILK.
TAILPLANE SECTION 6% THICK FLAT BOTTOM.
ALL TAIL RIBS 1/16".
TAIL SHEETED WITH 1/32
1" DIHEDRAL EACH TIP.
1/4" L.E.
1/8" T.E.
1/32" PLY
V.I.T. MOVEMENT 1/4" AT T.E.
FINS 1/8"
D/R.
PYLON 1/4" PLY
SEELIG TIMER.
1/8" PLY MOUNT.
SKID
COLLINS PAN.
MOTOR AT 45°
PORT TIP 1/8" WASHOUT. NO OTHER WARPS.
ALL DIMENSIONS IN INCHES.
1/16" SPRUCE SPARS. OUTER: 3/8 WIDE FULL LENGTH OF C.P. INNER: 1/4" WIDE TAPER OUT AFTER CENTRE 8 BAYS.
FIRST FOUR RIBS FROM CENTRE 1/8" BALSA.
ALL OTHER RIBS 1/16" BALSA.
1" DIHEDRAL.
4" LONG 8 S.W.G. WIRE JOINER IN BRASS TUBE.
WEBBING: FIRST SIX BAYS 1/16 SPRUCE FRONT & BACK. NEXT FIVE BAYS 1/16 BALSA FRONT & BACK. REMAINDER 1/16 BALSA CENTRAL.
5" DIHEDRAL WITH C.P. FLAT.
30.5
20
15
11
7.25
5.75
3.3
2.75
2.375
5
1.5
+1°
2
1

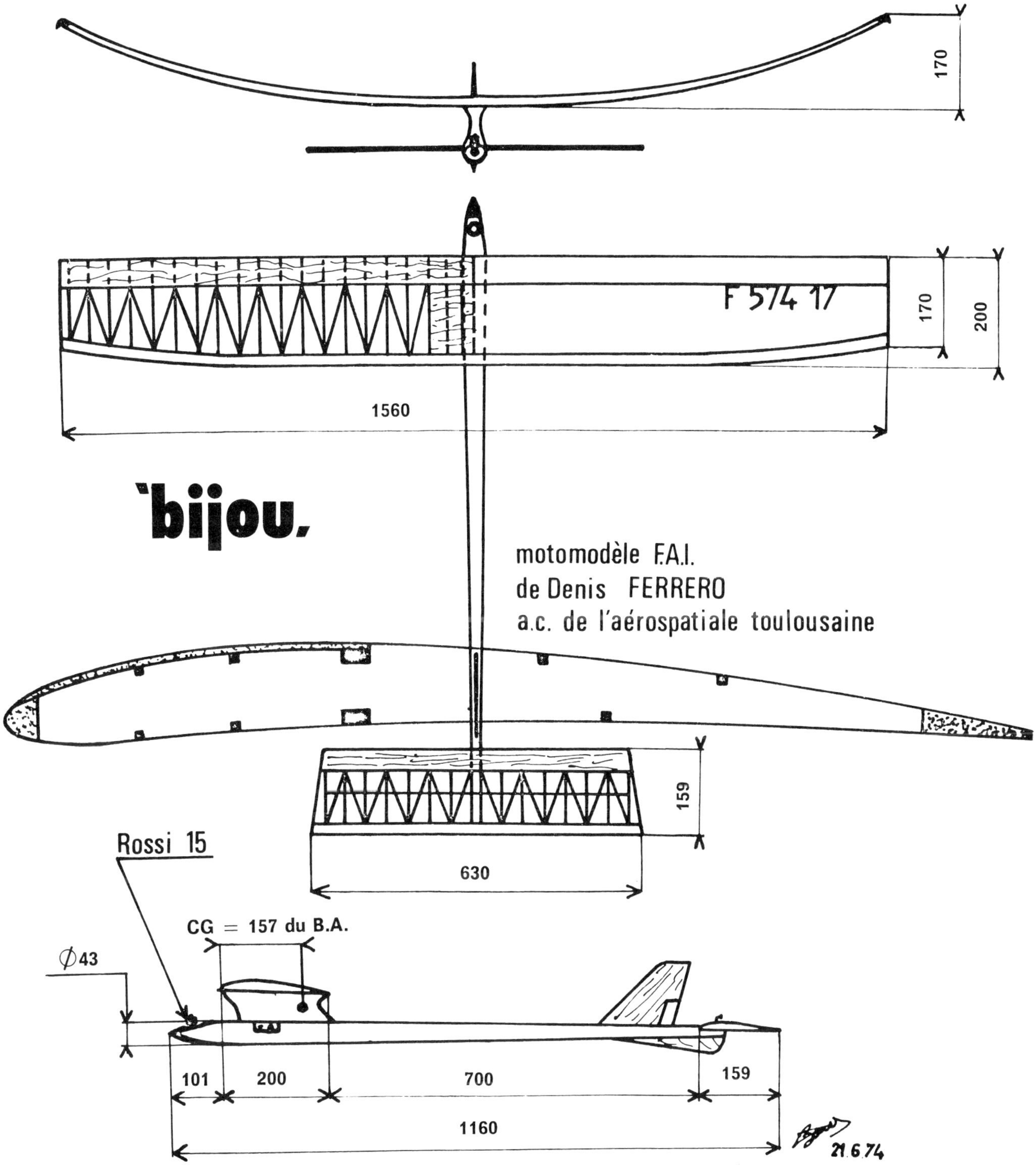

170
F 574 17
170
200
1560
'bijou.
motomodèle F.A.I.
de Denis FERRERO
a.c. de l'aérospatiale toulousaine
159
630
Rossi 15
CG = 157 du B.A.
⌀43
101
200
700
159
1160
21.6.74

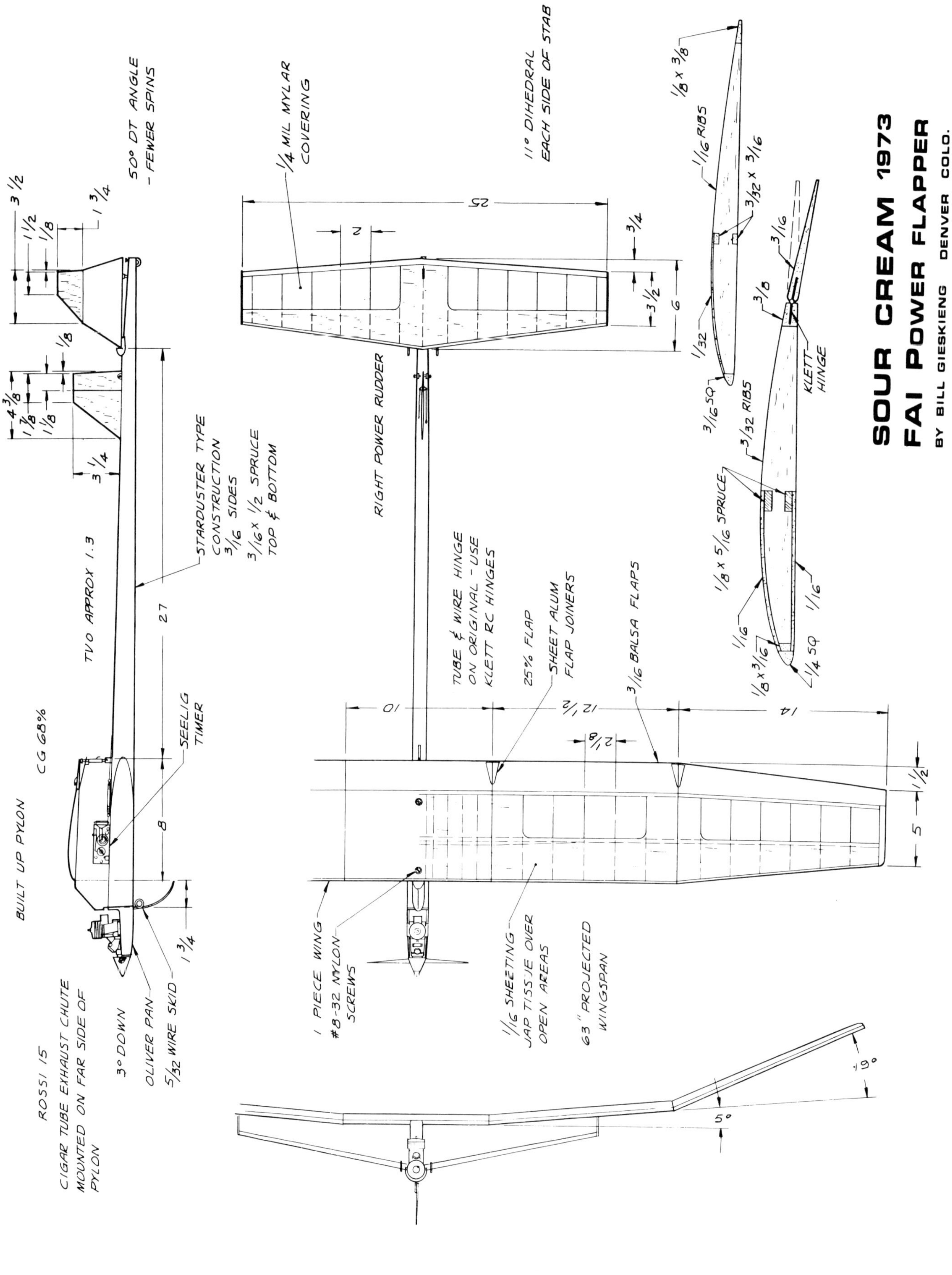
SOUR CREAM 1973
FAI POWER FLAPPER
BY BILL GIESKIENG DENVER COLO.
50° DT ANGLE - FEWER SPINS
1/4 MIL MYLAR COVERING
11° DIHEDRAL EACH SIDE OF STAB
BUILT UP PYLON
CG 68%
TVO APPROX 1.3
STARDUSTER TYPE CONSTRUCTION 3/16 SIDES 3/16 x 1/2 SPRUCE TOP & BOTTOM
SEELIG TIMER
RIGHT POWER RUDDER
ROSSI 15
CIGAR TUBE EXHAUST CHUTE MOUNTED ON FAR SIDE OF PYLON
3° DOWN
OLIVER PAN
5/32 WIRE SKID
1 PIECE WING
#8-32 NYLON SCREWS
1/16 SHEETING - JAP TISSUE OVER OPEN AREAS
63" PROJECTED WINGSPAN
TUBE & WIRE HINGE ON ORIGINAL - USE KLETT RC HINGES
25% FLAP
SHEET ALUM FLAP JOINERS
3/16 BALSA FLAPS
KLETT HINGE
1/8 x 5/16 SPRUCE
3/32 RIBS
1/16 RIBS
1/4 SQ
3/16 SQ
5°
19°

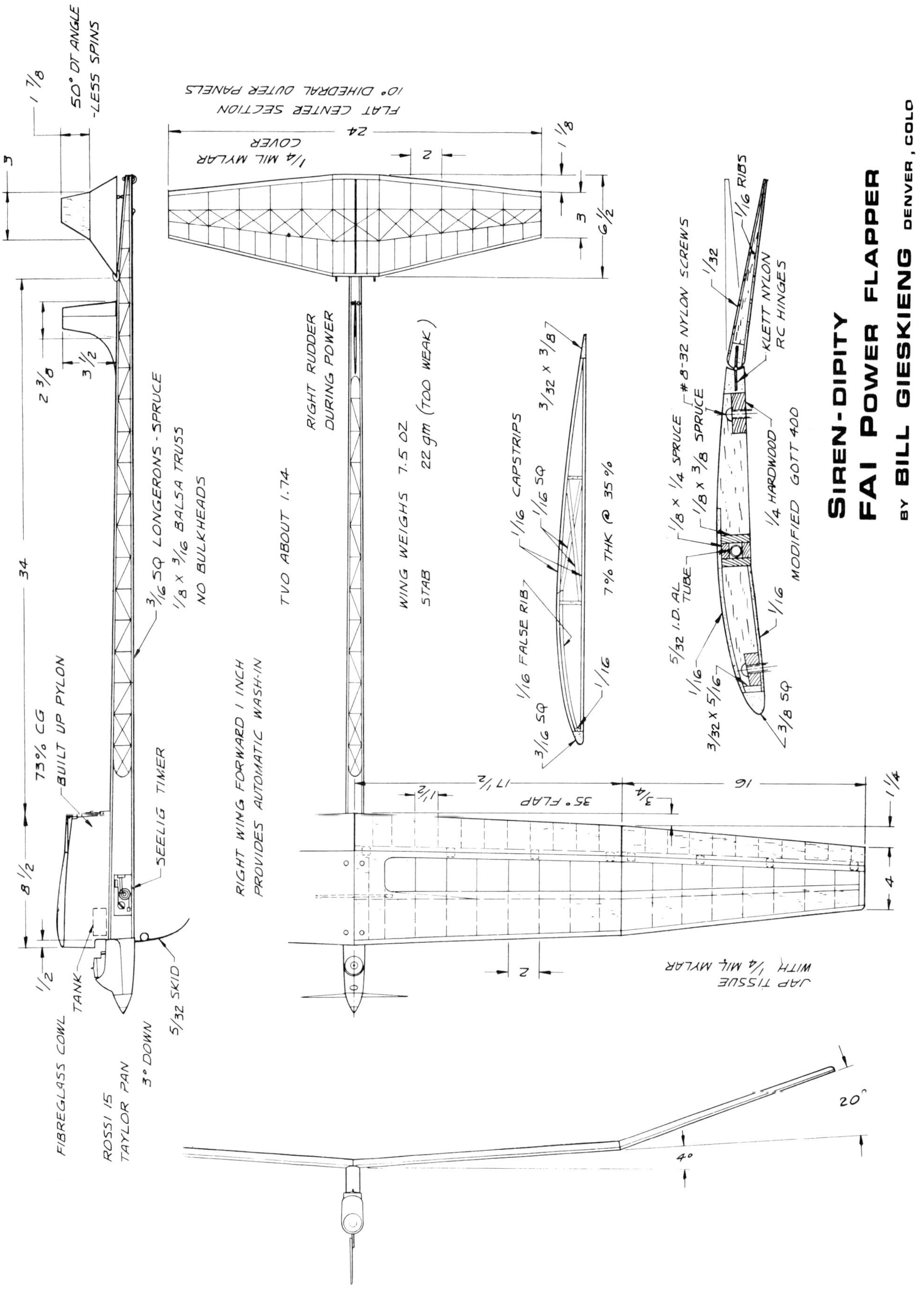

9 Gieskieng USA Scatter

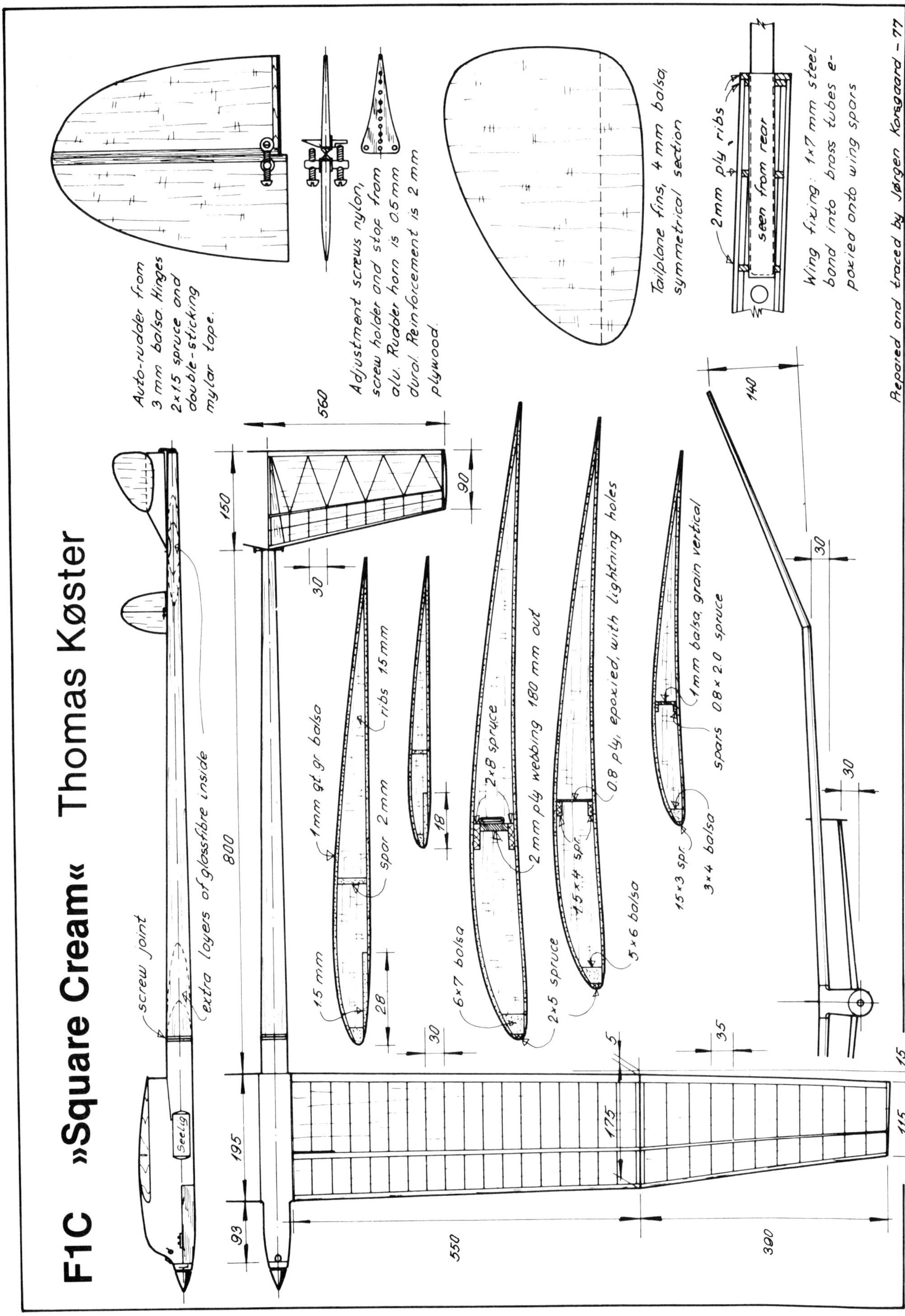
F1C »Square Cream« Thomas Køster
Auto-rudder from 3 mm balsa. Hinges 2×1.5 spruce and double-sticking mylar tape.
Adjustment screws nylon, screw holder and stop from alu. Rudder horn is 0.5 mm dural. Reinforcement is 2 mm plywood.
Tailplane fins, 4 mm balsa, symmetrical section
2mm ply ribs
seen from rear
Wing fixing: 1×7 mm steel bond into brass tubes epoxied onto wing spars
Prepared and traced by Jørgen Korsgaard - 77
screw joint
extra layers of glassfibre inside
Seelig
1 mm gt gr balsa
ribs 1.5 mm
spar 2 mm
1.5 mm
6×7 balsa
2×5 spruce
2×8 spruce
2 mm ply webbing 180 mm out
0.8 ply, epoxied, with lightning holes
1.5×4 spr.
5×6 balsa
1 mm balsa, grain vertical
spars 0.8×2.0 spruce
1.5×3 spr.
3×4 balsa
560
150
90
800
30
18
28
140
35
5
175
195
93
15
115
550
390

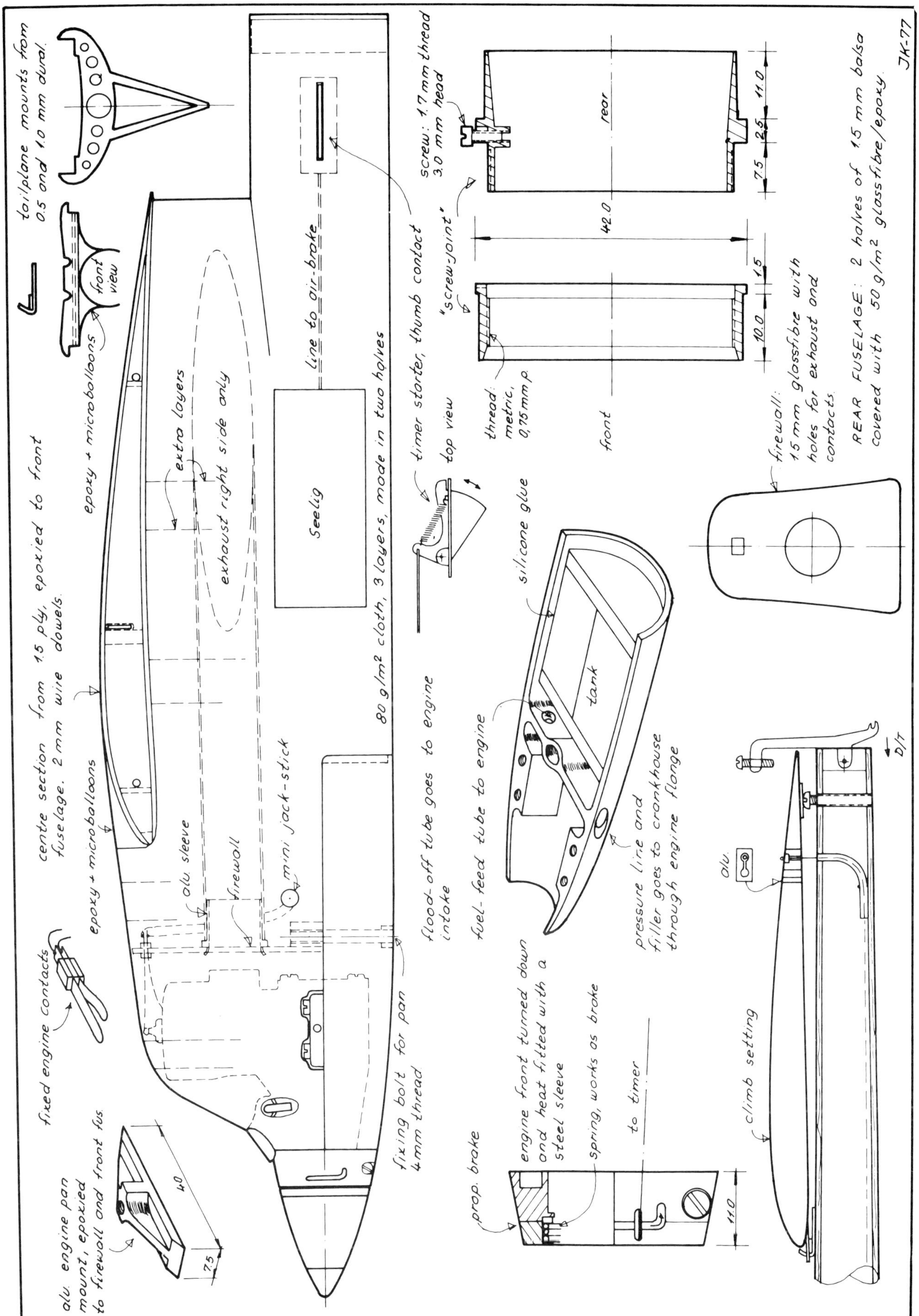

Koster *Square Cream* details

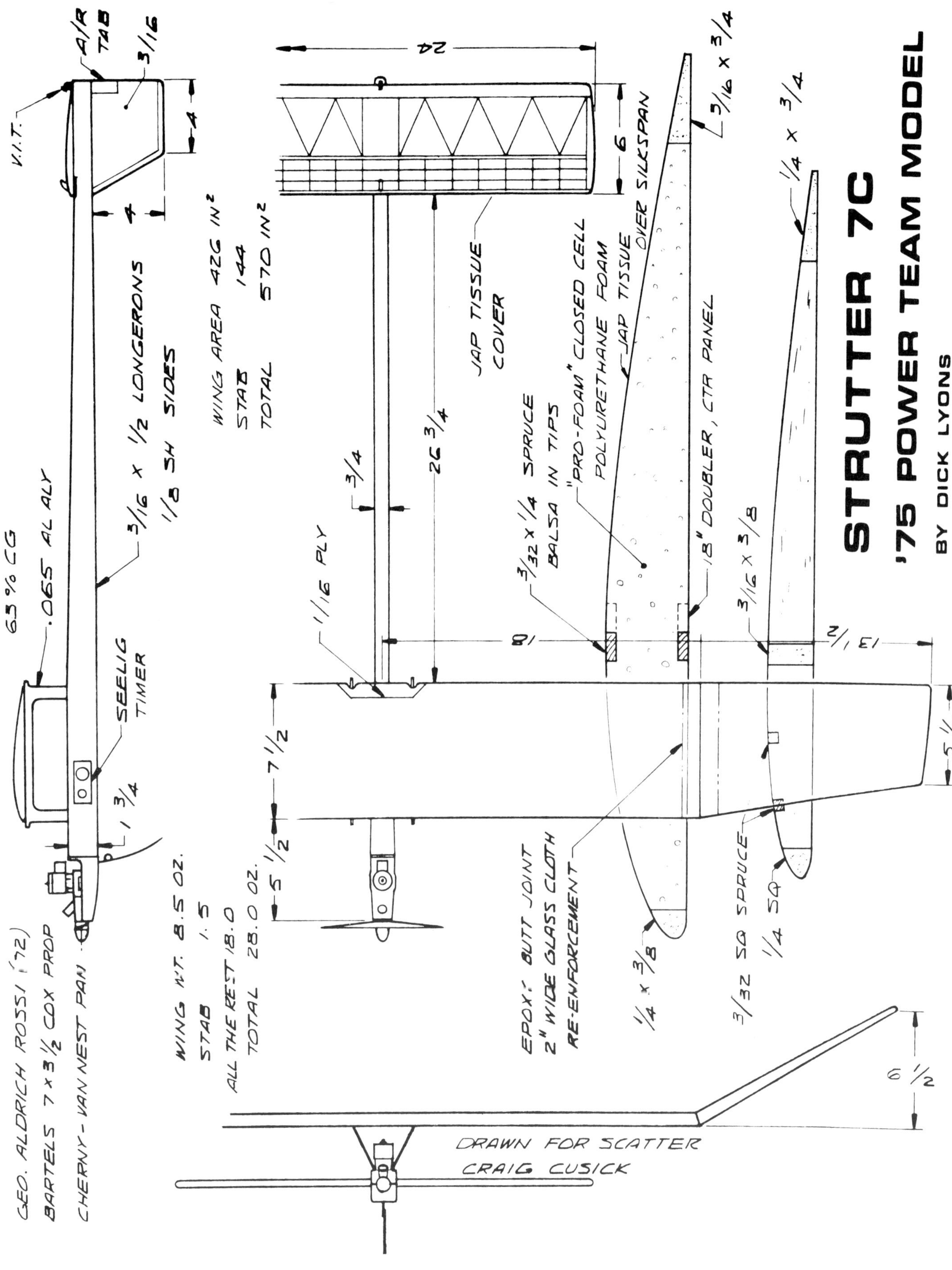
STRUTTER 7C
'75 POWER TEAM MODEL
BY DICK LYONS
A/R TAB
3/16
V.I.T.
4
4
63% CG
.065 AL ALY
3/16 × 1/2 LONGERONS
1/8 SH SIDES
SEELIG TIMER
1 3/4
GEO. ALDRICH ROSSI ('72)
BARTELS 7 × 3 1/2 COX PROP
CHERNY - VAN NEST PAN
WING WT. 8.5 OZ.
STAB 1.5
ALL THE REST 18.0
TOTAL 28.0 OZ.
WING AREA 426 IN²
STAB 144
TOTAL 570 IN²
24
6
JAP TISSUE COVER
26 3/4
3/4
1/16 PLY
18
7 1/2
5 1/2
5 1/2
3/16 × 3/4
OVER SILKSPAN
JAP TISSUE
"PRO-FOAM" CLOSED CELL POLYURETHANE FOAM
3/32 × 1/4 SPRUCE
BALSA IN TIPS
18" DOUBLER, CTR PANEL
1/4 × 3/4
3/16 × 3/8
13 1/2
EPOXY BUTT JOINT
2" WIDE GLASS CLOTH
RE-ENFORCEMENT
1/4 × 3/8
3/32 SQ SPRUCE
1/4 SQ
6 1/2
DRAWN FOR SCATTER
CRAIG CUSICK

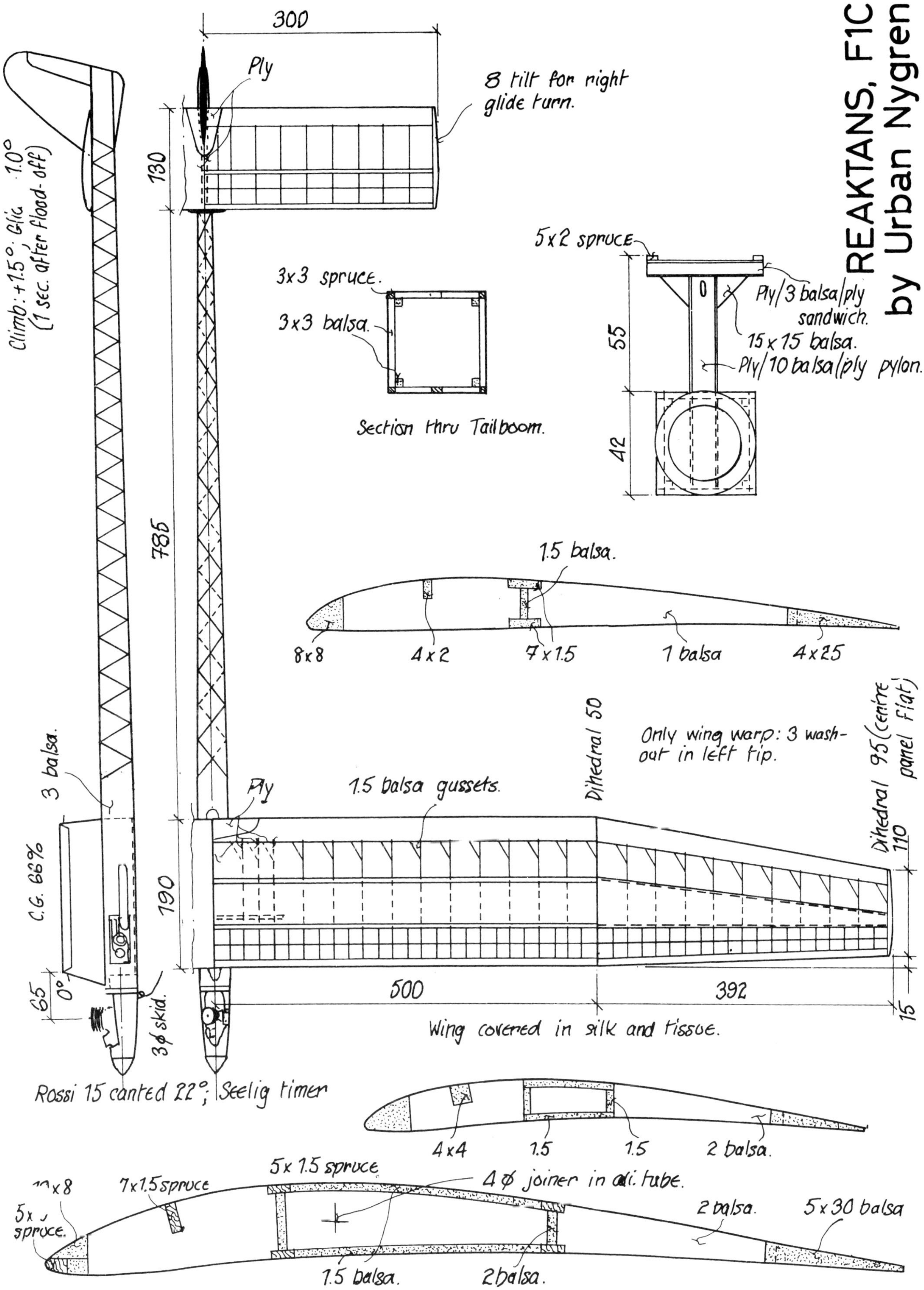

REAKTANS, F1C
by Urban Nygren
300
Ply
8 tilt for right glide turn.
130
Climb: +1.5°; Glide 1.0°
(1 sec. after flood-off)
5x2 spruce
3x3 spruce.
3x3 balsa.
Ply/3 balsa/ply sandwich.
15 x 15 balsa.
Ply/10 balsa/ply pylon.
55
42
Section thru Tailboom.
785
1.5 balsa.
8x8
4x2
7x1.5
1 balsa
4x25
Dihedral 50
Only wing warp: 3 wash-out in left tip.
Dihedral 95 (centre panel flat)
3 balsa.
Ply
1.5 balsa gussets.
110
C.G. 66%
190
65
0°
3ϕ skid.
500
392
15
Wing covered in silk and tissue.
Rossi 15 canted 22°; Seelig timer
4x4
1.5
1.5
2 balsa.
5x 1.5 spruce
4 ϕ joiner in ali. tube.
7x1.5 spruce
2 balsa.
5x30 balsa
1.5 balsa.
2 balsa.

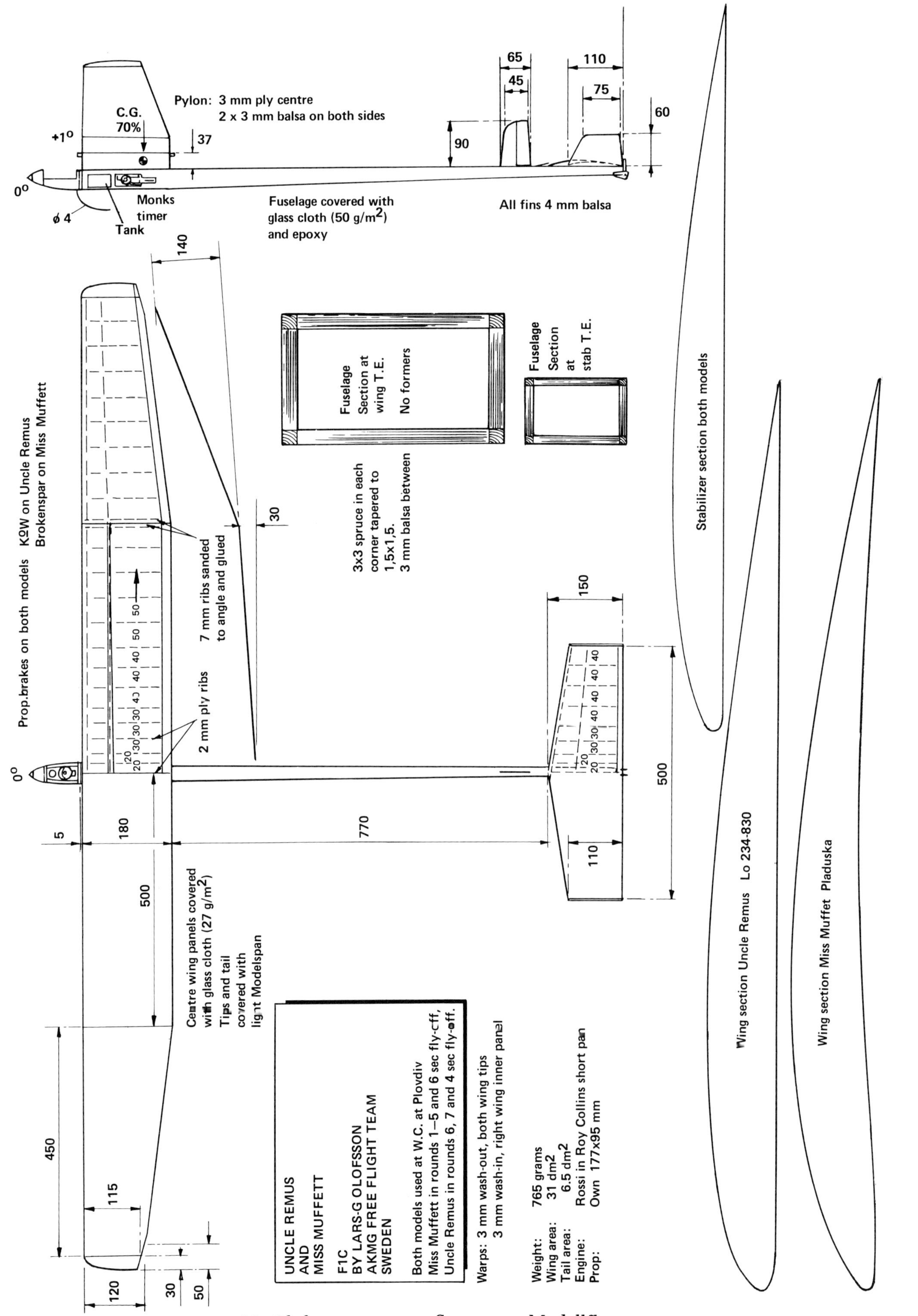

Pylon: 3 mm ply centre
2 x 3 mm balsa on both sides
C.G.
70%
+1°
37
0°
ϕ 4
Tank
Monks
timer
Fuselage covered with
glass cloth (50 g/m2)
and epoxy
90
65
45
110
75
60
All fins 4 mm balsa
140
Fuselage
Section at
wing T.E.
No formers
Fuselage
Section
at
stab T.E.
3x3 spruce in each
corner tapered to
1,5x1,5.
3 mm balsa between
Stabilizer section both models
Prop.brakes on both models
KΩW on Uncle Remus
Brokenspar on Miss Muffett
7 mm ribs sanded
to angle and glued
2 mm ply ribs
30
150
500
5
180
770
110
500
Centre wing panels covered
with glass cloth (27 g/m2)
Tips and tail
covered with
light Modelspan
450
115
120
30
50
Wing section Uncle Remus Lo 234-830
Wing section Miss Muffet Pladuska
UNCLE REMUS
AND
MISS MUFFETT
F1C
BY LARS-G OLOFSSON
AKMG FREE FLIGHT TEAM
SWEDEN
Both models used at W.C. at Plovdiv
Miss Muffett in rounds 1–5 and 6 sec fly-off,
Uncle Remus in rounds 6, 7 and 4 sec fly-off.
Warps: 3 mm wash-out, both wing tips
3 mm wash-in, right wing inner panel
Weight: 765 grams
Wing area: 31 dm2
Tail area: 6.5 dm2
Engine: Rossi in Roy Collins short pan
Prop: Own 177x95 mm

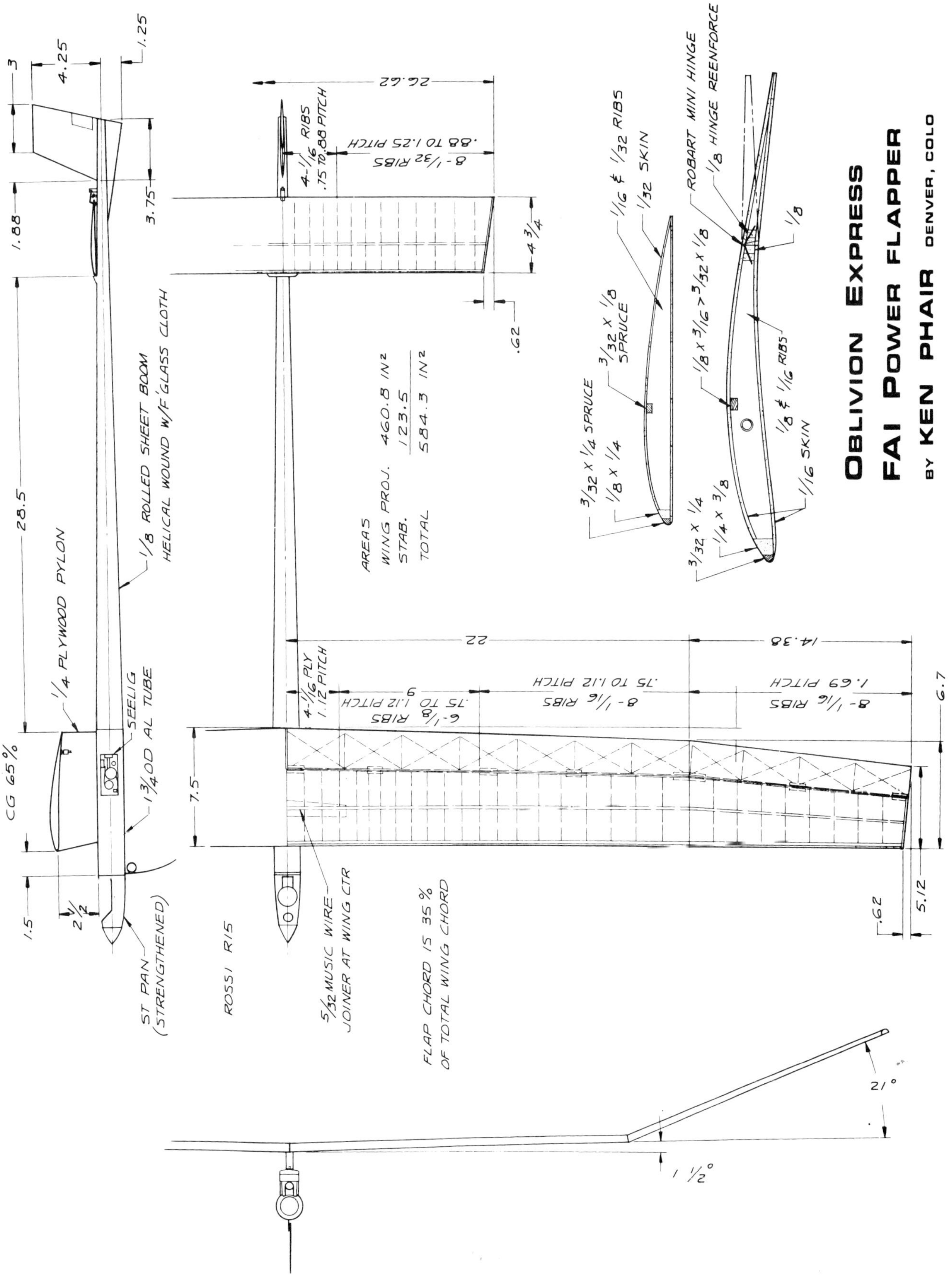
OBLIVION EXPRESS
FAI POWER FLAPPER
BY KEN PHAIR DENVER, COLO
1/4 PLYWOOD PYLON
1/8 ROLLED SHEET BOOM
HELICAL WOUND W/F GLASS CLOTH
SEELIG
13/4 OD AL TUBE
CG 65%
ST PAN (STRENGTHENED)
ROSSI R15
5/32 MUSIC WIRE JOINER AT WING CTR
FLAP CHORD IS 35 % OF TOTAL WING CHORD
AREAS
WING PROJ. 460.8 IN²
STAB. 123.5
TOTAL 584.3 IN²
ROBART MINI HINGE
1/8 HINGE REENFORCE
1/16 SKIN
1/32 SKIN
21°
1 1/2°

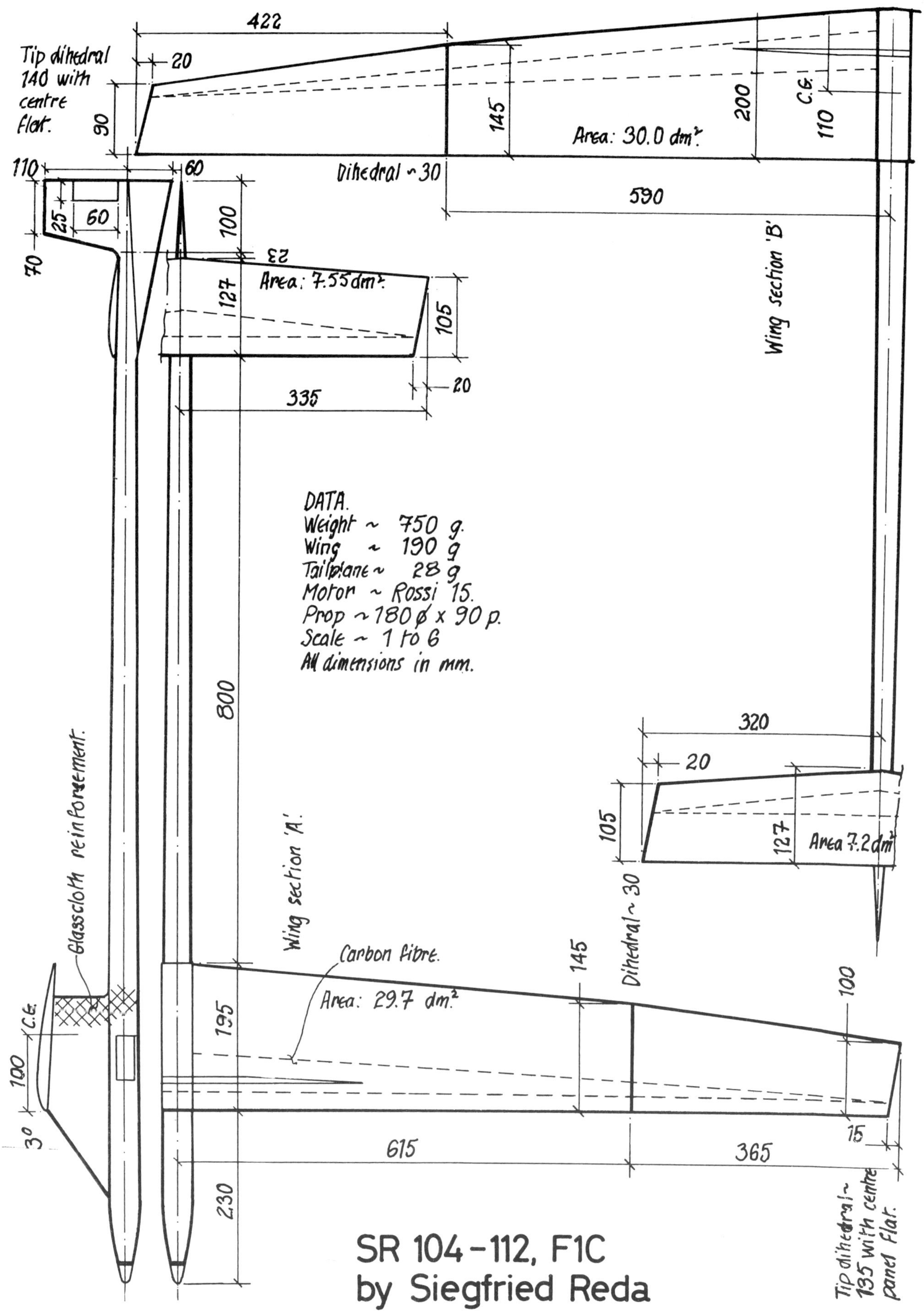

SR 104-112, F1C
by Siegfried Reda

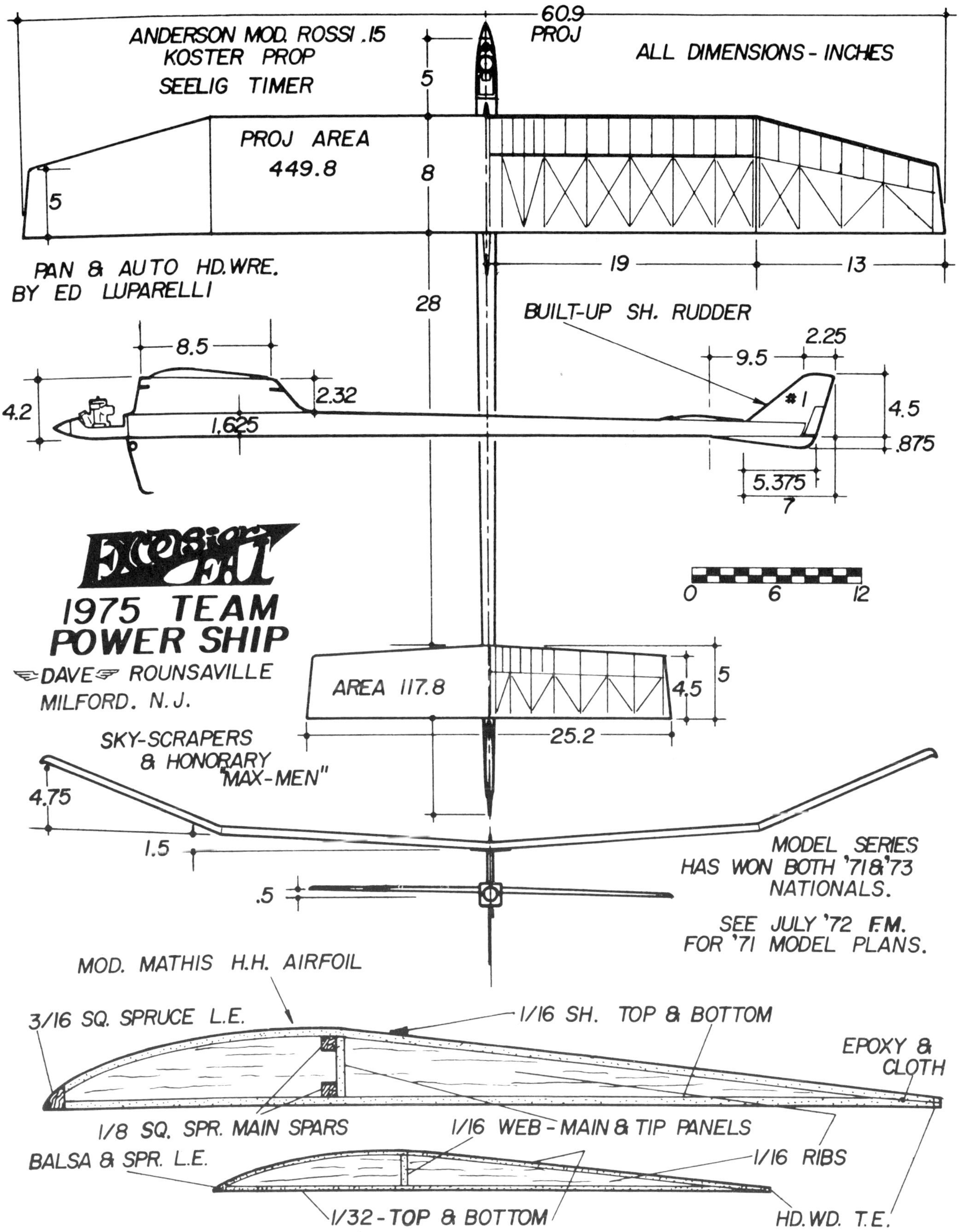
60.9
PROJ
ANDERSON MOD. ROSSI .15
KOSTER PROP
SEELIG TIMER
ALL DIMENSIONS - INCHES
PROJ AREA
449.8
5
8
5
19
13
PAN & AUTO HD.WRE.
BY ED LUPARELLI
28
BUILT-UP SH. RUDDER
8.5
2.25
9.5
4.2
2.32
1.625
#1
4.5
.875
5.375
7
0
6
12
1975 TEAM
POWER SHIP
DAVE ROUNSAVILLE
MILFORD. N.J.
AREA 117.8
4.5
5
25.2
SKY-SCRAPERS
& HONORARY
"MAX-MEN"
4.75
1.5
.5
MODEL SERIES
HAS WON BOTH '71&'73
NATIONALS.
SEE JULY '72 F.M.
FOR '71 MODEL PLANS.
MOD. MATHIS H.H. AIRFOIL
3/16 SQ. SPRUCE L.E.
1/16 SH. TOP & BOTTOM
EPOXY &
CLOTH
1/8 SQ. SPR. MAIN SPARS
1/16 WEB - MAIN & TIP PANELS
RIBS
BALSA & SPR. L.E.
1/16 RIBS
1/32 - TOP & BOTTOM
HD.WD. T.E.

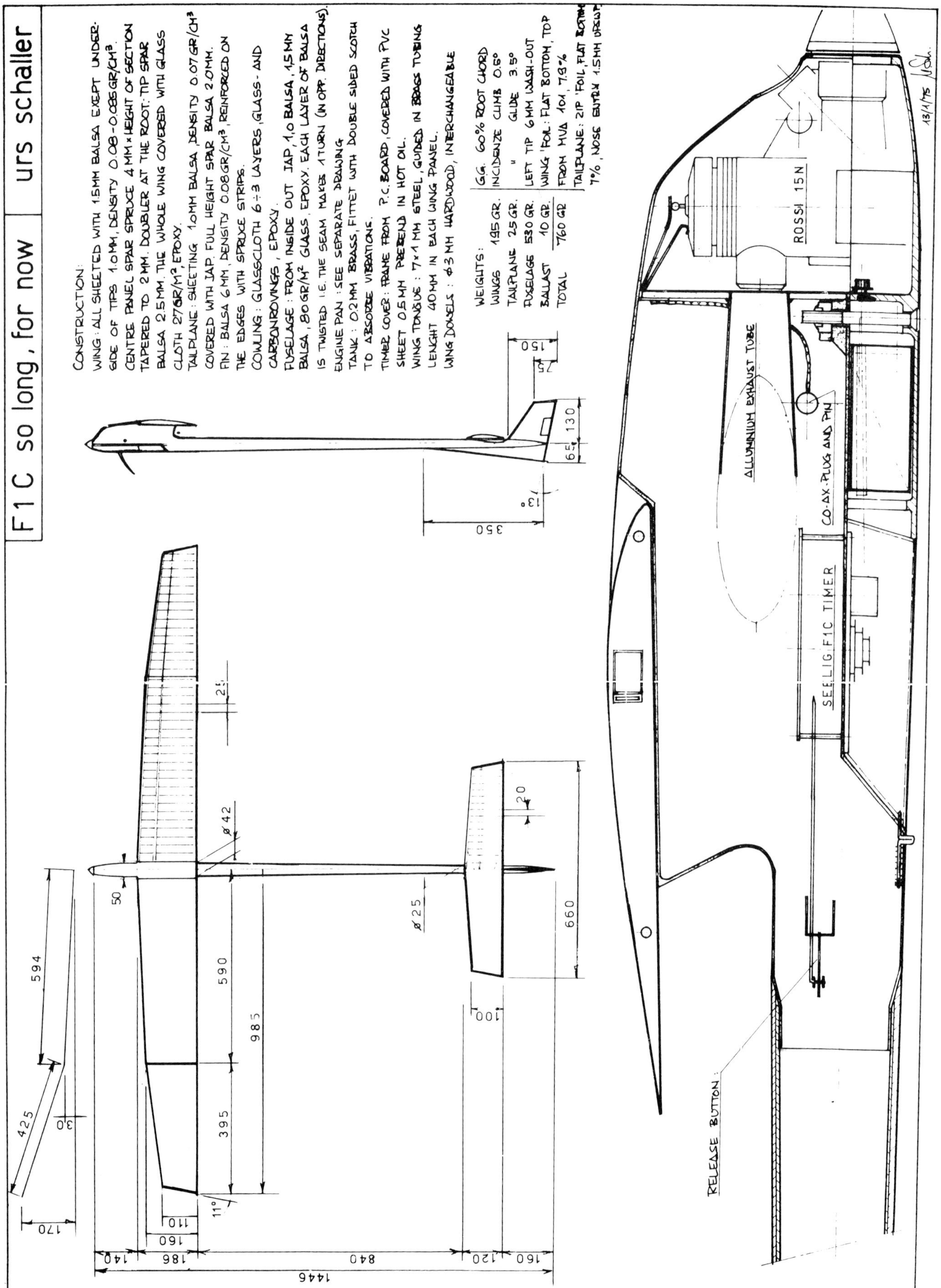
F1C so long, for now
urs schaller
CONSTRUCTION:
WING: ALL SHEETED WITH 1.5MM BALSA EXEPT UNDERSIDE OF TIPS 1.0MM, DENSITY 0.08-0.085GR/CM³.
CENTRE PANEL SPAR SPRUCE 4MM × HEIGHT OF SECTION TAPERED TO 2MM. DOUBLER AT THE ROOT. TIP SPAR BALSA 2.5MM. THE WHOLE WING COVERED WITH GLASS CLOTH 27GR/M², EPOXY.
TAILPLANE: SHEETING 1.0MM BALSA, DENSITY 0.07GR/CM³ COVERED WITH JAP. FULL HEIGHT SPAR BALSA 2.0MM.
FIN: BALSA 6MM, DENSITY 0.08GR/CM³, REINFORCED ON THE EDGES WITH SPRUCE STRIPS.
COWLING: GLASSCLOTH 6÷3 LAYERS, GLASS- AND CARBONROVINGS, EPOXY.
FUSELAGE: FROM INSIDE OUT JAP, 1,0 BALSA, 1,5MM BALSA, 80GR/M² GLASS, EPOXY. EACH LAYER OF BALSA IS TWISTED I.E. THE SEAM MAKES 1TURN (IN OPP. DIRECTIONS).
ENGINE PAN: SEE SEPARATE DRAWING
TANK: 0.2MM BRASS, FITTET WITH DOUBLE SIDED SCOTCH TO ABSORBE VIBRATIONS.
TIMER COVER: FRAME FROM P.C. BOARD, COVERED WITH PVC SHEET 0.5MM PREBEND IN HOT OIL.
WING TONGUE: 7×1 MM STEEL, GUIDED IN BRASS TUBING LENGHT 40MM IN EACH WING PANEL.
WING DOWELS: ø3MM HARDWOOD, INTERCHANGEABLE
WEIGHTS:
WINGS 195GR.
TAILPLANE 25GR.
FUSELAGE 530GR.
BALLAST 10GR.
TOTAL 760GR
G.G. 60% ROOT CHORD
INCIDENZE CLIMB 0.5°
" GLIDE 3.5°
LEFT TIP 6MM WASH-OUT
WING 'FOIL: FLAT BOTTOM, TOP FROM MVA 101, 7.9%
TAILPLANE: 21P 'FOIL, FLAT BOTTOM 7%, NOSE ENTRY 1.5MM DRAWN
425
594
170
30
140
186
160
110
50
11°
395
590
985
25
ø42
1446
840
ø25
120
100
160
20
660
350
13°
75
150
65
130
RELEASE BUTTON
ALLUMINIUM EXHAUST TUBE
ROSSI 15N
SEELIG F1C TIMER
CO-AX-PLUG AND PIN
13/1/75

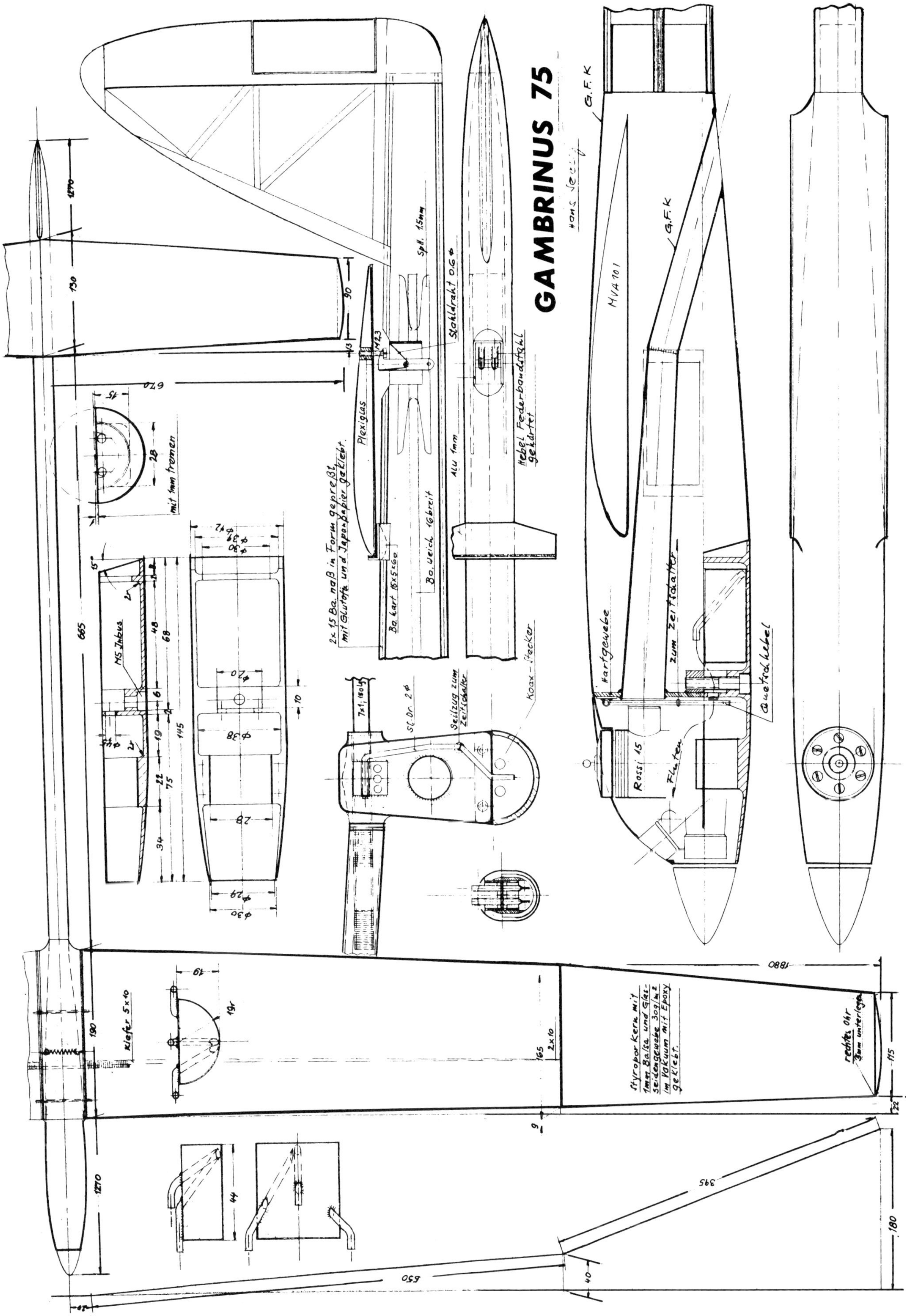

GAMBRINUS 75
G.F.K
MVA 101
Plexiglas
Rossi 15
Hartgewebe
Quertrudhebel
kiefer 5x10
MS Inbus
mit 1mm tronen

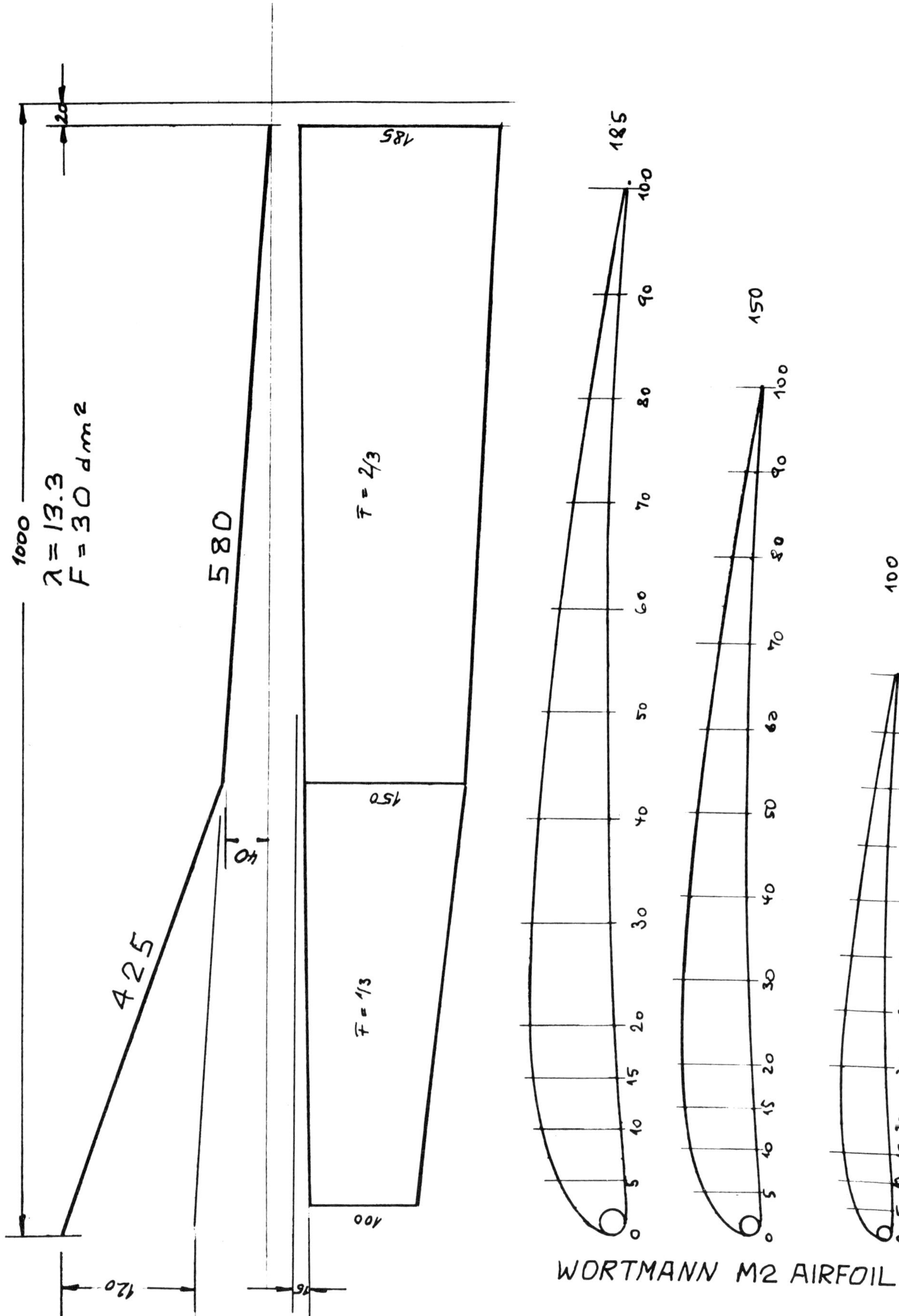

Seelig details

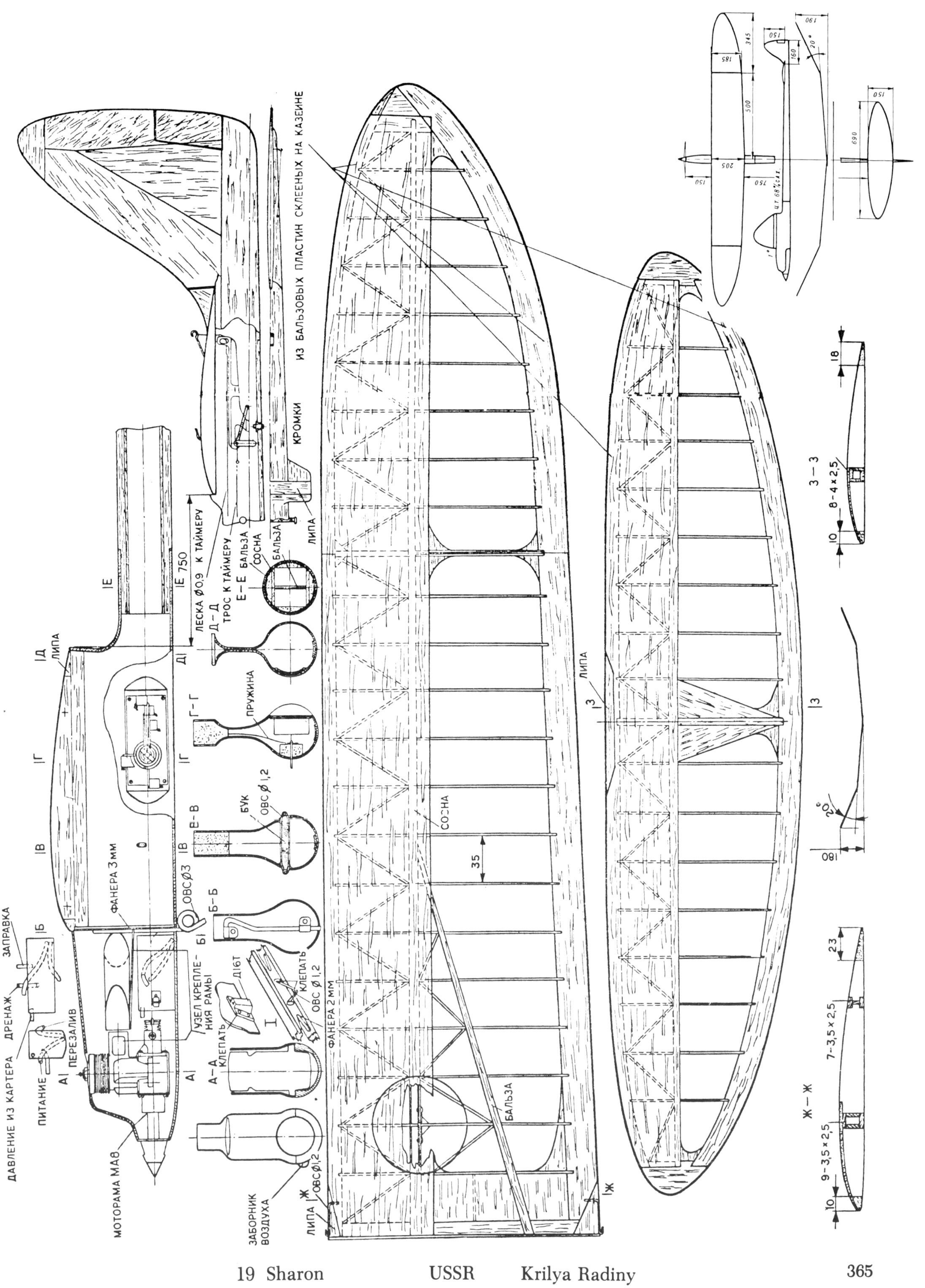

19 Sharon USSR Krilya Radiny

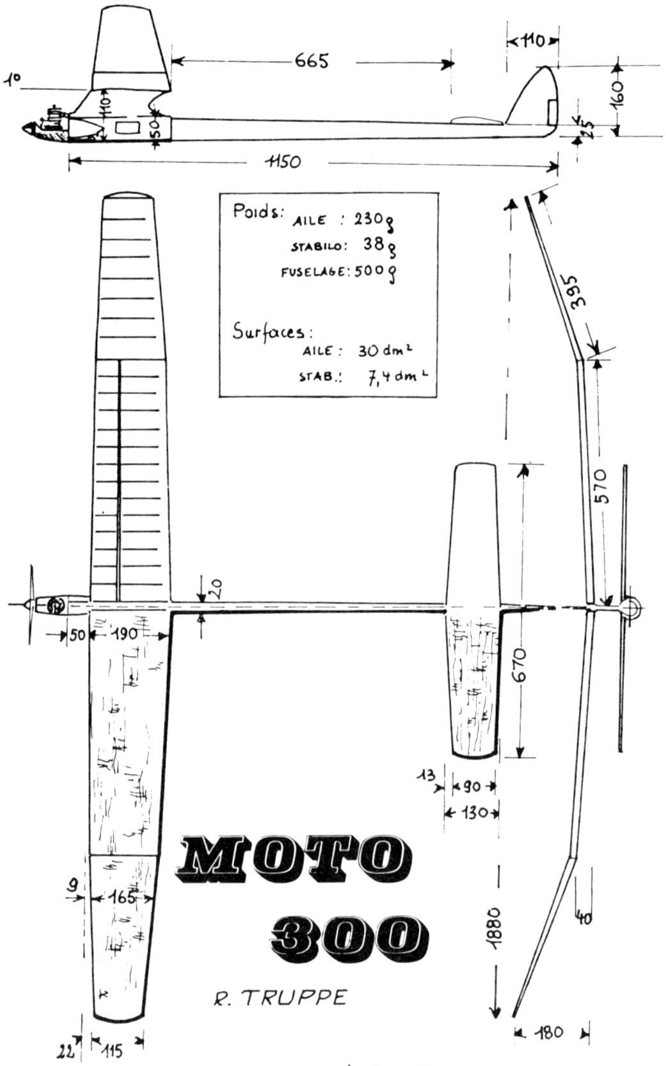

665
110
+1°
110
50
160
25
1150
Poids: AILE : 230 g
STABILO : 38 g
FUSELAGE : 500 g
Surfaces : AILE : 30 dm²
STAB. : 7,4 dm²
395
570
20
50
190
670
13
90
130
9
165
MOTO 300
R. TRUPPE
1880
40
22
115
180

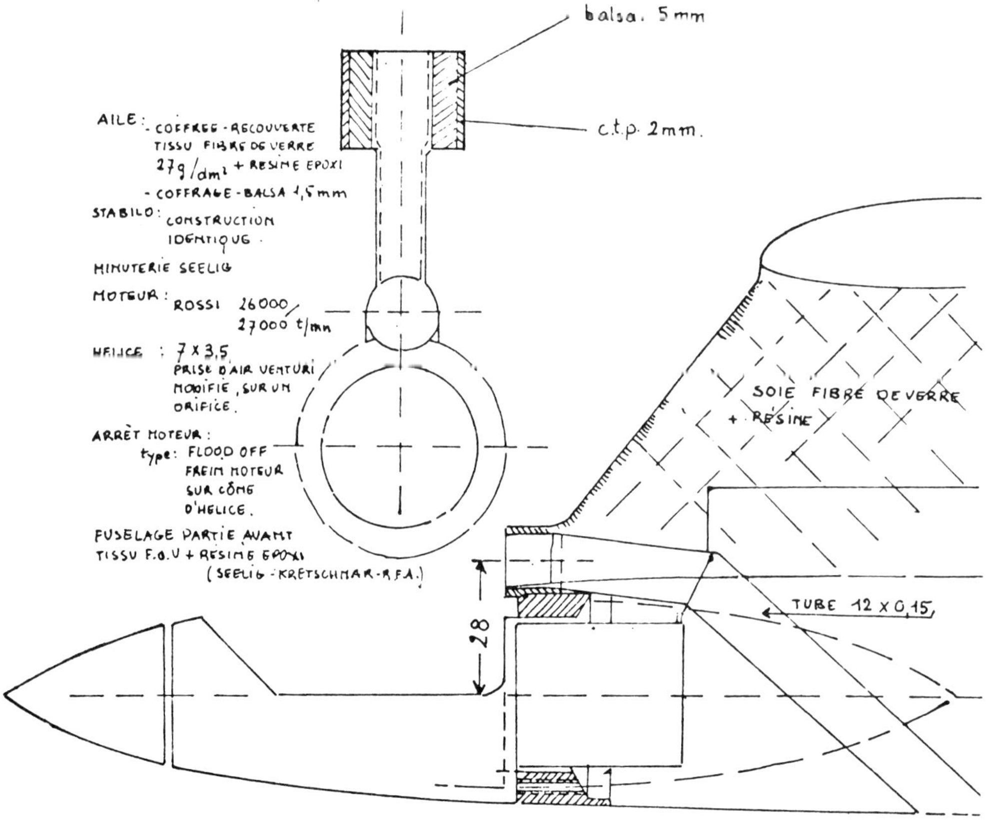

balsa. 5 mm
c.t.p. 2 mm.
AILE : - COFFRAGE - RECOUVERTE TISSU FIBRE DE VERRE 27 g/dm² + RESINE EPOXI
- COFFRAGE - BALSA 1,5 mm
STABILO : CONSTRUCTION IDENTIQUE.
MINUTERIE SEELIG
MOTEUR : ROSSI 26000/27000 t/mn
HELICE : 7 X 3,5
PRISE D'AIR VENTURI MODIFIE, SUR UN ORIFICE.
ARRÊT MOTEUR : type : FLOOD OFF FREIN MOTEUR SUR CÔNE D'HELICE.
FUSELAGE PARTIE AVANT TISSU F.d.V + RESINE EPOXI (SEELIG - KRETSCHMAR - R.F.A.)
SOIE FIBRE DE VERRE + RÉSINE
TUBE 12 X 0,15
28

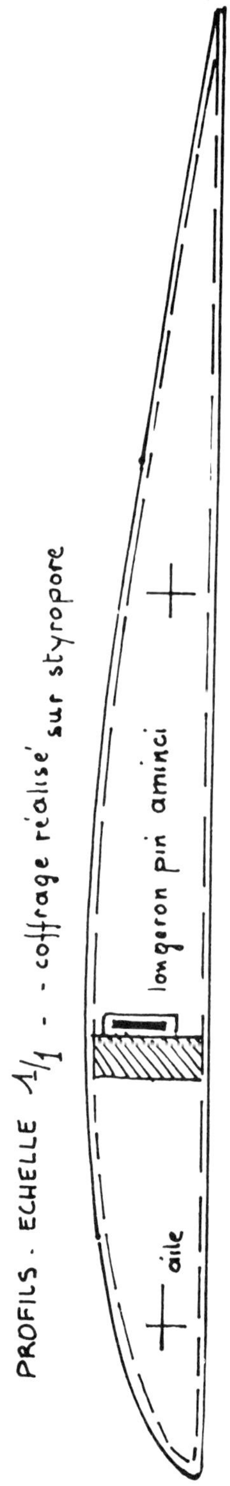

PROFILS . ECHELLE 1/1 - - coffrage réalisé sur styropore
longeron pin aminci
aile

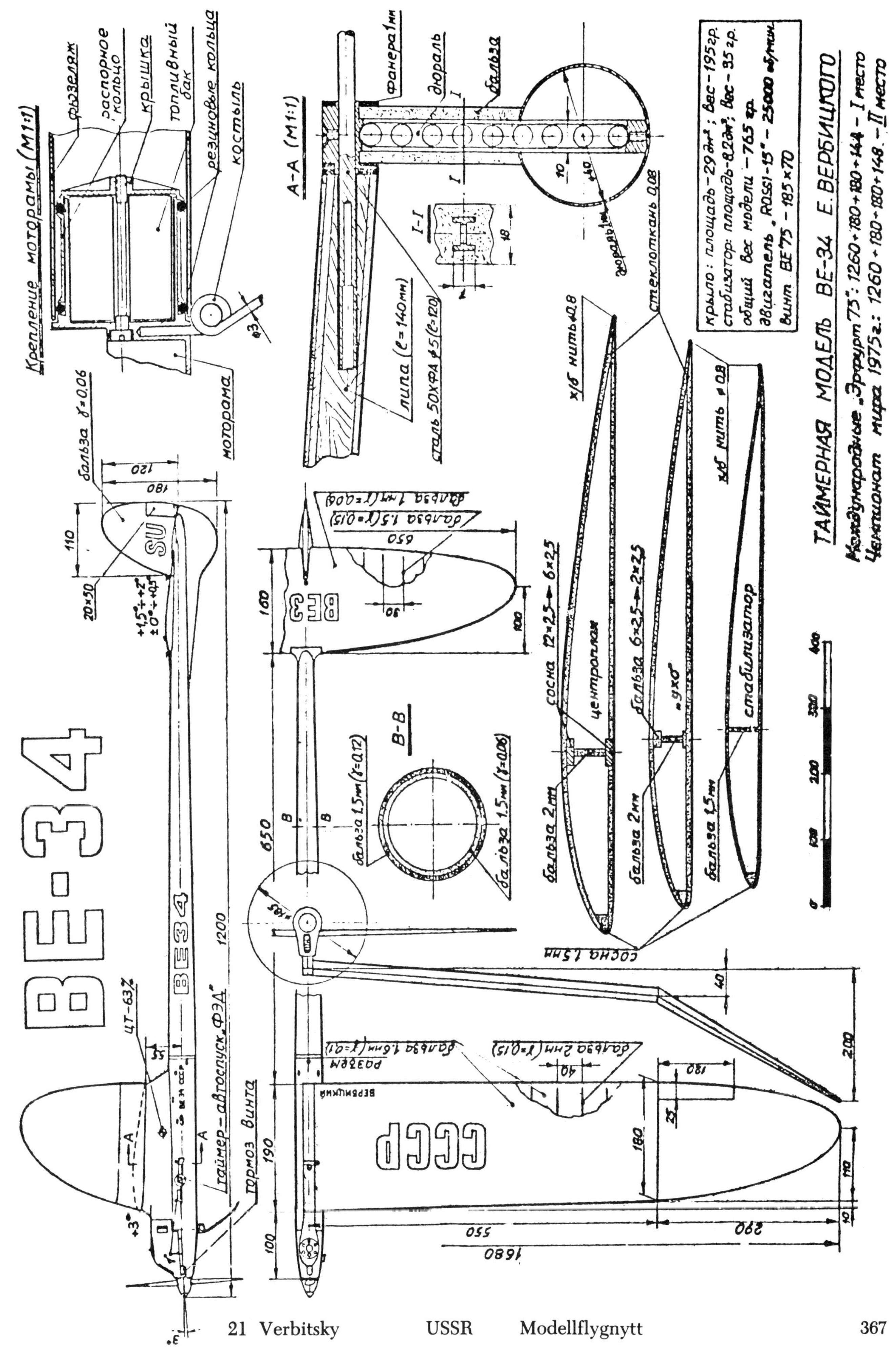
ВЕ-34
ТАЙМЕРНАЯ МОДЕЛЬ ВЕ-34 Е. ВЕРБИЦКОГО
СССР

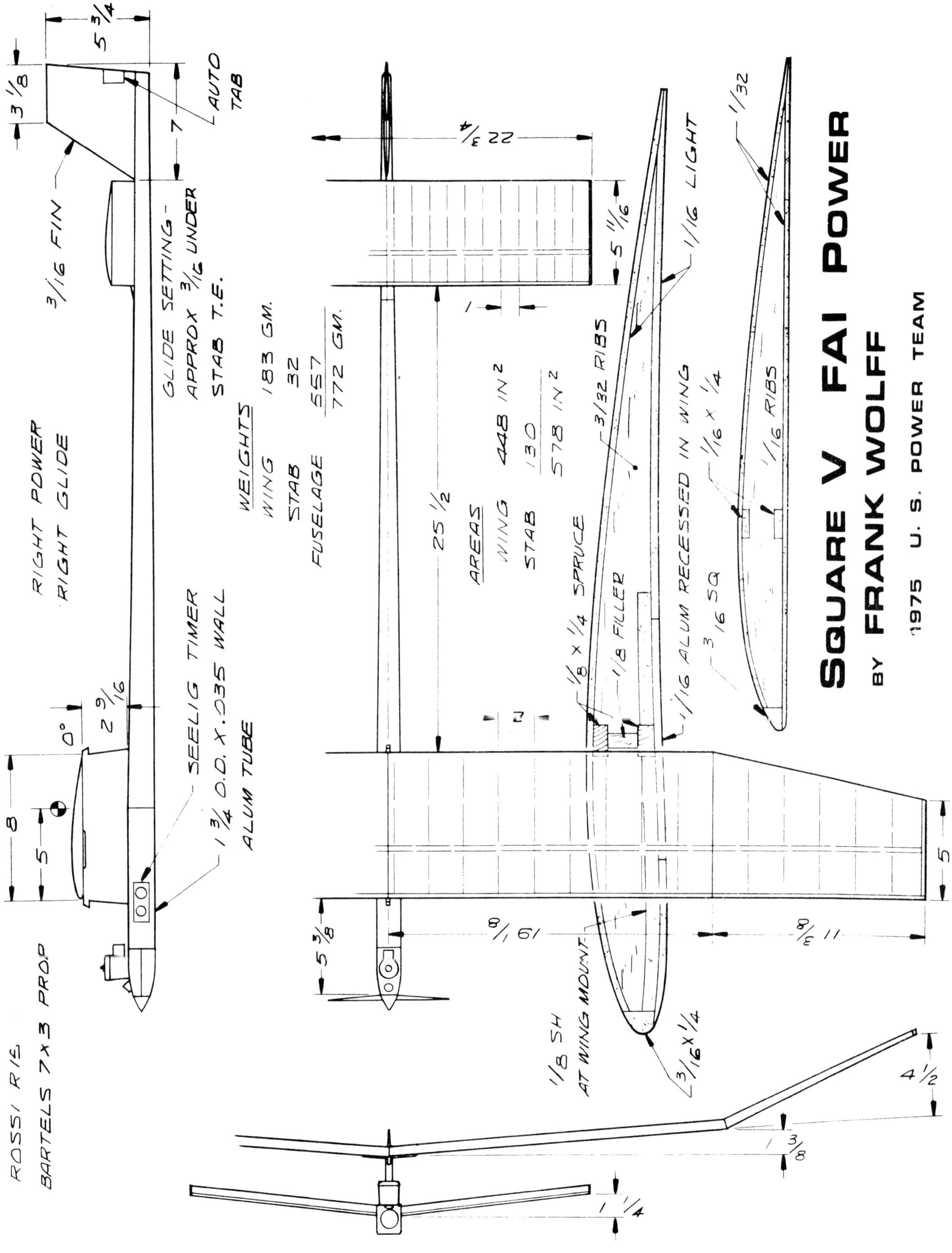
SQUARE V FAI POWER
BY FRANK WOLFF
1975 U. S. POWER TEAM
RIGHT POWER
RIGHT GLIDE
GLIDE SETTING-
APPROX 3/16 UNDER
STAB T.E.
WEIGHTS
WING 183 GM.
STAB 32
FUSELAGE 557
772 GM.
AREAS
WING 448 IN²
STAB 130
578 IN²
3/16 FIN
AUTO TAB
SEELIG TIMER
1 3/4 O.D. X .035 WALL
ALUM TUBE
ROSSI R15
BARTELS 7x3 PROP
3/32 RIBS
1/16 LIGHT
1/32
1/16 x 1/4
1/16 RIBS
1/16 ALUM RECESSED IN WING
3/16 SQ
1/8 x 1/4 SPRUCE
1/8 FILLER
1/8 SH
AT WING MOUNT
3/16 x 1/4

F1 D

F1D 1

Time Machine – Pete Andrews, Bogota, New Jersey

There isn't much doubt that Pete Andrews is a Grand Master of Indoor Free Flight. Pete was the first to do 30 minutes; first to do 40 minutes; and was World Champion in 1971. That winning model is shown in the drawing (thanks to Aeromodeller, Jan. '73).

Pete has these comments on the model. "This model design also holds the USA National FAI stick record of 41:43, set at Lakehurst Aug. 31, 1974. This exceeded my previous record of July 23, 1974, when I had three flights of 38:00, 38:22 & 39:12. The 41:43 flight was made in hanger No. 1 with a 22 in. dia., 32 in. pitch prop. A .059x19 in. loop of Pierelli was used with 2150 turns. Model weight was .0362 oz.

F1D 2

Rafter Banger - Laurie Barr, Bray, England

Laurie, as everyone knows is the prime mover behind the splendid indoor activity in England, concentrated with the use of the Cardington Air Sheds. And, of course, he consistently turns in topnotch performances with his own indoor models. A modeller for 30 years he was on the British Wakefield Team in 1969 and 1971, Indoor Team in 1972 and 1974 and has held numerous offices within the SMAE. He is the director and owner of A.G.M. Ltd., a company which makes scale industrial models.

The *Rafter Banger*, named for its natural proclivities, is reasonably conventional in design and well-built, of course. Laurie's best time at the '76 W/C was 36:54 official and 39:39, unfortunately unofficial. Near the end of the competition Laurie's last ditch efforts to catch Romak ended in frustration as he encountered bad air and rotten luck. Collision of the model with a top girder resulted in the model aimlessly wandering around, sans fin, for a number of minutes before it could be ballooned down.

F1D 3

1975 F1D by John Blount, Harlington, England

"Thanks for the note and pictures. I also didn't have too much time at Cardington to look around, however it's always nice to see old friends at these things. My main regret is that the Rumanians didn't send a team.

Model flown at the World Champs was my usual '76 model except that the wing section was a deep double-ender. Prop was a Holties 18x28 negative flare with a hard spar. Fin was placed on top of the tailplane. The philosophy is to slow up the model by using a slower (deeper) section. Even though the prop is quite small, RPM was kept low. Rubber weight used was also kept low at 1.1 to 1.2 g. The model has room for more development, and probably hasn't been flown to full potential in its present form. I don't believe that model planform is at all important, the whole problem is to match wing, prop and rubber together, and incorporate strength only where absolutely necessary. The weight of the model at the W/C was very close to 1 g before the first flight but increased later with the damp conditions."

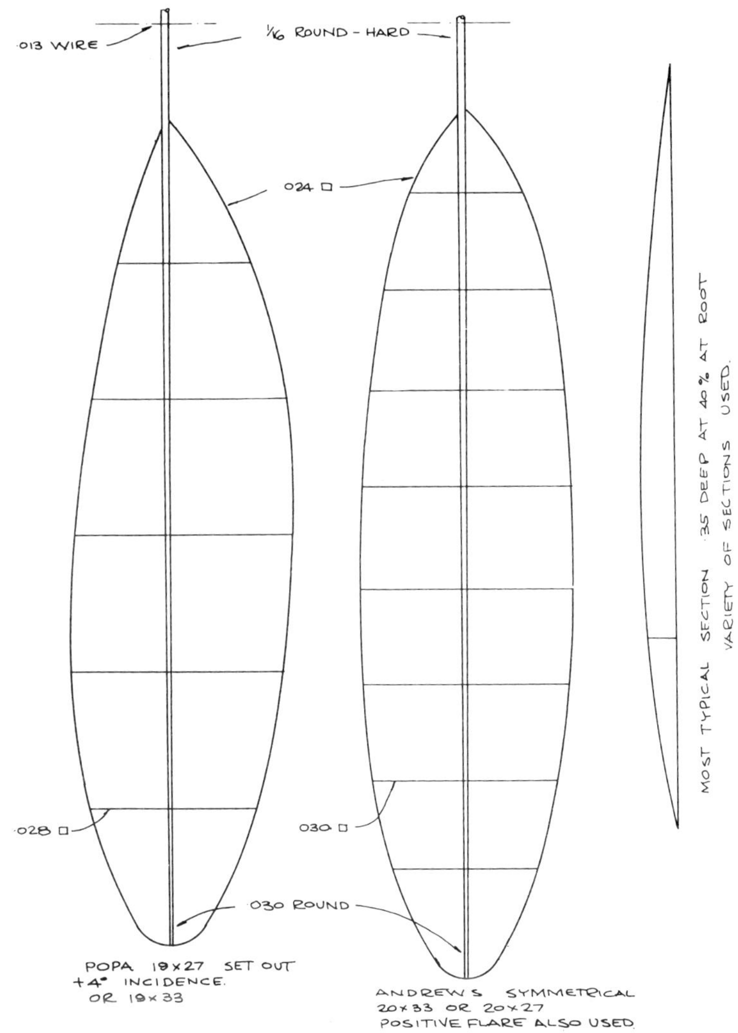

John Blount props.

John placed 9th in 1974 and 4th in 1976. He spends his working days as an airline captain.

F1D 4

1974 World Champion - Ryszard Czechowski, Krakow, Poland

The 1974 World Champs were held in the airship hangar at Lakehurst, New Jersey, famous site of much lighter-than-air activity in the 1920-1940 period. The Champion, Czechowski, has been an indoor flier for 10 of his 24 modelling years and has placed high in a number of national and international contests. He is a mechanical technician, having studied at the College of Krakow. His model is of conventional design but of excellent workmanship and, of more importance, expertly handled. In poor conditions his best time was 34:56, not the high time of the meet, but his matching flight of 34:53 gave him the highest total for the two.

F1D 5

World Record Holder - Dick Kowalski,
Warren, Michigan

Dick established a world's FAI absolute record for indoor, rubber-powered models (F1D-32) with the model shown in the drawing. That record is 50 min. 41 sec., flown in the Goodyear Aerospace Hangar in Akron, Ohio, on August 14, 1976. This is a larger model than permitted in F1D competition (limited to 90 cm span) but well within the area limit imposed on record attempts (1800 sq. in.) The exciting flight description is reprinted here, with thanks to *Indoor News and Views.*

Outdoor weather conditions during the earlier part of the day of this attempt were nearly ideal. Past experience has shown that days of moderate temperature, low wind velocity, partially obscured sky and low humidity can produce long indoor model flights in this hangar. The temperature this day was 80°F., the winds were 5 mph and the skies were partly cloudy. Although the humidity was generally low, a brief rain shower of 30 mins. duration fell on the hangar several hours before the attempt began.

Preflight preparations included a 44-min. first attempt and slight re-trimming thereafter to optimize the flight profile of the model for this attempt. Water drops, after the earlier rain shower, delayed this attempt most of the afternoon. Finally, at 6:30 p.m. conditions were judged to be adequate for the attempt. The rubber motor was carefully wound to 1950 turns using a 16:1 winder and was then fixed to the model with an initial torque measurement of .94 in. oz. The model was launched and began to climb rapidly. Although it exhibited some tendency to stall momentarily during its first orbit, it corrected itself easily as it quartered the slightly drifting air.

Climbing majestically, the model reached its peak altitude 9 min. and 15 sec. where it first contacted the rafters at a height of 183 ft. (55.8 m) above the floor of the hangar. The model continued in a shallow climb for another 10 min., 45 sec. but was stopped from gaining more altitude as it gently "tapped" the rafters once every 30 sec. At 21 min. of elapsed time it was just below the rafters where it cruised for nearly 9 min. Gradually descending at a very slow rate, the model drifted diagonally across the hangar until its final orbits were eclipsing some internal structures of the hangar. With nearly 50 min. already recorded on the stopwatches, it became necessary to steer the model away from these structures to avoid a collision with them. The model was at an altitude of 20 ft. (6.1 m). Using a helium filled balloon fixed to a line, the model was steered for three brief periods until its orbit was clear of the structures. The model finally landed in the clear, upon the hangar floor. The timekeepers reported 50 min. 41 sec. as the final record figure from the average of their watch readings.

In his note, Dick also acknowledged help by Erv Rodemsky with the record flight; Erv talked him into a few more turns in the rubber before launching the record flight.

F1D 6

Cat Walker - Jim Richmond,
Davidson, North Carolina

Jim's models have a distinctive tip shape that is easy to spot in the hangar. The competition at the 1976 World Champs kept an eye on it as Jim's potential was given maximum expectations. Jim wrote the following note after the fact. "Thanks much for the photos, we appreciated them very much. Enclosed is a drawing of the Catwalker — so named because of its ability to get me up on catwalks — something no other plane ever did — not once, but three times (Akron, West Baden and Cardington).

I think it's a good design in spite of problems at Cardington. The moisture in the air there expanded the wood causing my bracing to be so tight it warped parts. Also my ribs suffered from the same syndrome. They were made from straight strips and formed to shape, but the moisture caused them to straighten out again — almost flat. The large prop ran at 43 rpm on the 41:32 flight during the cruise. (at Akron)"

Jim has been modelling for 43 years (he started at an early age) and among his successes he can count the F1D World Championship earned in Rome, 1968.

F1D 7

Grand Gram - Bud Romak, Moraga, California

With tutors like Joe Bilgri to help out back home how can one go wrong? And at the 1976 World Champs Bud didn't go wrong, putting it all together with a certain precision. The first day's flying didn't produce any outstanding times for Bud but at the end of the day he had it all figured out as to what combination would be right the next day. And then Bingo! Two flights over 39 min. with the model straining at every joint, hanging through stalls at the start and then cruising near but out of the girders. Bud claims the model is really pretty simple, the drawing covers all the needed information.

You know, it's hard to tell how to be successful with an indoor model because we all do it the same — some are more lucky than others. The first thing I do is to look

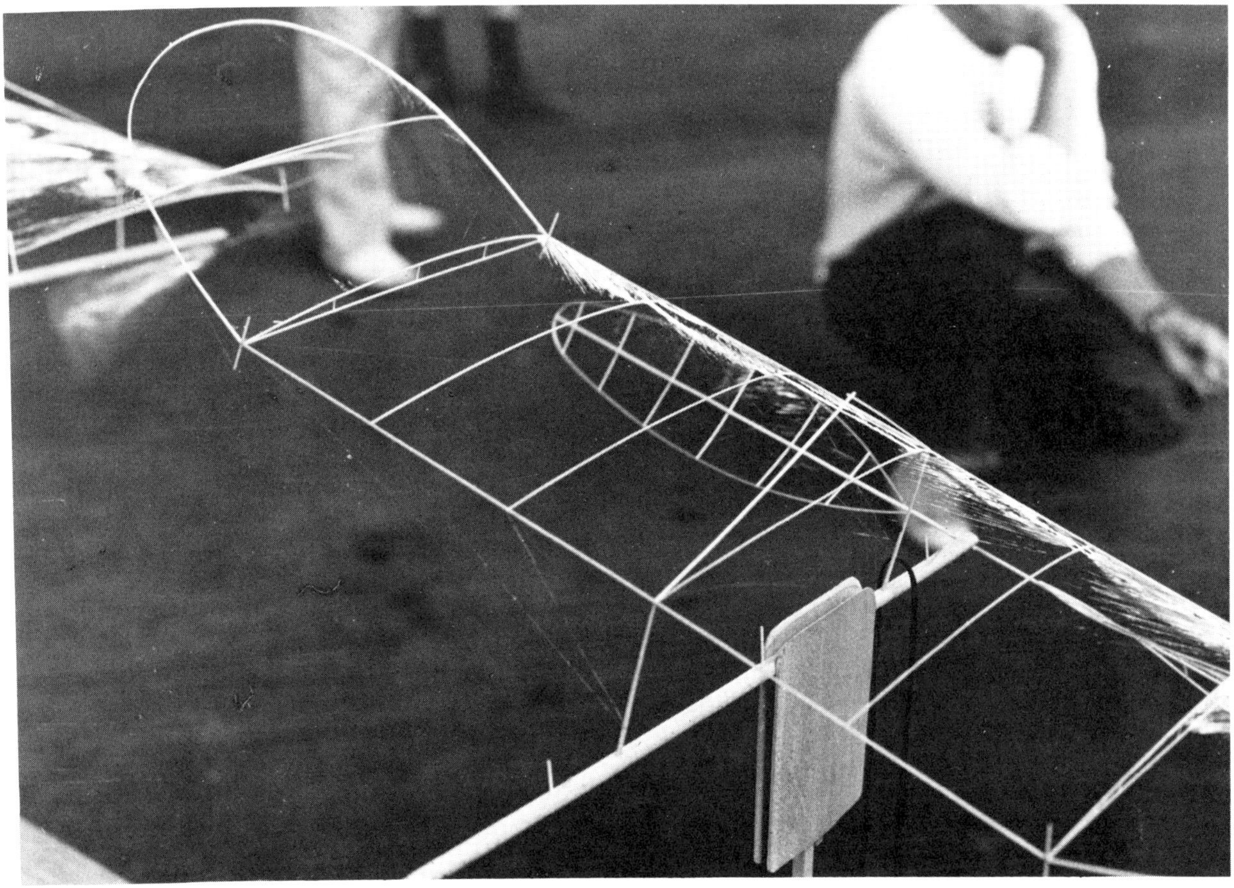

Bud Romak's 1976 World Champ F1D model details.

around and see what type of models are flown, worldwide. I look at my design and see where I can improve it, i.e., wing and prop design. I think the prop design is the most important. If you have good rubber and a good prop your troubles are solved.

After some days of scribbling, I will select a design and proceed to make the proper fixtures to build the models with. The next thing I do is pour the film.I always pour at least 30 to 40 sheets 14" x 30". I like to let the film age for at least four to six weeks. After storing the film in a safe place, I then select the balsa. Selecting the correct balsa can take hours. The motor stick, tail boom and wing spars are all very important. I usually cut enough spar stock and ribs for ten models. I make eight motor sticks, eight tail booms, eight stabs, four props and ten wings. This building process takes about two months.

Flying these models takes a lot of time. The main thing one must do is begin selecting the right prop and rubber combination. This can take at least four-full test sessions. I would say when you have completed all of your testing, you will find out that only four-out-of-the-eight models are really good and two-out-of-the-four are superior. Now you must go back to the building board and make at least four new models, making sure they are exactly the same as your four good ones.

Really, this indoor team selecting business can be really tough on you. You must be prepared to exclude yourself from all other extra activities for at least four months, and always be ready to build more wings and props. You must enjoy it to be successful and also have a patient understanding wife and family. When it comes to the indoor finals you had better be ready, i.e. have records on all of your models, props, rubber and torque meter readings. You can't do it without them.

F1D 8

Bucky '75/'76 - Bucky Servaites , Centerville, Ohio

The drawing of Bucky's model was made shortly after the '75 Team selection finals held in Lakehurst, and not many changes made for the '76 World Champs. Although demonstrating 38 mins. plus capability on other occasions the poor conditions at Cardington in '76 frustrated Bucky's efforts and his best flight there was 34:37. A smaller version of this model was flown by Bucky to a Class III FAI record on June 2, 1977, of 35:08. This was done in the Atrium of Northwood Institute, West Berlin, Indiana, that has a ceiling of 96 ft. 10 in.

F1D 9

Archaeopterix - Dieter Siebenmann,
Zurich, Switzerland

Dieter sent us a very nice drawing, done by Rene Butti, of the long tail moment arm models that he has designed and that were flown by the Swiss at the '76 World Champs. He also sent a detailed description of the aerodynamic theory behind this out-of-the-ordinary design. The theory has also been published in the October 1976 *Free Flight News* so we will not repeat it here. The following summary covers the main points.

"How can we improve the performance of an indoor F1D model? The tailplane lift is the easiest way to do this. So one must try to develop a model which can fly with a big, highly loaded tailplane. Moving back the center of gravity does not work because this only brings an unstable flight nearer the aperiodic pattern:

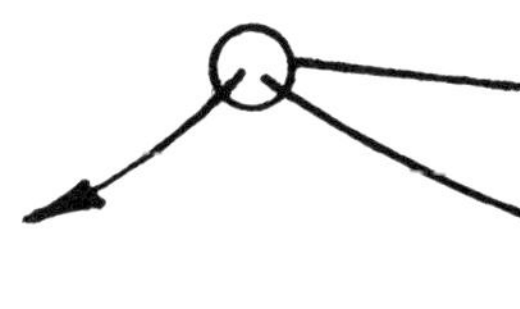

In order to find a way out, I connected the three formulae and computed the sinking speed which depended on the parameters: wing area, tailplane area in percent of wing area, and the fuselage length. These results led to the *Archaeopterix*. Its very long fuselage allows it to fly with a 50% tailplane at C_{lf} of 0.48 with ideal dynamic stability.

Sinking speed is theoretically decreased about 10% with this design compared to normal models. This could be decreased even more with a somewhat longer fuselage but structural weight becomes a problem."

This theory looks interesting and we look forward to further development. The initial attempt at Cardington in 1976 gave insufficient opportunity to develop the concept with working models, although a flight of 34:29 was made by Butti, and Siebenmann did 29:55.

F1D 10

Lucky Jim - Mike Thomas, Ontario, Canada

Lucky Mike flew *Lucky Jim* into just about every possible obstruction existent in the old Cardington Shed at the World Champs in 1976. But the model staggered on and on plowing through the pigeon droppings, and emerging with three flights over 30 mins to place 5th. Mike claims the model is just ordinary, nothing exotic. He made the wing tips a bit on the long side so that they would be a little more flexy on the burst, rather than the inboard panels. Prop was symmetrical outline because he couldn't make up his mind which way to flair it! Everything else he gives full credit to the recognized masters such as Kowalski (prop), Plotzki (covering), Triolo (wood selection), Barr (trim), Andrews (engineering) - say, that's quite a team!

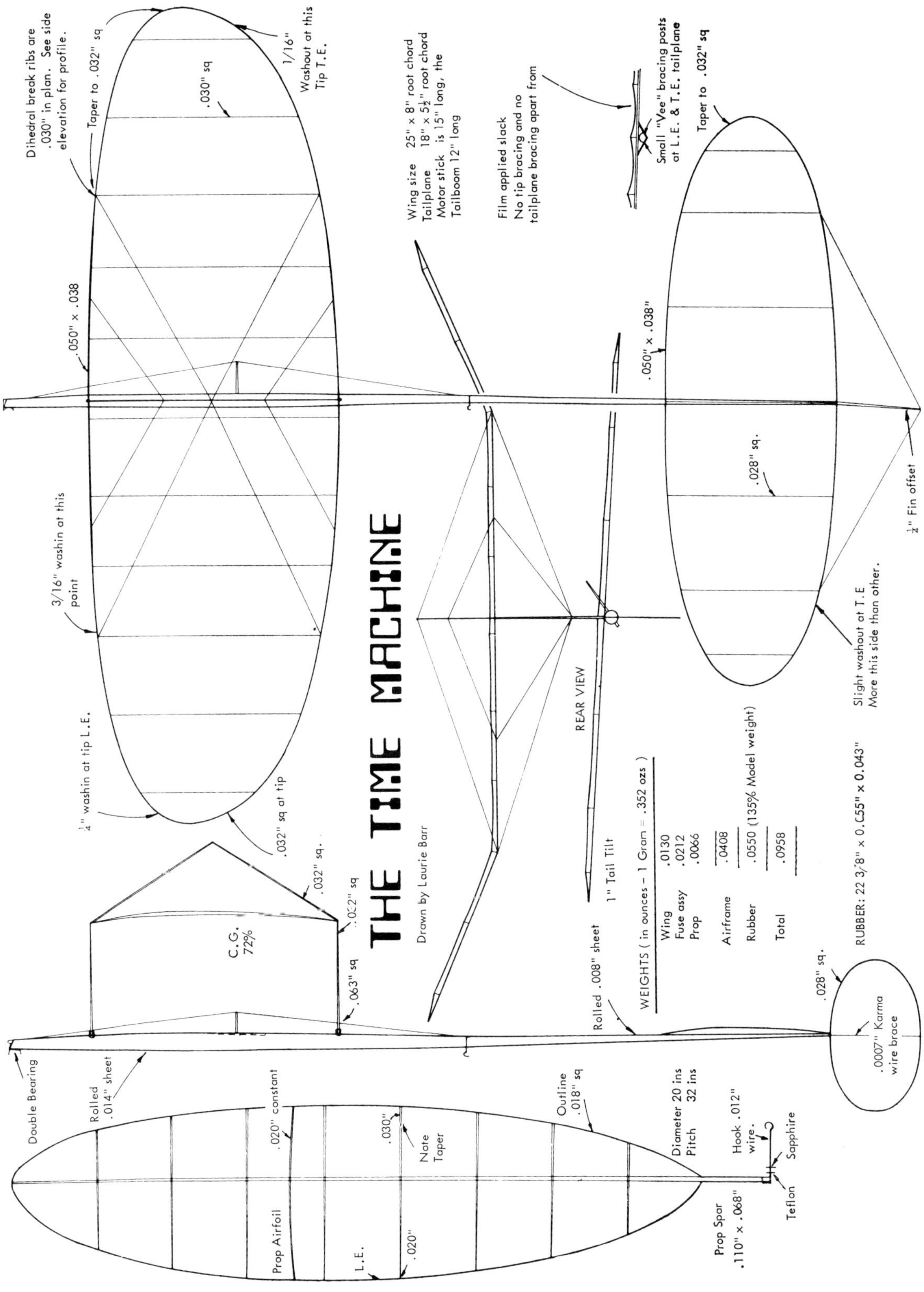
THE TIME MACHINE
Drawn by Laurie Barr
Dihedral break ribs are .030" in plan. See side elevation for profile.
Taper to .032" sq
.030" sq
1/16" Washout at this Tip T.E.
.050" x .038
3/16" washin at this point
¼" washin at tip L.E.
.032" sq at tip
Wing size 25" x 8" root chord
Tailplane 18" x 5½" root chord
Motor stick is 15" long, the Tailboom 12" long
Film applied slack
No tip bracing and no tailplane bracing apart from
Small "Vee" bracing posts at L.E. & T.E. tailplane
Taper to .032" sq
.050" x .038"
.028" sq.
¼" Fin offset
Slight washout at T.E. More this side than other.
REAR VIEW
1" Tail Tilt
WEIGHTS (in ounces – 1 Gram = .352 ozs)
Wing .0130
Fuse assy .0212
Prop .0066
Airframe .0408
Rubber .0550 (135% Model weight)
Total .0958
RUBBER: 22 3/8" x 0.055" x 0.043"
C.G. 72%
.032" sq.
.032" sq
.063" sq
Rolled .008" sheet
.028" sq.
.0007" Karma wire brace
Double Bearing
Rolled .014" sheet
.020" constant
.030"
Note Taper
Outline .018" sq
Diameter 20 ins
Pitch 32 ins
Hook .012" wire.
Sapphire
Teflon
Prop Spar .110" x .068"
Prop Airfoil
L.E.
.020"

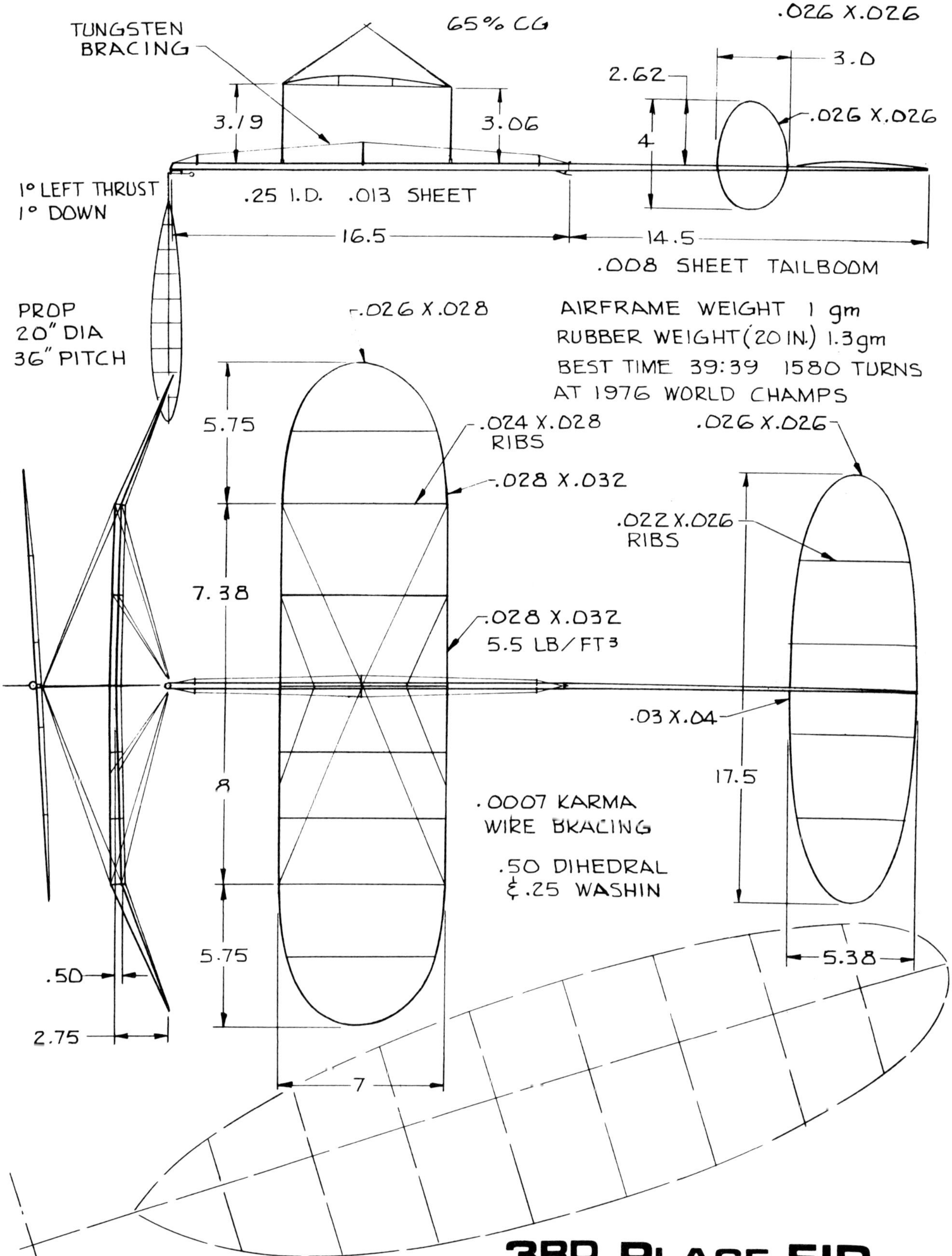

3RD PLACE FID 1976 INDOOR WORLD CHAMPS

BY LAURIE BARR

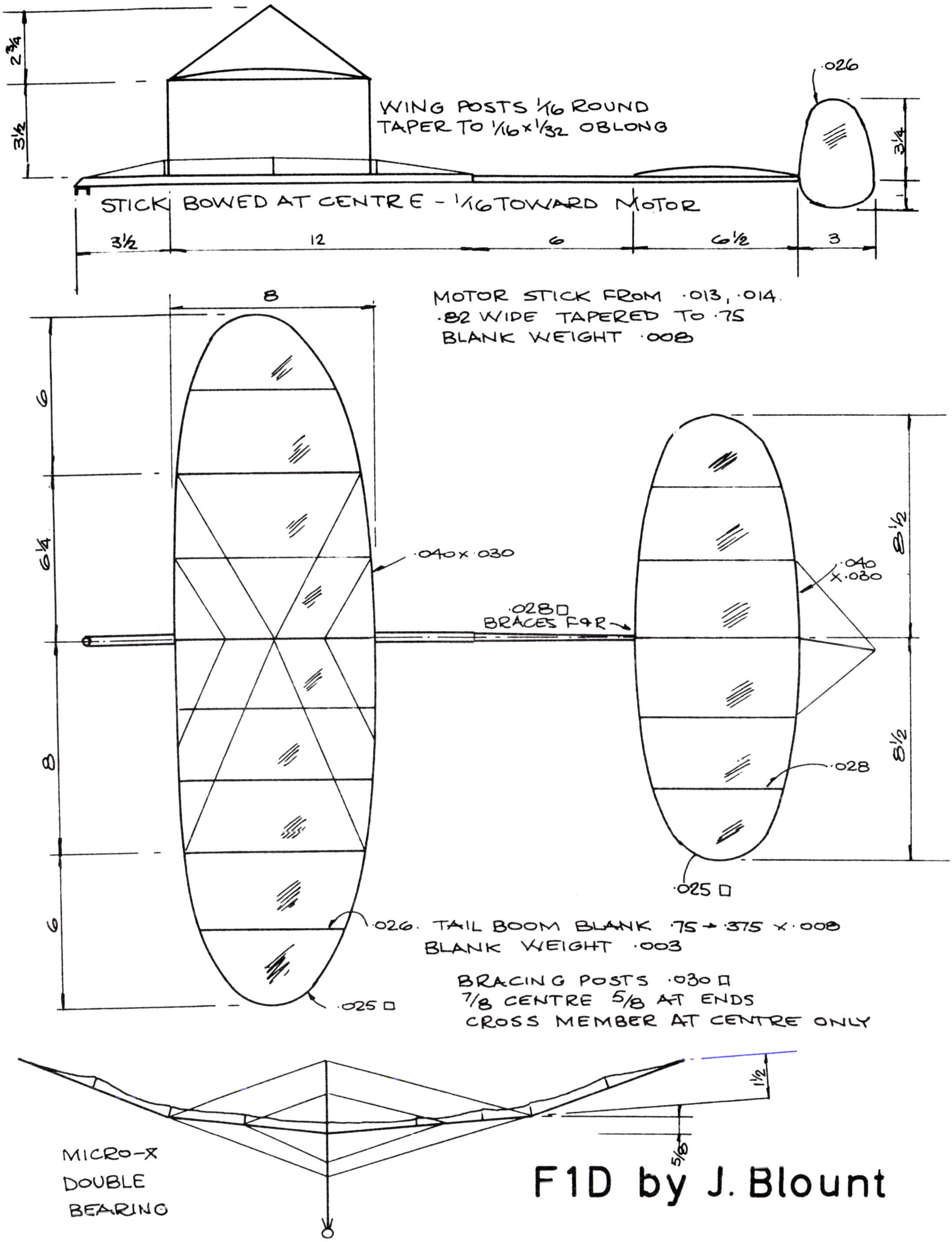

F1D by J. Blount

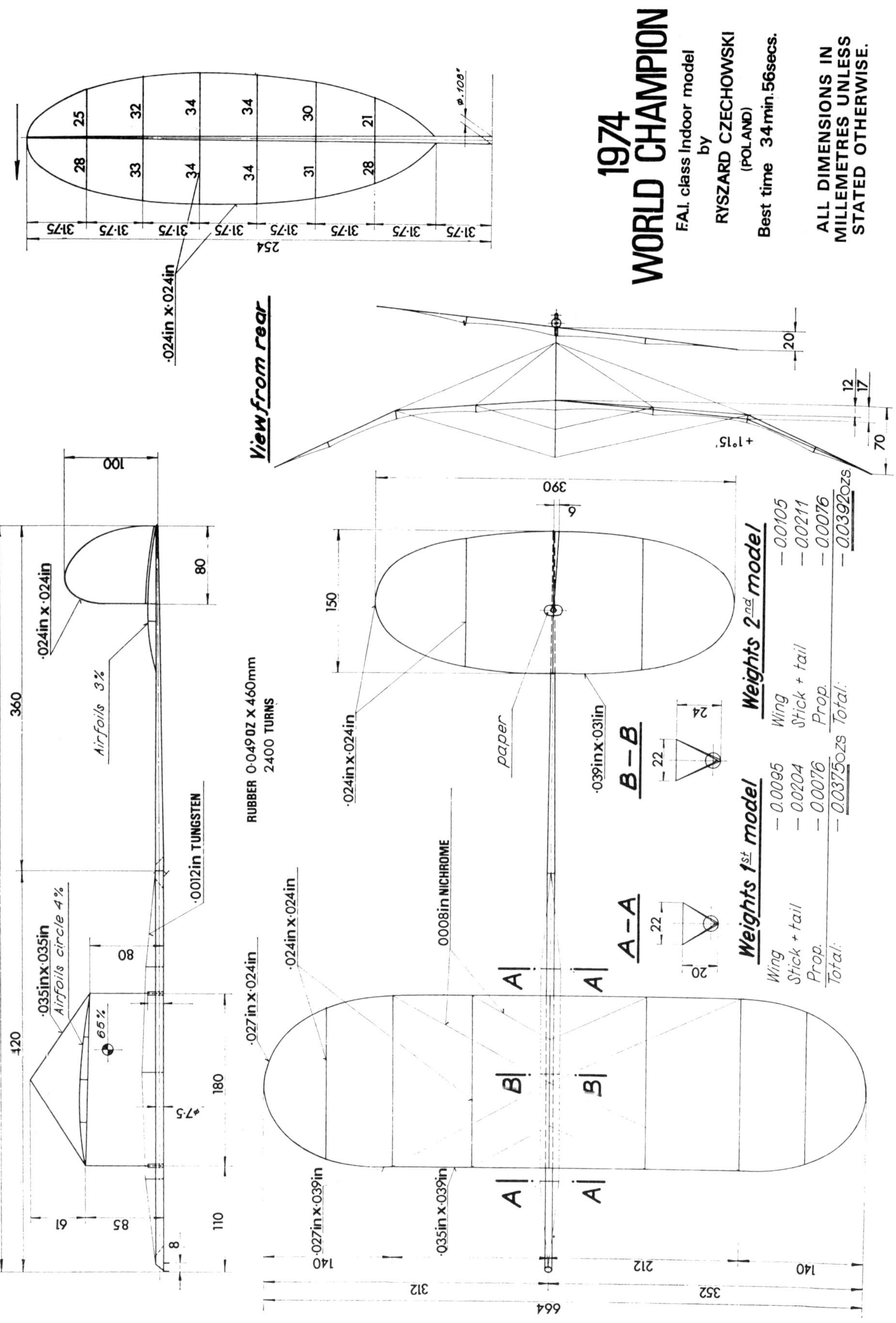
1974
WORLD CHAMPION
F.A.I. class Indoor model
by
RYSZARD CZECHOWSKI
(POLAND)
Best time 34min.56secs.
ALL DIMENSIONS IN
MILLEMETRES UNLESS
STATED OTHERWISE.
View from rear
RUBBER 0·049 OZ X 460mm
2400 TURNS
·0012in TUNGSTEN
·0008in NICHROME
Airfoils 3%
Airfoils circle 4%
paper
A–A
B–B
Weights 1st model
Wing – 0.0095
Stick + tail – 0.0204
Prop. – 0.0076
Total: – 0.0375ozs
Weights 2nd model
Wing – 0.0105
Stick + tail – 0.0211
Prop. – 0.0076
Total: – 0.0392ozs.

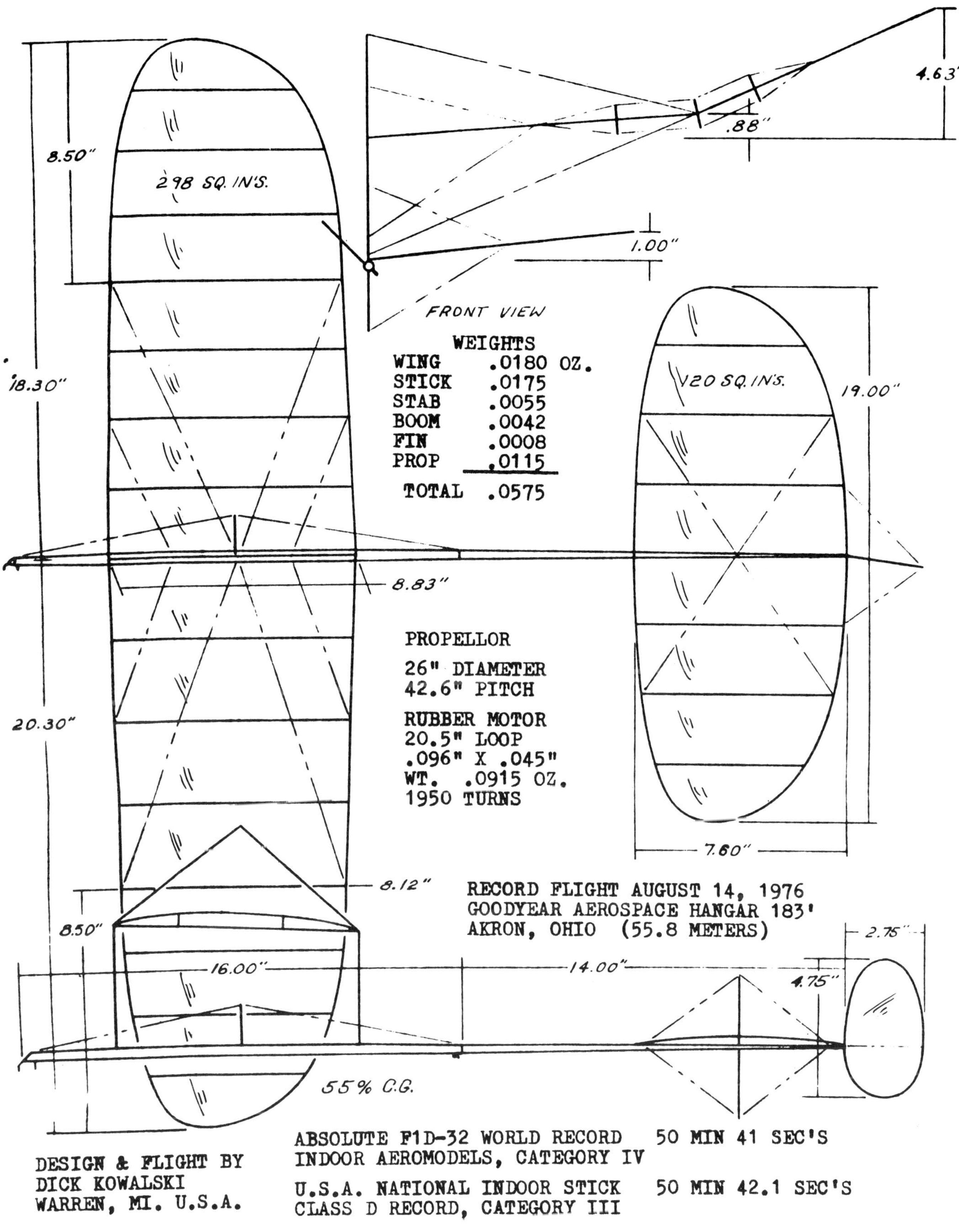

5 Kowalski USA Indoor News & Views

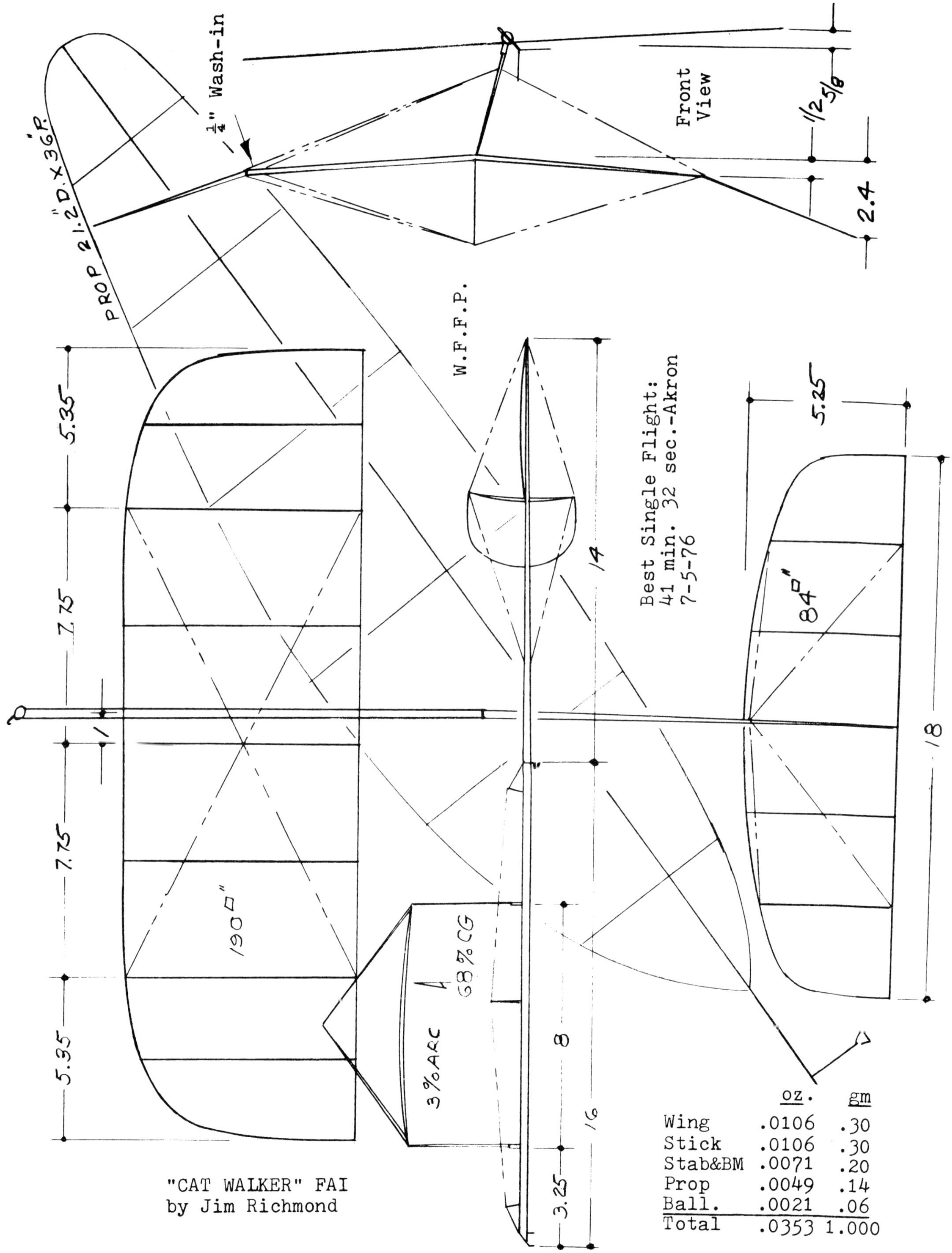

"CAT WALKER" FAI
by Jim Richmond

	oz.	gm
Wing	.0106	.30
Stick	.0106	.30
Stab&BM	.0071	.20
Prop	.0049	.14
Ball.	.0021	.06
Total	.0353	1.000

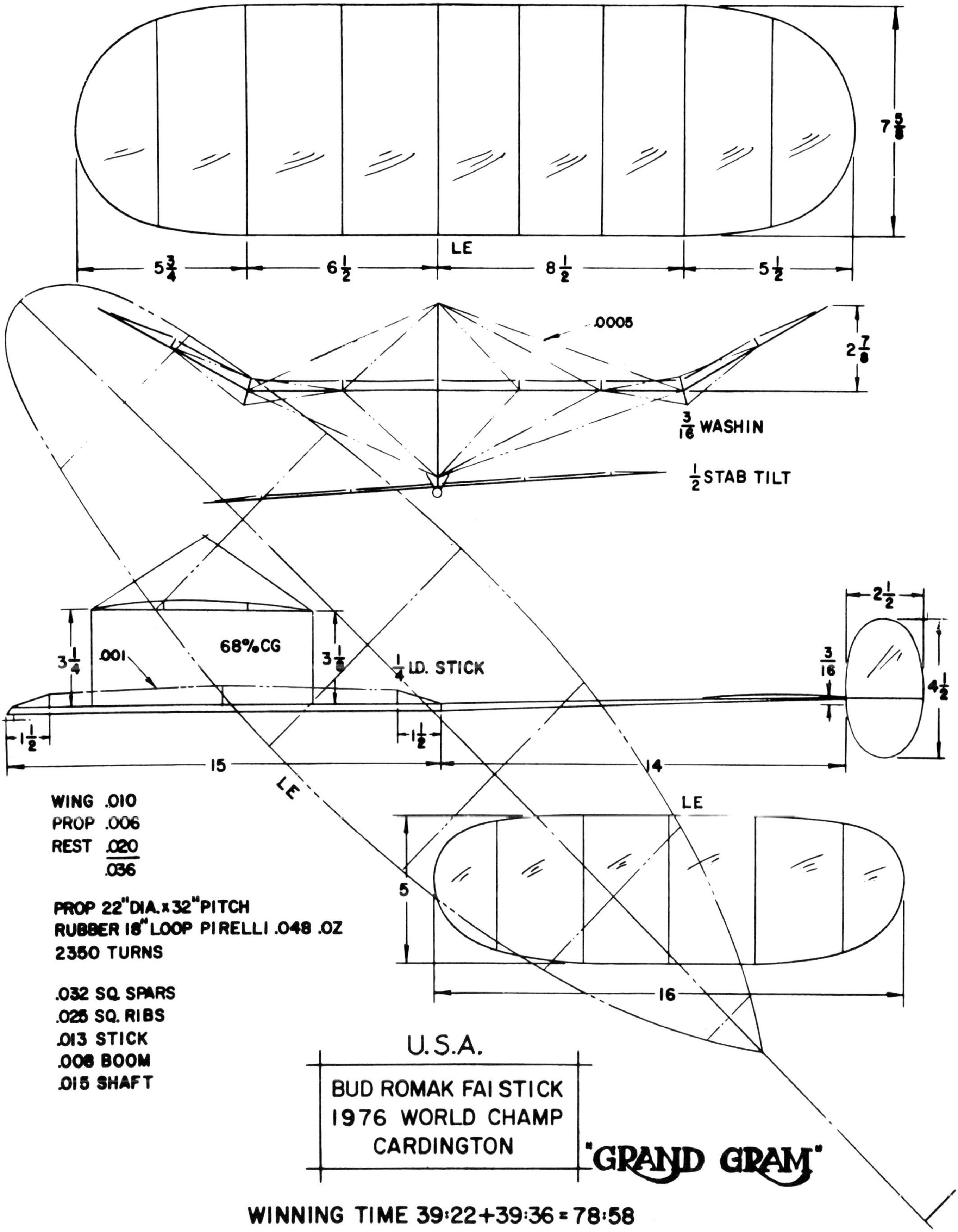

7 Romak USA Romak

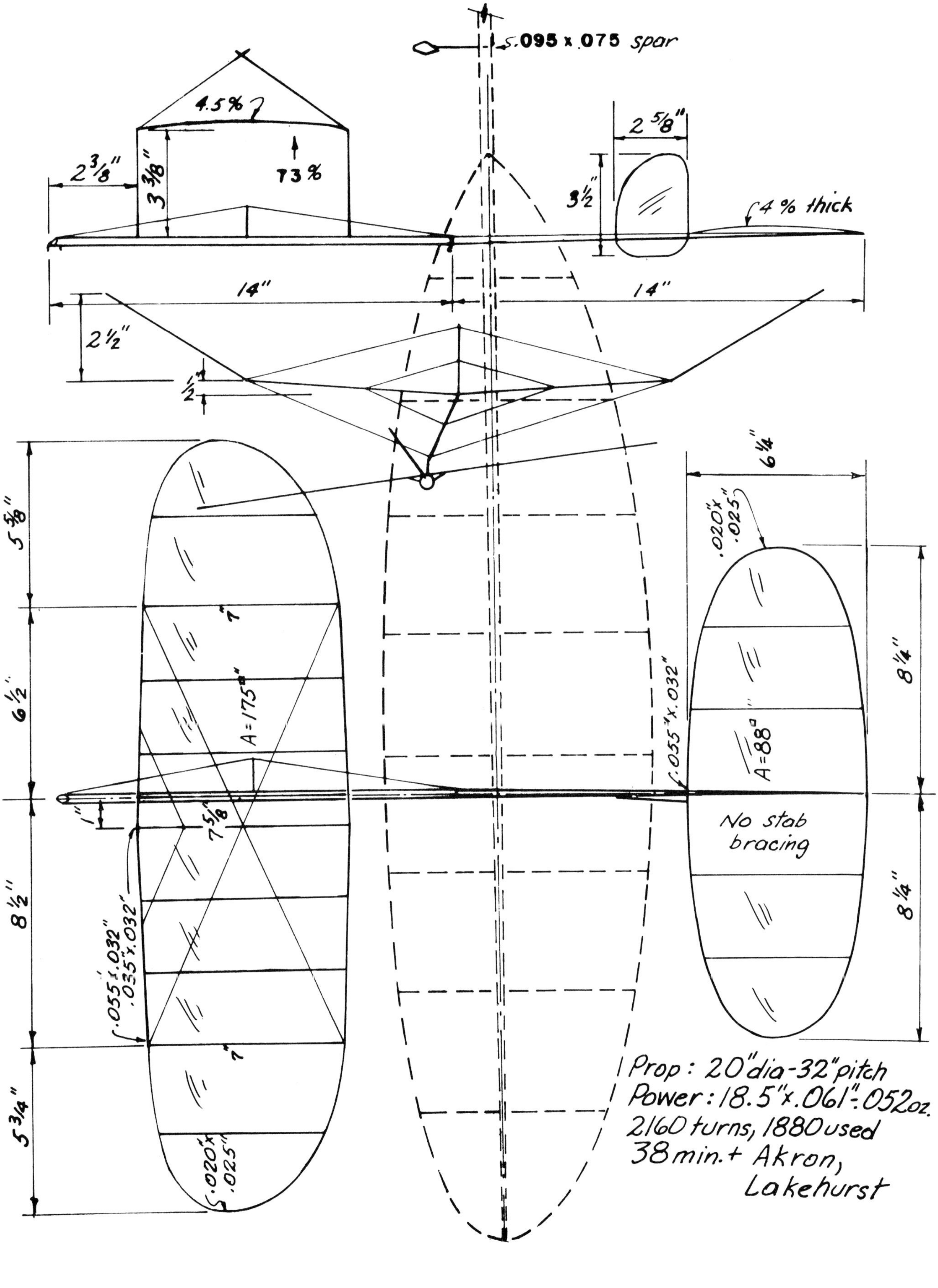
.095 x .075 spar
4.5%
73%
2 3/8"
3 3/8"
2 5/8"
3 1/2"
4% thick
14"
14"
2 1/2"
1/2"
5 5/8"
6 1/2"
8 1/2"
5 3/4"
A=175□"
7 5/8"
1"
.055"x.032"
.035"x.032"
.020"x .025"
6 1/4"
8 1/4"
8 1/4"
A=88□"
.055"x.032"
No stab bracing
Prop: 20"dia-32"pitch
Power: 18.5"x.061"-.052oz.
2160 turns, 1880 used
38 min.+ Akron, Lakehurst

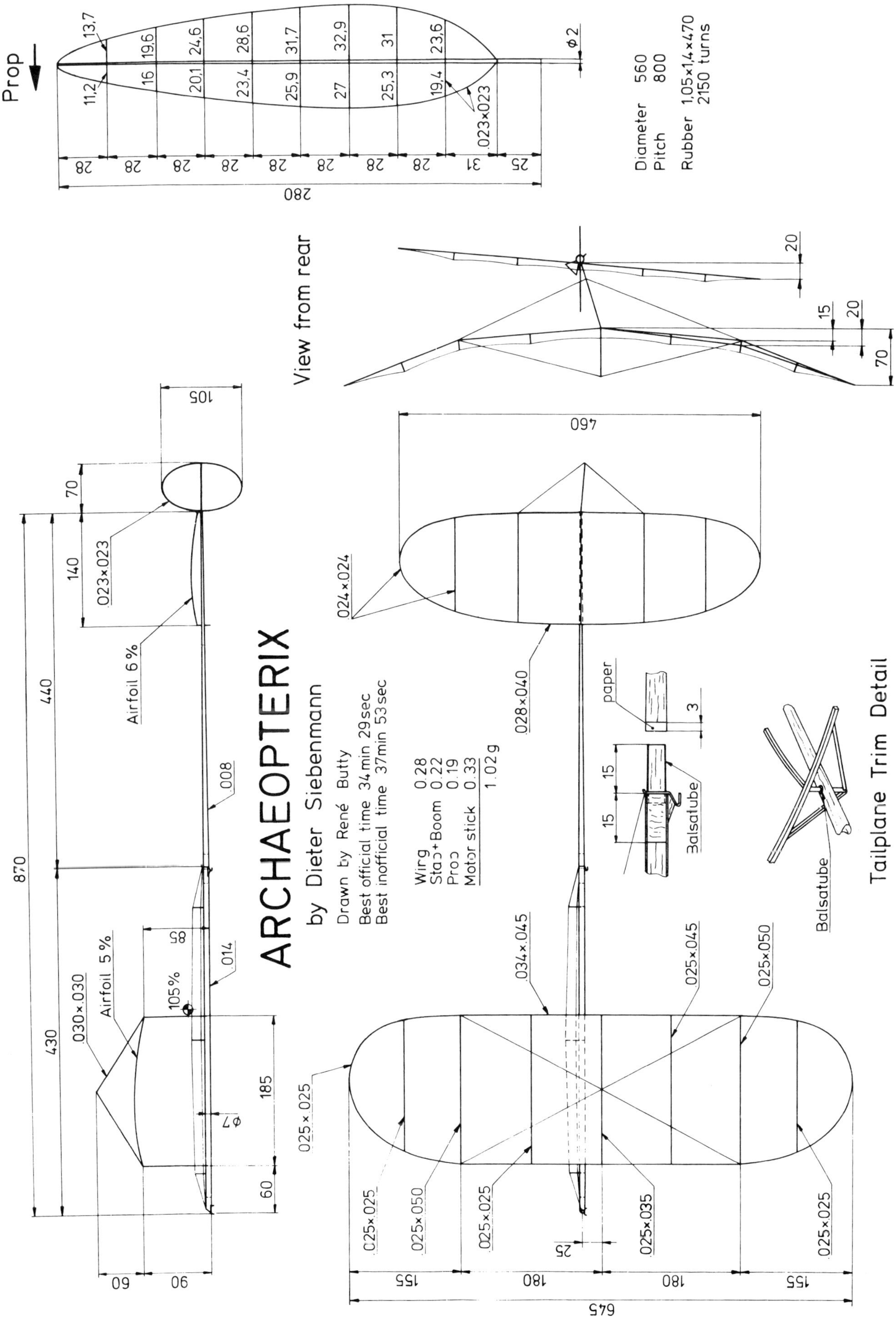
ARCHAEOPTERIX
by Dieter Siebenmann
Drawn by René Butty
Best official time 34min 29sec
Best inofficial time 37min 53sec
Wing 0.28
Stab+Boom 0.22
Prop 0.19
Motor stick 0.33
1.02 g
Diameter 560
Pitch 800
Rubber 1,05×1,4×470
2150 turns
Prop
View from rear
Tailplane Trim Detail
Balsatube
paper
Airfoil 6 %
Airfoil 5 %
105%

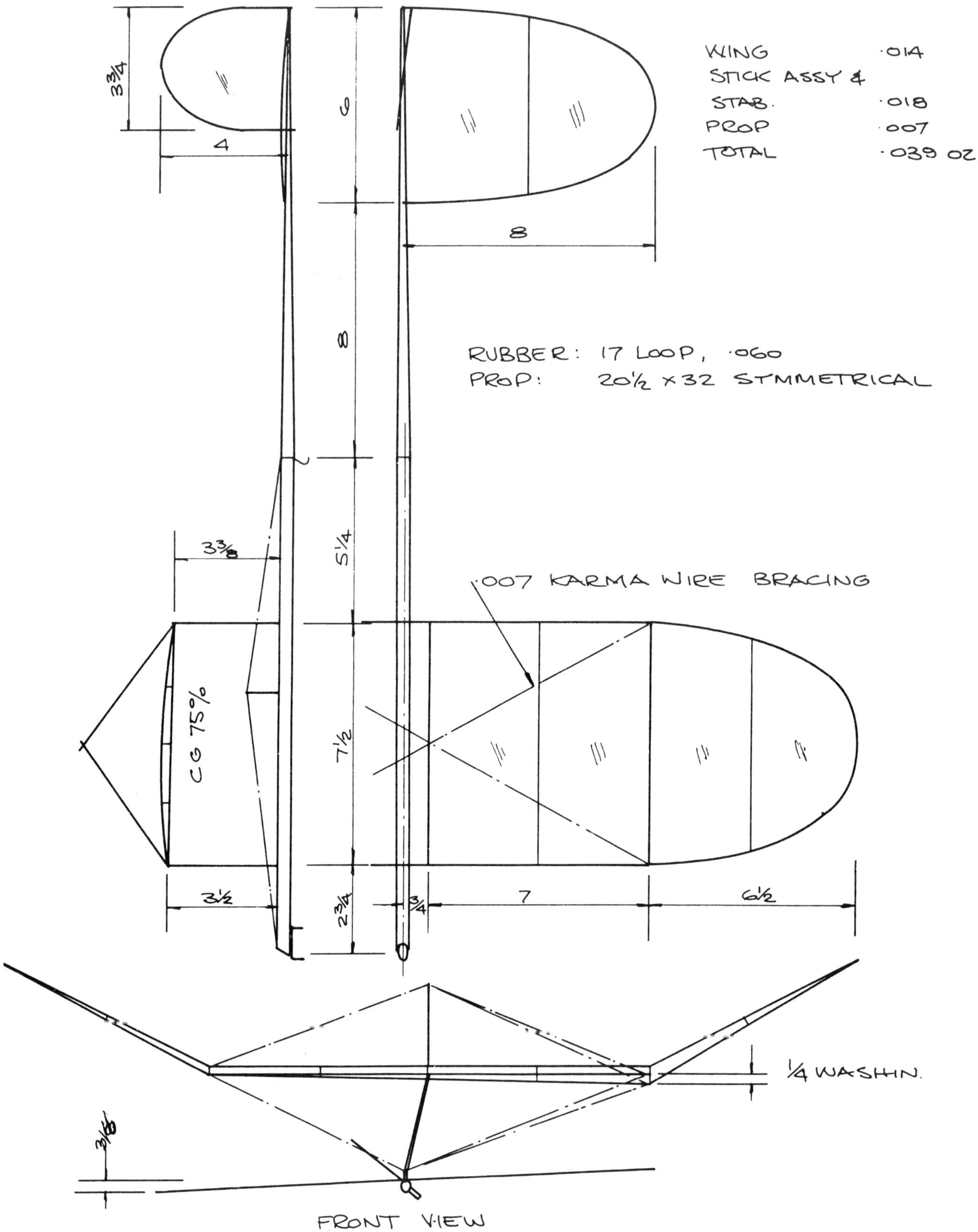

'LUCKY JIM' F.1.D. by Mike Thomas. Canada

F1 G

F1G1

Nike - G. Callegari, Milan, Italy

This model is the 1976 Italian Champion, and interestingly, uses the editors favorite airfoil, the MVA 342. Coupes seem to thrive on thin sections with well-distributed camber approaching a circular-arc shape. The Coupe cross-section rule was taken care of in deliberate fashion on this model by grafting a hatchet-head shaped pod onto the motor tube. Two-piece wing construction, although adding a little weight, really makes packing and travel a lot more convenient. The 40% CG location shown seems a bit forward, perhaps this is needed with the big wing and short tail moment? Ship placed second at the Turin, Italy, Coupe d'Hiver Challenge.

F1G2

Swallow - Ian Dowsett, Ruislip, England

Ian sent us a very nice plan for his latest Coupe model which is also shared with *Aeromodeller* (Jan. 1977). Swallow is on the "large-ish" side, reflecting the French trend and in fact the propellor is also patterned after Louis Dupuis' design. Ian is no slouch at design himself and the details of his model are quite good. The triple fins and dihedral tailplane are somewhat unusual and Ian claims that the ship is tolerant of bad launches and blustery weather. It can be built and flown as either an 80g or 100g model but special care must be exercised to keep the weight down on the lighter version.

F1G3

Louis' Coupe - Louis Dupuis, Chatellerault, France

The French are firmly committed to the 100g coupe concept and are masters at this deceptively simply class. Where others settle for quick building and somewhat sluggish performance, they have spent more time on clever design, large efficient structure and persistent flight training. Louis' model appears to be a scaled version of his Wakefield (F1B 4) including many of the construction features. Fuselage is an elliptic section moulded tube of balsa. Wings plug neatly in to faired root sections. Span is 1200 mm and wing area 14 dm^2. This is about 90% the size of a Wakefield, and powered with ¼ the rubber.

F1G4

Rubber Duckie - Dave Linstrum, Miami, Florida

Dave built *Rubber Duckie* in a London hotel room when he found his bathtub didn't have one. Stuck it together with "Hot-Stuff" and cajoled Pete Cameron into flying it proxy for him in the 1975 CdH International Challenge. The small size and flat-bottomed wing section preclude the still-air performance being very high but it would be useful in windy conditions. The design is more in line with what the proponents of the 80g rule perceive as an approach to simplification.

F1G5

F1G - Ing. Carlo Martegani, Fagnano, Italy

Carlo didn't send much description along with his Coupe drawing but he did include two good photos which we share with you here. This model was the winner of the 1977 Turin, Italy, International Challenge, against a nine-man fly-off.

MARTEGANI PHOTO

Bashful youngster holds Ing. Carlo Martegani's Coupe model.

MARTEGANI PHOTO

Now we show you Carlo and the underside of his 1977 Turin winner. Box fuselage, built-up tail boom. Two-piece wing for easy transport.

FIG 6

Les Trumeaux - George Matherat, Grenoble, France

Another example of the French "large" Coupe d'Hiver this model also sports a built-up motor tube box fuselage and a glass fiber tail boom. A number of models have recently been built using this design concept and they appear to be having some success. The original by Matherat has won the Cote d'Azure Cup, 1975 and placed second at the 1975 Aeromodeller Challenge (Halton, England). Since that time Georges has gone on to an even larger version, *Les Supertrumal* which has a projected span of 1505 mm and wing area of 19.40 dm.^{2}. That makes it bigger than a Wakefield.

FIG 7

Anglian Coupe - Ray Pavely, Ramsden Heath, England

This Coupe was built under the 100g rule and as such was able to carry a DT timer. Since reduction of the weight rules to 80g, the timer has been removed. Most of the design ideas have been taken from others and it is a good reliable design with no tricks. It is, in fact, subject of a "one design" club contest event.

This is the Coupe design of Matherat, built by Pierre Chaussebourg and flown proxy by Bob White in California.

FIG 8

Gannet Mk 6 - Stuart Savage, Canada

The *Gannet* was proxy flown by Ian Dowsett to a win at the British Nationals. In a continuing career in the hands of proxy fliers it placed ninth at the 100g International Challenge, 1977, this time being seen to by G. Dowsett, Jr. (Time marches on.) This model is an unpretentious capable machine that is easy to build, trim and fly.

FIG 9

Camira - Paul Van Leuven, Queensland, Australia

Fuselage of this Coupe is put together using 1/8 in. sq. longerons sided with 1/16 in. sheet balsa with the grain vertical to resist splitting. We have no contest performance record for the design but everything about it seems reasonable, although by French standards it is a "small" model.

FIG 10

Beau Coupe - Bob White, Monrovia, California

Beau Coupe is similar in construction and trim to Bob's Wakefields. Eight strands of 6mm Pirelli are used with a tension-prop stop. All-up structural weight is 90g so that a 10g ballast weight can be used for locating the best spot for the CG. Bob goes through a special routine with this and all his models. Rather than accept whatever CG comes out from the original design guess he carefully hand glides the model, trimming stab angle to get the best glide as he moves the weight back. When the stability reaches a limit, indicated by the glide becoming a dive, he moves the weight back forward and retrims the stab. How far forward he moves the weight depends on a lot of guesstimation but for Beau Coupe it is ½ inch.

Bob thought at one time that the 100g Coupe was a mistake and resulted in too sluggish a model, but he has changed his mind now and believes that the 100g rule is right.

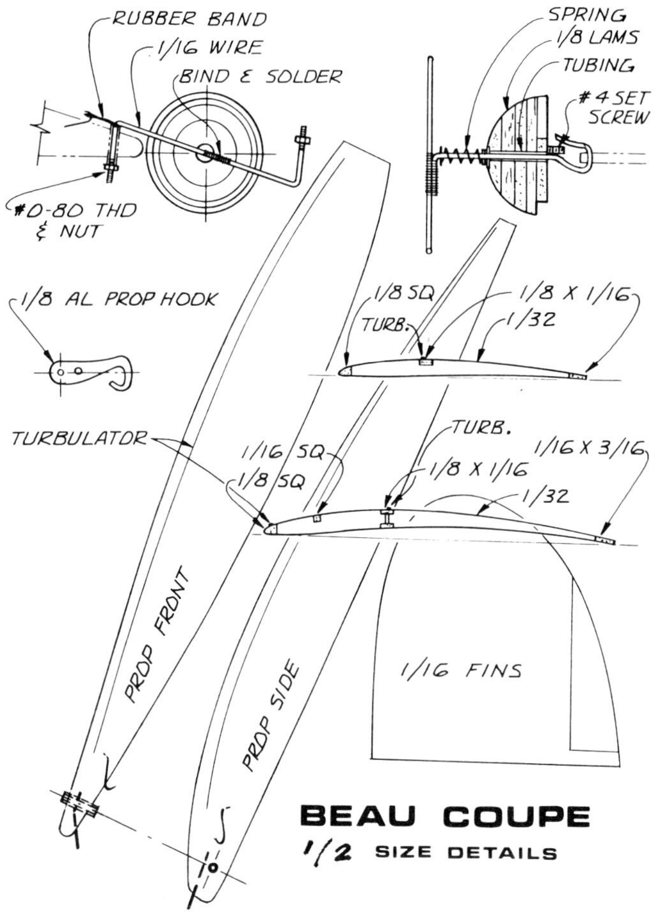

BEAU COUPE
1/2 SIZE DETAILS

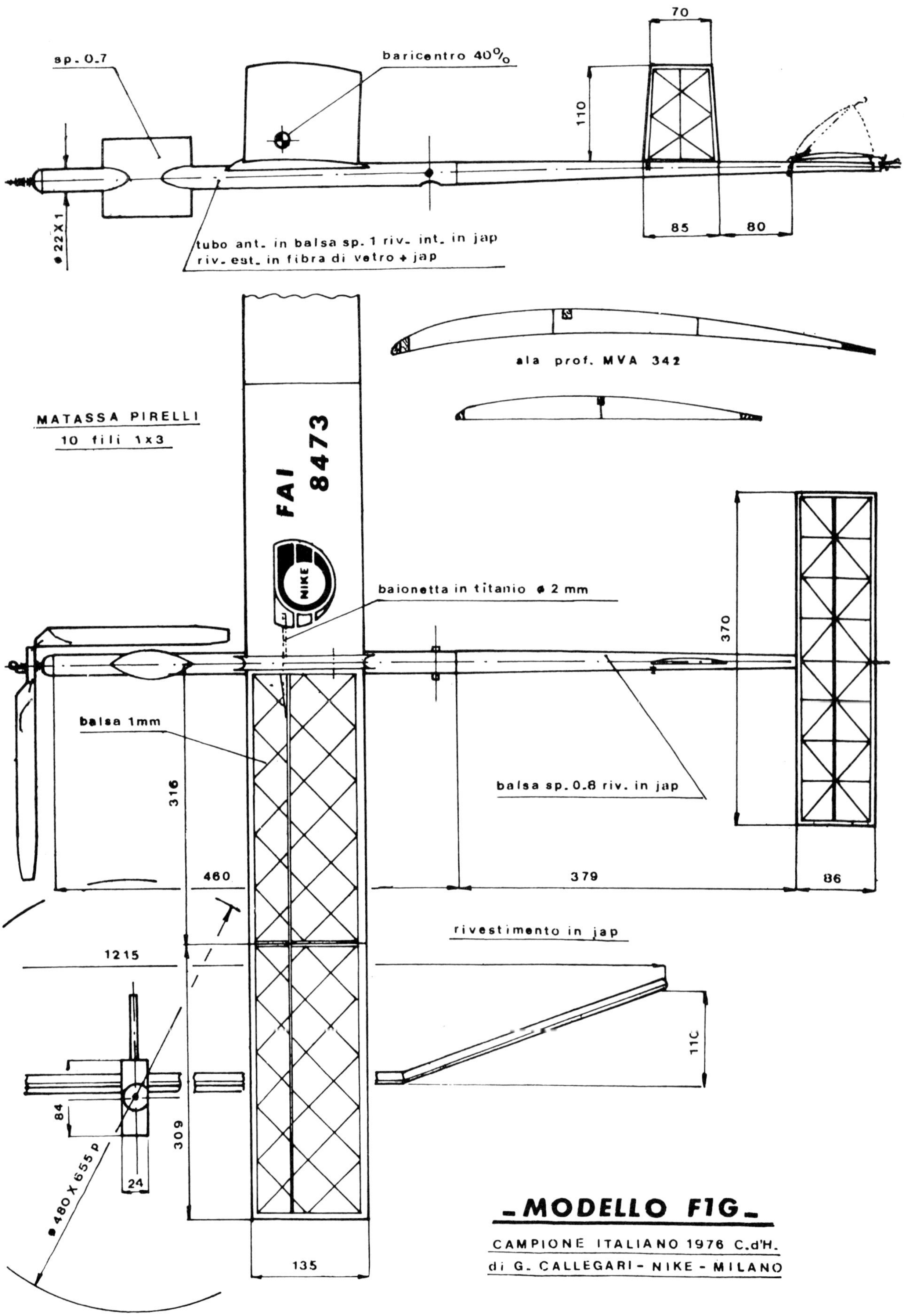

 1 Callegari Italy AVA

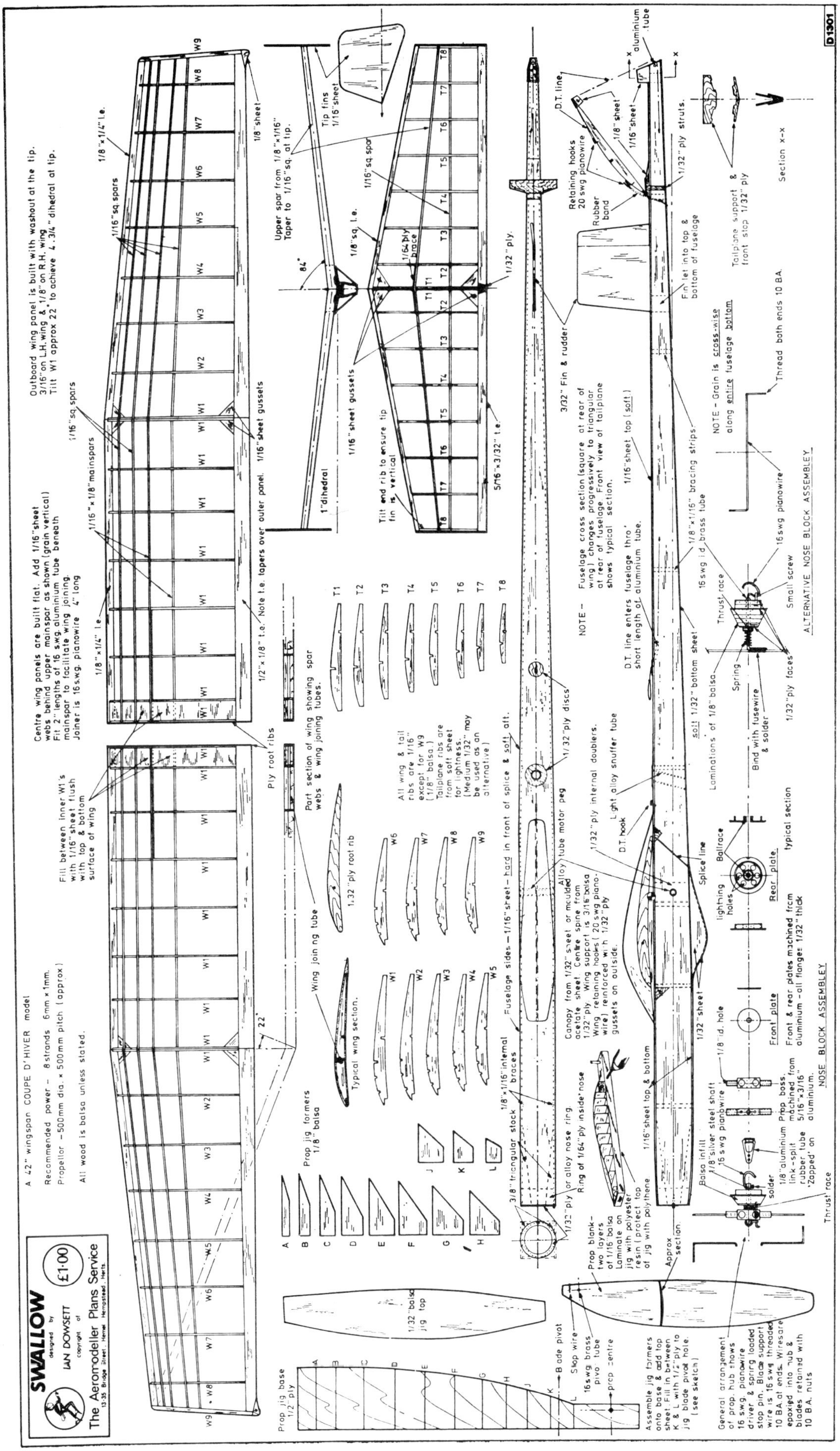

SWALLOW
designed by IAN DOWSETT
copyright of The Aeromodeller Plans Service
£1·00
D 1301
A 42" wingspan COUPE D'HIVER model
Recommended power - 8 strands 6mm x 1mm.
Propellor - 500mm dia. x 500mm pitch (approx)
All wood is balsa unless stated.
NOSE BLOCK ASSEMBLY
ALTERNATIVE NOSE BLOCK ASSEMBLY

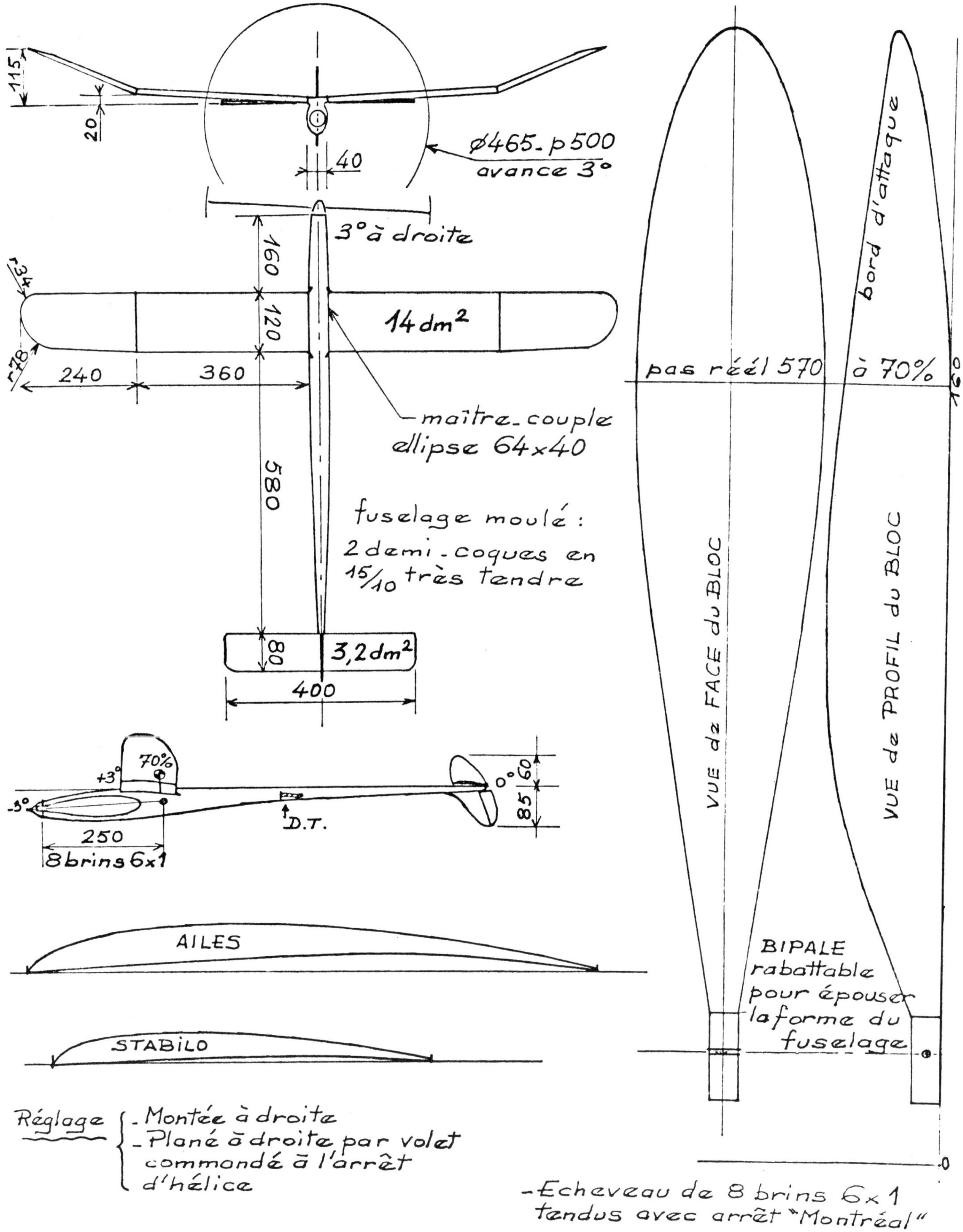
115
20
ø465 - p500
avance 3°
40
3° à droite
160
r34
120
14 dm²
r78
240
360
maître-couple
ellipse 64x40
580
fuselage moulé :
2 demi-coques en
15/10 très tendre
80
3,2 dm²
400
70%
+3°
-3°
0°
60
85
D.T.
250
8 brins 6x1
AILES
STABILO
pas réél 570
à 70%
bord d'attaque
VUE de FACE du BLOC
VUE de PROFIL du BLOC
BIPALE
rabattable
pour épouser
la forme du
fuselage
Réglage
- Montée à droite
- Plané à droite par volet
commandé à l'arrêt
d'hélice
- Echeveau de 8 brins 6x1
tendus avec arrêt "Montréal"

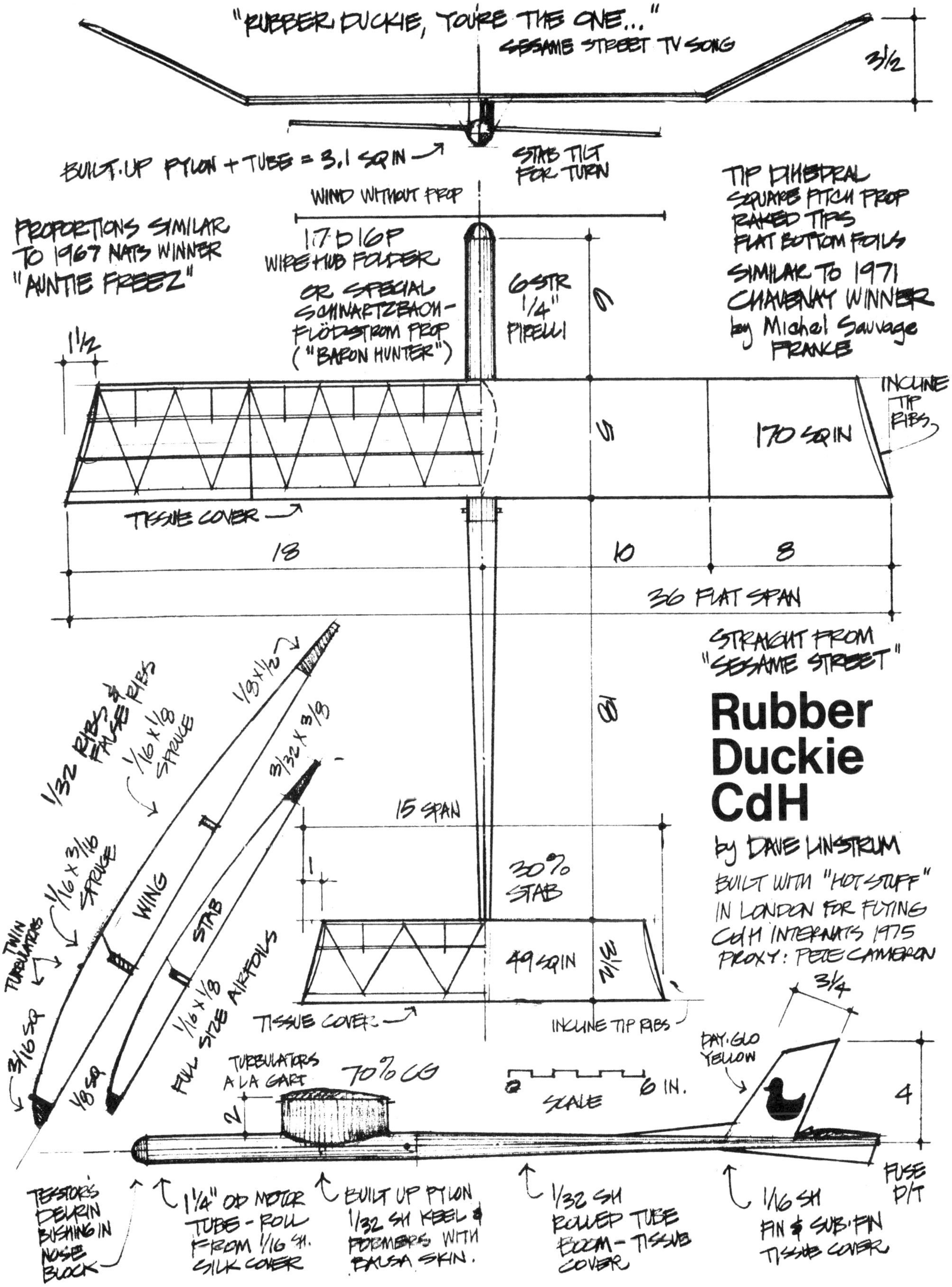
"RUBBER DUCKIE, YOU'RE THE ONE..."
SESAME STREET TV SONG
3½
BUILT-UP PYLON + TUBE = 3.1 SQ IN
STAB TILT FOR TURN
WIND WITHOUT PROP
TIP DIHEDRAL
SQUARE PITCH PROP
RAKED TIPS
FLAT BOTTOM FOILS
SIMILAR TO 1971 CHAVENAY WINNER by Michel Sauvage FRANCE
PROPORTIONS SIMILAR TO 1967 NATS WINNER "AUNTIE FREEZ"
17 D 16P
WIRE+HUB FOLDER
OR SPECIAL SCHWARTZBACH-FLÖDSTROM PROP ("BARON HUNTER")
6 STR 1/4" PIRELLI
1½
INCLINE TIP RIBS
170 SQ IN
TISSUE COVER
18
10
8
36 FLAT SPAN
STRAIGHT FROM "SESAME STREET"
Rubber Duckie CdH
by DAVE LINSTRUM
BUILT WITH "HOT STUFF" IN LONDON FOR FLYING CdH INTERNATS 1975
PROXY: PETE CAMERON
1/32 RIBS & FALSE RIBS
1/16 x 1/8 SPRUCE
1/8 x 1/16
3/32 x 3/8
1/16 x 3/16 SPRUCE
WING
STAB
TWIN TURBULATORS
1/16 x 1/8
FULL SIZE AIRFOILS
3/16 SQ
1/8 SQ
15 SPAN
30% STAB
49 SQ IN
TISSUE COVER
INCLINE TIP RIBS
3/4
DAY-GLO YELLOW
4
TURBULATORS A LA GART
70% CG
2
SCALE
6 IN.
FUSE D/T
TESTOR'S DELRIN BUSHING IN NOSE BLOCK
1¼" OD MOTOR TUBE - ROLL FROM 1/16 SH. SILK COVER
BUILT UP PYLON 1/32 SH KEEL & FORMERS WITH BALSA SKIN.
1/32 SH ROLLED TUBE BOOM - TISSUE COVER
1/16 SH FIN & SUB-FIN TISSUE COVER

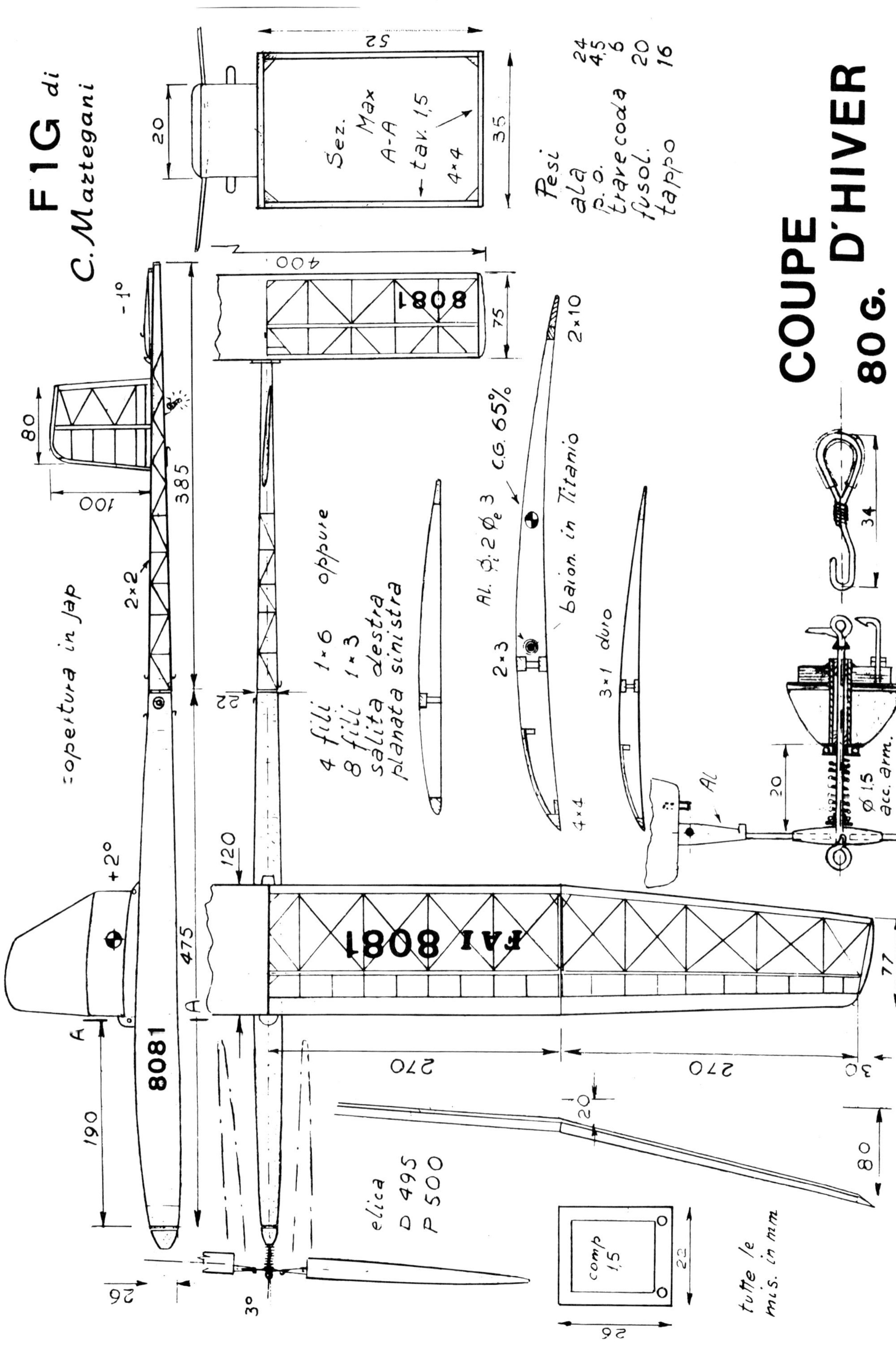

F1G di
C. Martegani
COUPE
D'HIVER
80 G.
Sez. Max A-A
tav. 1,5
4×4
Pesi
ala 24
p.o. 4,5
trave coda 6
fusol. 20
tappo 16
copertura in jap
4 fili 1×6 oppure
8 fili 1×3
salita destra
planata sinistra
C.G. 65%
Al. ϕ_i 2 ϕ_e 3
baion. in Titanio
3×1 duro
ϕ 1,5
acc. arm.
elica
D 495
P 500
comp 1,5
tutte le mis. in mm
FAI 8081
8081

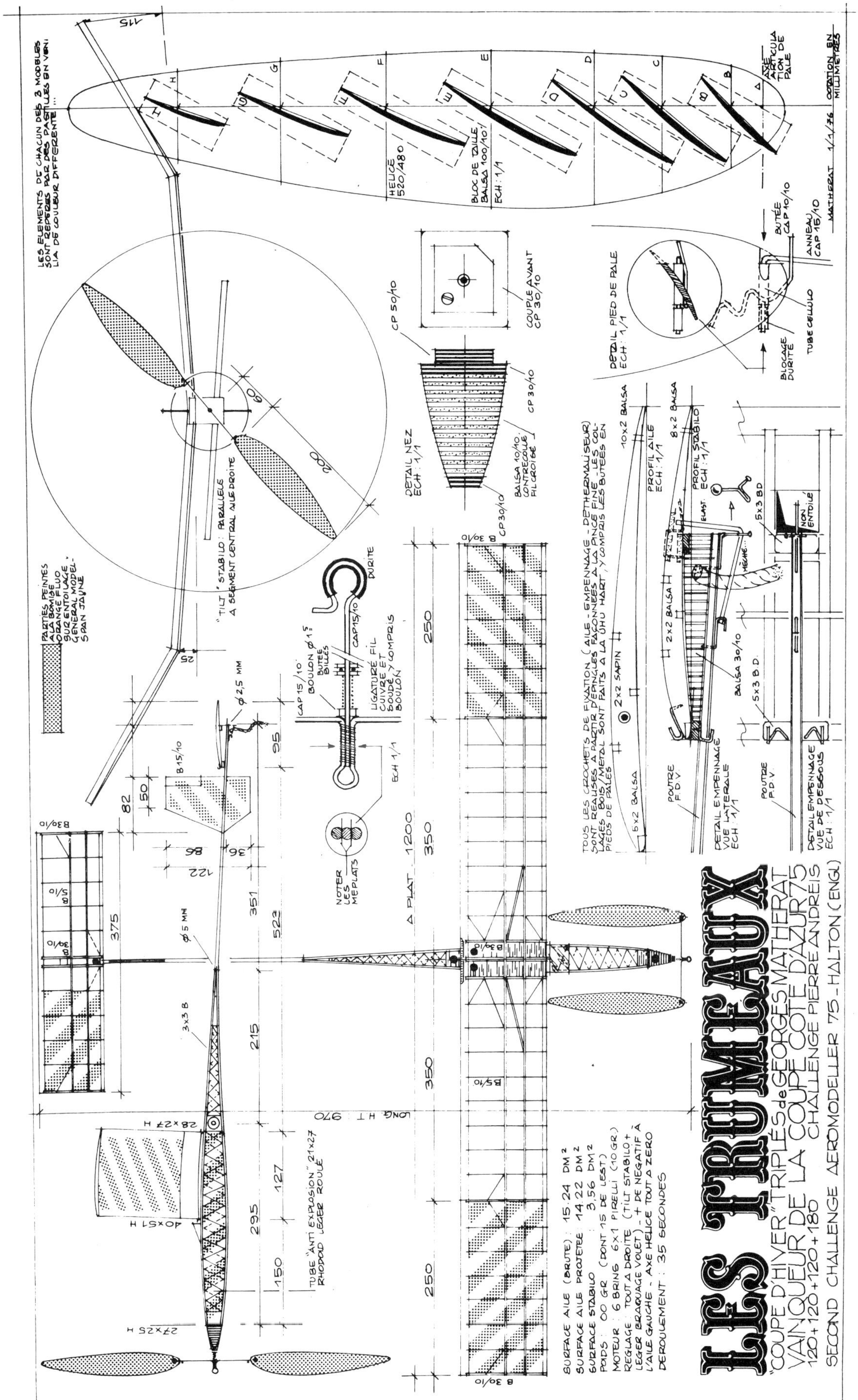

LES TRUMEAUX
"COUPE D'HIVER" TRIPLÉS de GEORGES MATHERAT
VAINQUEUR DE LA COUPE COTE D'AZUR 75
120+120+120+180 CHALLENGE PIERRE ANDREIS
SECOND CHALLENGE AEROMODELLER 75 - HALTON (ENGL)
SURFACE AILE (BRUTE) : 15.24 DM²
SURFACE AILE PROJETEE : 14.22 DM²
SURFACE STABILO : 3.56 DM²
POIDS : 00 GR (DONT 15 DE LEST)
MOTEUR : 6 BRINS 6x1 PIRELLI (10 GR)
DEROULEMENT : 35 SECONDES
HELICE 520/480
DETAIL NEZ ECH : 1/1
DETAIL PIED DE PALE ECH : 1/1
PROFIL AILE ECH : 1/1
PROFIL STABILO ECH : 1/1
DETAIL EMPENNAGE VUE LATERALE ECH : 1/1
DETAIL EMPENNAGE VUE DE DESSOUS ECH : 1/1
A PLAT 1200
LONG. H.T. 970

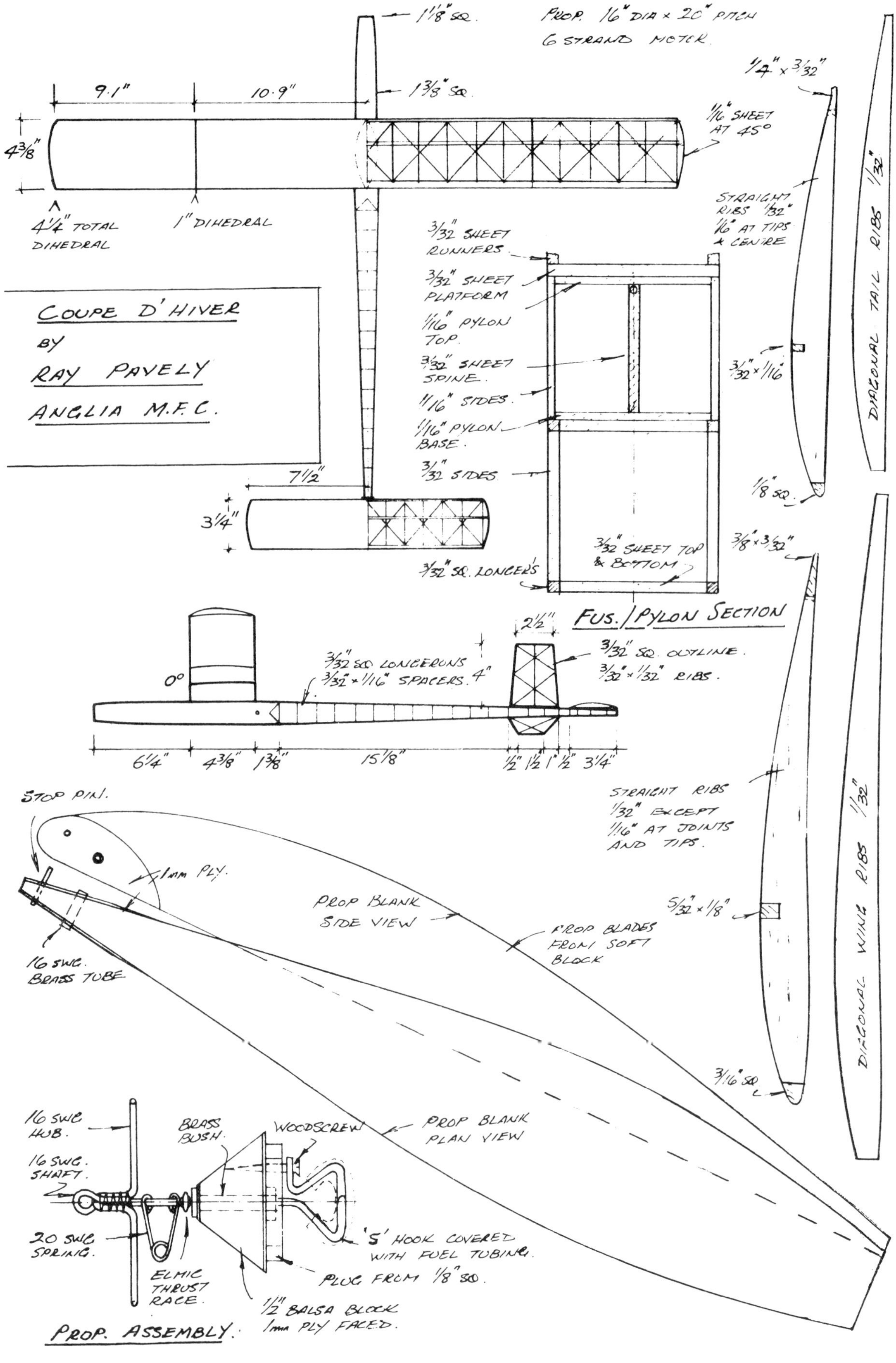

PROP. 16" DIA × 20" PITCH
6 STRAND MOTOR.
COUPE D'HIVER
BY
RAY PAVELY
ANGLIA M.F.C.
FUS./PYLON SECTION
PROP. ASSEMBLY.

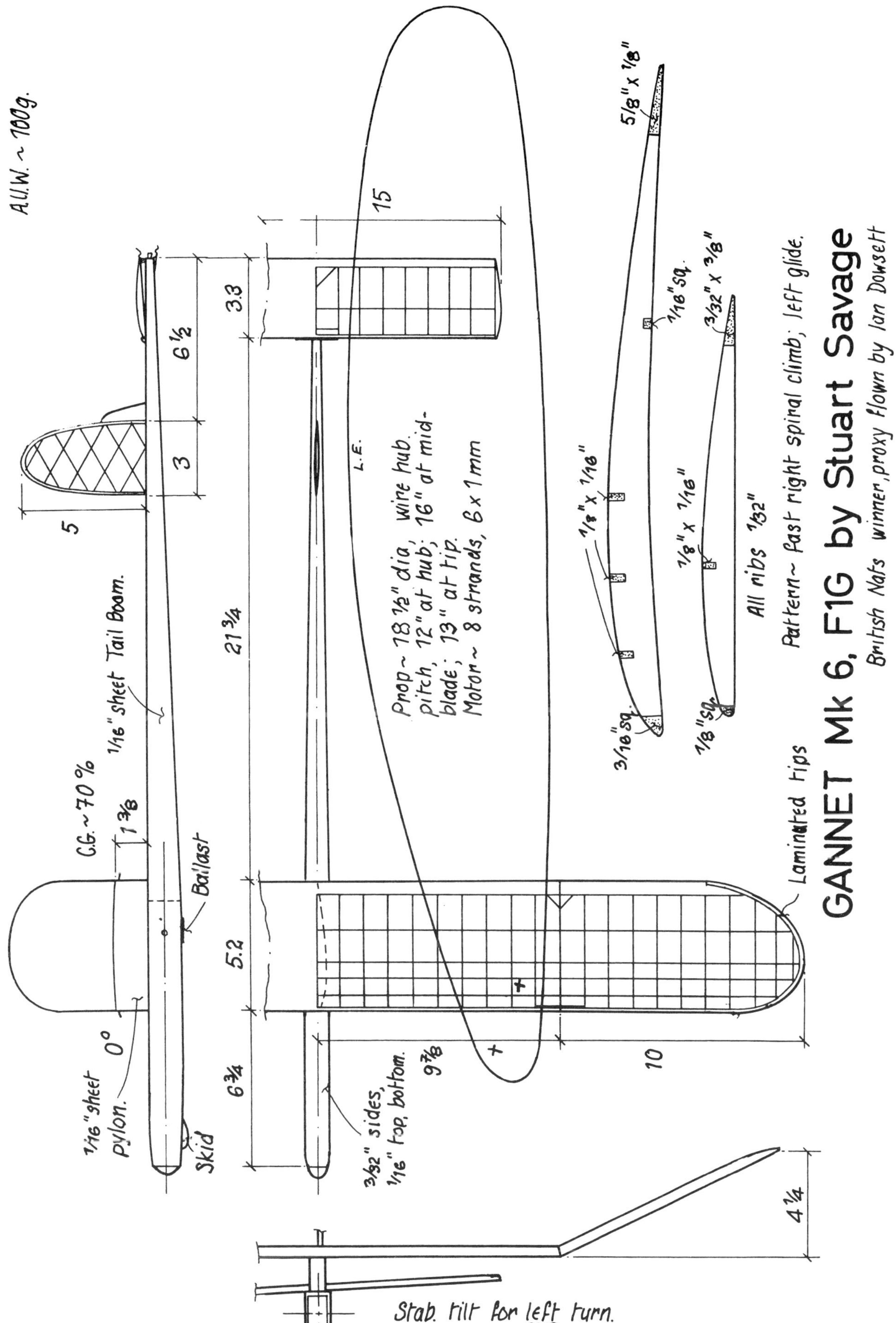
GANNET Mk 6, F1G by Stuart Savage
British Nats winner, proxy flown by Ian Dowsett
Pattern ~ fast right spiral climb; left glide.
A.U.W. ~ 100g.
Prop ~ 18 ½" dia, wire hub.
pitch, 12" at hub; 16" at mid-blade; 13" at tip.
Motor ~ 8 strands, 6 x 1 mm
All ribs 1/32"
Laminated tips
L.E.
C.G. ~ 70%
1/16" sheet Tail Boom.
1/16" sheet pylon.
Ballast
Skid
3/32" sides, 1/16" top, bottom.
Stab. tilt for left turn.
5/8" x 1/8"
1/16" sq.
3/32" x 3/8"
1/8" x 1/16"
1/8" x 1/16"
3/16" sq.
1/8" sq.
15
3.3
6 ½
3
5
21 ¾
1 3/8
0°
5.2
6 ¾
9 7/8
10
4 ¼

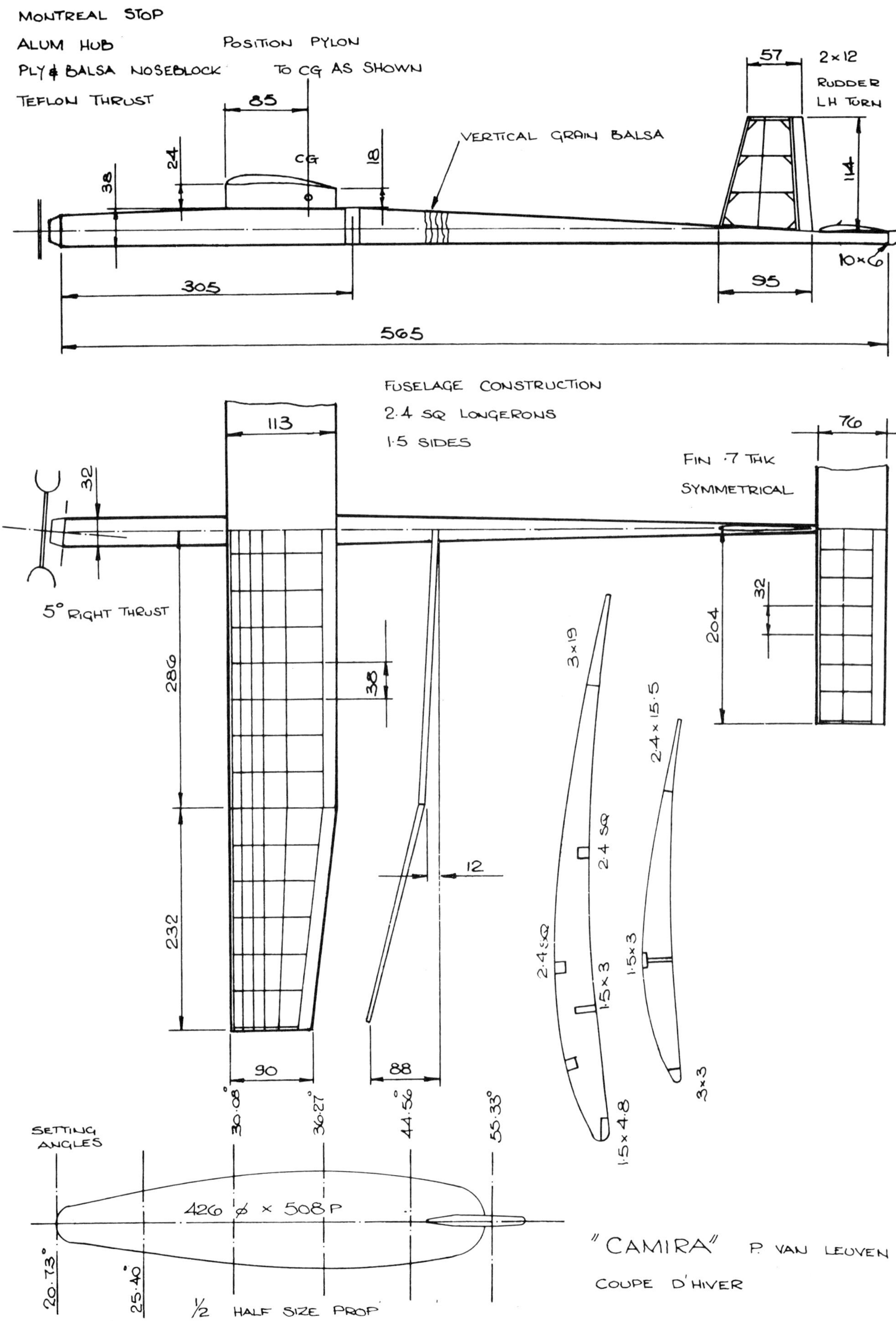
MONTREAL STOP
ALUM HUB
PLY & BALSA NOSEBLOCK
TEFLON THRUST
POSITION PYLON
TO CG AS SHOWN
VERTICAL GRAIN BALSA
CG
RUDDER
LH TURN
FUSELAGE CONSTRUCTION
2.4 SQ LONGERONS
1.5 SIDES
FIN ·7 THK
SYMMETRICAL
5° RIGHT THRUST
SETTING ANGLES
426 ⌀ × 508 P
½ HALF SIZE PROP
"CAMIRA" P. VAN LEUVEN
COUPE D'HIVER

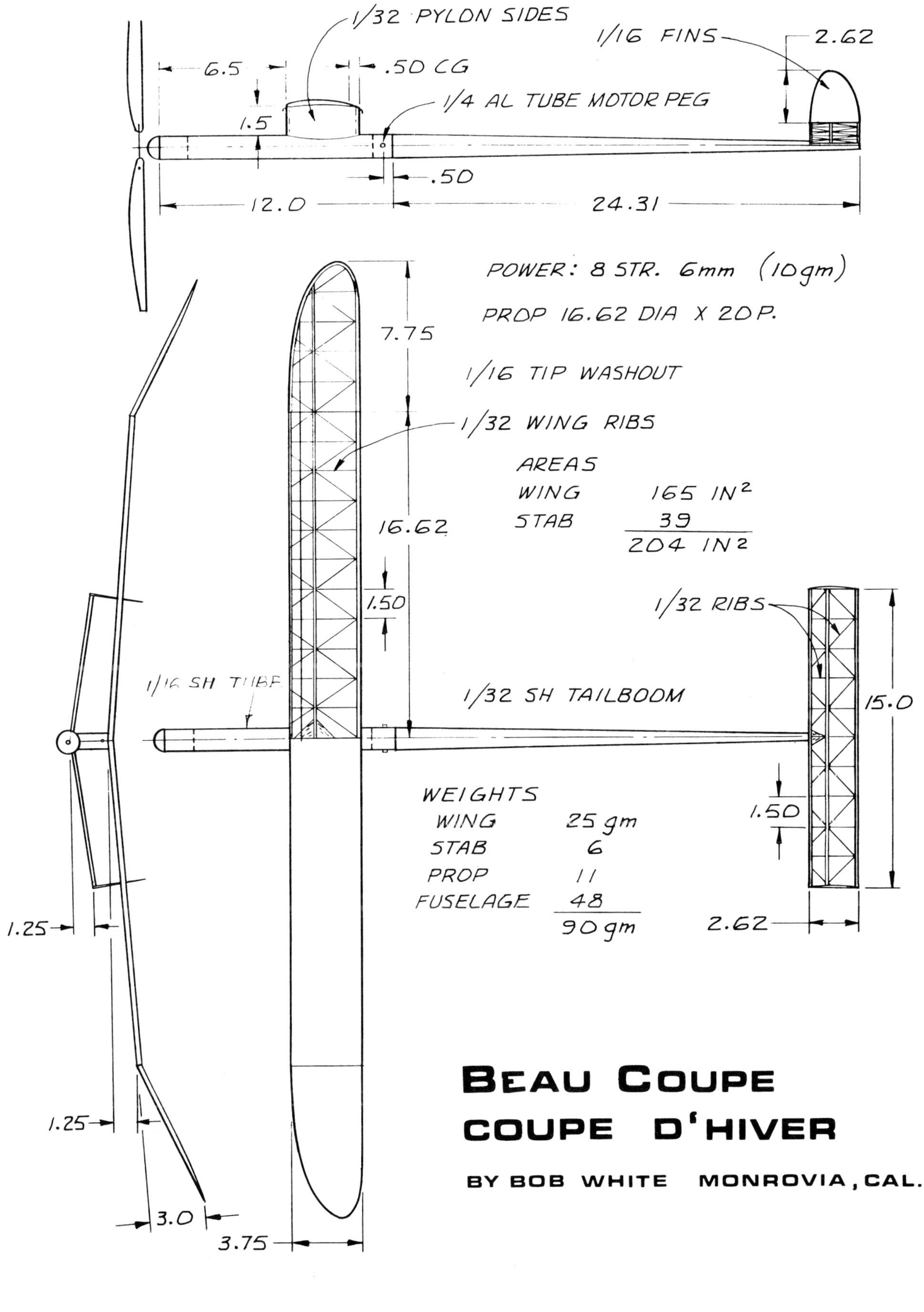
1/32 PYLON SIDES
1/16 FINS
2.62
6.5
.50 CG
1/4 AL TUBE MOTOR PEG
1.5
.50
12.0
24.31
POWER: 8 STR. 6mm (10gm)
PROP 16.62 DIA X 20 P.
7.75
1/16 TIP WASHOUT
1/32 WING RIBS
AREAS
WING 165 IN²
STAB 39
204 IN²
16.62
1.50
1/32 RIBS
1/16 SH TUBE
1/32 SH TAILBOOM
15.0
WEIGHTS
WING 25 gm
STAB 6
PROP 11
FUSELAGE 48
90 gm
1.50
1.25
2.62
1.25
3.0
3.75
BEAU COUPE
COUPE D'HIVER
BY BOB WHITE MONROVIA, CAL.

Taft, California.

9 Federation Aeronautique International

Organization

Aviators had no sooner figured out how to fly than it was decided that an international organization was needed. On June 10, 1905, at the Olympic Congress in Brussels, Belgium, Count de la Vaulx, then vice president of the Aero club of France; Fernand Jacobs, president of the Aero Club of Belgium; and Major Moedebeck of the German Airship League, presented their views on creating an Aero Federation. As a direct result the Olympic Congress passed the following resolution:

The congress recognizing the special importance of aeronautics, expresses the wish that there be formed in each country an association having the mission of regulating aeronautic sport and that there be formed later a world federation of aeronautics, including all the national associations, with a view to various needs and general regulations for the popularization of aeronautics as a science and a sport.

On October 12 of the same year a founders meeting was convened in Paris with representatives from Belgium, France, Germany, Italy, Spain, Switzerland, England and the United States attending. In two days of planning the FAI came into being with H.I.H. Prince Roland Bonaparte of France as the first president.

The FAI has grown steadily and now lists 64 member nations. Every form of aeronautics and astronautics including aeromodeling falls within its charter. The member nations also have their own national aero clubs which are affiliated to the FAI. The FAI can recognize only one national aero club per country. In the United States for example the National Aeronautics Association, chartered in 1922 (successor to the Aero Club of America) is the sole US representative of the FAI and as such, sanctions and certifies record flights and sponsors US teams participating in world sport aviation championships.

World Championships, Roskilde, Denmark, a Federation Aeronautique International Enterprise.

The NAA has a number of subdivisions that specialize in particular branches of sporting aviation. The aeromodeling division of NAA is the Academy of Model Aeronautics (AMA). Other countries have similar organizational arrangements such as the Royal Aero Club/Society of Model Aeronautical Engineers (SMAE) of England, and the Royal Danish Aeroclub/Danish Model Flying Union, for example.

Headquarters for the FAI is at 6 Rue Galilee, Paris, and is structured with a rather complex pyramidal committee system. Working down from the top we first come to the *International Committee for Sport Aviation* known as CASI. This group has overall jurisdiction over various procedural matters affecting the rules making of lower level special interest committees. The one we are interested in is the *Committee for International Aero Modeling* (CIAM). This committee meets at least once each year, usually in December, to fix the rules for competitions and world records for the future. The meeting also establishes the calendar for World Championships and other international events, elects officers and subcommittee chairmen, and conducts other business needing attention. The officers of the CIAM, a group referred to as the *Bureau*, meet in an interim meeting each year usually in April to conduct business in executive session. The CIAM also has a number of sub-committees including one for Free Flight. The chairman of the Free Flight subcommittee is elected by the full CIAM committee. he then appoints members to his subcommittee. Current Free Flight subcomittee chairman is Mr. Peter Allnutt, RR9, Brampton, Ontario, Canada.

Paris headquarters of the FAI at 6 Rue Galilee.

Le Grand Pub Lady Hamilton at l'Arc de Triomphe, around the corner from FAI headquarters.

Competition Rules

Indoor Rubber - (FAI)	1975	1976	1977
General			
Number of flights	6	6	6
Number of models	unlimited		
Maximum, secs	unlimited		
Scoring	Total of best two flights		
Indoor F1D			
Rubber weight, grams	unlimited		
Airframe weight, (min.) grams	1.0	1.0	1.0

Coupe d'Hiver (F1G)	1975	1976	1977
General			
FAI Status	provisional		yes
Number of flights	5	5	5
Number of models (max)	3	3	3
Maximum, secs	120	120	120
Fly-off max (progression)	30 sec/round		
Specifications CdH (F1G)			
Rubber weight, grams	*10*	*10*	*10*
Airframe weight, grams (FAI)	*70*	*70*	*70*
(M Bayet)	*90*	*90*	*90*
Fuselage cross section, cm^2	*20*	*20*	*20*

Outdoor Free Flight -(FAI)	1975	1976	1977
General			
Number of flights	7	7	7
Number of models (max.)	3	3	3
Maximum, secs	180	180	180
Nordic Glider F1A			
Towline length, meters	*50*	*50*	*50*
Pull test, kg	2	2	2
Total surface area, dm^2	*32/34*	*32/34*	32/34
Min. weight, grams	*410*	*410*	*410*
Fly – off, max, (progression)	60 sec/ round		
Wakefield F1B			
Rubber weight, grams	40	40	40
Total surface area, dm^2	17/19	17/19	17/19
Weight less motor, grams	190	190	190
Fly-off, max (progression)	60 sec/round		
Power F1C			
Engine displacement, max, cc	2.5	2.5	2.5
Fuel (glow) (methanol/castor)	80/20	80/20	80/20
Fuel (diesel)	unrestricted		
Engine run, sec	10	7	7
Loading, min/max, g/dm^2	20/50	20/50	20/50
Min. wt., g/cc displacement	300	300	300
Fly-off engine run, secs	8,6,4	7	7
Fly-off max	180	180	+ 60/rd.

Meeting Calendar *1975*

X Amsterdam Cup - 10/11 May - F1A, F1B, F1C * †
Rozendaalse Heath, Netherlands

Criterium du Nord - 17/18 May - F1A, F1B, F1C †
Mauberge, France

U.S. Free Flight Champs - 24/25/26 May - All events †
Taft, California, U.S.A.

United Kingdom Nationals - 24/26 May - Outdoor F/F
RAF Finningley, England

VII Eiffel Cup - 7/8 June - F1A, F1B, F1C
Drive, W Germany

United States Nationals - 3/10 July - All events
Lake Charles, Louisiana, U.S.A.

Munchen '75 - 5/6 July - F1A, F1B, F1C
Munich, W Germany

Indoor FAI - 14/15 July - F1D †
Brno, Czechoslovakia

Garcia Morato - 25/29 July - F1A, F1B, F1C
Alicante, Spain

World Championships - 15/20 Aug. - F1A, F1B, F1C * †
Plovdiv, Bulgaria

Hajdu Cup - 17/20 Aug. - F1D
Debrecen, Hungary

Criterium Pierre Trebod - 23/24 Aug. - F1A,F1B,F1C * †
Marigny-le-Grand, France

XI International Kolibri Cup - 28/31 Aug. - F1A, F1E
Voltendorf, Austria

Raba Cup - 20/21 Sept. - F1D
Per, Hungary

Free Flight News International - 25/26 Oct. - F1A, F1B, F1C
RAF Sculthorpe, England

WFFA Challenge - 8/9 Nov. - F1A, F1B, F1C †
Taft, California, U.S.A.

I Coupe d/Hiver International - 30 Nov. - F1G * †
RAF Halton, England

1976

XXXI Coupe d'Hiver MRA Anglo French Challenge - F1G
Gatinais/Montargis, France

Amsterdam Cup - 20/30 May - F1A, F1Bl, F1C * †
Rozendaalse Heide, Netherlands

US Free Flight Champs - 29/30/31 May - All events †
Taft, California, U.S.A.

Trans Tasman Challenge - 12/13/14 June - F1A, F1B, F1C
Amberly AFB, Brisbane, Australia

VIII Eiffel Cup - 19/20 June - F1A, F1B, F1C
Drove, W Germany

IX International Contest - 3/4 July - F1A, F1B, F1C
Munich, W Germany

Garcia Morato International - 16/19 July - F1A, F1B, F1C
Lerida, Spain

European F1C Champs - 28 July/2 Aug. - F1C †
Zagreb, Yugoslavia

United States Nationals - 2/7 Aug. - All events
Springfield, Ohio, U.S.A.

United Kingdom Nationals - 14/15 Aug. - Outdoor F/F
Little Rissington, England

Criterium Pierre Trebod - 21/22 Aug. - F1A, F1B, F1C * †
Marigny-le-Grand, France

FAI Indoor World Champs - 27/31 Aug. - F1D * †
RAE, Cardington, England

European F/F Champs - 10/12 Sept. -F1A, F1B, F1C †
Homburg/Saar, W Germany

WFFA Challenge - 20/21 Nov. - F1A, F1B, F1C * †
Taft, California U.S.A.

II Coupe d'Hiver International - 5 Dec. - F1G †
RAF Halton, England

XXX Australian Nationals - 28 Dec./3 Jan. - All events
Loxton, South Australia

1977

XXX Campeonato Nacional de aeromodelism - 7/10 April
Junin, Buenos Aires, Argentina

US Free Flight Champs - 28/29/30 May - All events * †
Taft, California, U.S.A.

Kratky cup - 9/12 June - F1A, F1B, F1C
Wiener Neustadt, Austria

SMAE/FFN FAI International - 26/27 June - F1A, F1B, F1C †
Bircham Newton, England

World Championships - 6/12 July - F1A, F1B, F1C * †
Roskilde, Denmark

Criterium Pierre Trebod - 16/17 July - F1A, F1B, F1C * †
Marigny-le-Grand, France

Canadian National - 16/24 July
Huron Park, Ontario, Canada

United States Nationals - 6/14 Aug. - All events
Riverside, California, U.S.A.

Journees Intl. de Vol Libre - 5/6/7 Aug. - F1A, F1B, F1C, F1G
Parthenay, France

Hajdu Cup - 18/20 Aug. - F1D
Debrecen, Hungary

Czech International 19/21 Aug. - F1A, F1B, F1C * †
Sezimovo Usti, Czechoslovakia

United Kingdom Nationals - 20/21 Aug. - Outdoor F/F
RAF Little Rissington, England

Free Flight - 19/20 Aug. - F1A, F1B
Ze'elim, Israel

XII Kolibri Cup - 25/28 Aug. - F1A, F1B, F1C, F1E
Herzogenburg/Weiner Neustadt, Austria

IX Intl. Eiffel Pokal - 27/28 Aug. - F1A, F1B, F1C
Zulprich, W Germany

Balkan Games - 25/Aug., 3/Sept. - F1A, F1B, F1C
Plovdiv, Bulgaria

European F/F Champs - 10/13 Sept.
Homburg-Miesau, W Germany

Indoor F/F - 7/9 Oct. - F1D
Nyiregyhaza, Romania

WFFA Challenge - 19/20 Nov. - F1A, F1B, F1C †
Taft, California, U.S.A.

III Australian Champs - 28 Dec./4 Jan. - All events
Camperdown, Victoria, Australia

* See story in *Competition Report*

† See results in *Competition Report*

Swedish Glider Cup World Championships Finale, Roskilde, Denmark, 1977.

10 Competition Record

Swedish Glider Cup.

The Year 1975

X Amsterdam Cup - 10/11 May - F1A, F1B, F1C
Rozendaalse Heath, Netherlands

A2 F.1.A.

	Land	1e	2e	3e	4e	5e	6e	7e	totaal
1. Baines	GB	180.	180.	180.	180.	180.	180.	162.	1242
2. Deubel	DL	161.	176.	180.	172.	177.	180.	180.	1226
3. Allnutt	CAN	180.	122.	180.	180.	180.	180.	180.	1202
4. Wilkening F.	DL	156.	164.	180.	180.	180.	180.	180.	1198
5. Erard J.	FR	180.	180.	180.	180.	112.	180.	180.	1192
6. Kappelhof	NL	180.	109.	180.	180.	180.	180.	180.	1189
7. Koster	DM	180.	180.	130.	180.	180.	180.	146.	1176
8. Breeman	NL	180.	172.	180.	180.	180.	101.	180.	1173
9. Verbree	NL	141.	128.	180.	180.	180.	180.	180.	1169
10. Erard	FR	180.	174.	180.	180	180.	108.	164.	1166
11. v.Eldik	NL	150.	180.	180.	180.	180.	180.	112.	1162
12. Lamers	NL	80.	180.	180.	180.	180.	180.	180.	1160
13. Wilkening H.	DL	180.	180.	180.	180.	158.	98.	156.	1132
14. Althof	DL	180.	134.	109.	180.	180.	180.	150.	1113
15. Warren	GB	93.	117.	180.	180.	180.	180.	180.	1110
16. v.Oostrom	NL	79.	180.	180.	180.	180.	180.	120.	1099
17. Wanngard	ZW	180.	45.	180.	146.	180.	180.	180.	1091
18. Grunnet	DM	180.	160.	54.	180.	180.	180.	144.	1078
19. Klink	DL	180.	104.	146.	180.	106.	180.	180.	1076
20. Steinacker	DL	28.	180.	155.	180.	164.	180.	180.	1067
21. Jansen	NL	123.	180.	180.	77.	180.	144.	180.	1064
22. Hohls	DL	55.	180.	108.	180.	180.	180.	180.	1063
23. Pinter	DL	145.	180.	115.	180.	180.	180.	77.	1057
24. de Ruwe	NL	180.	67.	180.	180.	114.	180.	144.	1045
25. Godinno	FR	128.	180.	57.	180.	180.	138.	180.	1043
26. Zeibig Dr.	DL	155.	118.	180.	180.	180.	76.	152.	1041
27. Brandenburg	NL	180.	97.	180.	177.	103.	180.	124.	1041
28. Baguley	GB	180.	86.	180.	176.	57.	180.	180.	1039
29. Brinks	NL	180.	180.	93.	109.	115.	180.	175.	1032
30. Peper	DL	59.	99.	180.	180.	180.	180.	144.	1022
31. Merkesteijn	NL	121.	59.	180.	180.	180.	112.	180	1012
32. LENDERING	NL	180.	180.	180.	180.	106.	114.	70.	1010
33. Dilly	GB	132.	101.	180.	180.	180.	125.	105.	1003
34. Muller	DL	130.	121.	180.	85.	126.	180.	180.	1002
35. Nocgue	FR	146.	59.	69.	180.	180.	180.	180.	994
36. Bünte	DL	53.	103.	180.	180.	166.	120.	180.	982
37. Dunkerbeck	DM	68.	156.	180.	107.	180.	105.	180.	976
38. Heidemann	DL	92.	96.	121.	180.	180.	111.	180.	960
39. Schallenberg H.G.	DL	180.	84.	180.	158.	112.	180.	58.	952
40. Gregorie	GB	180.	180.	128.	46.	180.	41.	180.	935
41. Lammerts	NL	171.	180.	76.	180.	71.	180.	71.	929
42. Fiks	NL	143.	33.	150.	57.	180.	180.	180.	923
43. Ferrou	FR	49.	180.	180.	180.	126.	26.	180.	921
44. Ottenhof	NL	69.	35.	73.	180.	180.	180.	180.	897
45. Litjens	NL	71.	129.	170.	180.	66.	94.	180.	890
46. De Boer	NL	180.	79.	92.	180.	180.	174.	48.	883
47. Bogaerts	BL	58.	83.	124.	180.	180.	135.	118.	878
48. Du Long	NL	115.	42.	180.	118.	180.	180.	47.	862
49. Masterman J.	GB	164.	124.	56.	180.	62.	180.	69.	835
50. Lammertink	NL	139.	180.	81.	170.	180.	69.	-	819
51. Kuhl	DL	48.	59.	79.	90.	180.	180.	180.	816
52. Müller B.	DL	155.	51.	180.	180.	75.	58.	102.	801
53. Müller G.	DL	66.	24.	80.	180.	180.	80.	180.	790
54. Jacobs	NL	99.	120.	135.	180.	106.	65.	75.	780
55. De Kruyff	NL	82.	45.	180.	35.	180.	99.	83.	704
56. v.Eede	NL	66.	-	-	180.	180.	180.	68.	674
57. Brouwer	NL	85.	32.	48.	180.	73.	55.	42.	515
58. Buge	DL	153.	180.	138.	-	-	-	-	471
59. Digby	GB	150.	180.	79.	-	-	-	-	409
60. Nielsen	DM	180.	-	145.	77.	-	-	-	402
61. Schaafsma	NL	151.	28.	178.	-	-	-	-	357
62. Schley	DL	90.	60.	105.	-	-	-	-	255
63. Piesk	DL	68.	23.	-	-	-	-	-	91
64. Gaull	DL	81.	-	-	-	-	-	-	81
65. Imgenberg	DL	46.	17.	-	-	-	-	-	63
66. Masterman P.	GB	58.	-	-	-	-	-	-	58
67. Weiland	DL	-	-	-	-	-	-	-	-

WAKEFIELD F.1.B.

1. Haacken	DL	156.	180.	180.	180.	180.	180.	180.	1236
2. Gialanella	I	177.	137.	180.	180.	180.	150.	180.	1184
3. Kongsberg	DM	180.	132.	180.	180.	180.	180.	140.	1172
4. Barnes	GB	90.	180.	180.	180.	180.	180.	180.	1170
5. Borchia	I	180.	114.	180.	180.	180.	180.	145.	1159
6. Marini	I	130.	180.	165.	124.	180.	180.	166.	1125
7. Boeren	DL	172.	180.	180.	180.	180.	132.	99.	1123
8. v. Hoorn	NL	137.	180.	180.	180.	78.	180.	180.	1115
9. Lonardi	I	165.	135.	101.	180.	180.	180.	150.	1091
10. De Ruyter	NL	83.	180.	180.	106.	180.	180.	168.	1077
11. Mönninghof	DL	180.	180.	107.	180.	158.	75.	180.	1060
12. Gaull	DL	164.	123.	138.	180.	129.	180.	126.	1040
13. Baguley	GB	100.	180.	72.	180.	180.	180.	93.	985
14. Manche	NL	73.	180.	180.	143.	68.	120.	156.	920
15. Ganbauge	DL	161.	50.	180.	180.	180.	63.	93.	907
16. Digby	GB	180.	78.	180.	142.	115.	132.	42.	869
17. Renk	DL	67.	162.	102.	180.	180.	-	-	691
18. Leveuvre	FR	84.	180.	180.	66.	180.	-	-	690
19. Durand	FR	105.	147.	109.	115.	180.	-	-	656
20. Masterman P.	GB	180.	3.	72.	50.	75.	-	-	380
21. Zeibig	DL	180.	119.	49.	-	-	-	-	348
22. Stöhr	DL	102.	-	-	-	-	-	-	102

POWER F.1.C.

									totaal	
1. Køster	DM	180.	180.	180.	180.	180.	180.	180.	1260	+3 x 180
2. Schwend	DL	180.	180.	180.	180.	180.	180.	180.	1260.	+ 2x 180 + 86
3. Broderac.J.	DL	180.	180.	180.	180.	180.	180.	180.	1260.	+ 180+166
4. Kuhl	DL	148.	180.	180.	180.	180.	165.	180.		1213
5. Maier E.	DL	130.	180.	180.	180.	180.	180.	180.		1210
6. Schallenberg	DL	180.	180.	180.	180.	180.	180.	98.		1178
7. Schley	DL	125.	180.	180.	180.	147.	180.	180.		1172
8. v.Uden	NL	180.	180.	102.	180.	155.	180.	180.		1157
9. Zeh	DL	122.	119.	180.	180.	180.	180.	180.		1141
10. Faux	GB	180.	155.	77.	104.	180.	180.	180.		1056
11. Borczewski E.	DL	23.	165.	180.	145.	180.	139.	180.		1012
12. Gröbl	DL	161.	-	-	157.	180.	180.	180.		858
13. Weber	DL	180.	146.	156.	180.	180.	-	-		842
14. Broderac. D.	DL	75.	101.	180.	180.	-	-	-		536
15. Imgenberg	DL	180.	123.	-	-	-	-	-		303
16. Heideman G.	DL	180.	-	-	-	-	-	-		180

BIJZ. TYPEN F.A.I.

1. De Kruyff	NL	78.	123.	71.	44.	130.	91.	93.	630
2. Leeuwangh	NL	56.	90.	102.	180.	34.	47.	52.	561
3. Jandt F.	DL	49.	91.	37.	74.	66.	49.	68.	334
4. Jandt P.	DL	30.	75.	77.	25.	15.	-	-	222

Criterium du Nord - 17/18 May - F1A, F1B, F1C
Mauberge, France

Glider F1A 71 flew

1 Brouez	F	180	180	180	180	180	180	180	=	1260	+240
2 Leeuwangh	NL	180	180	180	180	180	180	180	=	1260	+ 87
3 Allnutt	CDN	180	180	180	180	169	180	180	=	1249	
4 Depestel	B	180	180	180	180	148	180	180	=	1228	
5 Kleuskens	NL	168	180	180	180	104	180	180	=	1172	
6 De Boer	NL	180	180	180	180	82	180	180	=	1162	
7 Dorn	F	180	180	180	180	142	180	116	=	1158	
8 Schallenberg	BRD	180	180	180	180	110	147	180	=	1157	
9 Buge	BRD	175	180	180	118	180	180	120	=	1133	
10 Breeman	NL	180	180	180	180	180	180	45	=	1125	

Wakefield F1B 12 flew

1 Haacke	BRD	180	180	180	180	146	180	117	=	1163	+ 240
2 Graux	B	180	180	180	180	126	147	170	=	1163	+ 77
3 Helglais	F	180	180	180	180	162	142	115	=	1139	
4 Lefeuvre	F	106	180	180	180	145	180	86	=	1057	

Power F1C 15 flew

1 Simpson (NL)	USA	180	180	180	180	180	180	180	=	1260
2 D Broderac	BRD	180	180	150	180	180	180	180	=	1230
3 Stetz	BRD	180	145	180	180	180	180	180	=	1225
4 Seydel	BRD	180	180	180	173	180	138	180	=	1211
5 Schley	BRD	171	166	180	180	180	135	180	=	1192

U.S. Free Flight Champs - 24/25/26 May - All events
Taft, California, U.S.A.

30 Sec. Antique	
Al Hellman	13:18
Marc Tackett	13:05
Bruce Norman	12:08
Sal Taibi	11:18
Bob Longstreet	9:21

02 Replica	
Wade Wiley	15:00
Bruce Hannah, Jr.	14:20
Jim Farmer	14:09
Jim Persson	13:57
Rudy Calvo	12:00

OT Rubber	
Jim Persson	22:00
Bob Oslan	19:48
Gene Wallock	15:03
Jim Quinn	14:51
Vincent Costanzo	13:08

B Old Timer	
Bruce Norman	19:11
Greg Rasmussen	13:56
Al Rasmussen	12:16
Jerry Vernon	11:49
R.H. Osenga	11:37

C Old Timer	
Bob Oslan	13:23
Ron Martin	12:05
Bruce Norman	11:57
Walt Parker	11:26
Spiro Nicholaw	9:52

C Gas	
Bob Scully	38:57
Jon James	35:00
Bill Hunter	33:23
Scotty Harte	32:18
Ed Cadwell	31:35

B-C Junior	
Tim Young	15:58
Mark Kerzi	12:19
Scott Valentine	11:16
Steve Calhoun	9:58
Fred Calhoun	8:16

B Gas	
Ken Bauer	25:00
Glenn Schneider	25:00
Al Hotard	22:05
Red Johnson	17:30
Dave Lewis	17:19

Nite Flying	
Bob DeShields	19:26
Hulan Mathies	19:10
Jon James	18:54
Scotty Harte	18:47
Dick Myers	18:40

A Gas Open	
Guy Kirkwood	43:30
Mel Schmidt	30:00
Jack Moreland	27:51
Jim Haught	26:39
Richard Todd	21:42

A/1 Junior	
Bill Xenakis	12:29
Steve Kaiser	10:51
Jeff Livotto	10:46
Eric Dyer	10:16
David Bowser	8:18

A/1 Open	
Randy Secor	25:47
Jim Haught	23:42
Jim Farmer	21:51
Lee Hunt	21:31
Don McGhee	21:00

A Old Timer	
Bruce Norman	11:58
Jack Jella	11:30
Bruce Hannah	8:28
L. Nadolski	7:21
Spiro Nicholaw	6:31

Wakefield	
Walt Ghio	900
Jim Parker	892
Ed Wogulis	858
Bob Piserchio	834
Bob White	818

OHLG Jr.	
Eric Dyer	6:17
Steve Wittman	5:33
Marvin Miller	4:34
Steve Calhoun	3:51
Jeff Cunnyngham	3:45

1/2A Open	
Debbie Beron	31:26
Randy Secor	23:31
Jim Kelley	19:49
Dave Lewis	17:31
Howard Adamezuk	16:51

1/2A Jr.	
Eric Dyer	16:21
G. Voss	15:00
Bart Delbridge	13:14
Mark Kerzie	12:34
David Moyer	12:13

OHLG	
Dan Bragg	8:28
Bob DeShields	8:00
Ray Harper	7:37
Don Elliott	6:47
Ron Wittman	6:39

FAI Power	
Ed Carroll	1800
Reid Simpson	1713
Dave Parsons	894
Don McGhee	868
Walt Ghio	852

Indoor HLG	
Lee Hines	:60.5
Ron Wittman	:59.2
Bob DeShields	:56.6
Bob Meuser	:52.1
Randy Wallingford	:49.8

Penny Plane	
Clarence Mather	6:02
Earl Hoffman	6:01
Hal Cover	5:39
Dave Parsons	5:34
Fudo Takagi	5:16

Peanut	
C. Mather-Cougar	3
F. Takagi-Volksplane	8
M. Mulligan-J-3	8
H. Cover-Skycoupe	10
J. Person-Cougar	10

E-Z B	
Clarence Mather	6:41
Fudo Takagi	6:31
Earl Hoffman	6:28
Mark Valerius	4:00
Ken Bauer	2:18

A Gas Jr.	
Scott Valentine	16:42
Robert Langberg	13:27
Jack Moreland, Jr.	13:22
Jeff Cunnyngham	13:19
Fred Calhoun	12:24

Rubber Scale
Hal Cover
Brick Brickner
Mike Mulligan
Matt Johnson
Jim Persson

Payload	
Scott Valentine	7:18
Jim Haught	7:04
Dave Sbur, Jr.	5:51
Bill Valentine	4:20
Dave Sbur, Sr.	2:16

Rocket	
Jim Crocket	10:41
John Jennings	9:44
Ron Wittman	8:56
Irv Aker	7:41
Jim Haught	4:56

A-2 Nordic	
Jim Parker	1500
Jose Ramirez(MEXICO)	1322
Bob Isaacson	1319
Jose Doña	1207
Tom Hutchinson	1030

Unlimited Rubber	
Bob Tymcheck	1680
Del Adam	902
Jim Quinn	900
Irv Aker	879
Bob Piserchio	856

Unlimited Rubber Jr.	
Tim Young	647
Eric Dyer	620
Robert Langenberg	489
Bruce Armstrong	358
Walt Armstrong	354

Coupe D'Hiver	
Tom Medley	930
Irv Aker	734
Bob Tymchek	659
Al Hotard	575
Mark Valerius	546

D Gas	
Bill Hunter	22:40
Scotty Harte	22:08
Bill Moore	18:25
Tom Thompson	17:42
Fred Calhoun	14:11

Indoor HLG Jr.	
Steve Wittman	:45.4
Dan Sbur	:40.9
Walt Armstrong	:38.3
Tim Young	:37.5
Jim Laughlin	:36.1

Gas Scale
Bill Stroman
Bill Warner
Hal Cover
Gene Simpson
Brick Brickner

Grand Champion
Jim Haught

Senior	
Ken Bauer	378
Randy Secor	313
Jim Farmer	269
Jim Kelley	172
Bob Boyer	144

Junior	
Eric Dyer	520
Tim Young	503
Scott Valentine	441
Steve Calhoun	255
David Bowser	251

Open	
Bob DeShields	417
Mark Valerius	315
Jim Scarborough	313
Jack Moreland	309
Mel Schmidt	280

Teams	
PEARL #2; Moore, Simpson, Caldwell -	402
MAX MEN; Clark, Isaacson, White -	321
X; Elliott, Kelley, Valentine -	321
HOBBY FAIR; DeShields, Harte, Hunt -	319
DUST DEVILS; Kaiser, Moreland, Scarborough -	310

Indoor FAI - 14/15 July - F1D

Brno, Czechoslovakia

1	J Kalina	CZ	34.39	13.40	12.58	27.42	32.15	16.31	= 66.54
2	E Ciapala	PL	29.23	34.10	14.54	20.20	30.18	14.12	= 64.28
3	E Chlubny	CZ	9.15	23.50	12.04	31.02	32.26	10.58	= 63.28
4	R Czechowski	PL	26.56	23.24	29.14	28.35	30.16	23.12	= 59.30
7	L Barr	GB	7.56	4.17	7.10	25.22	31.55	13.15	= 57.17

World Championships - 15/20 Aug. - F1A, F1B, F1C

Plovdiv, Bulgaria

F1a Nordic

СВЕТОВНО ПЪРВЕНСТВО
ПО АВИОМОДЕЛИЗЪМ
Н. Р. БЪЛГАРИЯ, ПЛОВДИВ
15 20 АВГУСТ 1975 ГОДИНА

1	Tchop	SU	M	M	M	M	M	M	M	1260	+240	+300
2	Allnutt	CAN	M	M	M	M	M	M	M	1260	+240	+131
3	Henke	DDR	M	M	M	M	M	M	M	1260	+233	
4	Horeisi	CS	M	M	M	M	M	M	M	1260	+224	
5	Challin	FR	M	M	M	M	M	M	M	1260	+210	
6	Widensek	YU	M	M	M	M	M	M	M	1260	+203	
7	Fernandez	SP	M	M	M	M	M	M	M	1260	+191	
8	Korczak	POL	M	M	M	M	M	M	M	1260	+186	
9	Bucher	SCH	M	M	M	M	M	M	M	1260	+182	
10	Lepp	SU	M	M	M	M	M	M	M	1260	+182	
11	Gang	PRK	M	M	M	M	M	M	M	1260	+180	
12	Lustig	DDR	M	M	M	M	M	M	M	1260	+178	
13	Sillgren	FIN	M	M	M	M	M	M	M	1260	+172	
14	Isaenko	SU	M	M	M	M	M	M	M	1260	+171	
	Jeskosek	YU	M	M	M	M	M	M	M	1260	+171	
16	Vollbrecht	BDR	M	M	M	M	M	M	M	1260	+167	
17	Palmieri	ARG	M	M	M	M	M	M	M	1260	+162	
18	Voros	HUN	M	M	M	M	M	M	M	1260	+160	
19	Popa	ROM	M	M	M	M	M	M	M	1260	+156	
20	Broberg	SW	M	M	M	M	M	M	M	1260	+153	
21	Weiss	IS	M	M	M	M	M	M	M	1260	+152	
22	Ri	PRK	M	M	M	M	M	M	M	1260	+151	
	Fantham	GB	M	M	M	M	M	M	M	1260	+151	
	Totev	BG	M	M	M	M	M	M	M	1260	+151	
25	Martinez	CU	M	M	M	M	M	M	M	1260	+150	
26	Pop	ROM	M	M	M	M	M	M	M	1260	+149	
27	Ri Chul	PRK	M	M	M	M	M	M	M	1260	+144	
28	Zach	A	M	M	M	M	M	M	M	1260	+143	
29	Sodini	IT	M	M	M	M	M	M	M	1260	+141	
30	Lopez	SP	M	M	M	M	M	M	M	1260	+139	
31	Douglas	NEZ	M	M	M	M	M	M	M	1260	+130	
32	Isaacson	US	M	M	M	M	M	M	M	1260	+128	
33	Bardet	ARG	M	M	M	M	M	M	M	1260	+113	
34	Deubel	BDR	M	M	M	M	M	M	M	1260	+110	
35	Cooper	GB	M	M	M	M	M	M	M	1260	+109	
36	Odemark	NOR	M	M	M	M	M	M	M	1260	+104	
37	Buchwald	DAN	M	M	M	M	M	M	M	1260	+ 99	
	Jivkov	BG	M	M	M	M	M	M	M	1260	+ 99	
39	Dorn	FR	M	M	M	M	M	M	M	1260	+ 89	
40	Gregorie	NEZ	M	M	M	M	M	M	M	1260	+ 31	
41	Ovrutzky	IS	M	M	M	M	M	M	M	1260		
	Scvacsek	HUN	M	M	M	M	M	M	M	1260		

43	Kulmakko	FIN	179	M	M	M	M	M	M	1259
44	Wolf	DDR	M	M	172	M	M	M	M	1252
45	Tumaiti	IT	M	M	169	M	M	M	M	1249
46	Shmelik	A	M	M	M	M	M	M	163	1243
47	Petrescu	ROM	M	M	145	M	M	M	M	1225
	Dvorak	CS	M	M	M	145	M	M	M	1225
49	Massari	YU	M	M	M	M	136	M	M	1216
50	DeLuna	MEX	156	M	M	M	M	150	M	1206
51	DeBoer	HOL	M	165	M	139	M	M	M	1204
52	DeMello	CAN	M	M	127	M	M	M	176	1203
53	Lendering	HOL	M	M	M	134	M	165	173	1192
	Walters	US	M	M	M	M	M	112	M	1192
55	Kappelhof	HOL	M	M	M	M	M	109	M	1189
56	Echtjenkov	SU	M	M	M	M	M	104	M	1184
57	Abadjiev	BG	M	152	M	M	M	M	122	1174
58	Chancey	US	M	M	92	M	M	M	M	1172
59	Gronlund	DAN	M	M	98	M	M	M	171	1169
60	Schmidt	BDR	M	145	M	M	M	122	M	1167
	Olstad	NOR	M	M	M	M	87	M	M	1167
62	Haller	SCH	M	M	M	86	M	M	M	1166
63	Wallertin	SW	M	M	M	M	M	M	83	1163
64	Ramirez	MEX	M	M	M	M	79	175	179	1153
65	Gonzalez	CU	M	M	M	72	M	M	M	1152
	Spann	A	M	M	125	M	M	127	M	1152
67	Kubit	POL	M	M	M	M	85	160	M	1145
68	Diez	CU	M	M	M	M	M	64	M	1144
69	Grunnet	DAN	M	M	M	60	M	M	M	1140
70	Boone	GB	M	56	M	M	M	M	M	1136
71	Gurfein	IS	M	M	M	M	M	86	148	1134
72	Horvath	HUN	114	M	M	M	M	M	95	1109
73	Jarousseau	FR	M	M	M	M	33	M	175	1108
	Varetto	IT	M	M	M	M	97	M	111	1108
75	Sandahl	SW	M	M	M	131	M	74	M	1105
76	Leoni	ARG	M	M	M	M	63	138	M	1101
77	Hubert	CS	M	123	M	M	M	109	146	1098
78	Banasjuk	POL	91	M	M	M	M	M	102	1093
79	Larsen	NOR	M	M	M	82	M	M	81	1063
	Anderson	AUS	M	58	M	M	M	M	105	1063
81	Edwards	AUS	M	M	M	M	51	M	105	1056
82	Kamal	EG	M	M	M	35	M	M	114	1049
83	Tschuor	SCH	M	82	M	M	52	M	M	1034
84	Tahkapaa	FIN	120	M	55	M	91	M	M	986
85	Dimons	AUS	M	176	M	M	0	M	89	985
86	Soliman	EG	M	96	50	114	M	M	M	980
87	Garcia	SP	M	151	M	M	M	24	46	941
88	Thompson	CAN	M	M	107	77	M	M	26	930
89	Shezli	EG	81	142	89	M	M	M	61	913
90	Colmenares	MEX	48	178	M	118	28	73	M	805
91	Gomide	BRA	0	0	0	0	0	0	0	0
92	Charlier	BRA	0	0	0	0	0	0	0	0

F1b
Wakefield

1	Paik	PRK	M	M	M	M	M	M	M	1260	+240	+300	+303
2	White	US	M	M	M	M	M	M	M	1260	+240	+300	+281
3	Zachalmel	A	M	M	M	M	M	M	M	1260	+240	+300	+280
4	Kim	PRK	M	M	M	M	M	M	M	1260	+240	+300	+243
5	Oschatz	DDR	M	M	M	M	M	M	M	1260	+240	+300	+235
6	Artioli	IT	M	M	M	M	M	M	M	1260	+240	+300	+222
7	Vincze	ROM	M	M	M	M	M	M	M	1260	+240	+300	+214
8	Benedini	ARG	M	M	M	M	M	M	M	1260	+224		
9	McGillivray	CAN	M	M	M	M	M	M	M	1260	+212		
10	Hofsass	BRD	M	M	M	M	M	M	M	1260	+199		
11	Kmoch	YU								1260	+193		
12	Pecek	YU	M	M	M	M	M	M	M	1260	+173		
13	Thomas	CAN	M	M	M	M	M	M	M	1260	+171		
14	Kroon	HOL	M	M	M	M	M	M	M	1260	+147		
15	Rado	CS	M	M	179	M	M	M	M	1259			
16	Rashkov	BG	M	M	M	M	M	M	173	1253			
17	Rasmussen	DAN	M	M	M	172	M	M	M	1252			
18	Edwards	AUS	172	M	178	M	M	M	M	1250			
19	Zurad	POL	M	M	M	165	M	M	M	1245			
20	Klima	CS	M	M	M	M	M	M	160	1240			
21	Pentate	CU	152	M	M	M	M	M	M	1232			
22	Boiziau	FR	15C	M	M	M	M	M	M	1230			
23	Lapinski	POL	175	M	M	151	M	M	M	1226			
	Sallinen	FIN	M	M	M	M	M	146	M	1226			
25	Siebenmann	SCH	M	M	M	M	M	136	M	1216			
26	Liwenborg	SW	M	141	M	M	M	M	174	1215			
	Marquez	ARG	M	135	M	M	M	M	M	1215			
28	Valchev	BG	M	M	M	M	M	134	M	1214			
29	Reifler	SCH	M	132	M	M	M	M	M	1212			
30	Itzhak	IS	M	M	M	M	M	M	129	1209			
31	Pollard	GB	M	M	M	M	158	145	M	1203			
	Wlodarczik	POL	123	M	M	M	M	M	M	1203			
33	Cassi	IT	M	117	M	M	M	M	M	1197			
34	Kim Dong	PRK	160	M	M	135	M	M	M	1195			
	Gansli	SCH	M	M	M	M	M	115	M	1195			
	Stoianov	BG	M	M	M	M	M	115	M	1195			
37	Smitz	US	M	M	M	178	123	173	M	1194			
38	Silberg	SU	M	147	M	M	M	145	M	1192			
39	Mielitz	DDR	159	M	M	M	M	132	M	1191			
40	Jurov	SU	M	110	M	M	M	M	M	1190			
41	Gonzales	CU	157	M	M	132	M	M	M	1189			
42	Woodhouse	GB	M	103	M	M	M	M	M	1183			
43	Torgersen	NOR	152	M	M	M	M	127	M	1179			
44	Loffler	DDR	M	M	M	M	162	135	160	1177			
45	Nimptsch	BRD	M	129	M	M	M	146	M	1175			
	Purgai	HUN	111	164	M	M	M	M	M	1175			
47	Davis	US	M	M	110	M	M	M	164	1174			
48	Soininen	FIN	M	164	M	108	M	M	M	1172			
49	Cancelli	IT	M	119	M	M	M	M	149	1168			
	Gouverne	FR	M	M	M	M	117	155	176	1168			
51	Brand	IS	M	M	M	M	80	M	M	1160			
52	Leisner	BRD	M	121	M	M	M	136	M	1157			
53	Menendes	CU	141	M	M	155	M	M	139	1155			
54	Donde	IS	M	74	M	M	M	M	M	1154			
55	Depuis	FR	158	150	M	M	M	125	M	1153			
56	Reid	CAN	145	M	M	111	M	M	171	1147			
57	Korgard	DAN	M	126	M	M	115	M	M	1141			
58	Alujevic	YU	110	M	M	M	M	130	M	1140			
	Kongsberg	DAN	M	M	127	113	M	M	M	1140			
60	Reitterer	A	115	M	106	M	M	M	M	1121			
61	Linden	SW	151	89	158	M	M	M	M	1118			
62	Hoorn	Hol	130	167	98	M	M	M	M	1115			
63	Ribeiro	BRA	M	145	M	109	M	M	139	1113			
64	Kubes	CS	M	133	102	M	M	156	M	1111			
65	Wells	GB	M	108	108	M	157	M	M	1093			
66	Martin	A	M	M	M	117	143	135	152	1087			
	Sapeshni	SU	157	174	140	111	M	145	M	1087			
68	Prosser	AUS	143	120	M	165	117	M	M	1085			
69	Olstad	NOR	M	M	M	143	M	91	104	1058			
70	Zetterdahl	SW	0	M	M	M	M	140	M	1040			
71	Armesto	ARG	M	88	175	56	176	M	M	1035			
72	Ruyter	Hol	110	143	M	164	M	63	M	1020			
73	Pasanen	FIN	M	177	85	M	136	104	139	1001			
74	Jimenez	MEX	M	85	M	61	M	M	107	973			
75	Tongway	AUS	M	93	M	127	92	99	173	944			

F1c
Power

1	Olofsson	SW	M	M	M	M	M	M	M	1260	+180	+180	+158
2	Verbitski	SU	M	M	M	M	M	M	M	1260	+180	+180	+148
3	Burns	CAN	M	M	M	M	M	M	M	1260	+180	+180	+141
4	Reda	BRD	M	M	M	M	M	M	M	1260	+180	+180	+130
	Truppe	A	M	M	M	M	M	M	M	1260	+180	+180	+130
6	Schaller	SCH	M	M	M	M	M	M	M	1260	+180	+180	+128
7	Sin	PRK	M	M	M	M	M	M	M	1260	+180	+180	+119
8	Bjelayac	YU	M	M	M	M	M	M	M	1260	+180	+180	+115
9	Patek	CS	M	M	M	M	M	M	M	1260	+180	+180	+112
10	Meczner	HUN	M	M	M	M	M	M	M	1260	+180	+180	+111
11	Kraus	A	M	M	M	M	M	M	M	1260	+180	+180	+109
12	Denkin	BG	M	M	M	M	M	M	M	1260	+180	+180	+104
13	Englehard	DDR	M	M	M	M	M	M	M	1260	+180	+180	+101
	Kim Juna	PRK	M	M	M	M	M	M	M	1260	+180	+180	+101
15	Maurer	SCH	M	M	M	M	M	M	M	1260	+180	+180	+ 99
16	Iribarne	FR	M	M	M	M	M	M	M	1260	+180	+180	+ 89
17	Hartwanger	A	M	M	M	M	M	M	M	1260	+180	+180	+ 84
18	Goranov	BG	M	M	M	M	M	M	M	1260	+180	+180	+ 80
19	Lyons	US	M	M	M	M	M	M	M	1260	+180	+180	+ 77
20	Adlt	CS	M	M	M	M	M	M	M	1260	+180	+180	+ 69
21	Faux	GB	M	M	M	M	M	M	M	1260	+180	+180	
22	Schlachta	CAN	M	M	M	M	M	M	M	1260	+180	+180	
23	Mosirski	SU	M	M	M	M	M	M	M	1260	+180	+170	
24	Nigren	SW	M	M	M	M	M	M	M	1260	+180	+160	
25	Listrati	IT	M	M	M	M	M	M	M	1260	+180	+159	
26	Bonetto	ARG	M	M	M	M	M	M	M	1260	+180	+158	
27	Kibiki	NIP	M	M	M	M	M	M	M	1260	+180	+157	
28	Patek	CS	M	M	M	M	M	M	M	1260	+180	+149	
29	Benthin	DDR	M	M	M	M	M	M	M	1260	+180	+141	
30	Valdes	CU	M	M	M	M	M	M	M	1260	+180	+140	
31	Prosser	AUS	M	M	M	M	M	M	M	1260	+180	+122	
32	Krieg	DDR	M	M	M	M	M	M	M	1260	+180	+115	
33	Sharin	SU	M	M	M	M	M	M	M	1260	+180	+ 88	
34	Carlini	BRA	M	M	M	M	M	M	M	1260	+180	+ 11	
	Czizmapik	HUN	M	M	M	M	M	M	M	1260	+180		
	Kovacic	YU	M	M	M	M	M	M	M	1260	+180		
	Waltonen	FIN	M	M	M	M	M	M	M	1260	+180		
38	Savini	IT	M	M	M	M	M	M	M	1260	+174		
39	Rounsaville	US	M	M	M	M	M	M	M	1260	+169		
40	Miyamoto	NIP	M	M	M	M	M	M	M	1260	+148		
41	Jean	FR	M	M	M	M	M	M	M	1260			
	Masuda	NIP	M	M	M	M	M	M	M	1260			
43	Stetz	BRD	M	M	M	M	M	M	172	1252			
44	van Uden	HOL	165	M	M	M	M	M	M	1245			
45	Bortne	NOR	M	M	M	M	179	165	M	1244			
	Nielsen	DAN	M	170	174	M	M	M	M	1244			
47	DeLuna	MEX	M	163	M	179	M	M	M	1242			
48	Barbabella	IT	M	M	M	153	M	M	M	1233			
	Cuzzi	ARG	153	M	M	M	M	M	M	1233			
	Schenker	SCH	M	M	M	M	M	M	153	1233			
51	Oton	YU	M	M	M	139	M	M	M	1219			
	Kim Sung	PRK	M	M	M	M	139	M	M	1219			
53	Ferero	FR	M	M	M	142	M	M	175	1217			
54	Bailey	GB	M	M	133	M	M	M	M	1213			
55	Ochman	POL	M	M	M	127	M	M	M	1207			
56	Bauman	BRD	M	M	M	M	126	M	M	1206			
57	Akesson	SW	M	M	122	M	M	M	M	1202			
58	Harsfalvi	HUN	114	M	M	M	M	M	M	1194			
59	Isev	BG	M	M	M	110	M	M	M	1190			
60	Reus	HOL	118	M	M	M	M	171	M	1189			
61	Backx	HOL	175	M	119	M	173	M	M	1187			
62	Eggleston	CAN	M	M	M	M	M	103	M	1183			
63	Krzeminski	POL	M	M	M	146	130	M	M	1176			
64	Wolff	US	M	M	M	M	83	M	M	1163			
65	Collins	GB	91	M	M	157	M	M	M	1148			
66	Blanco	CU	144	122	147	M	M	M	M	1133			
67	Piatek	POL	M	M	M	M	42	M	M	1122			
68	Popa	ROM	M	M	M	M	M	32	M	1112			
69	Zito	ARG	M	M	M	M	0	M	M	1080			
70	Saukkonen	FIN	94	M	M	M	85	M	M	1079			
71	Luostrinen	FIN	40	93	M	M	119	135	M	927			
72	Jorgensen	DAN	162	M	M	155	0	0	M	857			
73	Koster	DAN	M	25	0	0	0	0	0	205			
74	Petrescu	ROM	0	0	0	0	0	0	0	0			
75	Vincze	ROM	0	0	0	0	0	0	0	0			
76	Torgersen	NOR	0	0	0	0	0	0	0	0			
77	Vela	MEX	0	0	0	0	0	0	0	0			

Team Results

1	USSR	SU	3780
2	North Korea	PRK	3780
3	East Germany	DDR	3772
4	Romania	R	3745
5	Yugoslavia	YU	3736
6	Bulgaria	BG	3694
7	West Germany	BRD	3687
8	Great Britain	GB	3656
9	Austria	A	3655
10	Israel	IL	3654
11	Hungary	H	3629
12	France	F	3628
13	USA	USA	3624
14	Argentina	ARG	3621
15	Italy	I	3617
16	Netherlands	NL	3585
17	Czechoslovakia	CS	3583
18	Danmark	DK	3569
19	Cuba	CU	3556
20	Sweden	S	3528
21	Finland	SF	3505
22	Poland	PL	3498
23	Norway	N	3490
24	Spain	E	3461
25	Switzerland	CH	3460
26	Canada	CDN	3393
27	Mexico	MEX	3164
28	Australia	AUS	3104
29	Egypt	ET	2942
30	New Zealand	NZ	2520

Team Results

1	North Korea	PRK	3715
2	Poland	PL	3674
3	Canada	CDN	3667
4	Bulgaria	BG	3662
5	Yugoslavia	YU	3660
6	East Germany	DDR	3628
	USA	USA	3628
8	Italy	I	3625
9	Switzerland	CH	3623
10	Czechoslovakia	CS	3610
11	West Germany	BRD	3592
12	Cuba	CU	3576
13	France	F	3551
	Great Britain	GB	3551
15	Danmark	DK	3533
16	Israel	IL	3523
17	Argentina	ARG	3510
18	USSR	SU	3569
19	Austria	A	3468
20	Finland	SF	3399
21	Netherlands	NL	3395
22	Sweden	S	3373
23	Australia	AUS	3279
24	Norway	N	2237
25	Romania	R	1260
26	Hungary	H	1175
27	Brazil	BR	1113
28	Mexico	MEX	973
29	Japan	J	854

Team Results

1	Austria	A	3780
2	Czechoslovakia	CS	3780
3	USSR	SU	3780
4	East Germany	DDR	3780
5	Japan	J	3780
6	Switzerland	CH	3753
	Italy	I	3753
8	North Korea	PRK	3739
	Yugoslavia	YU	3739
10	France	F	3737
11	Sweden	S	3722
12	West Germany	BRD	3718
13	Hungary	H	3714
14	Bulgaria	BG	3710
15	Canada	CDN	3703
16	USA	USA	3683
17	Great Britain	GB	3621
	Netherlands	NL	3621
19	Argentina	ARG	3573
20	Poland	PL	3505
21	Finland	SF	3266
22	Cuba	CU	2393
23	Danmark	DK	2306
24	Australia	AUS	1260
	Brazil	BR	1260
26	Norway	N	1244
27	Mexico	MEX	1242
28	Romania	R	1112

Criterium Pierre Trebod - 23/24 Aug. - F1A,F1B,F1C Marigny-le-Grand, France

PLANEURS - F1A

Pl.	Name	Nat.	1	2	3	4	5	6	7	Total
1	SEELIG P.	D	180	180	180	180	180	180	180	1260
										FLY OFF (240 - 166)
2	KRAUS W.	A	180	180	180	180	180	180	180	1260
										(240 - 153)
3	ALLNUTT P.	CDN	180	180	180	180	180	180	180	1260
										(166)
4	MAIWORM E.	D	180	180	180	180	180	180	180	1260
										(145)
5	VERBREE G.	NL	180	180	180	180	180	180	180	1260
										(142)
6	BARNES D.	GB	180	180	180	180	180	180	180	1260
										(139)
7	ALTHOFF H.	D	180	180	180	180	180	180	180	1260
										(121)
8	REYNDERS L.	B	180	180	180	180	180	179	180	1259
9	THOMPSON W.	CDN	180	180	180	180	180	180	178	1258
10	KIEHNLE U.	D	180	173	180	180	180	180	180	1253
11	DEL CERROCANTO A.	CUBA	180	180	180	180	171	180	180	1251
12	VAN ELDIK A.	NL	180	180	180	180	180	180	162	1242
13	GALICHET A.	F	180	180	180	180	180	156	180	1236
14	STIERLIN C.	CH	180	180	180	180	180	171	151	1222
15	SCHEUBLE P.	A	180	180	180	180	180	180	140	1220
15	HARTILL proxy SIMONS	USA	180	180	180	180	180	180	140	1220
17	CHAMPION R.	F	180	180	180	180	180	132	180	1212
18	HOHLS C.	D	180	180	180	180	180	126	180	1206
19	BAINES B.	GB	180	180	180	180	180	172	133	1205
20	DULOUT H.	F	180	180	180	180	180	180	123	1203
21	LANIER J.	F	180	180	180	180	170	180	128	1198
22	JORGENSEN P.	DK	140	180	180	180	153	180	180	1193
23	LEFEVER G.	GB	111	180	180	180	180	180	180	1191
24	STEWART P.	GB	180	180	180	180	180	180	110	1190
25	WANNGARD P.	S	180	160	180	180	180	180	127	1184
26	LAMMERTS H.	NL	180	180	180	180	180	155	128	1183
27	BAZILLON M.	F	180	180	180	180	180	180	99	1179
28	DE BOER P.	NL	180	180	180	180	180	119	157	1176
29	QVARNSTROM P.	S	180	180	180	180	180	156	113	1169
30	VAN MERKESTIJN P.	NL	115	180	180	180	180	180	151	1166
30	GODINHO J.	F	180	180	180	180	180	180	86	1166
32	MOTSCH H.	D	105	180	180	180	160	180	180	1165
33	GREAVES D.	GB	128	180	180	180	130	180	180	1158
34	ANDERSSON K.	S	140	180	180	125	172	180	180	1157
34	BLANCHET J.	F	128	129	180	180	180	180	180	1157
36	LELEUX J.	F	180	180	180	180	130	180	126	1156
36	RODRIGUEZ J.	CUBA	113	160	180	180	180	180	163	1156
38	CHALLINE J.P.	F	180	180	180	180	73	180	180	1153
39	SMITZ W.	USA	180	180	180	180	124	132	175	1151
39	BRAUN H.	L	180	116	180	180	180	180	135	1151
41	MADELIN G.	GB	180	125	180	180	157	180	148	1150
41	HIRLIMANN J.C.	F	180	180	180	180	180	94	156	1150
43	NORTH J.	GB	180	140	180	180	180	109	180	1149
44	BAROHIELLI G.	I	180	180	180	180	180	100	148	1148
44	DUNKERBECK E.	NL	180	180	180	180	132	116	180	1148
46	CHABOT J.M.	F	180	180	180	180	180	128	118	1146
47	KEIM P.	NL	162	180	180	180	180	93	170	1145
48	KRAFT K.	D	180	180	180	180	180	100	144	1144
48	BUNGART A.	D	180	180	180	141	180	180	103	1144
50	FERNANDEZ-ALMAGRO A.	E	180	180	180	180	100	169	150	1139
51	WILKENING F.	D	180	180	180	180	180	62	175	1137
52	DONNET J.	F	180	180	180	180	180	133	102	1135
53	HERNANDEZ-ABAB J.	E	141	180	180	180	180	93	180	1134
54	ALLAIS R.	F	124	180	180	180	180	139	150	1133
54	ROS B.	NL	180	180	180	180	83	180	150	1133
56	BASTOS C.	P	175	180	180	81	177	180	155	1128
57	CHMELIK H.	A	180	169	110	180	180	135	171	1125
58	GAULL S.	D	180	180	180	180	180	96	128	1124
59	SALZER K.	A	180	91	180	180	180	130	180	1121
60	THIES M.	L	178	180	180	180	180	96	121	1115
61	BOGAERTS J.	B	97	180	180	180	180	180	116	1113
62	LOPEZ GARCIA J.	E	180	180	104	180	174	180	108	1106
63	MATHERAT G.	F	180	149	180	105	180	180	125	1099
64	BRINKS G.	NL	65	180	180	180	132	180	180	1097
65	BRAUD L.	F	61	180	180	180	138	180	174	1093
66	REDA S.	D	167	180	180	180	37	180	167	1091
67	BERTHE R.	F	122	180	180	65	180	180	180	1087
68	BOUTILLIER B.	F	180	180	159	89	180	180	118	1086
69	PROCTOR K.	GB	180	150	180	180	114	99	180	1083
70	GREY T.	GB	180	163	180	180	85	180	114	1082
71	SAUER E.	D	180	156	180	180	180	74	131	1081
71	SCHWEND T.	D	102	180	180	180	180	120	139	1081
73	DE RUWE P.	NL	180	157	96	180	107	180	180	1080
74	SPAN R.	A	179	180	180	180	180	180	-	1079
75	LASSAIGNE J.	F	136	139	180	180	180	180	52	1067
76	WARREN M.	GB	180	110	180	136	180	87	180	1053
77	POLENZ P.	D	180	180	180	180	180	83	63	1046
78	KRAUTH M.	F	94	109	180	180	120	180	180	1043
79	NOUGE A.	F	180	96	63	180	180	180	158	1037
80	HAMELIN T.	F	36	180	180	180	180	180	96	1032
81	ROTH M.	CH	180	163	180	180	78	158	90	1029
82	SCHALLENBERG H.G.	D	114	89	180	180	129	180	153	1025
83	DUMONTET B.	F	180	64	180	180	145	107	158	1014
84	MANCHE L.	NL	105	180	180	180	120	119	121	1005
85	SCHLEY R.	D	180	75	180	180	57	180	150	1002
86	BUGE C.	D	180	180	119	180	141	71	120	991
87	ZWAHLEN A.	CH	180	180	180	103	85	103	157	988
88	SCHIMPITZ H.	D	180	180	84	41	180	180	132	977
89	IMGENBERG E.	D	80	180	112	180	180	112	132	976
90	OWENS P.	GB	180	180	180	180	52	32	166	970
91	MANG F.	A	68	180	180	180	180	180	-	968
92	HERBIET P.	F	128	127	180	180	98	172	82	967
93	MADORE J.C.	F	113	180	180	180	130	79	101	963
94	GARCIA SANCHEZ J.	E	175	3	180	180	64	180	180	962
95	MARTINEZ SIOTA M.	E	143	123	102	180	162	85	160	955
96	WACHTER R.	CH	110	67	180	180	125	180	101	943
97	RAPIN F.	F	180	180	180	60	83	79	180	942
98	FERREIRA J.	E	116	78	174	180	168	111	114	941
99	FUSS H.	A	180	114	180	71	80	180	133	938
100	NICHOLSON B.	GB	85	32	180	180	180	180	100	937
101	ROUX A.	F	145	180	180	31	172	92	132	932
102	PICOT D.	F	180	180	180	60	176	54	98	928
102	DIAZ E.	CUBA	180	95	108	69	180	125	171	928
104	DETON Y.	F	86	180	180	92	58	140	180	916
105	BARBARO J.P.	F	118	100	166	164	92	144	122	906
106	LENSI V.	I	-	-	180	180	180	180	180	900
107	TIMMERMANN W.	D	180	180	180	180	48	-	127	895
108	JIMENEZ C.	CUBA	90	180	25	180	180	101	133	889
109	KIMBER P.	GB	73	180	180	65	105	180	104	887
110	EDWARDS A.	AUS	54	180	180	180	180	111	-	885
111	DELCROIX J.	F	180	59	180	100	112	60	180	871
112	PICOT A.	F	177	178	145	101	36	127	101	865
113	MIHAI P.	F	180	180	180	21	144	100	57	862
114	PEPER H.	D	67	180	180	180	100	-	151	858
115	CHILD A.	GB	125	111	180	180	180	78	-	854
116	JELLIS P.	GB	126	180	135	180	95	48	72	836
117	FLINDT H.	D	105	89	180	180	180	-	100	834
118	WITTE H.	D	96	180	133	111	180	26	100	826
119	DEMEURE D.	B	180	106	121	136	54	123	101	821
120	FERROU J.L.	F	110	78	103	180	90	165	85	811
121	TILHAC J.M.	F	171	66	80	91	180	117	90	795
122	GLUE D.	GB	99	180	180	169	155	-	-	783
123	WILLIAMS J.	GB	180	180	29	47	57	139	148	780
124	BRAUD H.	F	104	23	180	87	180	13	180	767
125	KOSTER K.	DK	97	150	90	57	154	83	131	762
126	NEUMANN E.	D	180	108	180	180	65	45	-	758
127	MOTSCH H.	D	38	180	108	180	138	-	88	732
127	WILKENING H.	D	180	133	180	180	59	-	-	732
129	SCHANDEL A.	F	180	86	27	180	36	66	153	728
130	LUNDQVIST S.	S	89	180	36	180	122	-	95	702
131	FERREIRA C.	E.	-	39	96	172	180	68	105	660
132	CARREAU V.	F	53	72	39	180	109	84	117	654
133	THOMSON D.	GB	163	180	180	26	102	-	-	651
134	KOSTER T.	DK	180	110	180	180	-	-	-	650
135	HEIDEMANN G.	D	72	79	87	180	123	-	92	633
136	CHABAUD J.	F	88	78	60	144	91	46	102	609
137	JACK A.	GB	130	180	180	74	-	-	-	564
138	MONKS R.	GB	180	118	180	49	36	-	-	563
139	LEPAGE R.	F	113	180	180	22	29	-	-	517
140	MAIER E.	D	171	165	97	63	-	-	-	496
141	BUZZI G.	I	106	70	180	47	46	-	-	449
141	MOUCHARD A.	B	180	180	47	42	-	-	-	449
143	STEINACKER P.	D	139	32	78	180	-	-	-	429
144	JENSEN S.	DK	180	180	40	-	-	-	-	400
145	HENDERSON I.	GB	-	26	180	-	57	51	80	394
146	BARR A.	GB	90	180	108	-	-	-	-	378
147	ANDERSON D.	Aus	132	54	-	-	-	-	-	186
148	WOODHOUSE J.	GB	180	-	-	-	-	-	-	180

Criterium Pierre Trebod Aug. 23/24 1975

WAKEFIELDS - F1B

Pl.	Name	Nat.	1	2	3	4	5	6	7	Total
1	GREAVES D.	GB	180	180	180	180	180	180	180	1260
										FLY OFF (203)
2	LEPAGE P.	F	180	180	180	180	180	180	180	1260
										(156)
3	WHITE R.	USA	178	180	180	180	180	180	180	1258
4	UIGGIANO O.	RA	[illegible]	180	180	180	180	180	162	1242
5	BARNES J.	GB	180	180	180	153	180	180	180	1233
6	BENEDINI H.	RA	180	157	180	180	180	180	168	1225
6	ALLAIS R.	F	180	180	180	180	180	180	145	1225
8	WANTZENRIETHER J.	F	178	162	157	180	180	180	180	1217
9	MENENDEZ CG.	CUBA	180	180	180	180	180	180	135	1215
10	HIPPERSON D.	GB	180	180	180	145	160	180	180	1205
11	MARSELLA S.	CUBA	180	180	117	180	180	180	180	1197
12	RICO H.	CUBA	115	180	180	180	180	180	180	1195
13	MABILLE A.	B	113	180	180	180	180	180	180	1193
14	PISERCHIO R.	USA	180	180	148	180	139	180	180	1187
15	KOPPITZ A.	F	167	112	180	180	180	180	180	1179
16	PROCTER K.	GB	134	180	180	145	180	180	175	1174
17	BARR L.	GB	180	132	180	180	159	180	162	1173
17	KAYNES I.	GB	180	180	180	180	159	154	140	1173
19	MADELIN G.	GB	90	180	180	180	180	180	168	1158
20	SMITZ W.	USA	180	135	180	180	180	122	180	1157
21	PROSSER T.	AUS	128	180	123	180	180	180	180	1151
22	QVARSTROM P.	S	180	180	157	180	119	180	153	1149
23	LEFEVER G.	GB	180	180	180	180	123	180	124	1147
24	ZETTERDAHL J.	S	180	180	165	180	104	180	155	1144
25	KLINCK A.	D	143	180	180	180	100	180	180	1143
26	SANAVIO A.	I	136	180	180	180	179	180	105	1140
27	NOUGE A.	F	115	180	180	180	169	180	132	1136
28	BLANCHET J.	F	111	180	180	135	180	166	180	1132
29	PECCHIOLI P.	I	180	155	180	180	180	134	121	1130
30	RUYTER P.	NL	180	117	180	180	174	180	115	1126
31	CHAMPION R.	F	180	112	180	180	180	180	113	1125
32	MANCHE L.	NL	172	80	180	168	180	180	164	1124
33	DAVIS J.W	USA	180	106	163	180	180	180	130	1119
34	VANTOMME J.	B	180	99	180	180	180	180	116	1115
35	MONNINGHOFF P.	D	180	180	180	180	180	105	103	1108
36	TOMAS A.	S	174	111	180	180	134	180	148	1107
37	MARQUEZ R.	RA	180	107	180	180	143	131	180	1101
38	PRIOUX H.	F	180	110	180	180	137	160	141	1088
39	ARMESTO A.	RA	99	180	180	180	114	180	153	1086
40	HILLNBRECHT H.	D	180	119	180	180	95	180	145	1079
41	JACK A.	GB	105	180	180	147	108	180	172	1072
42	CHMELIK H.	A	140	84	124	180	180	180	180	1068
43	BOUTILLIER B.	F	130	180	180	180	87	151	159	1067
44	DURAND R.	F	131	178	180	64	180	180	151	1064

45	LUNDQVIST S.	S	126	180	180	180	143	180	74	1063
46	SODERSTROM B.	S	180	108	166	180	136	125	156	1051
47	LEFEUVRE M.	F	180	180	180	180	92	63	169	1044
48	COSTA C.	P	162	180	180	70	93	180	159	1024
49	GREY T.	GB	113	180	180	95	180	157	106	1011
50	BUSKELL J.	GB	102	130	180	180	123	180	108	1003
51	EDWARDS A.	AUS	180	175	146	180	180	137	-	998
52	DUCH M.	GB	85	180	125	131	180	180	158	979
53	BRAUD H.	F	150	114	94	148	180	180	108	974
54	RAPIN F.	F	180	145	180	107	66	180	103	961
55	DE SOUSA M.	P	180	180	180	180	54	168	14	956
56	DELCROIX J.	F	180	100	155	180	133	122	83	953
57	GRAUX A.	B	160	92	88	180	180	180	51	931
58	BASTOS C.	P	110	180	102	180	95	164	69	900
59	PILLER M.	F	88	108	180	97	180	85	135	873
60	CHAUSSEBOURG P.	F	140	180	158	180	180	-	-	838
61	GERMAIN Y.	F	151	86	75	63	142	140	178	835
61	MADORE J.C	F	128	102	180	47	180	68	130	835
63	ANDERSSON K.	S	180	180	104	180	180	6	-	830
64	SEGRAVE M.	CDN	118	73	180	180	38	90	106	785
65	NEUMANN P.	D	5	100	180	94	180	113	82	754
66	ELLIOTT N.	GB	180	58	93	180	94	75	69	750
67	RUIZ PENEZ A.	CUBA	159	114	45	180	4	180	66	748
68	NEW R.	GB	54	92	117	114	129	42	109	657
69	ELLIOTT R.	GB	102	180	180	180	-	-	-	642
70	DUPUIS L.	F	180	180	124	114	-	-	-	598
71	BOGAERTS J.	B	-	91	109	132	91	64	72	559
72	ROMCZYK D.	F	78	84	180	115	76	-	-	533
73	MATHERAT G.	F	159	125	65	-	-	-	-	349
74	EKENDAHL T.	S	105	56	79	60	-	-	-	300
75	BOGART B.	USA	44	89	78	67	-	-	-	278
77	NEUMANN E.	D	147	51	-	-	-	-	-	198
78	GAULL S.	D	62	55	-	-	-	-	-	117

Criterium Pierre Trebod
Aug. 23/24 1975

MOTOMODELES - F1C

1	KRAUS W.	A	180	180	180	180	180	180	180	1260 (180 - 180 - 180 - 147)
2	SEYDEL S.	D	180	180	180	180	180	180	180	1260 (180 - 180 - 180 - 132)
3	TRUPPE R.	A	180	180	180	180	180	180	180	1260 (180 - 180 - 180 - 126)
4	LANDEAU A.	F	180	180	180	180	180	180	180	1260 (180 - 180 - 152)
5	REDA S.	D	180	180	180	180	180	180	180	1260 (180 - 180 - 134)
6	COLLINS R.	GB	180	180	180	180	180	180	180	1260 (180 - 180 - 113)
7	BUSKEL P.	GB	180	180	180	180	180	180	180	1260 (180 - 180 - 87)
8	FERRERO D.	F	180	180	180	180	180	180	180	1260 (180 - 159)
9	BARBARO J.P.	F	180	180	180	180	180	180	180	1260 (180 - 144)
10	ALLEN J.	GB	180	180	180	180	180	180	180	1260 (180 - 142)
11	MONKS R.	GB	180	180	180	180	180	180	180	1260 (180 - 82)
12	BARTHEL F.	I	180	180	180	180	180	180	180	1260 (180 - 0)
12	IMGENBERG U.	D	180	180	180	180	180	180	180	1260 (180 - 0)
12	WOLFF F.	USA	180	180	180	180	180	180	180	1260 (180 - 0)
15	BAILEY R.	GB	180	180	180	180	180	180	180	1260 (134)
16	JEAN M.	F	180	180	180	180	180	180	180	1260 (0)
16	LINDHOLM H.	S	180	180	180	180	180	180	180	1260 (0)
18	PYMM D.	GB	180	175	180	180	180	180	180	1255
19	QUINTERO A.	CUBA	180	180	180	180	173	180	180	1253
20	SCHWEND T.	D	180	180	180	180	180	180	170	1250
21	SCHALLENBERG H.	D	180	180	180	180	180	180	165	1245
22	BOND P.	GB	180	180	180	180	180	180	163	1243
23	LIMA REIS A.	P	180	180	180	180	180	156	180	1236
24	ROUNSAVILLE P.	USA	180	180	162	180	180	180	172	1234
25	SOUVETON J.C.	F	180	145	180	180	180	180	180	1225
26	EGGLESTON B.	CDN	180	180	137	180	180	180	180	1217
26	NOA B.	CUBA	180	180	180	137	180	180	180	1217
28	SJOMAN O.	S	180	180	135	180	180	180	180	1215
29	JOHNSON R.	GB	180	134	180	180	178	180	176	1208
30	SOARES M.	P	180	126	180	180	180	180	180	1206
31	SEELIG H.	D	180	180	180	122	180	180	180	1202
32	SCHLEY R.	D	180	180	114	180	180	180	180	1194
33	PLATT H.	D	180	117	180	180	180	174	180	1191
34	SCHLACHTA F.	CDN	180	180	180	180	159	128	180	1187
35	LELEUX J.	F	180	163	163	147	180	180	155	1169
36	MAURICIO Z.	RA	180	180	87	180	180	180	180	1167
37	FAUX K.	GB	180	180	180	180	180	96	155	1151
38	HARRIS P.	GB	180	180	175	180	180	108	134	1137
39	BONETTO J.	RA	180	167	154	180	180	120	154	1135
40	ZEH A.	D	180	161	101	174	135	180	180	1111
41	SAUER K.	D	180	173	180	32	180	180	180	1105
42	CUZZI O.	RA	180	89	180	180	180	180	107	1096
43	CHILD A.	GB	154	116	180	180	180	113	164	1087
44	GIAMPIERI V.	I	180	180	105	82	180	170	180	1077
45	BOUTILLIER B.	F	180	23	180	180	180	180	150	1073
46	CHILTON F.	GB	51	157	180	180	180	180	139	1067
47	SCHALLER U.	I	180	180	130	180	180	166	-	1016
48	ROUX A.	F	173	180	180	180	30	-	-	743
49	MAIER E.	D	180	180	180	45	106	-	-	691
50	RAPIN F.	F	70	129	88	85	105	-	160	637
51	CHELIUS T.	D	65	-	180	180	99	-	-	524
52	CHILTON A.	GB	83	127	180	-	-	-	-	390
53	JORGENSEN P.	DK	180	180	-	-	-	-	-	360
54	WEBER A.	D	180	-	-	-	-	-	-	180

Free Flight News International - 25/26 Oct. -
F1A, F1B, F1C RAF Sculthorpe, England

F1A Glider (Ronytube Trophy)

1	A Cordes	180	180	180	180	180	=	900 + 240
2	J H Cooper	180	180	180	180	180	=	900 + 194
3	J B Spooner	180	180	180	180	180	=	900 + 180
4	G Madelin	180	180	180	180	180	=	900 + 173
5	B Baines	180	180	180	180	180	=	900 + 159
6	A G Jack	180	180	180	180	180	=	900 + 140
7	M Fantham	180	180	180	180	180	=	900 + 112
8	J Abbey	180	180	180	174	180	=	894
9	J Hanson	180	173	180	180	180	=	893
10	B Halford	180	180	180	169	180	=	889
11	D H Thomson	180	180	180	180	164	=	884
12	D Barnes	161	180	180	180	180	=	881
13	B D Silcocks	176	158	180	180	180	=	874
14	E B Jones	180	140	180	180	180	=	860
15	M J Woodhouse	149	155	180	180	180	=	844
16	J E Billam	121	180	180	180	180	=	841
17	C Batty	105	180	180	180	180	=	825
18	P M Stewart	180	99	180	180	180	=	819
19	D Glue	110	166	180	180	163	=	799
	C P Williams	98	161	180	180	180	=	799
21	M B Cowley	154	180	131	180	150	=	795
22	M Coomes	180	67	180	180	180	=	787
23	P A Owens	132	180	180	96	180	=	768
24	M Dilly	180	73	144	180	180	=	757
	T J Punter	132	143	122	180	180	=	757
26	C T James	180	120	93	149	180	=	722
27	J O'Donnell	137	180	180	38	180	=	715
28	R Cedar	105	53	180	180	180	=	698
29	M C Warren	180	105	81	146	178	=	690
30	A le Vey	180	180	120	125	83	=	688
31	S Marriott	126	180	0	180	180	=	666
32	M Prickett	180	88	180	161	42	=	651
33	I V Dowsett	138	79	138	114	180	=	649
34	J Powell	110	67	180	180	93	=	630
35	D Oldfield	180	88	180	180	0	=	628
36	G le Vey	51	90	180	122	180	=	623
37	M C Gregorie NZ	180	22	164	85	171	=	622
38	I B Bracken	66	180	132	58	180	=	616
39	P G Jubb	95	97	180	60	180	=	612
40	J Williams	114	86	180	93	103	=	576
41	L Moore	142	57	168	115	77	=	559
42	B Nicholson	90	93	180	70	110	=	543
43	P Hawkins	180	84	114	129	32	=	539
44	J Ashby	-	-	180	139	180	=	499
45	G J Lefever	150	153	123	-	-	=	426
46	D Yates	180	98	-	-	-	=	278
47	T J Grey	116	115	-	-	-	=	231
48	E P Drew	180	0	0	-	-	=	180
49	R J Latimer	28	120	-	-	-	=	148
	J P Masterman	23	125	-	-	-	=	148
	No. of maxes	24	19	32	26	31		1
	No. of full houses		11	10	8	7		1

F1B Wakefield (RAFMAA Trophy)

1	J Barnes	180	180	180	180	180	=	900 + 208
2	R C Pollard	180	180	180	180	180	=	900 + 161
3	C P Williams	180	180	180	180	180	=	900 + 107
4	G J Lefever	173	176	180	180	180	=	889
	(A R Wells)	180	180	180	160	180	=	860
5	J H Cooper	160	180	180	160	180	=	860
6	A G Jack	167	145	180	180	180	=	852
7	T J Punter	108	180	180	180	180	=	828
8	M Duce	180	180	180	114	170	=	824
9	J E Billam	134	180	180	137	180	=	811
10	J van Rij NL	180	138	136	170	180	=	804
11	J B Spooner	180	180	108	55	180	=	803
12	D Hipperson	180	148	114	177	180	=	799
13	S Marriott	145	120	180	159	180	=	784
14	R L Elliott	180	162	100	151	180	=	773
15	J H Gunn	81	180	171	158	180	=	770
16	B L Halford	180	110	175	105	180	=	750
17	J O'Donnell	149	127	146	147	180	=	749
18	C T James	134	180	180	70	180	=	744
19	I Taylor	146	180	180	115	116	=	737

20	B Kershaw	82	169	180	180	119	= 730
21	T J Grey	106	180	116	90	180	= 672
22	M J Woodhouse	112	119	100	180	128	= 639
23	M A Dixon	112	180	103	61	180	= 636
	D Greaves	180	120	129	180	27	= 636
25	B V Rowe	153	21	162	119	176	= 631
26	R C Monks	100	180	116	-	-	= 396
27	K Proctor	144	89	67	-	-	= 300
28	G Madelin	97	-	-	-	-	= 97
	No. of maxes	11	15	14	9	20	
	No. of full houses		6	5	3	3	

F1C Power (Hans Seelig Trophy)

1	R C Monks	-	180	180	180	180	= 720 + 240
2	S C Screen	-	180	180	180	180	= 720 + 185

3	P K Bond	-	180	180	177	180	= 717
4	K A Faux	-	180	180	103	147	= 610
5	R Johnson	-	180	100	141	180	= 601
6	P R Harris	-	180	101	176	133	= 590
7	F G Chilton	-	180	160	90	142	= 572
8	R A Collins	-	180	180	0	180	= 540
9	M B Cowley	-	107	118	180	132	= 537
10	R Baggott	-	180	0	180	165	= 525
11	J F Allen	-	107	153	89	174	= 523
12	P Rowledge	-	180	0	-	-	= 180
13	R L Bailey	-	74	-	-	-	= 74
14	J E Billam	-	0	-	-	-	= 0
	No. of maxes		10	5	4	5	1
	No. of full houses			5	2	2	1

NB: All nationalities, British, except where noted.

WFFA Challenge - 8/9 Nov. - F1A, F1B, F1C
Taft, California, U.S.A.

F1A NORDIC (WFFA Annual)

		1	2	3	4		5	6	7	8	9	10	
1.	Lee Hines	M	M	M	M	PARTY SAT. NITE	135	M	M	M	M	M	1755
2.	Bob Isaacson	M	095	M	M		M	M	M	M	M	M	1715
3.	Jim Trego	M	173	M	M		102	M	M	M	M	M	1715
4.	Don Zink	M	M	M	M		M	M	M	M	M	82	1702
5.	Bob Norton	M	M	043	M		114	M	M	M	M	M	1597
6.	Bill Gibbons	137	146	M	131		142	M	M	M	131	M	1587
7.	Jose Dona	M	136	116	096		155	M	M	M	M	M	1583
8.	Juan Livotto	071	116	177	140		149	M	M	M	165	M	1538
9.	Jim Wilson	M	-	M	106		141	M	M	M	M	M	1507
10.	Ed Wogulis	057	M	M	106		M	M	M	M	M	71	1494
11.	Val Sidlauskus	161	137	73	130		M	M	M	M	M	-	1421
12.	Gary Medley	M	100	M	-		M	M	M	M	45	M	1405
13.	Hector Diez	M	076	M	150		100	172	M	M	M	-	1398
14.	Bob Critchlow	M	153	056	170		93	M	M	M	90	103	1385
15.	Tom Hutchinson	M	97	74	129		M	M	M	M	58	-	1258
16.	Craig Cusick	M	M	80	M		77	67	15	M	58	15	1032
17.	Ray Harper	77	61	56	73		90	-	M	46	23	61	673
18.	Ken Bauer	M	155	-	135		-	-	-	-	-	-	470
19.	Walt Ghio	M	-	-	-		-	-	-	-	-	-	180

F1B WAKEFIELD (WFFA Annual)

		1	2	3	4		5	6	7	8	9	10	
1.	Al Hotard	M	M	159	M	PARTY SAT. NITE	M	M	M	M	M	M	1779
2.	Larry Jordan	M	157	M	M		170	M	M	143	M	171	1721
3.	Jim Quinn	M	M	112	M		153	M	M	M	M	M	1705
4.	P. Tsiknopoulos	131	M	M	M		M	M	M				1677
5.	Val Sidlauskas	093	M	150	M		168	M	M	M	M	M	1671
6.	Walt Ghio	M	123	M	M		M	M	M	105	M	M	1668
7.	Bob White	118	145	M	M		M	140	M	M	M	M	1663
8.	Bob Tymchek	134	101	151	M		176	161	M	M	145	M	1588
9.	Bill Gibbon	M	089	144	M		M	131	M	M	M	112	1556
10.	Les Dewitt	127	145	M	M		M	124	M	81	M	98	1475
11.	Bill Wiley	M	91	129	146		108	M	79	137	M	M	1410
12.	Bill Bogart	M	081	145	122		116	M	97	M	118	M	1399
13.	Dick Gildersleeve	M	M	M	112		-	-	-	-	-	-	652
14.	Frank Newquist	097	111	104	119		-	-	-	-	-	-	431
15.	Hugo Sandroni	038	103	113	086		-	-	-	-	-	-	340
16.	Tom Medley	-	-	-	-		133	-	-	-	-	-	133

F1C POWER (WFFA Annual)

		1	2	3	4		5	6	7	8	9	10	
1.	Dick Myers	M	M	M	M	PARTY SAT NITE	M	M	M	M	M	M	1800
2.	Reid Simpson	M	173	M	M		M	M	M	M	M	M	1793
3.	Roger Simpson	M	M	M	154		M	M	M	M	M	M	1774
4.	Ken Oliver	M	M	M	M		M	M	151	M	M	M	1771
5.	Dave Parsons	M	136	M	M		M	M	M	M	M	M	1756
6.	Ronnie Young	M	M	M	171		98	M	M	M	M	M	1709
7.	Bruce Hannah	137	123	M	M		M	M	M	M	M	M	1700
8.	Doug Galbreath	M	126	M	161		M	145	M	M	163	M	1675
9.	Ken Phair	132	M	M	M		133	132	M	M	M	M	1657
10.	Larry Sicuranza	109	138	142	124		145	M	M	5	0	0	1023

I Coupe d/Hiver International - 30 Nov. - F1G
RAF Halton, England

80gr (Aero Modeller Trophy)

1.	L. Dupuis (F)	120	120	120	120	120	600
2.	G. Matherat (F)	120	120	106	120	120	586
3.	H. Tubbs	120	120	70	120	120	550
4.	D. Hipperson	120	120	120	78	108	546
5.	J. Cooper	120	120	120	70	115	543
6.	A. Wells	79	100	120	120	120	539
7.	S. Marriot	120	120	83	109	90	522
8.	R. White (USA) (Proxy P.Chaussebourg)	120	120	110	97	72	519
9.	J. O'Donnell	120	120	87	94	97	918
10.	G.C. Hart	115	106	120	118	54	513
11.	R. Elliott	120	99	87	120	77	505
12.	B. Boutillier (F)	120	120	76	90	98	504
13.	P. Cameron	120	117	68	120	78	503
14.	P.R. Bixby	89	120	120	82	80	491
15.	T. Linstrum (USA) Proxy P. Cameron)	120	90	91	107	71	479
16,	P. Chaussebourg (F)	120	120	97	71	68	476
17.	I.W. Kaynes	80	120	100	90	80	470
18.	I. Dowsett	77	120	88	116	60	461
19.	D. Greaves	120	72	120	66	79	457
20.	B.V. Rowe	120	120	95	120	-	455
21.	A. Roux (F)	85	78	110	120	56	449
22.	R. Fleetwood	105	120	54	98	77	444
23.	Fudo Takagi (USA) (Proxy R. Johnson)	87	116	90	57	86	436
24.	R. Coleman	120	55	75	120	62	432
25.	R. Johnson	68	72	120	84	87	431
26.	S. Millett (F) (Proxy A. Roux)	120	100	62	73	52	407
27.	W. Houghton	62	89	77	120	53	401
28.	P. Freebrey	115	78	83	63	58	397
29.	G. Ferer	56	73	88	60	120	387
30.	A. Crisp	120	68	66	61	68	373
31.	T. Taylor	86	114	56	56	60	372
32.	M. Cowley	107	43	78	99	43	370
33.	J. Wright	86	101	64	120	96	367
34.	S. Taylor	52	77	65	90	46	330
35.	D. Roche	77	97	51	58	42	325
36.	D. Bell	47	43	120	37	65	312
37.	Foster	79	62	61	58	48	308
38.	R. Moore	44	38	67	72	52	273
39.	B. Cox	69	75	64	-	-	208
40.	L. Ranson	120	11	-	-	-	121
41.	Mdme C. Dupuis	120	-	-	-	-	120

100gr (M.R.A. Challenge)

1.	B. Boutillier (F)	120	120	120	360 + 113	
2.	L. Dupuis (F)	120	120	120	360 + 88	
3.	F. Monts (USA) (Proxy J.O'Donnell)	104	79	120	303	
4.	L. Dupuis (F)	83	84	120	287	
5.	D. Greaves	99	92	79	270	
6.	G. Matherat (F)	70	120	62	252	
7.	G. Matherat (F)	120	120	-	240	
8.	R. Moore	60	56	120	236	
9.	R. White (USA) (Proxy P. Chaussebourg)	71	78	85	234	
10.	A. Wells	76	78	65	219	
11.	D. Linstrum (USA) (Proxy P. Chaussebourg)	101	56	57	214	
12.	R. Moore	77	60	61	208	
13.	C. Sotich (USA) (Proxy D. Goodwin)	99	62	46	207	
14.	P. Chaussebourg (F)	60	69	76	205	
15.	S. Millett (F) (Proxy A. Roux)	68	66	69	203	
16.	A. Roux (F)	73	65	61	199	
17.	S. Savage (CANADA) (Proxy I. Dowsett)	52	85	58	195	
18.	P. Boutillier (F)	71	60	63	194	
19.	P. Freebrey	55	54	70	179	
20.	R.A. Twomey	57	47	65	169	
21.	D. Roche	47	120	-	167	
22.	I. Sutton	85	63	-	140	
23.	S. Savage (CANADA) (Proxy I. Dowsett)	71	71	-	142	
24.	L. Ranson	66	-	-	66	
25.	R.A. Twomey	44	-	-	44	
26.	I. Dowsett	39	-	-	39	

The Year 1976

Amsterdam Cup - 20/30 May - F1A, F1Bl, F1C
Rozendaalse Heide, Netherlands

Klasse F1A - A2

1	Van Dijk	NL	180	180	180	180	120	120	120	1080
2	Salzer	A	162	162	156	180	120	115	120	1015
3	Lendcring	NL	180	180	171	120	120	120	120	1011
4	Bünte	D	180	180	160	129	120	110	120	999
5	Van Rinsum	NL	180	170	119	180	120	120	104	993
6	Wilke	D	180	142	180	158	85	120	120	985
7	Hirlimann	F	180	180	165	180	60	89	120	974
8	Van Eldik	NL	180	180	158	119	95	120	120	972
9	De Ruwe	NL	180	180	144	180	120	46	120	970
10	Marriott	GB	180	180	180	168	120	33	108	969
11	Ehrlich	A	180	177	86	180	120	114	89	946
12	Spann	A	180	180	180	112	120	120	46	938
13	Higgins	NL	130	177	155	180	120	120	50	932
14	Litjens	NL	180	180	125	180	98	120	46	929
15	Noque	F	180	90	156	170	120	120	68	904
16	Wanngard	S	180	172	125	118	120	99	80	894
17	Peper	D	180	111	63	180	120	120	119	893
18	Madelin	GB	180	180	138	180	76	76	62	892
19	Williams	GB	180	71	90	180	120	120	120	881
20	Erard, H.	F	180	152	180	180	32	100	51	875
21	Verbree	NL	180	180	122	93	93	77	120	865
22	Lammertink	NL	180	105	148	143	120	83	76	855
23	Lamers	D	180	156	180	115	70	120	29	850
24	Althoff	D	180	84	160	152	120	29	120	845
25	Jack	GB	180	69	152	97	120	105	120	843
26	Dulout	F	180	165	119	145	100	120	7	836
27	Klink, D.	D	180	180	70	122	120	65	97	834
28	Klink, H.	D	108	180	55	180	92	120	94	829
29	Dams	NL	180	180	46	101	120	120	72	819
30	Baines	GB	180	180	130	52	120	78	56	796
31	Stewart	GB	50	120	180	112	120	120	78	780
32	Breeman	NL	180	108	43	107	86	120	120	764
33	Van Eede	NL	90	135	102	150	76	75	120	748
34	Gilmore	GB	180	71	136	40	82	115	120	744
35	Erard, J.	F	180	124	118	62	32	120	91	727
36	Cowley	GB	180	135	137	180	80	---	---	712
37	Ferrou	F	180	141	70	180	32	49	53	705
38	Büge	D	180	180	111	136	56	26	---	689
39	Le Leux	F	180	146	127	130	66	27	---	676
	Greaves	GB	180	180	92	113	111	---	---	676
41	Griessenburg, F.	D	180	180	130	180	---	---	---	670
42	Litjens, P.	NL	180	180	92	180	36	---	---	668
43	Kreetz	NL	180	115	180	163	27	---	---	665
44	Moust	NL	180	122	112	124	120	---	---	658
45	Schley Sr.	D	178	93	153	60	37	32	103	656
46	Godinho	F	180	84	103	180	---	63	37	647
47	Fleischhacker	A	62	180	166	77	62	38	44	629
48	Schley Jr.	D	180	180	---	26	82	50	99	617
49	Müller, B.	D	180	180	85	49	120	---	---	614
50	Schallenberg, G.	D	180	180	72	180	---	---	---	612
51	Brinks	NL	180	65	128	180	49	---	---	602
52	Wilkening	D	180	139	97	180	---	---	---	596
53	Baguley	GB	180	180	54	93	79	---	---	586
54	De Boer	NL	180	148	153	104	---	---	---	585
55	Kappe	D	180	139	107	112	---	45	---	583
56	Arns	NL	180	180	12	69	99	42	---	582
57	Warren	GB	180	180	105	67	45	---	---	577
	Walleneit	D	97	180	120	180	---	---	---	577
59	Fiks	NL	180	180	90	79	45	---	---	574
60	Mohr	D	180	121	80	180	---	---	---	561
61	Carrasco	D	126	180	54	63	120	---	---	543
62	Dilly	GB	127	180	126	90	---	---	---	523
63	Jansen	NL	103	112	85	64	120	37	---	521
64	Lehmann	D	180	73	76	132	44	---	---	505
65	Brandenburg	NL	17	164	180	95	45	---	---	501
66	Müller, G.	D	180	110	180	25	---	---	---	495
67	Müller, I.	D	180	180	133	---	---	---	---	493
68	Hacken	NL	180	171	45	90	---	---	---	486
69	Van Bragt	NL	180	180	115	---	---	---	---	475
70	Van den Dungen	NL	54	169	70	180	---	---	---	473
71	Moss	NL	180	83	122	86	---	---	---	471
72	Hawkins	GB	180	180	97	---	---	---	---	457
73	Griessenburg, A.	D	28	180	---	180	64	---	---	452
74	Schiller, K.	D	96	160	72	38	41	---	---	407
75	Bogaerts	B	168	82	---	144	---	---	---	394
76	Proctor	GB	159	88	82	60	---	---	---	389
77	Flindt	D	85	180	---	58	65	---	---	388
78	Coenen	NL	180	167	6	---	---	---	---	353
79	La Peyre	F	180	50	99	---	---	---	---	329
80	Schiller	D	27	132	115	42	---	---	---	316
81	Schnittger	NL	57	175	68	---	---	---	---	300
82	Dunkerbeck	DM	126	92	64	13	---	---	---	295
83	Lammerts	NL	91	90	96	---	---	---	---	277
84	Imgenberg	D	85	80	90	---	---	---	---	255
85	Thomson	GB	68	180	---	---	---	---	---	248
86	Kappelhoff	NL	154	---	76	---	---	---	---	230
87	Hohls	D	123	94	---	---	---	---	---	217

Klasse F1B - Wakefield

1	Greaves	GB	180	180	172	117	120	120	120	1009
2	Van Hoorn	NL	180	180	151	102	120	120	120	973
3	Kaynes	GB	180	180	140	178	55	91	120	944
4	Pollard	GB	180	137	158	126	80	120	120	921
5	Jack	GB	179	160	167	75	120	105	.96	902
6	Jakobsen	DM	142	177	135	54	120	120	117	865
	Ter Velde	NL	180	140	92	180	66	120	87	865
8	Bailey	GB	180	180	142	66	84	120	92	864
9	Manche	NL	180	180	174	78	57	111	78	858
10	Van Rij	NL	180	156	123	135	120	45	78	837
11	Marriott	GB	178	120	128	96	85	120	97	824
12	Korsgaard	DM	180	180	147	110	61	110	34	822
13	Merkestijn	NL	180	99	83	116	86	49	107	720
14	Haacken	D	180	162	163	85	68	---	---	658
15	Madelin	GB	180	137	139	132	---	---	---	588
16	Mönninghoff	D	121	159	90	90	120	---	---	580
17	Baguley	GB	180	141	129	83	---	---	---	533
18	Boeren	D	180	179	127	3	---	---	---	489
19	Reike	D	130	180	138	---	---	---	---	448
20	Ganbauge	D	24	130	115	66	---	---	---	335
21	Lefevre	GB	9	---	---	---	---	---	---	9
22	Proctor	GB	4	---	---	---	---	---	---	4

Klasse F1C - Power

1	Køster	DM	180	180	180	180	120	120	120	1080 (1[illegible]
2	Huijben	NL	180	180	180	180	120	120	120	1080 (1[illegible]
3	Broderac, J.	D	137	180	180	180	120	120	120	1037
4	Zeh	D	180	180	180	180	120	89	99	1028
5	Jørgenson	DM	180	180	180	127	120	---	120	907
6	Chilton	GB	132	152	140	110	89	120	51	794
7	Seydel	D	177	180	180	109	---	---	---	646
8	Heidemann, T.	D	180	158	120	152	---	---	---	610
9	Friedrich	D	180	180	171	75	---	---	---	606
10	Imgenberg	D	180	180	180	---	---	---	---	540
	Seidelmayer	D	180	180	180	---	---	---	---	540
12	Broderac, D.	D	94	132	180	86	---	---	---	492
13	Kuhl	D	180	153	150	---	---	---	---	483
14	Borczewski	D	180	147	51	45	---	55	---	478
15	Stetz	D	180	180	87	---	---	---	---	447
16	Faux	GB	130	180	130	---	---	---	---	440
17	Cowley	GB	180	21	95	---	---	---	---	296
18	Heidemann, G.	D	180	111	---	---	---	---	---	291

Bijzondere modellen

1	Leeuwangh	NL	86	63	95	44	20	36	40	384
2	Jandt	D	67	98	90	34	25	---	1	315
3	De Kruijff	NL	45	58	---	42	39	120	---	304

US Free Flight Champs - 29/30/31 May - All events
Taft, California, U.S.A.

PAYLOAD

Terry Thorkildsen	8:06
Ron Wittman	7:11
Scott Valentine	5:42
Bill Valentine	4:37

1/2A GAS, JUNIOR

Eric Dyer	36:11
Penny Johnson	20:00
Jeff Cunnyngham	12:19
Guy Mennano	11:35
Fred Calhoun	8:27

1/2A GAS OPEN

Guy Kirkwood	37:35
Blen Christensen	36:01
Andrew Barron	32:26
Mike Kerzie	20:56
Randy Weiler	17:11

A-GAS JUNIOR

Parker McQuawn	16:37
Flaxen Johnson	15:00
Jeff Cunnyngham	14:03
Eric Dyer	10:16
Keith Morgan	9:28

A-GAS

Randy Weiler	62:02
Nick Cacci	28:00
Jim Scarborough	25:34
Russ Backer	22:50
Bob Scully	22:50

JR B/C GAS

Jeff Cunnyngham	25:53
Eric Dyer	12:07
Keith Morgan	11:44
Daniel Hannah	11:04
Fred L. Calhoun	5:44

B GAS

Mark Valerius	46:55
Cliff Tanaka	31:12
Ray Sahlberg	25:00
Bob Scully	21:40
Nick Cacci	20:00*
Bruce Hannah, Jr	20:00*

*Tie broken by flyoff.

C - GAS

Doug Galbreath	70:00
Randy Secor	42:26
Bill Hunter	33:30
Chuck Thompson	24:36
Tom Scully	24:19

D GAS

Raymond Smith	25:00
Ken Bauer	21:52
Lyman Armstrong	20:00
Bob Coombs	16:31
Steve Calhoun	15:00

FAI POWER

Reid Simpson	1579
Al Bissonette	1578
Don McGhee	1224
Dick Myers	1166
Bruce Hannah	1144

OUTDOOR PEANUT

Warner
Mather
Mulligan
Oldenkamp
Moss

SCALE RUBBER

Cover
Baren
Moss
C. Haught
D. Haught

INDOOR PEANUT

M. Mulligan
Bill Warner
Bill Stroman
Hal Cover
Larry Moss

EASY B

Clarence Mather	600
Earl Hoffman	549
Mark Valarius	433
Ken Bauer	301
Keith Morgan	228

PENNYPLANE

Clarence Mather	404
Earl Hoffman	362
Bob Meuser	349
John Magnus	320
Bill Xenakis	181

OLD-TIMER RUBBER

Bob Oslan	15:00
Ray Berens	12:25
Wade Wiley	11:42
Irv Acker	10:40
Russ Backer	10:36

020 REPLICA

Bob Oslan	11:27
Jack Jella	11:00
Thomas Keppler	9.37
Brick Brickner	9:00
Bruce Hannah Jr	8:43

A-B CABIN

Hal Cover	13:51
Bob Oslan	12:14
Bruce Chandler	11:47
Bob Dittmer	11:25
Ralph Prey	10:34

30 SEC. ANTIQUE

Wade Wiley	15:00
Fred Emmert	14:07
J.W. Savage	13:28
Leslie Norman	10:53
Jerry Otis	10:17

C - OLDTIMER CABIN

Fred Emmert	14:30
T.C. O'Meara	13:28
Leslie Norman	11:19
J.W. Savage	11:16
Spiro Nicholan	9:45

C - OLDTIMER PYLON	
Bruce Norman	15:00
Ron Martin	14:09
Bob McBride	13:31
Tony Blizzard	12:01
Russel Adamzuk	11:52

A/B OT PYLON	
Bruce Norman	15:00
Leslie Norman	15:00
Sal Taibi	13:36
Otto Bernhart	12:53
Larry Boyer	12:44

COUPE D'HIVER	
Bill Davis	600
Jim Quinn	569
Tom Medley	567
Merwin Bristol	559
Rich Rohrke	557

WAKEFIELD	
Les DeWitt	1236
Bob Piserchio	1214
Jim Quinn	1202
Bob Tymchek	1183
Irv Aker	1125

UNLIMITED RUBBER, OPEN	
Geo. Xenakis	1680
Jim Quinn	1142
Bob Piserchio	1053
Andy Faykun	1050
Mark Valarius	996

INDOOR HANDLAUNCH GLIDER	
Bob Meuser	62.0
Ron Wittman	58.5
Bob Boyer	54.9
Bob De Shields	54.5
Phedon Tsiknopoulous	54.3

INDOOR HLG JUNIOR	
Steve Wittman	52.8
Jim Geraghty	37.9
Eric Dyer	37.8
Steve Mounsey	24.4
Bret Payne	17.6

A/1 JUNIOR	
Ed Wogulis	736
Jim Geraghty	696
Eric Dyer	670
Dean Koesling	596
Danny Diez	580

A/1 GLIDER, SR OPEN	
Bill Xenakis	1440
Bob DeShields	1071
Jack Moreland	967
Bob Critchlow	890
Russ Backer	859

A/2 TOWLINE GLIDER	
Bob Isaacson	1636
Craig Cusick	1605
Tom Hutchinson	1242
George Xenakis	1208
Wayne Drake	1200

NIGHT	
Vic Cunnyngham Jr	25:00
Bill Hunter	24:51
Jon James	24:44
Hulan Mathies	20:00
Jack Moreland	19:46

HLG JUNIOR	
Fernando Diez	360
Danny Diez	296
Steve Wittman	274
Bruce Armstrong	265
Marvin Miller	219

HLG OPEN	
Bill Blanchard	871
Randy Secor	455
Bill Xenakis	390
George Batiuk	353
Jon James	349

ROCKET	
Irv Aker	12:00
Harry Steinmetz	7:06
Jim Crocket	6:27
Bill Haught	6:19
Andrew Barron	5:00

UNLIMITED RUBBER, JUNIOR	
David Steinmetz	807
Greg Richardson	444
Van Richardson	427
Eric Dyer	247
Jim Geraghty	173

SCALE GAS
Brickner
Cover
Warner
Stroman
Moss

GRAND CHAMPION; Chuck Broadhurst Tr.
Bruce Hannah Jr.

HIGH TIME: Ocie Randall Trophy
Doug Galbreath

JUNIOR SWEEPSTAKES
Eric Dyer

OPEN SWEEPSTAKES
Irv Aker

TEAM
Uncle Sal's Flying Circus:
Sal Taibi, Bob White, Lee Hines

SENIOR SWEEPSTAKES
Andrew Barron

European F1C Champs - 28 July/2 Aug. - F1C Zagreb, Yugoslavia

1	A Denkin	BG									1260 + 188
2	G Barbarella	I									1260 + 183
3	W Kraus	A									1260 + 137
4	O Velunsek	JU	180	158	180	180	180	180	180	=	1238
5	U Schaller	CH	180	180	180	180	180	157	180	=	1237
6	S Reda	BRD	180	180	180	180	145	180	180	=	1225
7	R Friedrich	BRD	180	180	180	180	142	180	180	=	1222
	J Szecsenyi	H	180	180	142	180	180	180	180	=	1222
9	K Engelhardt	DDR	180	137	180	180	180	180	180	=	1217

Team results

1	German Fed. Rep.	BRD	3576 seconds
2	Hungary	H	3534
3	Jugoslavia	JU	3479
4	Czechoslovakia	CZ	3444

Criterium Pierre Trebod - 21/22 Aug. F1A, F1B, F1C Marigny-le-Grand, France

1976 12 Critérium Pierre Trébod

PLANEURS F1A

1	GRUNNET P.	DK	180	180	180	180	180	180	180	1260
2	BAINES B.	GB	169	180	180	180	180	180	180	1249
3	SPANN R.	A	165	180	180	180	180	180	180	1245
4	DE BOER P.	NL	161	180	180	180	180	180	180	1241
5	LEPAGE R.	F	164	180	180	180	180	175	180	1239
6	DIBBETS P.	NL	154	180	180	180	180	180	176	1230
7	LEBEE J.	F	147	180	180	180	173	180	180	1220
8	PAILLOU P.	F	180	180	180	180	180	180	134	1214
9	KRAUTH M.	F	129	180	180	180	180	180	180	1209
10	MADELIN G.	GB	128	180	180	180	180	180	180	1208
11	WILKENING H.	D	125	180	180	180	180	174	180	1199
12	LELEUX J.	F	141	156	180	180	180	180	180	1197
13	JACK A.	GB	115	180	180	180	180	180	180	1195
14	PIQUER J.	F	113	180	180	180	180	180	180	1193
15	GREAVES D.	GB	112	180	180	180	180	180	180	1192
16	BUNGART A.	D	111	180	180	180	180	180	180	1191
17	CHALLINE J.P.	F	180	180	180	180	107	180	180	1187
17	ALTHOFF H.	D	159	180	180	180	180	128	180	1187
19	AIMELET F.	F	115	180	180	180	180	180	171	1186
20	FLINDT H.	D	174	180	180	180	180	180	104	1178
21	COOPER J.	GB	153	178	180	123	180	180	180	1174
22	PEPER H.	D	173	180	180	180	180	180	96	1169
23	GAENSLI F.	CH	098	180	168	180	180	180	180	1166
24	MULLER J.	D	149	180	180	116	180	180	180	1165
25	HUMPHREY	CDN	145	112	170	180	180	180	180	1147
26	GODINHO J.	F	157	180	180	180	180	084	180	1141
27	COWLEY M.	GB	140	111	180	180	180	180	168	1139
28	HORCICKA V.	A	145	180	180	180	118	180	145	1128
29	SALZER K.	A	155	072	180	180	180	180	180	1127
30	BOGAERTS J.	B	138	180	180	084	180	180	180	1122
31	BERTHE R.	F	168	176	180	180	126	180	106	1116
32	MANCHE L.	NL	146	180	180	180	154	090	180	1110
33	RASMUSSEN P.	DK	126	180	180	180	180	180	081	1107
34	BATTY C.	GB	130	180	180	075	180	180	180	1105
34	CEDAR R.	GB	099	180	180	180	106	180	180	1105
34	BROBERG H.	S	180	180	147	180	112	180	126	1105
34	BACHMANN C.	CH	135	180	151	180	177	180	102	1105
38	CORDES A.	GB	165	158	061	180	180	180	180	1104
39	REDA S.	D	130	068	180	180	180	180	180	1098
40	DRAPEAU JL	F	085	180	180	117	175	180	180	1097
41	WILKENING F.	D	116	140	180	120	180	180	180	1096
42	MICHENAUD L.	F	091	180	180	180	134	180	147	1092
43	CHAUSSEBOURG P.	F	109	180	080	180	180	180	180	1089
44	MONICA A.	P	106	150	180	176	112	180	180	1084
45	JENSEN S.	DK	097	180	115	180	180	180	148	1080
46	SEELIG P.	D	155	180	124	100	100	100	075	1074
47	ROS B.	NL	107	180	180	180	123	112	180	1072
48	DUMONTET B.	F	106	180	180	180	180	082	156	1064
49	BOUTILLIER B.	F	129	078	180	180	126	180	180	1053
50	PROCTOR R.	GB	126	127	180	180	077	180	180	1050
51	ALLAIS R.	F	103	113	178	180	180	180	113	1047
52	KONGSTAD K.	DK	169	120	037	180	180	180	180	1046
53	NICHOLSON B.	GB	102	180	163	143	180	093	180	1041
53	STEWART P.	GB	080	153	180	180	180	180	088	1041
55	ABREU J.	P	093	180	094	180	180	180	133	1040
56	DILLY M.	GB	115	180	180	180	180	127	076	1038
57	REYNDERS L.	B	124	180	180	091	095	180	180	1030
58	DULOUT H.	F	147	180	180	180	097	113	132	1029
59	[illegible]INKS G.	NL	131	180	180	180	084	085	180	1020
60	DEMEURE C.	B	129	079	135	180	173	180	141	1017
61	COIN C.	F	034	159	180	180	145	180	135	1013
62	VAN HAUDENARD M.	B	139	180	180	144	062	124	180	1009
63	KAPPE B.	D	152	180	180	077	174	085	152	1000
64	TILHAC JM	F	090	125	117	180	180	180	126	998
65	LAMMERTS H.	NL	130	180	115	180	107	101	180	993
66	GARCIA E.	P	102	145	131	082	180	171	180	991
67	GLUE D.	GB	116	180	180	130	083	120	180	999
68	BUGE C.	D	126	180	180	180	106	111	100	983
68	RAPIN F.	F	092	180	075	180	180	172	104	983
70	HARTILL W.	USA	145	180	085	079	132	180	180	981
71	WITTE H.	D	148	102	143	180	180	180	040	973
72	DUPUIS L.	F	043	127	082	180	180	180	180	972
73	WANNGARD P.	S	072	180	065	180	180	180	112	969
74	DELCROIX J.	F	125	180	100	180	049	147	180	961
75	DETON Y.	F	115	180	180	180	083	094	126	958
76	THOMSON D.	GB	092	180	180	130	092	098	180	952
76	CHAMPION R.	F	134	100	081	180	180	097	180	952
78	GUFFENS J.	B	122	075	180	180	180	034	180	951
79	OWENS P.	GB	118	104	180	180	180	066	122	950
80	GAULL S.	D	127	180	180	119	120	034	180	940
81	HAIDEMANN T.	D	135	180	118	066	180	180	078	937
82	HACKEN A.	NL	116	102	180	081	180	032	180	921
83	PILLER M.	F	100	180	139	180	050	088	180	917
84	KEIM P.	NL	114	180	054	180	060	144	180	912
85	GOISMIER J.	F	045	033	180	110	180	180	180	908
86	KIEHNLE U.	D	125	064	180	091	084	180	180	904
87	FOURNIER JM	F	077	120	180	180	180	078	079	894

88	VAN EEDE T.	NL	124	055	180	132	180	081	138	890
89	BREEMAN C.	NL	094	180	180	087	125	112	100	878
90	IMGENBERG E.	D	117	100	123	180	160	092	100	872
91	HOHLS C.	D	133	180	180	093	180	103	-	869
91	GALAN JIMENEZ J.	E	141	180	108	053	072	135	180	869
93	PRISON H.D.	D	171	097	180	119	047	138	111	863
94	NOCQUE G.	F	137	180	180	137	180	041	-	855
95	ROUX A.	F	122	125	115	081	180	180	045	848
96	EKKELENKAMP W.	NL	103	088	043	130	180	180	116	840
97	BARBARO J.P.	F	106	180	086	180	084	092	108	836
98	HAMELIN T.	F	105	123	172	180	062	047	135	824
99	DUCE M.	GB	087	180	053	130	141	043	180	814
100	TEJERO-CASTELL L.	E	072	150	110	180	058	057	180	807
101	MARRIOTT S.	GB	104	180	180	083	085	085	088	805
101	CASIER C.	F	126	180	056	048	074	180	141	805
103	BASTOS C.	P	103	180	078	180	-	072	180	793
104	BACHMANN GO	CH	099	078	059	136	180	158	058	768
105	GAILCHET A.	F	116	180	180	180	103	-	-	749
106	LITJENS J.	NL	135	113	180	083	115	080	033	739
107	BIXBY P.	GB	030	077	097	163	145	127	099	738
108	TOMCZYK D.	F	125	122	180	120	180	-	-	727
109	HERZOG A.	B	021	060	118	180	123	059	162	723
110	NOUGE A.	F	137	180	180	081	137	-	-	715
111	NEUMANN E.	D	060	093	129	028	114	180	100	704
111	GREY T.	GB	106	130	118	092	180	078	-	704
113	LOPES A.	P	079	107	053	180	053	180	037	689
114	WILKE J.	D	134	180	180	180	-	-	-	674
115	BRAWN D.	GB	119	047	147	060	080	101	095	649
116	SCHANDEL A.	F	100	180	114	050	098	065	-	607
117	MADORE J.C.	F	110	140	180	104	046	-	-	580
118	ALLNUTT P.	CDN	163	180	180	052	-	-	-	575
119	KOSTER K.	DK	156	180	040	180	-	-	-	556
120	BROCHARD G.	F	110	113	180	087	052	-	-	552
121	BACHMANN GU	CH	091	081	025	043	081	175	055	551
122	DE MELLO	CDN	128	180	180	023	-	-	-	511
123	HIRLIMANN J.C.	F	122	180	109	060	-	-	-	471
124	LYANT R.	F	104	115	016	094	-	029	107	465
125	WILLIAMS J.	GB	180	083	080	113	-	-	-	456
126	HAWKINS P.	GB	105	085	060	093	099	-	-	442
127	RODRIGUEZ-GARRIDO S.	E	033	054	085	080	066	044	079	441
128	GOUDEAU C.	F	122	078	180	060	-	-	-	440
129	MATHERAT G.	F	121	141	156	-	-	-	-	418
130	SODERSTROM B.	S	152	180	059	-	-	-	-	391
131	BAZILLON M.	F	163	180	-	-	-	-	-	343
132	LASSAIGNE J.	F	125	120	-	-	-	-	-	245
133	HUBLER H.	D	071	135	--	--	-	-	-	206
134	BRODARAC D.	D	148	-	-	-	-	-	-	148
135	MONKS R.	GB	145	-	-	-	-	-	-	145
136	JELLIS P.	GB	030	-	-	-	-	-	-	-

W A K E F I E L D S   F 1 B

1	GOUVERNE E.	F	180	180	180	180	180	180	180	1260
2	DUPUIS L.	F	180	180	180	165	180	180	180	1245
3	DELCROIX J.	F	147	180	180	180	180	180	180	1227
4	GAENSLI F.	CH	180	135	180	180	180	180	180	1215
5	BENEDINI H.	RA	180	120	180	180	180	180	180	1200
6	MATHERAT G.	F	180	178	180	180	111	180	180	1189
7	PRIOUX H.	F	180	172	180	180	150	180	127	1169
7	CHAMPION R.	F	180	175	180	180	094	180	180	1169
9	KERSHAW B.	GB	151	109	180	180	180	180	180	1160
10	BARNES J.	GB	158	180	180	170	095	180	180	1143
11	BOUTILLIER B.	F	180	180	170	065	180	180	180	1135
12	PECCHIOLI P.	I	137	140	150	180	180	180	164	1131
13	WITHE R. (Proxy CHAUSSEBOURG)	USA	136	180	163	180	103	180	180	1122
14	SILZ B.	D	149	180	180	143	180	180	090	1102
15	GREY T.	GB	180	118	149	180	180	180	114	1101
16	ORTHWEIN M.	D	116	090	180	180	154	180	180	1080
17	VAN MERKESTIJN P.	NL	176	180	180	180	101	147	113	1077
18	SANAVIO A.	I	180	180	180	180	159	101	091	1071
19	MADELIN G.	GB	123	146	180	180	180	078	180	1067
20	NOUGE A.	F	161	180	180	180	129	180	026	1036
21	RASMUSSEN P.	DK	082	180	180	180	131	180	095	1028
22	MARRIOTT S.	GB	129	149	155	148	180	124	102	987
23	GREAVES D.	GB	180	141	140	058	180	180	099	978
24	ALLAIS R.	F	127	180	180	180	180	126	-	973
25	COSTA C.	P.	123	070	180	180	180	103	104	940
26	SODERSTROM B.	S	140	105	180	180	062	180	069	916
27	ROBERTS	CDN	127	180	081	180	160	180	-	908
28	HELMBRECHT H.	D	143	180	023	165	037	175	180	903
29	ROHRER E.	CH	136	166	088	180	025	101	180	876
30	DUCE M.	GB	133	056	140	180	180	180	-	869
31	CHAUSSEBOURG P. (Proxy WETTERBERG)	F	149	097	080	180	161	114	085	866
32	AMBROSO G.	F	180	123	180	061	104	180	-	828
33	HENDERSON I.	DK	180	084	096	129	048	147	140	824
34	BASTOS C.	P	113	180	055	106	070	140	150	814
35	GARCIA E.	P	056	095	180	088	180	080	123	812
36	DE KRUYFF G.	NL	078	180	-	180	088	098	180	804
37	RAPIN F.	F	180	061	093	180	066	119	090	789
38	NEUMANN E.	D	150	046	180	163	180	-	-	719
39	JAKOBSEN E.	DK	180	102	069	065	180	060	-	656
40	GILLIVRAY	CDN	155	180	180	121	-	-	-	636
40	MANCHE L.	NL	144	180	132	180	-	-	-	636
42	WOODHOUSE E.	GB	157	180	180	091	-	-	-	608
43	WANNGARD P.	S	173	104	109	180	-	-	-	566
44	HIPPERSON D.	GB	180	180	113	076	-	-	-	549
45	KAYNES I.	GB	122	145	180	096	-	-	-	543
46	GRAUX A.	B	105	180	180	072	-	-	-	537
47	GOUDEAU C.	F	102	180	180	045	-	-	-	507
47	NEGLAIS J.C.	F	180	180	147	-	-	-	-	507
49	GAULL S.	D	112	180	180	-	-	-	-	472
50	SAUBUSSE R.	F	096	050	180	063	066	-	-	455
51	CASIER C.	F	128	139	180	-	-	-	-	447
52	BOGAERTS J.	B	114	145	150	-	-	-	-	409
53	RIGAUT I.J.	F	105	075	138	081	-	-	-	399
54	KOPPITZ A.	F	139	172	043	-	-	-	-	354
55	REYNDERS L.	B	095	046	180	-	-	-	-	321
56	LEPAGE P.	F	117	112	084	-	-	-	-	313
57	MADORE J.C.	F	075	070	114	049	-	-	-	308
58	VANTOMME J.	B	178	083	035	-	-	-	-	296
59	COGNET G.	F	142	086	040	-	-	-	-	268
60	DURAND R.	F	126	133	-	-	-	-	-	259
61	BATTY C.	GB	097	158	-	-	-	-	-	255
62	PILLER M.	F	134	105	-	-	-	-	-	239
63	PROCTOR K.	GB	129	050	-	-	-	-	-	179
64	GOISMIER J.	F	076	061	-	-	-	-	-	137

M O T O M O D E L E S   F 1 C

1	FERRERO D.	F	180	169	180	180	180	180	180	1249
2	HUBLER H.	D	180	168	180	180	180	180	180	1248
3	STETZ H.	D	180	180	180	180	180	180	163	1243
4	FAUX K.	GB	180	180	180	180	180	180	158	1238
5	MONKS R.	GB	180	180	180	180	180	180	141	1221
6	IMGENBERG U.	D	180	180	180	180	180	137	180	1217
7	SCHALLER U.	I	180	141	180	180	180	171	180	1212
8	ZITO M.	RA	154	127	180	180	180	180	180	1181
9	KOSTER T.	DK	180	180	180	180	180	100	180	1180
10	LINDHOLM H.	S	180	180	180	081	170	180	180	1151
11	BRODARAC J.	D	150	180	094	180	180	180	180	1144
12	BARBARO J.P.	F	094	146	180	180	180	180	180	1140
13	SEELIG H.	D	180	124	112	180	180	180	180	1136
14	ZEH R.	D	174	180	180	180	116	180	104	1114
15	HEIDEMANN G.	D	180	180	151	180	074	180	116	1061
16	JOHNSON R.	GB	120	112	180	180	180	100	180	1052
17	SOARES P.	P	165	096	149	180	180	160	102	1032
18	SAUER K.H.	D	180	180	180	180	103	138	057	1018
19	REDA S.	D	180	180	180	180	180	105	-	1005
20	JEAN M.	F	180	180	180	180	127	156	-	1003
20	HARTWAGNER F.	A	180	150	180	180	156	056	101	1003
22	COWLEY M.	GB	180	180	180	180	051	126	-	897
23	JACK A.	GB	-	167	180	180	108	165	065	865
24	FRIEDRICH R.	D	180	180	180	180	081	-	-	801
25	COLLINS R.	GB	163	177	180	025	105	097	-	747
26	BURROWS L.	GB	115	097	084	102	120	135	091	744
27	ILLAN ESPINOSA M.	E	158	076	-	092	143	088	180	737
28	LANDEAU A.	F	180	180	180	180	-	-	-	720
29	SCHILLING G.	D	008	180	147	064	176	139	-	714
30	MAIER E.	D	171	180	180	180	-	-	-	711
31	JORGENSEN P.	DK	160	152	180	180	-	-	-	672
32	SEYDEL S.	D	118	180	180	180	-	-	-	658
33	ZIMMER C.	F	-	180	175	180	-	-	-	535
34	HEIDEMANN T.	D	180	156	180	006	-	-	-	522
35	ROUX A.	F	175	052	180	-	-	-	-	407
36	BRODARAC D.	D	177	180	027	-	-	-	-	384
37	PYMM D.	GB	180	129	065	-	-	-	-	374
38	SOARES M.	P	132	025	180	-	-	-	-	337
39	REIS A.	P	131	098	-	-	-	-	-	229
40	BERNI G.	I	068	-	-	-	-	-	-	068

FAI Indoor World Champs

- 27/31 Aug. - F1D

RAE, Cardington, England

F.A.I. 76 CARDINGTON
INDOOR MODEL
WORLD CHAMPIONSHIPS

1	P Romak	USA	34.54	29.23	32.48	39.22	39.36	-	=	78.58
2	E Ciapala	PL	35.45	36.18	30.19	0.46	29.32	2.48	=	72.03
3	L Barr	GB	34.30	36.54	17.44	32.05	13.45	11.03	=	71.24
4	J Blount	GB	25.25	35.42	34.05	18.05	34.44	24.58	=	70.26
5	M Thomas	CDN	32.19	35.32	33.18	23.14	28.20	27.51	=	68.50
6	K Rybecki	CZ	34.35	33.39	31.22	34.08	7.17	15.49	=	68.43
7	R Green	GB	33.10	34.33	33.06	30.12	33.53	18.49	=	68.26
8	B Servaites	USA	34.37	14.27	10.15	32.34	23.48	33.41	=	68.18
9	J Richmond	USA	9.37	31.32	36.29	20.53	31.41	17.28	=	68.10
10	A de Mello	CDN	33.22	34.43	11.12	29.34	4.30	28.51	=	68.05
11	B Felsted	AUS	27.25	31.47	31.37	32.18	20.23	35.45	=	68.03
12	J Kalina	CZ	32.45	33.34	7.25	32.42	9.33	23.17	=	66.19
13	T Strasberger	JU	27.30	23.34	29.28	29.21	36.51	-	=	66.19
14	J MacGillivray	CDN	29.50	36.06	29.02	26.20	12.10	6.27	=	65.56
15	V Kmoch	JU	26.28	30.31	34.58	26.03	29.39	19.41	=	65.29
16	E Chlubny	CZ	28.57	36.04	24.28	28.46	11.39	-	=	65.01
17	R Butti	CH	23.25	29.40	22.33	5.40	34.29	14.15	=	64.09
18	W Wetzel	BRD	30.04	31.01	18.39	32.17	13.32	-	=	63.33
19	L Gabrijel	JU	26.34	33.41	13.00	1.12	1.28	29.22	=	63.03
20	S Kujawa	PL	29.02	32.45	16.56	8.29	28.26	5.15	=	61.47
21	E Liem	NL	26.03	8.55	25.05	-	22.53	34.12	=	60.15
22	S Pontan	S	17.07	19.14	28.09	11.32	20.23	31.35	=	59.44
23	Y Banba	J	28.48	29.26	30.14	21.55	29.20	14.08	=	59.40

24	P Migani	I	26.14	7.35	33.15	25.49	8.30	13.46	=	59.29
25	F Migani	I	24.38	19.49	28.22	30.01	7.30	18.30	=	58.23
26	D Siebenmann	CH	26.20	29.55	27.03	26.31	27.46	28.14	=	58.09
27	C Cotugno	I	28.53	29.10	24.54	16.15	22.01	7.17	=	58.03
28	R Czechowski	PL	10.05	30.24	8.45	23.51	8.51	27.23	=	57.47
29	C Wolthoorn	NL	25.54	26.42	11.38	28.39	22.36	26.19	=	55.21
30	K Vogler	BRD	17.25	28.25	15.14	23.33	26.19	17.32	=	54.44
31	P Nore	SF	27.42	19.22	17.52	20.25	25.47	3.42	=	53.29
32	H Enomoto	J	22.40	25.43	22.03	24.01	4.27	14.45	=	49.44
33	H Raulio	SF	22.35	20.53	24.43	20.31	5.23	4.29	=	47.18
34	K Nottelmann	BRD	21.36	18.39	19.40	19.31	25.33	5.47	=	47.09
35	S-O Linden	S	12.26	1.38	7.20	12.45	34.06	5.17	=	46.51
36	H Odagiri	J	20.39	22.29	-	19.04	23.10	15.05	=	45.39
37	H Erofejeff	SF	18.42	19.19	21.08	23.23	18.42	0.16	=	44.31
38	W Beekmeyer	NL	16.21	21.56	9.10	19.54	9.24	3.01	=	41.50
39	P Sodersten	S	16.36	17.34	21.45	18.47	15.13	-	=	40.32
40	G Cognet	F	14.03	16.07	17.39	7.41	19.38	1.07	=	37.17
41	W Heise	CH	14.58	13.37	16.39	1.15	0.53	14.21	=	31.37

Team results

1	United States	USA	215.26
2	Great Britain	GB	210.16
3	Canada	CDN	202.51
4	Czechoslovakia	CZ	200.03
5	Jugoslavia	JU	194.51
6	Poland	PL	191.37
7	Italy	I	175.55
8	German Fed. Rep.	BRD	165.26
9	Netherlands	NL	157.26
10	Japan	J	155.03
11	Switzerland	CH	153.55
12	Sweden	S	147.07
13	Finland	SF	145.18
14	Australia	AUS	68.03
15	France	F	37.17

Ernie Kopecky Trophy for the best single individual flight: P 'Bud' Romak.

European F/F Champs - 10/12 Sept. -F1A, F1B, F1C Homburg/Saar, W Germany

F1A Glider

1	G Zach	A	180	180	180	180	180	180	164	=	1244
2	P Williams	GB	180	180	180	180	180	180	121	=	1201
3	W Kraus	A	180	180	178	99	180	178	180	=	1175
4	G Zojceski	JU	180	180	180	132	177	180	142	=	1171
5	B Leskosek	JU	155	180	155	180	180	180	106	=	1136
6	V Lustig	DDR	180	180	143	180	66	180	180	=	1109
7	A Dolder	CH	180	180	80	172	147	180	118	=	1057
8	H-J Wolf	DDR	180	144	180	59	131	180	180	=	1054
9	S Vinco	I	180	95	162	94	175	180	150	=	1036
10	M Fantham	GB	180	180	80	129	179	106	180	=	1034
11	J Müller	BRD	180	78	180	169	95	180	146	=	1028
12	S Nevzat	T	180	138	180	125	119	180	94	=	1016
13	O Ben-David	IL	157	136	180	180	180	180	-	=	1013
14	G Totev	BG	144	180	180	63	180	157	102	=	1006
15	A Vidensek	JU	180	55	165	115	180	180	121	=	996
16	I Sundstedt	S	180	180	180	48	180	107	95	=	970
17	P Lendering	NL	145	180	85	14	180	180	180	=	964
18	I Weiss	IL	77	180	137	86	180	180	114	=	954
19	B Baines	GB	180	100	180	58	180	180	71	=	949
20	A van Eldik	NL	180	65	180	180	30	180	131	=	946
21	A Bucher	CH	170	78	180	-	148	180	180	=	936
22	N Wallertin	S	67	180	170	180	128	119	88	=	932
23	H-W Althoff	BRD	54	180	180	71	165	180	91	=	921
24	W Gerlach	BRD	180	85	61	124	83	180	180	=	893
25	W Haase	DDR	124	180	60	117	145	158	105	=	889
26	N Nikoloz	BG	180	180	73	38	165	180	60	=	876
27	G Verbree	NL	180	112	93	68	56	180	180	=	869
28	H Chmelik	A	180	91	66	103	111	180	113	=	844
29	G Tschuor	CH	96	110	180	112	96	180	67	=	841
30	I Yosipovitch	IL	180	112	113	80	80	180	59	=	804
31	P Vassil	BG	96	151	60	28	180	180	68	=	763
32	S Lundequist	S	153	133	168	62	180	-	16	=	712
	E Vörös	H	180	10	112	-	180	180	50	=	712
34	I Kovacs	H	79	121	117	80	125	141	25	=	688
35	M Ledocq	B	71	65	158	76	95	155	50	=	670
36	G Stranieri	I	96	103	89	111	63	71	95	=	628
37	I Aslan	T	45	180	43	39	75	107	75	=	564
38	J Leleux	F	40	-	74	75	168	150	51	=	558
39	K Kösegil	T	60	56	112	27	98	33	92	=	478
40	L Fülöp	H	57	32	125	63	92	55	46	=	470
41	V Lensi	I	137	106	180	-	-	-	-	=	423
42	D Demeure	B	180	180	59	-	-	-	-	=	419
	no of maxes		22	18	15	7	14	26	8		
	no of full houses			10	4	2	2	2	-		

Team results, F1A

1	Yugoslavia	JU	3303
2	Austria	A	3263
3	Great Britain	GB	3184

Team results, F1B

1	Italy	3330	
2	DDR	3320	(East Germany)
3	BRD	3198	(West Germany)

F1B Wakefield

1	R Artioli	I	125	180	180	176	170	180	180	=	1191
2	B Söderström	S	180	180	122	174	160	180	180	=	1176
3	L Stoianov	BG	180	180	180	180	158	107	175	=	1160
4	B Kroon	NL	180	180	178	170	165	159	120	=	1152
5	R Schlesinger	BRD	180	180	180	174	164	122	140	=	1140
6	E Balarini	I	157	130	180	180	132	180	180	=	1139
7	M Gialanella	I	154	180	104	180	180	161	163	=	1122
	A Oschatz	DDR	180	180	168	98	136	180	180	=	1122
9	R Brand	IL	141	180	180	108	143	180	180	=	1112
10	E Mielitz	DDR	180	90	180	179	113	180	180	=	1102
	A Wells	GB	146	134	180	102	180	180	180	=	1102
12	H van Hoorn	NL	180	131	152	180	118	180	160	=	1101
13	E Sebahattin	T	134	147	120	158	180	180	180	=	1099
14	M Kapetanovic	JU	180	153	172	112	120	180	180	=	1097
15	J Löffler	DDR	180	129	91	156	180	180	180	=	1096
16	H Zachelmel	A	180	71	164	119	180	180	180	=	1074
17	A Busch	BRD	153	180	122	180	63	180	180	=	1058
18	P-T Skjulzstad	N	180	127	166	111	180	139	151	=	1054
19	L Purgai	H	180	180	125	102	157	180	114	=	1038
20	J Zetterdahl	S	180	180	94	151	163	132	135	=	1035
21	D Siebenmann	CH	95	180	180	100	119	180	180	=	1034
22	A Haagen	A	155	180	137	100	101	180	177	=	1030
23	W Eggimann	CH	180	180	180	56	102	148	172	=	1018
24	A Donde	IL	137	180	180	78	82	180	180	=	1017
25	G Walker	GB	98	180	180	103	180	86	180	=	1007
26	K Leissner	BRD	82	180	180	76	122	180	180	=	1000
	A Sanavio	I	114	92	97	157	180	180	180	=	1000
	K Andersson	S	84	180	163	91	145	170	167	=	1000
29	D Pecek	JU	105	180	117	104	180	121	180	=	987
30	I Kaynes	GB	116	180	118	90	107	180	180	=	971
31	A Graux	B	180	180	120	111	126	68	180	=	965
32	A Hakensson	S	180	118	157	106	42	180	177	=	960
33	K Jusufbasic	JU	72	63	180	167	169	127	175	=	953
34	M Bülent	T	180	88	98	115	111	180	180	=	952
35	A Borchia	Sard	180	118	96	38	170	180	166	=	948
36	E Reitterer	A	137	180	179	173	59	47	150	=	925
37	H Reifler	CH	159	121	60	95	123	179	180	=	917
38	S-T Voltchev	BG	98	180	100	78	164	180	95	=	895
39	A Szeri	H	87	78	130	75	144	180	180	=	874
40	J Pietchev	BG	77	180	91	120	155	70	179	=	872
41	T Akca	T	171	44	121	136	82	180	137	=	871
42	M Varadi	H	116	180	31	79	-	180	130	=	716
43	H van Rij	NL	131	163	63	77	70	95	90	=	689
44	I Den Itzhak	IL	61	112	106	27	118	106	112	-	642
45	C Ledocq	B	-	107	81	-	-	-	83	=	271
	no of maxes		18	25	13	5	9	27	23		
	no of full houses			9	3	1	-	-	-		

WFFA Challenge - 20/21 Nov. - F1A, F1B, F1C Taft, California U.S.A.

			1	2	3	4	5	6	7	8	9	10	Total
F1A	1	Lee Hines	M	M	M	165	171	173	M	M	M	M	1769
	2	Ed Wogulis	M	M	M	134	M	M	M	M	160	M	1734
	3	Jim Wilson	M	M	M	137	123	M	M	M	145	M	1665
	4	Gary Baughman	M	M	M	124	153	M	M	M	M	115	1652
	5	Hector Diez	62	167	M	M	153	M	M	118	M	M	1580
	6	Craig Cusick	M	120	173	50	161	95	M	173	M	M	1492
	7	Juan Livotto	M	M	M	M	164	79	M	109	147	54	1453
	8	Bob Norton	M	M	M	142	160	157	M	63	123	53	1418
	9	Bill Gibbons	115	80	89	M	146	142	126	M	66	M	1304
	10	Jeff Livotto	M	36	M	79	80	M	75	60	94	36	1000
	11	Don Zink	-	-	-	-	-	M	M	M	149	M	869
		Maxes	8	6	8	3	1	6	9	6	4	7	
		Flights	10	10	10	10	10	11	11	11	11	11	
F1B	1	Les DeWitt	M	M	M	M	M	M	M	M	M	159	1779
	2	Val Sidlauskas	M	168	142	M	M	M	M	M	M	148	1718
	3	Jim Quinn	M	M	M	119	M	120	M	M	M	M	1679
	4	Bob White	M	M	133	M	M	160	110	M	M	M	1663
	5	Walt Ghio	123	174	138	152	156	M	M	M	M	M	1643
	6	Bill Humbert	M	M	M	M	160	M	M	134	M	83	1637
	7	Bob Tymchek	M	133	130	139	M	M	M	M	140	M	1622
	8	Bob Piserchio	177	M	104	M	M	173	84	M	158	M	1596
	9	Irv Aker	M	110	M	145	106	M	136	M	M	161	1558
	10	Bill Gibbons	M	M	71	129	M	110	M	M	173	122	1505
	11	Larry Jordan	152	92	M	149	156	-	-	-	-	-	729
	12	Bill Wiley	142	149	41	-	-	-	-	-	-	-	332
		Maxes	8	6	5	5	7	6	7	9	7	5	
		Flights	12	12	12	11	11	10	10	10	10	10	
F1C	1	Ed Carroll	M	M	M	M	M	M	M	M	M	M	1800
	2	Doug Galbreath	M	177	M	M	M	M	M	M	M	M	1797
	3	Bruce Hannah	176	M	M	M	M	M	M	M	M	M	1796
	4	Dick Myers	M	164	M	M	M	M	M	M	M	M	1784
	5	Reid Simpson	M	M	M	M	M	M	M	M	133	M	1753
	6	Hardy Brodersen	M	132	M	M	M	M	176	M	M	M	1748
	7	Dave Bogdanyi	151	M	M	149	M	M	M	M	M	M	1740
	8	Ronny Young	M	M	M	M	M	M	M	M	M	114	1734
	9	Ken Oliver	M	129	M	M	M	144	M	M	M	M	1713
	10	Bill Hartill	M	M	M	151	152	M	M	M	110	-	1493
	11	Frank Newquist	174	M	M	M	140	-	-	-	-	-	854
	12	Dave Parsons	168	M	M	M	94	-	-	-	-	-	802
	13	Jon James	131	120	M	131	133	-	-	-	-	-	695
	14	Al Vela	77	35	M	-	-	-	-	-	-	-	292
		Maxes	8	8	14	10	9	9	9	10	8	8	
		Flights	14	14	14	13	13	10	10	10	10	9	

Rounds 1 to 5, Nov 20, 1 to 6PM. Rounds 6 to 10 Nov 21, 8AM to 1PM
Weather - Wind 2 to 5MPH, completely overcast with low clouds, and temperature not exceeding 59 degF (15 degC)

II Coupe d'Hiver International - 5 Dec. - F1G
RAF Halton, England

100 gram event: MRA Challenge; 46 flew

1	G Matherat (F)	120	113	120	= 353
2	B Boutillier (F)	120	120	105	= 345
3	B Boutillier (F)	115	105	120	= 340
4	A Roux (F)	103	120	108	= 331
5	B Cox				327
6	S Marriott				296
7	I Kaynes				293
8	G Matherat (F)				291

80 gram event: Aero Modeller Trophy; 55 flew

1	D Hipperson	109	120	120	120	120	= 589
2	M Dilly	120	120	104	120	119	= 583
3	J O'Donnell	118	111	120	120	108	= 577
4	S Marriott	116	120	120	86	120	= 562
5	G Matherat (F)						553
6	B Boutillier (F)						546
7	R Godden						543
8	R Fleetwood						535

The Year 1977

US Free Flight Champs - 28/29/30 May - All events
Taft, California, U.S.A.

H.L.G. S.O.	
Bill Blanchard	925
Dick Mathies	890
Greg Sussex	691
Ron Wittman	677
Bob De Shields	630

C Cabin O.T.	
Sal Taibi	23:01
Bruce Norman	18:13
Leslie Norman	17:22
Fred Emmert	15:00
Larry Boyer	9:18

A-B Cabin O.T.	
Bruce Norman	14:52
Wade Wiley	11:30
Phil McCary	10:50
L. Nadolski	10:39
Leslie Norman	10:26

½A Gas S.O.	
Nick Jones	1500
Ed Weiler	1380
Bob Boyer	1360
Steve Shibuyo	1325
Debbie Beron	1259

A Gas Jr.	
Joey Foster	25:00
Scott Adikes	16:25
Todd Kerzie	10:42
Jeff Cunnyngham	10:31
Bobby Morgan	9:40

B-C Gas Jr.	
Scott Adikes	15:17
Jeff Cunnyngham	13:35
Chuck Nadolski	3:42

F.F. Scale Rubber
Clarence Mather
Hal Cover
Dave Haught
Mike Mulligan
Clarence Haught

A-2 Glider	
Lee Hines	1620
Hector Diaz	1526
Mike Szura	1223
Jim Wilson	1180
Dennis Mihora	1175

Rocket	
Bill Booth, Jr.	12:12
Ed Eliot	9:28
Richard Wood	8:51
Irv Aker	8:50
Jim Haught	7:38

Indoor Peanut Scale
Mike Mulligan
Jim Pearson
Bill Warner
C.C. Johnson
Mark Valerius

Unlimited Rubber S.O.	
Jim Quinn	3300
Bob White	2333
Mac MacClintock	1277
Andy Faykun	853
John Patton	803

WAKEFIELD RUBBER	
Carl Drake	1211
Bob Piserchio	1161
Irv Aker	1125
Bob White	1123
Walt Ghio	1122

Novice Penny Plane	
Earl Hoffman	403
Bob Meuser	385
Bill Vanderbeek	349
Cezar Banks	348
Paul Andrade	340

Outdoor Peanut Scale
Hal Cover
Mike Mulligan
Clarence Mather
Bill Warner
Bill Stroman

A-1 Glider S.O.	
Bill Xenakis	1119
Lee Hunt	824
Bob Sunberg	816
Clarence Haught	799
Scott Valentine	790

NITE A.M.A. Gas	
Hulan Mathies	68:31
Randy Archer	33:47
Bruce W. Hannah	29:42
Mike Morgan	29:31
Jon James	25:00

F.F. Scale Gas
Bill Warner
Hal Cover
Gene Simpson
Bill Stroman
Dave Haught

A-1 Glider Jr.	
Jason Kendy	678
Joey Foster	516
Dennis Kramer	435
Jeff Cunnyngham	381
Fransisco Barragan	365

Payload	
Ed Eliot	10:16
Bob Meuser	9:55
Jim Haught	7:14
Bill Gunther	7:06
Steve Wittman	6:39

E.Z.B.	
Earl Hoffman	537
Cezar Banks	469
Clarence Mather	436
Jean Andrews	367
Fudo Takagi	359

H.L.G. Jr.	
Eric Dyer	460
Dave Turgeon	459
Joey Foster	413
Steve Wittman	413
Eric Altizer	318

O.T. Rubber	
Barnett Kernoff	14:47
Bob Oslan	14:09
Andy Faykun	13:52
Jim Quinn	13:17
Ernst Johnson	12:31

30 Sec. Antique O.T.	
Larry Boyer	18:27
Bruce Norman	17:03
L. Nadolski	12:26
Wade Wiley	11:26
Fred Emmert	11:19

½A Gas Jr.	
Bobby Morgan	1867
Keith Morgan	1706
Jeff Cunnyngham	1320
Scott Adikes	1313
Parker McQuewn	863

B Gas	
Randy Secor	49:23
Bob Scully	30:00
Mark Valerius	28:58
Mike Kerzie	28:56
Joe Pinto	27:16

D Gas	
Fred Calhoun Jr.	25:00
John Travino	17:24
Fred Calhoun Sr.	17:22
Doug Galbraith	14:45
Bob Vinson	14:34

Unlimited Rubber Jr.	
Jeff Cunnyngham	1260
Mike Davis	708
Terry Legenza	614
Keith Morgan	460
Van Richardson	385

Indoor H.L.G. S.O.	
Ron Wittman	63.3
Bob De Shields	61.6
Bob Boyer	58.9
Lee Hines	57.3
Phedon Tsikmpolos	55.3

C Pylon O.T.	
Bob Mc Bride	12:41
Jerry Vernon	12:04
Cliff Silva	11:40
Cris Cristenson	11:34
Bruce Norman	11:08

A-B Pylon	
Bob Oslan	35:00
Leslie Norman	33:39
Jack Jella	19:58
Larry Clark	16:52
Cliff Silva	13:54

020 Rep. O.T.	
Jack Jella	18:00
Lee Hunt	15:00
Bruce W. Hannah	11:15
L. Nadolski	8:55
Ron Wittman	8:31

A Gas S.O.	
Guy Kirkwood	31:19
Hulan Mathies	25:00
Bill Xenakis	24:09
Paul Andrade	20:00
Randy Secor	17:02

C Gas	
Doug Galbraith	38:25
Fred Calhoun, Jr	25:00
Hulan Mathies	24:21
Irv Aker	18:38
John Sparling	17:45

FAI Power	
Roger Simpson	2700
Mike Achterberg	2590
Bill Gieskieng	2091
Ronnie Young	1782
Bill Lovins	1256

Coupe d'Hiver	
Irv Aker	1213
John Lenderman	1054
George Batick	883
Bill Davis	838
Bill Humbert	720

Indoor H.L.G. Jr.	
Steve Wittman	47.5
David Turgon	33.5
Ed Wogulis	32.8
Joey Foster	32.6
Jeff Cunnyngham	18.4

SWEEPSTAKES

Open	Hulan Mathies
Senior	Scott Valentine
Junior	Jeff Cunnyngham

GRAND CHAMPION Irvin Aker, Sr.

SMAE/FFN FAI International - 26/27 June - F1A, F1B, F1C Bircham Newton, England

F1A Glider 28 flew; 7 x 180 seconds

1	G Madelin	Crookham	180	180	180						
1	G Madelin	Crrokham	180	180	172	180	164	135	180	=	1191
2	A Cordes	Whitefield	111	147	162	135	180	180	180	=	1095
3	I Bracken	St Albans	68	180	180	131	129	180	180	=	1048
4	J Walton	York	180	180	180	67	142	178	109	=	1036
5	J Bogaerts	Belgium	165	180	180	88	180	71	164	=	1028
6	D Barnes	Vulcans	91	151	160	180	175	88	180	=	1025
7	B Baines	RAFMAA	80	180	115	180	97	180	180	=	1012
8	A Jack	Tynemouth	92	178	131	122	102	180	180	=	985

F1B Wakefield 16 flew; 7 x 180 seconds

1	J O'Donnell	Whitefield	180	180	180	164	165	180	169	=	1218
2	H Gunn	Bristol & W	180	180	180	180	119	131	135	=	1105
3	A Jack	Tynemouth	178	92	110	179	180	143	180	=	1062
4	K Proctor	York	160	180	180	165	113	162	43	=	1003
5	S Marriott	Biggles	108	140	100	177	116	180	180	=	1001

F1C Power 5 entered, 4 flew; 7 x 180 seconds

1	L Burrows	Blackheath	180	159	180	164	176	149	180	=	1188
2	F Chilton	Crookham	180	174	174	161	146	106	154	=	1095
3	R Moore	CM	-	123	165	110	103	97	180	=	778

World Championships - 6/12 July - F1A, F1B, F1C * † Roskilde, Denmark

* WORLD FREE FLIGHT CHAMPIONSHIP *

ROSKILDE LUFTHAVN 6. - 12. JULI 1977 - DANSK MODELFLYVEFORBUND

INDIVIDUAL FLIGHT F1A

PLACING	NAME	COUNTRY	1. P	2. P	3. P	4. P	5. P	6. P	7. P	8. P	9. P	TOTA
1	KOSTADIN ABADJIEV	BG	180	180	180	180	180	177	180			1257
2	ANDRES LEPP	USSR	180	180	180	180	180	180	164			1244
3	WERNER KRAUS	A	180	180	154	180	180	180	180			1234
4	RI SONG CHAN	DPRK	165	167	180	180	180	180	180			1232
5	LIONEL BRAUD	F	167	180	180	180	180	160	180			1227
6	IVAN HOREJSI	CSSR	180	180	180	165	180	180	151			1216
7	KANG JONG SIK	DPRK	180	165	164	180	180	180	164			1213
8	VICTOR CHOP	USSR	180	180	180	180	170	180	137			1207
9	HERBERT SCHMIDT	D	180	180	157	180	180	180	142			1199
10	RI CHOL	DPRK	180	126	180	180	180	180	154			1180
11	KLAUS D THORMANN	DDR	180	180	180	180	180	100	180			1180
12	VICTOR ISAENKO	USSR	180	180	103	180	180	174	172			1169
13	JOHN COOPER	GB	162	123	176	180	180	157	180			1158
14	JEAN-LUC DRAPEAU	F	180	152	180	180	180	180	103			1155
15	JAMES WALTERS	USA	162	169	180	180	169	140	151			1151
16	PAVEL DVORAK	CSSR	180	180	180	171	180	133	126			1150
17	GEORGIJE ZOJCESKI	YU	180	180	180	89	180	180	151			1140
18	RUDIGER SPANN	A	180	180	180	142	180	132	142			1136
19	LEO REYNDERS	B	111	180	180	180	180	180	122			1133
20	CHARLES MARKOS	USA	155	178	180	180	180	110	143			1126
21	JUHA TITOFF	SF	116	174	180	180	180	180	115			1125
22	BO JANSSON	S	161	116	180	180	180	180	124			1121
23	ANTON VIDENSEK	YU	168	180	180	121	120	180	170			1119
24	ANTON BUCHER	CH	180	180	119	180	180	180	92			1111
25	HANS-JURGEN WOLF	DDR	156	180	180	120	180	167	126			1109
26	ROBERT SIFLEET	USA	147	180	160	180	149	107	180			1103
27	MILOSLAV POKORNY	CSSR	178	144	123	180	180	180	115			1100
28	GIORA HERTZBERG	IL	180	130	107	164	180	180	159			1100
29	GOTTFRIED ZACH	A	170	180	180	180	77	157	151			1095
30	DIETMAR HENKE	DDR	152	180	180	114	180	107	180			1093
31	SVEIN A LARSEN	N	148	158	180	153	151	131	160			1081
32	SVEIN O OLSTAD	N	180	163	165	180	56	148	171			1063
33	LENSI VALDEMARO	I	99	180	180	180	180	112	131			1062
34	MICHEL BERNISSON	F	180	180	61	115	174	172	176			1058
35	PER GRUNNET	DK	180	117	180	180	148	73	180			1058
36	ILAN WIESS	IL	180	133	180	180	127	131	123			1054
37	GERARD FIKS	NL	173	180	180	180	180	54	107			1054
38	HAKAN BROBERG	S	163	163	180	113	177	128	128			1052
39	RAIMO KATAJAMAKI	SF	156	173	152	180	106	119	166			1052
40	JENS MULLER	D	146	122	172	180	89	161	180			1050
41	M DILLY P ROS DOUGLAS	NZ	155	166	156	180	180	102	106			1045
42	BRANKO LESKOSEK	YU	179	180	180	180	80	131	115			1045
43	PETER OTTE	DK	135	180	164	159	180	98	118			1034
44	DAVID SIMONS	AUS	180	119	143	180	115	152	144			1033
45	LUIS COLMENARES GAMEZ	MEX	180	105	154	176	68	180	157			1020
46	ANDY CRISP	GB	139	165	90	163	150	180	128			1015
47	ANDRE DOLDER	CH	154	164	97	180	180	93	142			1010
48	GEORGES TSCHUOR	CH	106	104	180	174	180	180	81			1005
49	GERARD BRINKS	NL	139	116	145	173	180	121	124			998
50	WALTER PALMIERI	RA	132	112	180	180	137	145	107			993
51	JOHN BAILEY	GB	180	113	120	88	180	162	142			985
52	VICENTE MOLFINO	RA	129	128	180	108	180	143	112			980
53	FINN BJERRE	DK	74	129	180	156	174	151	97			961
54	HANS WILHELM ALTHOFF	D	127	162	37	180	180	180	94			960
55	MARTIN GREGORIE	NZ	87	180	180	180	96	159	63			945
56	GORD MACKENZIE	CDN	158	180	148	53	180	137	75			931
57	MIKE MCMAHON	CDN	148	78	146	151	129	180	93			925
58	JACQUES FOULON	B	114	126	104	180	168	108	119			919
59	NEIL MURRAY	EAT	158	180	175	119	54	115	117			918
60	MAX BOCCARDO	AUS	151	180	126	69	141	180	66			913
61	ITZHAK YOSIFOVITCH	IL	180	140	178	106	54	133	118			909
62	JQRN QDEMARK	N	109	143	148	180	87	93	120			880
63	JOSE LUIS R BARBOSA	MEX	97	86	135	180	82	180	116			876
64	ANTONIO A DE LA ESCOSURA	MEX	107	155	148	75	48	177	165			875
65	KUSHIBIKI TAKASHI	J	134	84	97	180	115	140	111			861
66	YOSHIKAWA HIROSHI	J	100	94	106	145	146	158	109			858
67	VINCO SABBADINI	I	104	100	147	81	109	126	66			845
68	HANY MYNA	EAZ	106	143	115	129	180	93	72			838
69	MICHEL LEDOCQ	B	121	118	136	125	125	69	132			826
70	FATHALLA KAMAL	EAZ	151	142	92	180	65	91	101			822
71	PIET LENDERING	NL	104	121	102	180	102	82	130			821
72	ANDY DEMELLO	CDN	74	109	180	32	141	122	158			816
73	BJQRN SQDERSTRQM	S	125	172	105	180	92	51	86			811
74	GINO STRANIERI	I	121	0	176	178	121	134	78			808
75	FELIPE HERNANDEZ	RA	131	173	180	180	85	55	0			804
76	F NUTINI P W NUTINI	BR	124	73	167	112	128	48	96			748
77	QUEIROZ P LILIANE	BR	55	113	77	180	98	123	96			742
78	LINEU DE CAIRES CHARLIER	BR	146	78	2	65	176	152	110			729
79	JUKKA SILLGREN	SF	136	149	96	83	43	116	94			717
80	ABBAS FATHY	EAZ	69	105	50	111	69	178	111			693
81	RONALD MAGILL	NZ	138	65	131	54	156	101	44			689
82	JAMES GARAY	AUS	19	54	30	0	0	0	0			103

* WORLD FREE FLIGHT CHAMPIONSHIP *

ROSKILDE LUFTHAVN 6. - 12. JULI 1977 - DANSK MODELFLYVEFORBUND

INDIVIDUAL FLIGHT F1B

PLACING	NAME	COUNTRY	1. P	2. P	3. P	4. P	5. P	6. P	7. P	8. P	9. P	TOTAL
1	KIM DONG SIK	DPRK	180	173	180	180	180	180	180			1253
2	SERGEY SAMOKISH	USSR	180	180	180	180	165	180	180			1245
3	BAIK CHANG SON	DPRK	171	180	174	180	180	175	180			1240
4	SHIBACHI MASABUMI	J	180	180	155	180	172	180	180			1227
5	JEAN-CLAUDE NEGLAIS	F	169	180	133	180	180	180	180			1202
6	FRANTISEK RADO	CSSR	127	180	180	170	180	180	180			1197
7	GIOVANNI CASSI	I	140	180	145	180	180	180	180			1185
8	WERNER NIMPTSCH	D	180	180	180	132	180	145	180			1177
9	ALBRECHT OSCHATZ	DDR	166	180	172	180	109	180	180			1167
10	EDY MAURI	I	86	180	153	190	180	180	180			1139
11	ERNST REITTERER	A	180	164	155	145	136	177	180			1137
12	KJELL LIWENBORG	S	122	159	180	180	111	180	180			1112
13	JOSEF KLIMA	CSSR	180	153	180	152	124	152	160			1101
14	MATSUMOTO ISAO	J	180	180	130	176	144	107	180			1097
15	KIM IN SOL	DPRK	180	155	173	103	156	135	180			1082
16	HANS ZACHHALMEL	A	150	180	180	102	108	180	180			1080
17	RON POLLARD	GB	137	156	125	120	180	180	180			1078
18	MIRSAD KAPETANOVIC	YU	123	180	147	180	120	180	144			1074
19	WILLARD SMITZ	USA	116	147	138	180	180	134	177			1072
20	IGOR ZILJBERG	USSR	144	173	88	180	180	180	115			1060
21	PAWEL WLODARCZYK	PL	177	58	116	180	174	180	175			1060
22	LJUBOMIR PETROV	BG	111	180	117	147	180	129	180			1044
23	ANDRZEJ POCZOBUT	PL	101	157	162	145	169	168	139			1041
24	KALEVI KOSKINEN	SF	104	98	140	147	180	180	180			1029
25	ANTONIO SANAVIO	I	112	180	180	133	135	151	134			1025
26	ROBERT PISERCHIO	USA	110	173	114	180	142	119	180			1018
27	EGON MIELITZ	DDR	180	0	180	118	180	180	180			1018
28	DIETER SIEBENMANN	CH	36	158	124	154	180	180	180			1012
29	PER TH SKJULSTAD	N	174	97	129	67	180	180	180			1007
30	KAZIMIERZ IAPINSKI	PL	109	117	162	180	109	180	148			1005
31	ISHII HIDEO	J	119	89	180	130	113	180	180			991
32	ARCANGEL ARMESTO	RA	180	180	180	85	2	180	180			987
33	RONALD MAGILL	NZ	101	168	124	174	105	180	120			972
34	PAUL VAN LEUVEN	AUS	110	80	99	170	180	149	180			968
35	ALFREDO HERBON	RA	83	139	159	136	90	180	180			967
36	WALTER GHIO	USA	167	106	103	150	80	180	180			966
37	PENTTI AALTO	SF	136	109	180	117	116	123	180			961
38	JOACHIM LQFFLER	DDR	137	136	134	83	105	180	180			955
39	SIQUEIRA P L S MATUD	BR	138	110	101	134	180	180	108			951
40	ALLEN EDWARDS	AUS	126	141	159	91	152	153	119			941
41	KNUT ANDERSON	S	85	102	180	105	180	180	105			937
42	OLE TORGERSEN	N	113	116	180	180	116	107	119			931
43	VICTOR ROSHONOK	USSR	100	100	143	127	157	142	151			920
44	ROLANDD SCHIESINGER	D	153	86	128	139	61	180	172			919
45	JAN ZETTERDAL	S	161	118	137	127	133	114	129			919
46	JACK MCGILLIVRAY	CDN	103	180	180	100	120	81	151			915
47	JACK REID	CDN	31	133	163	93	163	149	180			912
48	JOSEF LIBRA	CSSR	148	180	180	2	64	144	180			898
49	IBRAHIM SINANAGIC	YU	136	79	87	149	127	180	140			898
50	TIM TEN VELDE	NL	90	111	157	157	180	105	94			894
51	PETER WILLIAMS	GB	130	108	119	117	180	117	116			887
52	OSSI KILPELAINEN	SF	145	157	107	80	112	175	101			877
53	ERIK JACOBSEN	DK	180	111	147	72	145	137	85			877
54	MICHAEL WOODHOUSE	GB	115	130	129	91	171	144	95			875
55	JOSIP FRANIC	YU	20	55	180	122	145	180	167			869
56	BERNARD BOUTILLIER	F	78	92	156	98	84	180	180			868
57	PETER RASMUSSEN	DK	50	90	77	180	104	180	180			861
58	BERT KROON	NL	108	110	106	68	180	142	133			847
59	MIKE THOMAS	CDN	68	116	146	93	104	133	180			840
60	JQRGEN KORSGAARD	DK	119	150	140	81	90	106	146			832
61	ITZHAK BEN-ITZHAK	IL	70	98	102	116	173	123	144			826
62	KLAUS LEISSNER	D	93	143	95	76	114	134	166			821
63	HENK VAN HOORN	NL	157	38	114	134	78	180	97			798
64	ANDRE GRAUX	B	120	106	59	82	80	180	165			792
65	HERBERT CHMELIK	A	180	71	32	136	78	139	148			784
66	MARIO BLITZMAN	RA	91	102	141	132	103	105	98			772
67	PAULO SOLON RIBEIRO	BR	91	167	91	73	119	50	179			770
68	REUVEN BRAND	IL	145	67	69	81	80	136	180			758
69	RAUL A DE LA CADENA	MEX	98	62	114	81	96	116	180			747
70	HANS REIFLER	CH	130	63	60	172	32	180	99			736
71	ERNESTO JIMENEZ DIAZ	MEX	7	55	154	176	59	164	109			724
72	GUNTER WALTER ALBRECHT	BR	71	152	61	80	180	89	47			680
73	ERICH ROHRER	CH	145	125	91	106	26	104	72			669
74	PETER TH RAUTENBACH	EAT	73	80	116	102	100	103	88			662
75	PETER SMITH	AUS	116	104	71	38	82	79	86			576
76	RENE ALLAIS	F	26	95	65	139	5	61	132			523
77	ARIE DONDE	IL	3	77	94	64	11	150	61			460
78	KALED EL-SHAZLY	EAZ	0	80	119	53	0	0	0			252
79	HESHAM MEHRES	EAZ	49	15	2	3	0	0	0			69
80	STEINER HESTHAGEN	N	59	1	0	0	0	0	0			60

* WORLD FREE FLIGHT CHAMPIONSHIP *

ROSKILDE LUFTHAVN 6. - 12. JULI 1977 - DANSK MODELFLYVEFORBUND

TEAM FLIGHT F1A

PLACE	NO	PARCIPANT			COUNTRY	1. P	2. P	3. P	4. P	5. P	6. P	7. P	TOTA
1	H119	153	154	155	DPRK	525	458	524	540	540	540	498	3625
2	H130	184	185	186	USSR	540	540	463	540	530	534	473	3620
3	H128	178	179	180	CSSR	538	504	483	516	540	493	392	3466
4	H133	194	195	196	A	530	540	514	502	437	469	473	3465
5	H111	130	131	132	F	527	512	421	475	534	512	459	3440
6	H109	124	125	126	DDR	488	540	540	414	540	374	486	3382
7	H129	181	182	183	USA	464	527	520	540	498	357	474	3380
8	H116	144	145	146	YU	527	540	540	390	380	491	436	3304
9	H131	188	189	190	D	453	464	366	540	449	521	416	3209
10	H125	171	172	173	GB	481	401	386	431	510	499	450	3158
11	H123	165	166	167	CH	440	448	396	534	540	453	315	3126
12	H113	136	137	138	IL	540	403	465	450	361	444	400	3063
13	H108	121	122	123	DK	389	426	524	495	502	322	395	3053
14	H120	156	157	158	N	437	464	493	513	294	372	451	3024
15	H126	174	175	176	S	449	451	465	473	449	359	338	2984
16	H110	127	128	129	SF	408	496	428	443	329	415	375	2894
17	H103	107	108	109	B	346	424	420	485	473	357	373	2878
18	H112	133	134	135	NL	416	417	427	533	462	257	361	2873
19	H101	101	102	103	RA	392	413	540	468	402	343	219	2777
20	H117	147	148	149	MEX	384	346	437	431	198	537	438	2771
21	H114	139	140	141	I	356	360	503	439	410	372	275	2715
22	H118	150	151	152	NZ	380	411	467	414	432	362	213	2679
23	H106	115	116	117	CDN	380	367	474	236	450	439	326	2672
24	H132	191	192	193	EAZ	326	390	257	420	314	362	284	2353
25	H104	110	111	112	BR	325	264	246	357	402	323	302	2219
26	H102	104	105	106	AUS	350	353	299	249	256	332	210	2049
27	H115	142	143	0	J	234	178	203	325	261	298	220	1719
28	H105	113	0	0	BG	180	180	180	180	180	177	180	1257
29	H127	177	0	0	EAT	158	180	175	119	54	115	117	918

* WORLD FREE FLIGHT CHAMPIONSHIP *

ROSKILDE LUFTHAVN 6. - 12. JULI 1977 - DANSK MODELFLYVEFORBUND

TEAM FLIGHT F1B

PLACE	NO	PARCIPANT			COUNTRY	1. P	2. P	3. P	4. P	5. P	6. P	7. P	TOTAL
1	H219	250	251	252	DPRK	531	508	527	463	516	490	540	3575
2	H214	238	239	240	I	338	540	478	493	495	511	494	3349
3	H215	241	242	243	J	479	449	465	486	429	467	540	3315
4	H230	282	283	284	USSR	424	453	411	487	502	502	446	3225
5	H228	276	277	278	CSSR	455	513	540	324	368	476	520	3196
6	H209	223	224	225	DDR	483	316	486	381	394	540	540	3140
7	H221	257	258	259	PL	387	332	440	505	452	528	462	3106
8	H229	279	280	281	USA	393	426	355	510	402	433	537	3056
9	H233	291	292	293	A	510	415	367	383	322	496	508	3001
10	H226	272	273	274	S	368	379	497	412	424	474	414	2968
11	H231	285	286	287	D	426	409	403	347	355	459	518	2917
12	H210	226	227	228	SF	385	364	427	344	408	478	461	2867
13	H216	244	245	246	YU	279	314	414	451	392	540	451	2841
14	H225	269	270	271	GB	382	394	373	328	531	441	391	2840
15	H201	201	202	203	RA	354	421	480	353	195	465	458	2726
16	H206	214	215	216	CDN	202	429	489	286	387	363	511	2667
17	H211	229	230	231	F	273	367	354	417	269	421	492	2593
18	H208	220	221	222	DK	349	351	364	333	339	423	411	2570
19	H212	232	233	234	NL	355	259	377	359	438	427	324	2539
20	H202	204	205	206	AUS	352	325	329	299	414	381	385	2485
21	H223	263	264	265	CH	311	346	275	432	238	464	351	2417
22	H204	208	209	210	BR	300	429	253	287	479	319	334	2401
23	H213	235	236	237	IL	218	242	265	261	264	409	385	2044
24	H220	254	255	256	N	346	214	309	247	296	287	299	1998
25	H217	247	248	0	MEX	105	117	268	257	155	280	289	1471
26	H205	211	212	213	BG	111	180	117	147	180	129	180	1044
27	H218	249	0	0	NZ	101	168	124	174	105	180	120	972
28	H203	207	0	0	B	120	106	59	82	80	180	165	792
29	H227	275	0	0	EAT	73	80	116	102	100	103	88	662
30	H232	288	289	290	EAZ	49	95	121	56	0	0	0	321

* WORLD FREE FLIGHT CHAMPIONSHIP *

ROSKILDE LUFTHAVN 6. - 12. JULI 1977 - DANSK MODELFLYVEFORBUND

INDIVIDUAL FLIGHT F1C

PLACING	NAME	COUNTRY	1.-7.P	8.P	9.P	1o.P	TOTAL
1.	THOMAS KØSTER	DK	126o	24o	3oo	34o	214o
2.	ANDRAS MECZNER	H	126o	24o	3oo	3o1	21o1
3.	EVGENY VERBITSKY	USSR	126o	24o	3oo	276	2o76
4.	URS SCHALLER	CH	126o	24o	3oo	275	2o75
5.	DAVE SUGDEN	CDN	126o	24o	3oo	259	2o59
6.	SIEGFRIED REDA	D	126o	24o	3oo	253	2o53
7.	OTON VELUNSEK	YU	126o	24o	3oo	247	2o47
8.	SILVANO LUSTRATI	I	126o	24o	3oo	235	2o35
9.	KIM JONG HUI	DPRK	126o	24o	295		1795
1o.	BRUNO FIEGL	I	126o	24o	266		1766
11.	MIKE BURNS	CDN	126o	24o	253		1753
12.	STAFFORD SCREEN	GB	126o	24o	197		1697
13.	STEEN AGNER	DK	126o	24o	186		1686
14.	SERGEY SHARIN	USSR	126o	237			1497
15.	GIAN P. BARBABELLA	I	126o	235			1495
16.	FRANZ HARTWAGNER	A	126o	231			1491
17.	MARTYN COWLEY	GB	126o	221			1481

18	CENEK PATEK	CSSR	180	180	180	180	180	180	180	206	1466
19	SIN SANG GOL	DPRK	180	180	180	180	180	180	180	179	1439
20	THOMAS MCLAUGHLAN	USA	180	180	180	180	180	180	180	155	1415
21	ORESTES CUZZI	RA	180	180	180	180	180	180	180	94	1354
22	REINHARD FRIEDRICH	D	180	180	180	180	180	180	180	32	1292
23	VACLAV PATEK	CSSR	180	180	180	180	166	180	180	0	1246
24	ALEIN ROUX	F	180	180	180	180	180	180	161	0	1241
25	LADISLAV KOVACIC	YU	180	180	170	180	164	180	180	0	1234
26	MICHEL JEAN	F	180	180	180	180	152	180	180	0	1232
27	FRANK SCHLACHTA	CDN	180	180	180	180	180	180	148	0	1228
28	TOR BORTNE	N	180	180	180	155	180	180	171	0	1226
29	OSZKAR MACZKO	H	180	180	180	180	144	180	180	0	1224
30	CHARLES P MARTIN	USA	180	180	180	180	180	141	180	0	1221
31	JANOS SZECSENYI	H	180	152	180	180	164	180	180	0	1216
32	REINHARD TRUFFE	A	180	180	180	180	180	131	180	0	1211
33	ALEXANDER DENKIN	BG	180	180	180	180	123	180	180	0	1203
34	KIM SONG HAK	DPRK	164	160	180	156	180	180	180	0	1202
35	AL BISSONNETTE	USA	180	139	180	180	180	162	180	0	1201
36	VALENTIN MOZYRSKY	USSR	180	180	180	180	116	180	180	0	1196
37	WALTER NUTINI	BR	180	180	180	180	180	133	163	0	1196
38	EOLO CARLINI	BR	166	180	180	177	154	171	164	0	1192
39	MAURICIO ZITO	RA	180	150	180	180	180	141	180	0	1191
40	HORST KRIEG	DDR	180	180	180	132	180	180	158	0	1190
41	OSWALD EBNER	A	180	180	173	180	180	150	137	0	1180
42	GERHARD FISCHER	DDR	180	180	180	179	96	180	180	0	1175
43	PETER MAURER	CH	180	180	123	180	159	162	180	0	1164
44	BEDRICH KRYCER	CSSR	180	180	180	180	180	180	76	0	1156
45	MIRKO BJELAJAC	YU	180	180	180	180	146	138	148	0	1152
46	RICHARD JOHNSON	GB	180	180	180	180	180	105	136	0	1141
47	NIELS CHR CHRISTENSEN	DK	180	180	180	180	118	164	138	0	1140
48	KLAUS ENGELHARDT	DDR	180	180	180	180	180	103	135	0	1138
49	KOSTADIN ABADJIEV	BG	180	180	179	166	180	56	180	0	1121
50	REIJO SAUKKONEN	SF	113	180	167	135	180	177	157	0	1109
51	DE ALMEIDA PC R C [illegible]ANCE	BR	117	152	165	180	180	169	100	0	1063
52	JOSE LUIS DE LUMA SOLANO	MEX	114	157	180	180	180	80	166	0	1057
53	REIMA LUOSTARINEN	SF	143	62	179	180	168	180	134	0	1046
54	HOWARD GOSTELOW	AUS	0	161	171	180	180	180	168	0	1040
55	KOITABASHI ISAMU	J	180	180	149	180	0	180	169	0	1038
56	KJELL JOHANSSON	S	160	164	156	180	176	69	118	0	1023
57	MICHAEL J PETTIGREW	AUS	180	41	180	180	180	149	107	0	1017
58	FRANZ BAUMANN	D	180	166	180	180	180	124	0	0	1010
59	RUEDI SCHENKER	CH	127	162	180	172	61	174	128	0	1004
60	LARS-G OLOFSSON	S	180	180	180	180	180	5	55	0	960
61	HANS LINDHOLM	S	0	180	180	180	180	111	122	0	953
62	ALBERTO RUIZ VELA	MEX	167	174	180	180	0	81	5	0	787
63	MICHEL IRIBARNE	F	180	180	180	180	50	0	0	0	770
64	WILLIAM EAST	AUS	143	11?	172	116	120	5	0	0	672
45	VESA LAPPALAINEN	SF	180	0	0	140	96	0	0	0	416

* WORLD FREE FLIGHT CHAMPIONSHIP *

ROSKILDE LUFTHAVN 6. - 12. JULI 1977 - DANSK MODELFLYVEFORBUND

TEAM FLIGHT F1C

PLACE	NO	PARCIPANT			COUNTRY	1. P	2. P	3. P	4. P	5. P	6. P	7. P	TOTAL
1	H312	332	333	334	I	540	540	540	540	540	540	540	3780
2	H305	313	314	315	CDN	540	540	540	540	540	540	508	3748
3	H316	341	342	343	DPRK	526	520	540	516	540	540	540	3722
4	H327	372	373	374	USSR	540	540	540	540	476	540	540	3716
5	H325	366	367	368	H	540	512	540	540	488	540	540	3700
6	H326	369	370	371	USA	540	499	540	540	540	483	540	3682
7	H324	363	364	365	CSSR	540	540	540	540	526	540	436	3662
8	H322	357	358	359	GB	540	540	540	540	540	465	496	3661
9	H307	319	320	321	DK	540	540	540	540	478	524	498	3660
10	H330	381	382	383	A	540	540	533	540	540	461	497	3651
11	H314	336	337	338	YU	540	540	530	540	490	498	508	3646
12	H328	375	376	377	D	540	526	540	540	540	484	360	3530
13	H308	322	323.	324	DDR	540	540	540	491	456	463	473	3503
14	H303	307	308	309	BR	463	512	525	537	514	473	427	3451
15	H320	351	352	353	CH	487	522	48?	532	400	516	488	3428
16	H310	328	329	330	F	540	540	540	540	382	360	341	3243
17	H323	360	361	362	S	340	524	516	540	536	185	295	2936
18	H302	304	305	306	AUS	323	318	523	476	480	334	275	2729
19	H309	325	326	327	SF	436	242	346	455	444	357	291	2571
20	H301	301	302	303	RA	360	330	360	360	360	321	360	2451
21	H304	310	312	0	BG	360	36?	359	346	303	236	360	2324
22	H315	339	340	0	MEX	281	331	360	360	180	161	171	1844
23	H317	344	0	0	N	180	180	180	155	180	180	171	1226
24	H313	335	0	0	J	180	180	149	180	0	180	169	1038

Criterium Pierre Trebod - 16/17 July - F1A, F1B, F1C Marigny-le-Grand, France

THE WORLD CONTEST MARIGNY 77 13me CRITERIUM PIERRE TREBOD

PLANEURS - F1A

1	MOTSCH H.	D	1260	(199)	
2	LELEUX J.	F	1260	(143)	(160)
3	BAINES B.	GB	1260	(143)	(156)
4	REYNDERS L.	B	1260	(139)	
5	PICOT A.	F	1260	(132)	
6	BLEUR H.	CH	1260	(120)	
7	BARNES D.	GB	1260	(115)	
8	BAZILLON M.	F	1260	(113)	
9	OWENS P.	GB	1260	(107)	
10	PICOT D.	F	1260	(78)	

11	PALMIERI W.	RA	172	180	180	180	180	180	180	1252
12	BACHMANN C.	CH	175	180	180	180	180	172	180	1257
13	CHAUSSEBOURG P.	F	166	176	180	180	180	180	180	1242
14	HOHLS C.	D	180	180	180	160	170	180	180	1230
15	GALICHET A.	F	149	180	180	180	180	180	180	1229
15	LAMMERTS H.	NL	149	180	180	180	180	180	180	1229
17	SIMMS W.	GB	136	180	180	180	180	180	180	1216
18	TOMCZYK D.	F	180	180	180	180	180	135	180	1215
18	LASSAIGNE J.	F	180	180	135	180	180	180	180	1215
18	SIMONS D.	AUS.	180	180	135	180	180	180	180	1215
21	BUISSON G.	F	180	116	180	180	180	180	180	1196
22	WILKENING H.	D	125	180	180	170	180	180	180	1195
23	GRUNNET P.	DK	152	140	180	180	180	180	180	1192
24	EIMAR B.	S	180	110	180	180	180	180	180	1190
25	DRAPEAU J.L.	F	180	161	180	180	180	180	126	1187
26	MOLFINO V.	RA	180	180	180	180	179	180	106	1185
27	MARTINEZ A.	CUBA	149	180	180	180	129	180	180	1178
28	COIN C.	F	123	180	180	180	180	180	154	1177
29	GOUDEAU C.	F	180	180	180	180	180	142	130	1172
30	KIEHNLE U.	D	180	137	180	134	180	180	180	1171
31	HERZOG A.	B	180	180	180	180	180	180	088	1168
32	GODINHO J.	F	107	180	180	180	180	180	160	1167
33	CHAMPION R.	F	180	109	180	180	180	180	156	1165
34	ALLAIS R.	F	180	180	180	076	180	180	180	1156
35	VAN HAUDENARD M.	B	075	180	180	180	180	180	180	1155
36	BREEMAN C.	NL	180	180	138	116	180	180	180	1154
37	SALZER K.	A	180	180	180	180	180	180	072	1152
38	ALTHOFF H.	D	069	180	180	180	180	180	180	1149
39	BUETTY R.	CH	090	180	180	180	156	180	180	1146
39	WARREN M.	GB	180	145	180	180	101	180	180	1146
41	MURRAY N.	ZA	180	180	180	180	064	180	180	1144
42	SMITZ W.	USA	146	180	180	097	180	180	180	1143
43	HERNANDEZ F.	RA	062	180	180	180	180	180	180	1142
44	LEBEE J.	F	180	180	180	134	162	180	125	1141
45	VISONNEAU C.	F	126	180	180	180	180	180	112	1138
46	DILLY M.	GB	180	180	057	180	179	180	178	1134
47	LEPAGE R.	F	150	180	180	180	180	180	076	1126
48	JACK A.	GB	091	180	180	180	180	180	134	1125
48	PARKER J.	USA	136	125	180	180	180	144	180	1125
50	PEPER H.	D	180	119	180	180	180	180	103	1122
51	NOUGE A.	F	180	180	136	180	180	127	135	1118
51	BACHMANN G.	CH	107	180	176	180	180	136	159	1118
53	MAIWORM E.	D	067	180	180	148	180	180	180	1115
54	DE BOER P.	NL	180	119	180	105	180	180	170	1114
55	GREAVES D.	GB	132	180	180	180	180	147	113	1112
56	BOUTILLIER B.	F	158	180	099	168	180	143	180	1108
56	CLERMONT A.	NL	168	180	101	180	180	168	131	1108
58	MADORE JC	F	061	180	180	180	180	180	145	1106
59	SIEBENMANN D.	CH	180	056	180	180	149	180	180	1105
60	KAYNES I.	GB	180	180	180	132	180	101	150	1103
61	MADELIN G.	GB	180	100	180	180	180	180	099	1099
62	HATCHECK B.	USA	051	180	180	180	180	180	147	1098
63	REIFLER H	CH	136	180	180	099	180	180	140	1095
64	BJERRE F.	DK	104	105	180	180	180	180	161	1090
65	MARRIOTT S.	GB	113	180	180	076	180	180	180	1089
66	LITJENS P.	NL	180	116	180	073	180	176	180	1085
67	CHABOT J-M	F	147	180	180	180	151	110	134	1082
68	DUMONTET B.	F	133	180	180	180	132	090	180	1075
68	AIMELET F.	F	180	180	180	105	180	180	070	1075
70	SJOLUND P.	S	180	028	169	155	180	180	180	1072
70	WILLIAMS J.	GB	180	132	180	180	180	112	108	1072
72	LITJENS J.	NL	148	180	180	116	103	180	164	1071
73	LAUREAU J.P.	F	115	180	180	180	071	180	164	1070
74	HALLER W.	CH	055	126	180	180	180	180	168	1069
75	DETON Y.	F	180	180	081	180	180	092	175	1068
76	GONZALEZ J.	CUBA	145	180	180	071	180	131	180	1067
77	BESNARD J.	F	128	180	180	180	180	093	122	1063
78	WITTE H.	D	144	180	075	121	180	180	180	1060
79	RASMUSSEN J.	DK	180	097	180	153	171	097	180	1058
80	COWLEY M.	GB	149	180	180	100	115	180	153	1057
81	LANEURIE J.P.	F	180	180	090	180	113	180	124	1047
82	DUNKERBECK E.	NL	080	175	180	071	180	180	180	1046
83	PAILLOU P.	F	180	163	180	095	055	180	180	1033
84	GREGORIE M.	NZ	099	180	180	107	158	180	126	1030
85	LEDOCQ M.	B	180	180	085	060	158	180	180	1023
86	MULLER J.	D	100	180	180	180	107	180	087	1014
87	BOGAERTS J.	B	099	086	180	180	180	180	107	1012
88	BLEUER K.	CH	130	126	180	063	180	180	147	1006
88	FAUX K.A.	GB	077	137	180	180	180	105	147	1006
90	VAN DIJK M.H.	NL	180	180	091	081	180	180	113	1005
90	ALLNUTT P.	CDN	106	180	086	180	180	180	093	1005
92	GOISMIER J.	F	146	180	162	180	180	090	064	1002
93	MAC MAHON M.	CDN	080	180	110	180	180	111	153	994

94	BIXBY P.	GB	127	180	077	180	180	066	180	990
95	DULOUT H.	F	173	099	180	180	180	082	096	990
96	LAFEUILLE P.	F	180	071	180	121	131	180	123	986
97	SEELIG P.	D	087	097	101	160	180	180	180	985
98	VAN MERKESTYN P.	NL	180	080	180	180	104	114	137	975
99	TIDSTRAND P.	S	180	061	180	153	180	046	148	948
100	PILLER M.	F	123	113	180	112	180	114	125	947
101	JELLIS P.	GB	120	086	180	110	180	180	085	941
102	VAN EEDE T.	NL	075	099	095	131	180	180	170	930
103	BUGE H.C.	D	088	121	180	180	180	[illegible]	120	928
104	COUVARD H.	F	100	043	180	180	180	111	123	917
105	PROCTOR K.	GB	105	178	092	110	108	180	138	911
106	MARILIER T.	F	171	132	180	057	039	140	180	899
107	SEREN P.	D	104	050	160	121	153	180	121	889
108	FUSS H.	A	180	076	180	060	180	180	030	886
109	ROTH M.	CH	070	135	180	180	082	107	124	878
110	KAPPE B.	D	110	180	180	118	--	100	180	868
111	LAMARCHE P.	F	--	169	102	112	151	158	175	867
112	EKKELENKAMP W.	NL	124	180	047	078	167	180	088	864
113	SUEN E.	CUBA	153	102	180	012	053	180	180	860
114	GERARD P.	F	180	062	084	079	161	180	106	852
115	MAROT J.	F	180	065	180	089	068	154	096	832
115	FRIEBIS W.	USA	131	085	115	133	180	092	096	832
117	CARREAU V.	F	047	096	180	090	180	133	091	817
118	HIRLIMANN J.C.	F	021	091	075	180	180	080	175	802
119	BROCHARD G.	F	094	180	107	061	047	180	132	801
120	HAWKINS P.	GB	117	085	180	068	088	090	164	792
121	BUREAU L.	F	140	086	146	101	180	076	059	788
122	MATHERAT G.	F	107	070	063	138	053	180	171	782
123	WAECHTER R.	CH	072	180	180	071	028	104	141	776
124	TILHAC J.M.	F	180	180	180	180	--	-	--	720
125	DUPUIS L.	F	123	095	050	088	059	123	180	718
126	CABRERA C.	CUBA	062	180	092	120	150	045	053	702
127	IMGENBERG E.	D	119	098	180	080	078	-	-	555
128	FRIEBIS J.	USA	116	065	150	079	--	081	058	549
129	MICHELIN F.	F	060	083	--	046	086	069	172	516
130	CASIER C.	F	085	108	024	068	089	040	050	466
131	IMGENBERG R.	D	107	137	180	--	--	--	--	424
132	DAHLMANS J.	NL	180	180	--	--	--	--	--	360
133	HEIDEMANN T.	D	180	--	--	--	--	--	--	180
134	GAENSLI F.	CH	086	075	--	--	--	--	--	161
135	ROTTELEUR D.	F	102	--	--	--	--	--	--	102
136	JENSEN S.	DK	099	--	--	--	--	--	--	099
137	POLLARD R.	GB	061	--	--	--	--	--	--	061

WAKEFIELD

1	NEGLAIS J.C.	F			1260		(240)			
2	PROCTOR K.	GB			1260		(159)			
3	TAYLOR I.	GB	147	180	180	180	180	180	180	1227
4	CHAUSSEBOURG P.	F	180	180	180	180	180	146	180	1226
4	ALLEN C.	USA	180	180	180	156	180	170	180	1226
6	MARRIOTT S.	GB	180	180	180	180	178	180	135	1213
7	BLITZMAN M.	RA	180	123	180	180	180	180	180	1203
8	ZETTERDAHL J.	S	139	180	180	180	180	149	180	1188
8	SIEBENMANN D.	CH	180	180	108	180	180	180	180	1188
8	ROWSELL D.	CDN	180	180	180	180	108	180	180	1188
11	NOUGE A.	F	180	137	180	150	180	180	180	1187
12	GOUVERNE E.	F	180	180	180	180	180	180	101	1181
13	RASMUSSEN P.	DK	180	169	114	180	180	180	175	1178
14	MENENDEZ M.	CUBA	180	180	180	180	180	180	095	1175
15	BENEDINI H.	RA	180	132	180	180	152	170	180	1174
16	POLLARD R.	GB	178	180	180	180	180	087	180	1165
17	DAVIS J.	USA	180	126	180	163	180	151	180	1160
18	ARIESTO A.	RA	180	156	180	100	180	180	180	1156
19	ROHRER B.	CH	180	147	160	180	180	122	180	1149
20	RUYTER P.	NL	180	174	160	180	174	180	094	1142
21	WOODHOUSE J.	GB	180	180	180	180	180	180	055	1135
22	REIFLER H.	CH	180	180	180	180	157	070	180	1127
23	GIBBONS B.	USA	180	180	180	180	066	180	158	1124
24	CHAMPION R.	F	161	120	180	180	180	120	180	1121
25	GAENSLI F.	CH	180	152	180	180	180	180	061	1113
26	MAC GILLIVRAY	CDN	180	127	180	180	180	142	115	1104
27	BOUTILLIER B.	F	173	180	180	071	142	177	180	1103
28	RUIZ A.	CUBA	150	104	180	152	180	180	140	1086
29	DUCE M.	GB	040	180	153	180	171	180	180	1084
30	HELMBRECHT H.	D	180	180	180	180	091	116	144	1071
31	WIDELL K.	DK	175	114	180	180	180	057	180	1066
31	EIMAR B.	S	111	085	180	090	180	180	180	1066
33	BUSCH A.	D	143	165	180	180	180	126	080	1054
34	MONTPLAISIR C.	USA	177	180	063	147	180	114	165	1026
35	ORTHWEIN M.	D	180	180	157	142	163	099	097	1018
36	ALLAIS R.	F	180	180	044	163	180	078	180	1005
37	BRAUCHLE P.	D	180	120	104	180	180	041	180	985
38	VIGGIANO O. C.	RA	180	118	180	044	180	099	180	981
39	KOPPITZ A.	F	148	135	103	144	164	172	096	962
40	PRIOUX H.	F	180	083	110	133	138	117	180	941
41	GREAVES D.	GB	159	167	066	180	142	090	130	934
42	NIKITENKO F.	F	074	128	180	180	180	100	075	917
43	MERITTE A.	F	180	096	180	072	180	108	090	906
44	LEPAGE P.	F	180	093	159	180	075	112	097	896
45	SCHUPPE B.	D	111	180	180	180	089	083	064	887
46	PENATE S.	CUBA	180	164	180	110	--	116	120	870
47	ERICSSON L.	S	180	055	180	180	040	106	110	851
48	AKER I.	USA	180	128	122	130	110	180	-	850
49	MOREJON I.	CUBA	096	180	074	069	157	145	100	821
50	PISERCHIO R.	USA	180	180	180	180	099	--	--	819
51	DURAND R.	F	099	110	096	172	095	180	054	806
52	DUPUIS L.	F	180	180	180	087	127	--	--	754
53	KAYNES I.	GB	180	079	180	180	106	--	--	725
54	ZWAHLEN A.	CH	129	154	075	180	082	-	051	671
55	TOMCZYK D.	F	144	095	180	140	082	-	--	641
56	SMITZ W.	USA	180	123	119	180	-	-	-	602
57	CASIER C.	F	088	180	093	057	180	-	-	598
58	MATHERAT G.	F	138	180	180	098	--	-	-	596
59	BUISSON G.	F	105	141	140	180	-	-	-	566
60	MADELIN G.	GB	147	180	149	076	--	--	-	552
61	WHITE R.	USA	180	118	180	--	--	-	-	478
62	VAN MERKESTYN P.	NL	180	116	067	093	-	-	-	456
63	SEGRAVE M.	USA	113	150	180	--	-	-	-	443
64	MINASSIAN J.	USA	180	157	092	--	-	-	--	429
65	HIPPERSON D.	GB	180	127	097	--	-	-	-	404
66	COGNET G.	F	170	098	132	--	--	-	-	400
67	BUSKELL J.	GB	180	089	060	067	--	-	-	396
68	MICHELIN F.	F	080	052	180	064	--	-	-	376
69	LEBEE J.	F	074	053	045	180	-	-	-	352
70	GRAUX A.	B	072	068	138	057	--	-	-	335
71	PILLER M.	F	090	057	042	--	--	--	-	189

MOTOMODELES - F1 C

1	COWLEY M.	GB			1260			(240)		
2	HEIDEMANN T.	D			1260			(186)		
3	CUZZI O.	RA			1260			(150)		
4	HARTWAGNER F.	A			1260			(119)		
5	ROUNSAVILLE D.	USA	180	170	180	180	180	180	180	1250
6	KOSTER T.	DK	180	184	164	180	180	180	180	1244
7	BOND P.	GB	180	180	180	159	180	180	180	1239
8	GIESKING M.	USA	178	180	128	180	180	180	180	1206
9	SCHLACHTA F.	CDN	180	180	180	180	180	166	138	1204
10	JEAN M.	F	180	180	180	178	180	124	180	1202
11	JACK A.	GB	177	180	180	166	134	180	180	1197
12	FAUX H.	GB	180	162	180	180	180	180	129	1191
13	BOUTILLIER B.	F	174	180	180	180	156	138	180	1188
14	MARTIN C.	USA	180	176	180	180	180	180	110	1186
15	SEELIG H.	D	180	180	180	136	180	147	180	1183
16	WEBER A.	D	180	180	180	180	180	070	180	1150
17	LINDHOLM H.	S	089	180	180	180	177	180	158	1144
18	IMGENBERG U.	D	180	174	095	180	152	180	180	1141
19	TALOUR C.	F	180	157	180	180	113	140	180	1130
20	STUTZ H.	D	180	180	180	180	180	139	089	1128
21	ZEH R.	D	180	152	180	180	135	134	165	1126
22	CHILTON F.	GB	180	180	180	180	180	101	103	1104
23	FERRERO D.	F	094	180	180	180	180	145	121	1080
24	HUEBLER H.	D	180	173	180	175	180	180	-	1068
25	ZITO M.	RA	144	180	135	162	180	139	120	1060
26	JOHNSON R.	GB	165	132	180	074	179	140	180	1050
27	BLANCO M.	Cuba	168	180	150	148	123	180	100	1049
28	HEIDEMANN G.	D	164	124	165	180	180	101	109	1023
29	LANDEAU A.	F	180	180	180	164	130	180	-	1014
30	COLLINS R.	GB	180	180	180	180	106	180	-	1006
31	SCHILLING G.	D	082	180	100	180	141	128	180	991
32	BURROWS L.	GB	175	180	106	180	180	-	119	940
33	MAURER P.	CH	180	180	180	025	180	123	058	926
34	CASTILLO L.	Cuba	166	125	180	155	180	065	054	925
35	PLATT H.	D	179	144	163	110	078	120	074	868
36	VALDES A.	Cuba	157	177	098	180	180	065	-	857
36	JOHANSSON K.	S	180	156	088	060	095	098	180	857
38	SEYDEL S.	D	127	180	180	180	180	-	-	847
39	ROUX A.	F	166	180	180	180	23	-	-	729
40	BISSONNETTE A.	USA	180	172	180	180	-	-	-	712
41	MAC LAUGHLAN T.	USA	180	180	180	152	-	-	-	692
42	SAUER K.	D	139	119	180	125	-	-	-	563
43	EAST W.	AUS	055	114	180	180	-	-	-	529
44	BUSKELL P.	GB	171	166	--	-	-	-	-	337
45	GROSSL S.	D	120	-	063	-	-	-	-	183

REGLEMENT

Les épreuves du Critérium International Pierre Trébod, réservées aux aéro-modèles des classes F 1 A, F 1 B et F 1 C, sont organisées conformément au code Sportif, Sections 1 à 4 de la F.A.I.

Des groupes de 6 ou 7 concurrents seront formés et se verront attribués un poste de chronométrage, avec un chronométreur différent pour chaque round.

Les vols débuteront en principe à 7 h 30, chaque round durant 1 h, les fly-off ayant lieu à partir de 17 h 30.

Le Critérium est doté de coupes individuelles attribuées aux trois premiers de chaque catégorie. Dans chaque classe, une coupe-challenge est mise en compétition. Cette coupe-challenge étant attribuée définitivement au concurrent qui l'aura gagnée trois années consécutives ou non.

Le challenge Jean MAGNIETTE récompensera la meilleure nation son attribution sera calculée de la façon suivante : addition des temps réalisés par le premier concurrent de chaque nation dans chaque catégorie (F 1 A, F 1 B, F 1 C).

Naturellement, les accompagnateurs désirant participer au repas de clôture pourront le faire moyennant la somme de 35,00 F et l'inscription obligatoire sur le bulletin d'engagement des compétiteurs.

Comme l'an passé, les commerçants viendront sur le terrain, au camping afin que vous puissiez effectuer vos achats de nourriture.

Le bar sera, bien sûr, à votre disposition.

Czech International 19/21 Aug. - F1A, F1B, F1C Sezimovo Usti, Czechoslovakia

FAI Glider F1A 80 entries, 74 flew

1	- Preuss	DDR	1260 + 183								
2	P Dvorak	CZ	1260 + 177 + 227								
3	W Korczak	PL	1260 + 177 + 178								
4	B Johansson	S	1260 + 173								
5	A Barta	CZ	1260 + 170								
6	M Klima	CZ	1260 + 164								
	S Kubit	PL	1260 + 164								
8	I Veselka	CZ	1260 + 161								
9	P Körnhofer	CZ	1260 + 153								
10	L Vaculik	CZ	1260 + 148								
11	Z Bosak	CZ	1260 + 135								
12	E Sauer	DDR	1260 + 132								
13	- Hofrichter	CZ	1260 + 127								
14	J Orel	CZ	180	180	180	180	180	180	179	=	1259
15	J Janu	CZ	180	180	180	180	180	180	173	=	1253
16	Z Dudacek	CZ	180	180	180	180	180	180	172	=	1252
Maxes			51	61	56	58	62	56	29		
Full houses				45	36	33	30	28	13		

FAI RUBBER F1B 35 entries, 32 flew

1	P Chaussebourg	F	1260 + 240 + 207									
2	J Klima	CZ	1260 + 240 + 120									
3	E Mielitz	DDR	1260 + 190									
4	J Libra	CZ	1260 + 171									
5	J Löffler	DDR	180	180	142	180	180	180	180	=	1222	
6	L Mucha	CZ	180	180	180	180	180	180	141	=	1221	
7	- Terlanda	CZ	180	180	180	180	180	180	137	=	1217	
8	H Kucharski	PL	180	180	135	180	180	180	180	=	1215	
Maxes			20	23	15	16	18	22	17			2
Full houses				14	9	7	6	6	4			2

FAI Power F1C 33 entries, 29 flew

1	H Lindholm	S	1260 + 240									
2	H Krieg	DDR	1260 + 180									
3	V Patek	CZ	1260 + 28									
4	B Krycer	CZ	180	180	180	180	180	180	176	=	1256	
5	H Antoni	DDR	180	180	180	180	180	165	180	=	1245	
6	J Blazek	CZ	180	180	180	180	163	180	180	=	1243	
7	- Schmeling	DDR	180	180	180	180	180	180	154	=	1234	
Maxes			17	20	22	15	13	19	10			1
Full houses				13	13	10	7	6	3			1

WFFA Challenge - 19/20 Nov. - F1A, F1B, F1C Taft, California, U.S.A.

WFFA 1977 NORDIC F1A

		SAT.					SUN.					
	CONTESTANT	1	2	3	4	5	6	7	8	9	10	TOTAL
1	MATT GEWAIN	180	180	180	180	180	157	172	180	180	180	1769
2	JUAN LIVOTTO	180	180	180	180	180	179	139	180	180	180	1758
3	DON ZINK	180	180	180	180	164	180	145	180	180	180	1749
4	BOB ISAACSON	180	180	127	180	180	180	180	180	180	180	1747
5	LEE HINES	76	180	180	180	180	180	180	180	180	180	1696
6	DICK MEYERS	180	180	180	180	115	146	141	180	180	180	1662
7	GAIL GEWAIN	180	180	180	180	120	158	130	180	180	180	1658
8	JIM WILSON	54	180	180	180	180	180	158	180	180	180	1652
9	PETER ALLNUTT	180	141	180	80	180	151	66	180	180	180	1518
10	MIKE SZURA	180	129	84	0	180	158	180	180	180	180	1451
11	HECTOR DIEZ	119	180	180	121	180	10					790
12	BRUCE HANNAH	180	180	110	68	128						676
13	JEFF LIVOTTO	180	180	180								540
14	JON JAMES	180	180									360

WFFA 1977 WAKEFIELD F1B

		SAT.					SUN.					
	CONTESTANT	1	2	3	4	5	6	7	8	9	10	TOTAL
1	BOB WHITE	180	180	180	180	180	180	180	180	117	180	1737
2	WALT GHIO	180	180	180	180	180	165	125	180	180	180	1730
3	JOE FOSTER	180	180	180	152	180	180	146	180	165	180	1721
4	VAL SIDLAUSKAS	180	180	140	104	180	167	180	168	180	180	1659
5	JIM QUINN	99	175	134	180	180	150	156	180	180	180	1614
6	BOB DODDS	180	180	180	170	129	144	134	113	161	180	1571
7	IRV AKER	180	121	180	136	111	115	180	180	180	180	1563
8	TOM MEDLEY	180	180	107	180	150	115	166	180	91	180	1529
9	BILL WILEY	180	150	127	180	152	147	139	180	92	180	1527
10	HUGO SANDRONI	180	180	180	180	22	150	147	180	63	180	1482
11	AL HOTARD	180	99	180	84	106	163	134	180	121	126	1473
12	VINCE COSTANZO	180	180	86	117	145	118	147	84	180	180	1417
13	BILL BOGART	68	180	132	108	172	140	135	180	158	112	1385
14	LARRY JORDAN	180	62	145	114	144	120	108	115			988
15	BRUCE HANNAH	180	180									360
	LES DEWITT	180	180									360
17	RAY HARPER	180										180

WFFA 1977 FAI POWER F1C

		SAT.					SUN.					
	CONTESTANT	1	2	3	4	5	6	7	8	9	10	TOTAL
1	ROGER SIMPSON	180	180	180	180	180	180	180	180	180	180	
									240	300	175	2515
2	RONNY YOUNG	180	180	180	180	180	180	180	180	180	180	
									240	300	143	2483
3	KEN OLIVER	180	173	180	180	180	180	180	180	180	180	1793
4	WALT GHIO	180	180	180	169	180	180	180	180	180	180	1789
5	ED CARROLL	180	158	180	180	180	180	180	180	180	180	1778
6	HARDY BRODERSON	180	180	180	180	180	180	180	180	180	111	1731
7	MIKE ACHTERBERG	180	180	180	180	175	180					1075
8	BILL HARTILL	180	180	180	142	180						862
9	JERRY CLARK	121	180	170								471